Nuclear Magnetic Resonance of Liquid Crystals

NATO ASI Series

Advanced Science Institutes Series

A series presenting the results of activities sponsored by the NATO Science Committee, which aims at the dissemination of advanced scientific and technological knowledge, with a view to strengthening links between scientific communities.

The series is published by an international board of publishers in conjunction with the NATO Scientific Affairs Division

A	Life Sciences	Plenum Publishing Corporation
B	Physics	London and New York
C	Mathematical and Physical Sciences	D. Reidel Publishing Company Dordrecht, Boston and Lancaster
D	Behavioural and Social Sciences	Martinus Nijhoff Publishers
E	Engineering and Materials Sciences	The Hague, Boston and Lancaster
F	Computer and Systems Sciences	Springer-Verlag
G	Ecological Sciences	Berlin, Heidelberg, New York and Tokyo

Series C: Mathematical and Physical Sciences Vol. 141

Nuclear Magnetic Resonance of Liquid Crystals

edited by

J. W. Emsley
Chemistry Department, The University of Southampton,
Southampton, U.K.

D. Reidel Publishing Company

Dordrecht / Boston / Lancaster

Published in cooperation with NATO Scientific Affairs Division

Proceedings of the NATO Advanced Study Institute on
Nuclear Magnetic Resonance of Liquid Crystals
San Miniato, Italy
July 26-August 7, 1983

Library of Congress Cataloging in Publication Data

NATO Advanced Study Institute on Nuclear Magnetic Resonance of Liquid Crystals (1983 : San Miniato, Italy)
Nuclear magnetic resonance of liquid crystals.

(NATO ASI series. Series C, Mathematical and physical sciences ; vol. 141)
"Proceedings of the NATO Advanced Study Institute on Nuclear Magnetic Resonance of Liquid Crystals, San Miniato, Italy, July 26–August 7, 1983"-T.p. verso.
Includes index.
1. Liquid crystals–Analysis–Congresses. 2. Nuclear magnetic resonance spectroscopy–Congresses. I. Emsley, J. W. (James William) II. Title. III. Series: NATO ASI series. Series C, Mathematical and physical sciences ; no. 141.
QD923.N37 1983 548′.9 84–18140

ISBN-13: 978-94-009-6519-5 e-ISBN-13: 978-94-009-6517-1
DOI: 10.1007/978-94-009-6517-1

Published by D. Reidel Publishing Company
P.O. Box 17, 3300 AA Dordrecht, Holland

Sold and distributed in the U.S.A. and Canada
by Kluwer Academic Publishers,
190 Old Derby Street, Hingham, MA 02043, U.S.A.

In all other countries, sold and distributed
by Kluwer Academic Publishers Group,
P.O. Box 322, 3300 AH Dordrecht, Holland

D. Reidel Publishing Company is a member of the Kluwer Academic Publishers Group

CONTENTS

PREFACE

The liquid crystalline state has been known for about a century and has been studied by many techniques. Nuclear magnetic resonance has been used to study mesophases for thirty years, but it has been in very recent years that advances in this form of spectroscopy have led to a rapid growth in its applications to the study both of liquid crystals and of solutes dissolved in them. It has become apparent that no other method of studying liquid crystals can yield such a wealth of data and it is unrivalled as a means of probing the behaviour of the molecules in mesophases. There has also been a steady increase in the study of the shape of small molecules dissolved in liquid crystals via the analysis of their NMR spectrum. In fact, the study of solutes was until recently regarded as a separate activity to the study of liquid crystals themselves, but this artificial division arose only from the gap between the large amount of information that could be derived from the spectrum of a small molecule and the rather meagre data set obtainable from the spectra of liquid crystals. This gap has, however, narrowed and it is now possible to derive a very detailed picture of the structure and orientational ordering of the large molecules typical of those which form liquid crystals. There has also been a rapid growth of interest in the liquid crystalline state. This has been stimulated in part by their importance in electro-optic display devices, in the recognition of their role in living systems, and the parallel that exists between these biological systems and the lyotropic phases formed by soaps and surfactants. The parallel growth of research on liquid crystals and on NMR spectroscopy stimulated the organisation of a NATO Advanced Study Institute to consider the fundamental theory and applications of the use of NMR to study mesophases. The ASI brought together theoreticians and experimentalists during the two weeks July 26th to August 7th, 1983, and the lectures which form the basis of this monograph were given to an audience of chemists, physicists and biologists. It was a very stimulating experience which I hope these detailed accounts of the lectures will convey.

The ASI was organised by myself, being mainly responsible for constructing the programme of lectures, and Dr Carlo Alberto Veracini of the Istituto di Chimica Fisica, Universitá di Pisa, who worked extremely hard and efficiently to make all the practical arrangements for the ASI. It was held at "i Cappuccini", Centro Studi della Casa di Risparmio di San Miniato, whose staff helped to make the event pleasurable, as well as stimulating.

The ASI was made possible by a grant from the Scientific Affairs Division of NATO, and I wish to record my appreciation of the support given by them to both lecturers and students.

My task in editing these articles has been made easier by the careful sub-editing of Graham Johnson and the accurate typing of Sally Johnson.

University of Southampton, England J.W. Emsley
August 1984

QUANTITATIVE DESCRIPTION OF ORIENTATIONAL ORDER: RIGID MOLECULES

Claudio Zannoni

Istituto di Chimica Fisica
Universita'
Viale Risorgimento, 4, 40136 Bologna, Italy

LOOKING FOR A COMPLETE DESCRIPTION

The problem we consider here is that of giving a microscopic description of a certain anisotropic mesophase. For a real system, which may be typically composed of a number of molecules of the order of 10^{23}, the microscopic description we are looking for is a statistical one. In this sense the complete information (i.e. the state of the system) is represented by the information necessary to calculate all the average properties of interest (1 - 3).

To be more specific we consider a system of N molecules in a certain state of aggregation. We shall assume the molecules of interest to be classical, rigid, particles with centre of mass at position r and orientation ω. Our molecules could be solute molecules, dissolved in a given solvent or they could be constituents of a one component system. The assumption of rigidity means that three parameters, often taken to be Euler angles (4) are sufficient to specify the molecular orientation. Quite often an assumption of molecular cylindrical symmetry is also made. Notice, however, that real molecules are normally neither cylindrically symmetric nor rigid, for they can have, for example, flexible chains or rings that can rotate with respect to each other. We shall have therefore to expect that, beyond a certain level of sophistication, features like deviation from cylindrical symmetry and flexibility will have to be taken into account. We shall discuss flexibility in the next chapter and describe here a formalism for rigid particles.

The state of the system at a given instant of time is

J. W. Emsley (ed.), Nuclear Magnetic Resonance of Liquid Crystals, 1–34.

represented for our purposes by its configuration i.e. by the set of positions and orientations ω_i of all the particles. The orientation of each rigid particle can be specified in terms of the angles (α, β, γ) defined following Rose (4) convention (cf. Appendix). The enormous number of positional and orientational coordinates specifying the various configurations is fortunately unnecessary if we are only interested in calculating average properties. The average, observable value of an n-particle property i.e. of a property which depends collectively on the positions and orientation of n particles can be written as

$$\langle A\{X^n\}\rangle = \{(N - n)!/N!\}\int\{dX^n\}A(\{X^n\})\ P^{(n)}\ (\{X^n\})\ . \qquad (1)$$

Here we have introduced the probability of finding n particles out of the given N in the range $X_1+dX_1, X_2+dX_2\ldots.,\ X_n+dX_n$, which can be written for a system of N particles in a volume V at temperature T (canonical conditions) as the n-particle distribution

$$P^{(n)}(\{X^n\}) = (N!/(N-n)!Z_N)\int\{dX^N_{n+1}\}\exp\ (-U(\{X^N\})/kT)\ , \qquad (2)$$

where, for economy of notation we have used X to indicate the six variables (r,ω) and the curly brackets to denote collectively N variables (2). Thus $\{X^N\} \equiv (X_1,X_2,\ldots,X_N)$ and, similarly, $\{dX^N\} \equiv dX_1.dX_2\ldots.dX_N$. We have also introduced the notation $\{dX^N_m\} \equiv dX_m dX_{m+1}\ldots.dX_N$. Each volume element dX is equivalent to $dr d\omega$, where $dr = dr_x dr_y dr_z$ and $d\omega = d\alpha \sin\beta d\beta d\gamma$. In Eq. (2) k is the Boltzmann constant and the normalization factor Z_N is called the configurational partition function. To complete the remarks about notation we mention that in the absence of possible ambiguities we shall use only one integration sign to indicate the, possibly multiple, integration over all the variables whose volume elements appear. Integration over coordinates is extended to the sample volume V and to the usual domains $0 \leq \alpha \leq 2\pi$, $0 \leq \beta \leq \pi$ and $0 \leq \gamma \leq 2\pi$ for Euler angles. Notice that $P^{(n)}$ is not normalized to one but to the number of n-plets that can be formed by choosing n variables out of N i.e.

$$\int\{dX^n\}P^{(n)}(\{X^n\}) = N!/(N-n)!\ . \qquad (3)$$

In a number of techniques including magnetic resonance we do not need to concern ourselves very much with n-particle quantities and with the general n-particle distribution but mainly with the one- and two-particle distributions, or, as they are often called, the singlet and the pair distributions. These can be used to define the canonical ensemble average $\langle A\rangle$ of any property depending on position and orientation of one or two particles respectively. Thus

$$\langle A(X_1)\rangle = (1/N)\int dX_1\ A(X_1)\ P^{(1)}(X_1) \tag{4}$$

and

$$\langle A(X_1,X_2)\rangle = (1/N(N-1))\int dX_1\ dX_2\ A(X_1,X_2)\ P^{(2)}(X_1,X_2). \tag{5}$$

For example the dipolar coupling between two nuclei in the same molecule is a single particle property (5). The similar coupling between two nuclei belonging to different molecules is an example of a pair property. Inspection of Eqs. (4) and (5) shows that formal alternative definitions of $P^{(1)}$ and $P^{(2)}$ are (1)

$$P^{(1)}(r_1,\omega_1) = N\langle\delta(r_1-r_1')\delta(\omega_1 - \omega_1')\rangle \tag{6}$$

and

$$P^{(2)}(r_1,\omega_1;r_2,\omega_2)=N(N-1)\langle\delta(r_1-r_1')\delta(\omega_1-\omega_1')\delta(r_2-r_2')\delta(\omega_2-\omega_2')\rangle. \tag{7}$$

where $\delta(a-b)$ is a Dirac delta function and the integration implied in the ensemble average $\langle\ldots\rangle$ is over the primed variables. The singlet and pair distribution can be rewritten in a convenient way using the Fourier integral representation of the positional delta function

$$\delta(r) = (2\pi)^{-3}\int dk\ \exp\ (ik.r) \tag{8}$$

and the expansion of the angular delta function in generalized spherical harmonics or Wigner rotation matrices $D^L_{m,n}(\omega)$ (4)

$$\delta(\omega-\omega') = \sum\ ((2L+1)/8\pi^2)\ D^L_{m,n}(\omega)\ D^{L*}_{m,n}(\omega'). \tag{9}$$

The functions $D^L_{m,n}(\omega)$ are of great importance in the description of ordering. We shall give their definition and some of their important properties in the Appendix. The sum in Eq. (9) runs over L, m and n; we shall normally take the summations on the right hand side of an equation to run over all indices not appearing on the left hand side. Replacing the delta functions in Eqs. (6,7) gives some general expansions for the probability distributions. As an example

$$P^{(1)}(r_1,\omega_1) = N\int dk \sum c_{Lmn}(k)\ \exp(ik.r_1)\ D^L_{m,n}(\omega_1). \qquad (10)$$

The expansion coefficients

$$c_{Lmn}(k) = ((2L+1)/64\pi^5)\langle \exp(-ik.r_1')\ D^{L*}_{m,n}(\omega_1')\rangle, \qquad (11)$$

afford a systematic way of approximating the singlet distribution. We shall identify them with the positional-orientational order parameters for the system (2).

For a uniform system like an ordinary isotropic fluid or a nematic (but not a smectic) the physical properties are invariant under translation and the interaction energy $U(\{X^N\})$ depends only on relative distances. Therefore we can write

$$P^{(1)}(r_1,\omega_1) = \rho f(\omega_1) \qquad (12)$$

and

$$P^{(2)}(r_1,\omega_1;r_2,\omega_2) = \rho^2 f(\omega_1)\ f(\omega_2)\ g^{(2)}(r_{12},\omega_1,\omega_2), \qquad (13)$$

where $\rho=N/V$ is the number density and $f(\omega_1)$ is a purely orientational singlet distribution normalized to unity:

$$f(\omega_1) = \langle\delta(\omega_1 - \omega_1')\rangle \qquad (14a)$$

$$= (V/Z_N)\int \{dX_2^N\}\exp\ (-U\{X^N\})/kT) \qquad (14b)$$

This orientational distribution is of particular importance here since it can be used to express any single particle orientational property

$$\langle A(\omega_1)\rangle = \int d\omega_1\ f(\omega_1)\ A(\omega_1). \qquad (15)$$

Substitution in Eq. (9) yields the formal expansion

$$f(\omega_1) = (1/8\pi^2)\sum(2L+1)\ \langle D^{L*}_{m,n}\rangle D^L_{m,n}(\omega_1). \qquad (16)$$

The set of averaged Wigner orientation matrices $\langle D^L_{m,n}\rangle$ allows a complete characterization of $f(\omega)$. The quantities $\langle D^L_{m,n}\rangle$ are called orientational order parameters (2,6-7). Notice that a purely orientational distribution formally analogous to $f(\omega)$ can be introduced for nonhomogeneous phases as well by integrating $P^{(1)}(r,\omega)$ over positions.

For an isotropic molecular fluid the singlet and pair distribution can be simply written as

$$P^{(1)}(r_1,\omega_1) = \rho/8\pi^2 \tag{17}$$

and

$$P^{(2)}(r_1,\omega_1;r_2,\omega_2) = (\rho/8\pi^2)^2\ g^{(2)}(r_{12},\omega_1,\omega_2). \tag{18}$$

The singlet and pair distributions introduced in this way can be used to obtain expressions for the relevant thermodynamic quantities such as energy, specific heat, free energy, pressure etc. (2). Here we shall concentrate on the description of orientational order and its relation to NMR. Since single particle properties are possibly more important in NMR and certainly more accessible experimentally we shall concentrate here on the singlet distribution.

Instead of following a very formal approach starting from Eq. (10) we shall begin from a very simple case, that of cylindrical molecular symmetry. This will allow us to discuss many of the relevant concepts while introducing later on asymmetric molecules as molecules deviating from this uniaxial symmetry. For example the significance and usefulness of rank four order parameters is most conveniently discussed for cylindrical molecules.

A SIMPLE CASE: CYLINDRICAL MOLECULES IN A UNIAXIAL PHASE

Let us consider a uniform anisotropic fluid like a nematic (6-9). We are interested by now only in the single particle orientational distribution. This is a function of the two polar angles giving the molecular orientation of the particle in question i.e.

$$f(\omega) = f(\alpha,\ \beta)$$

if our molecules have cylindrical symmetry. The detailed form of $f(\omega)$ is of course unknown, but some constraint imposed on it by symmetry can nevertheless be easily taken into account. We know that experiments at least in a nematic and in a smectic A are consistent with a symmetry of the mesophase uniaxial around the director (6). If we choose this direction as our Z axis this means that rotating the sample about Z no observable property will change. Thus the probability for a molecule to have orientation $\omega = (\alpha,\ \beta)$ should be the same whatever the angle α.

More concisely

$$f(\alpha, \beta) = f(\beta)/2\pi. \tag{19}$$

Another experimental finding is that turning the aligned sample upside down changes nothing. Thus we should have

$$f(\beta) = f(\pi - \beta). \tag{20}$$

This is quite reasonable if we think of the molecules of interest as spherocylinders or other cylindrically symmetric objects where head and tail are not distinguishable. It is also convenient to renormalize our distribution, since we have only one angle to deal with. Thus we take

$$\int_0^\pi d\beta \sin\beta f(\beta) = 1 \tag{21}$$

since there is a certainty of finding the molecule at some angle β. The full singlet orientational distribution can be produced by computer simulation (10), and in principle it is contained in certain X-ray (11) and neutron scattering (12) measurements. In practice it is extremely difficult to get complete information on the distribution $f(\omega)$. A useful approach is, however, that of trying to approximate $f(\omega)$ in terms of a set of quantities that we can obtain from experiment. We need for this a set of functions orthogonal when integrated over $d\cos\beta$. Such a set of functions is that of Legendre polynomials (13) $P_L(\cos\beta)$, (cf. Appendix) for which we have

$$\int_0^\pi d\beta \sin\beta \, P_L(\cos\beta) \, P_N(\cos\beta) = 2\delta_{LN}/(2L+1) \tag{22}$$

The explicit form of the first few Legendre polynomials is given in Table 1, while in Fig. (1) we show a graph of $P_2(\cos\beta)$ and $P_4(\cos\beta)$ versus $\cos\beta$. Legendre polynomials have the useful property that $P_L(\cos\beta)$ is an even function of $\cos\beta$ if the rank L is even and an odd function if L is odd. Thus

$$P_L(\cos\beta) = (-)^L P_L(-\cos\beta) \tag{23}$$

Since in particular $\cos(\pi - \beta) = -\cos\beta$ this means that in writing our orientational distribution which is even in $\cos\beta$ (see Eq. (20)) in terms of $P_L(\cos\beta)$ functions only even L terms will need to be retained. Thus we can write

Table 1. *The explicit form of the Legendre polynomials $P_L(\cos\beta)$ for $L = 0$ to 6.*

$P_0(\cos\beta) = 1$	$P_1(\cos\beta) = \cos\beta$
$P_2(\cos\beta) = (3\cos^2\beta - 1)/2$	$P_3(\cos\beta) = (5\cos^3\beta - 3\cos\beta)/2$
$P_4(\cos\beta) = (35\cos^4\beta - 30\cos^2\beta + 3)/8$	$P_5(\cos\beta) = (63\cos^5\beta - 70\cos^3\beta + 15\cos\beta)/8$
$P_6(\cos\beta) = (231\cos^6\beta - 315\cos^4\beta + 105\cos^2\beta - 5)/16$	

$$f(\beta) = \sum_L f_L P_L (\cos\beta) \; ; \; L \text{ even.} \tag{24}$$

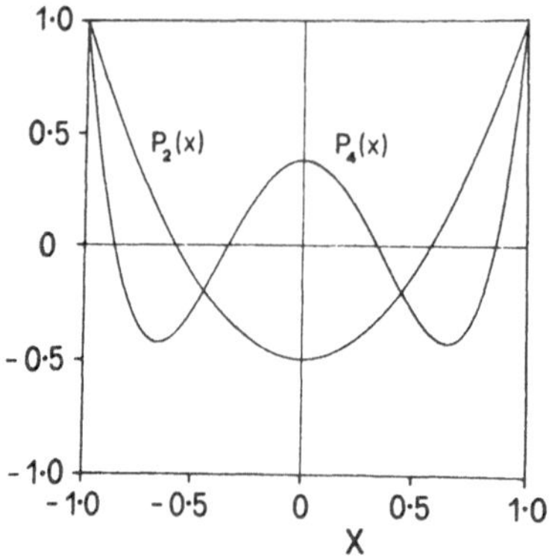

Figure 1. *The second and fourth rank Legendre polynomials $P_2(x)$ and $P_4(x)$ plotted as a function of $x = \cos\beta$.*

Multiplying both sides of Eq. (24) by $P_N(\cos\beta)$ and integrating over $\sin\beta d\beta$:

$$\int_0^\pi d\beta f(\beta)\, P_N\, (\cos\beta)\, \sin\beta = \sum_L f_L \int_0^\pi d\beta P_L\, (\cos\beta)\, P_N\, (\cos\beta)\, \sin\beta, \tag{25}$$

we find the coefficients in Eq. (24) as

$$f_N = (2N + 1) \langle P_N \rangle / 2, \tag{26}$$

where we have used the notation

$$\langle P_N \rangle = \int_0^\pi d\beta \sin\beta P_N\, (\cos\beta) f(\beta). \tag{27}$$

Eq. (24) is a special case of the general formal expansion Eq. (16) in view of Eq. (A6).

The averages of Legendre polynomials $\langle P_N \rangle$ represent our set of orientational order parameters. The knowledge of the (infinite) set of $\langle P_N \rangle$ would completely define the distribution. We can write

$$f(\beta) = \tfrac{1}{2} + (\tfrac{5}{2})\langle P_2 \rangle P_2(\cos\beta) + (\tfrac{9}{2})\langle P_4 \rangle P_4(\cos\beta) + \ldots \tag{28}$$

The first non trivial term contains the second rank order parameter

$$\langle P_2 \rangle = 3(\langle \cos^2\beta \rangle - 1)/2 \tag{29}$$

which corresponds to the S order parameter introduced by Zwetkoff (14). It is easy to see that $\langle P_2 \rangle$ has the properties that we would intuitively expect an order parameter to possess. For a system of perfectly aligned molecules where, $\beta = 0$ or π for every molecule, $\langle P_2 \rangle = 1$. At the other extreme, for a completely disordered system such as an ordinary isotropic fluid we have

$$\langle \cos^2\beta \rangle = \int_0^\pi d\beta \sin\beta \cos^2\beta \Big/ \int_0^\pi d\beta \sin\beta = 1/3 \tag{30}$$

and therefore for a disordered system we find $\langle P_2 \rangle = 0$. On going from an ordered to a disordered system the order parameter jumps discontinuously to zero if the transition is of the first order type (15), i.e. if it is associated with an entropy jump. At a continuous, second order transition the change in order parameter is instead a smooth function of temperature. Usually there are also pretransitional variations in the isotropic phase. We shall see later on that the second rank order parameter $\langle P_2 \rangle$ is proportional to the anisotropy in various experimentally measurable properties. This anisotropy can therefore be used to monitor orientational phase transitions.

Quite similarly to what we have just said the higher order parameters $\langle P_4 \rangle$, $\langle P_6 \rangle$, etc. are respectively one for complete order and zero for an isotropic system. We may wonder therefore if there is an advantage in considering more than one order parameter. That this is the case becomes apparent if we refer to Fig. (1) where the angular variation of the Legendre polynomials $P_2(\cos\beta)$ and $P_4(\cos\beta)$ is shown as a function of $\cos\beta$. As we can deduce from Fig. (1), if we measure $\langle P_2 \rangle$ and find that $\langle P_2 \rangle > 0$, this will mean that between 0 and 90 degrees the majority of molecules have a cylindrical axis orientation β between zero and the so called magic angle, $\beta \cong 54.7$ degrees, or $\cos^{-1}(\tfrac{1}{3})^{\frac{1}{2}}$, the zero of $P_2(\cos\beta)$. If, on the other hand, $\langle P_2 \rangle < 0$, then we may expect that on average molecules will have an orientation giving a negative P_2 (e.g. β between 54.7 and 90 degrees). Let us now consider the Legendre polynomial P_4 and its average $\langle P_4 \rangle$. The zeros of P_4 in the first quadrant fall at $\beta \sim 30.5$ and $70.1°$.

Suppose we have now measured $<P_2>$ and $<P_4>$ with some experimental technique. If $<P_2>>0$ and $<P_4>>0$ then the distribution of orientations will be such that the majority of molecules has a long axis orientation between 0 and 30.5 degrees. An example of distribution function of this type is reported in Fig. (2). We might, however, find a positive $<P_2>$,

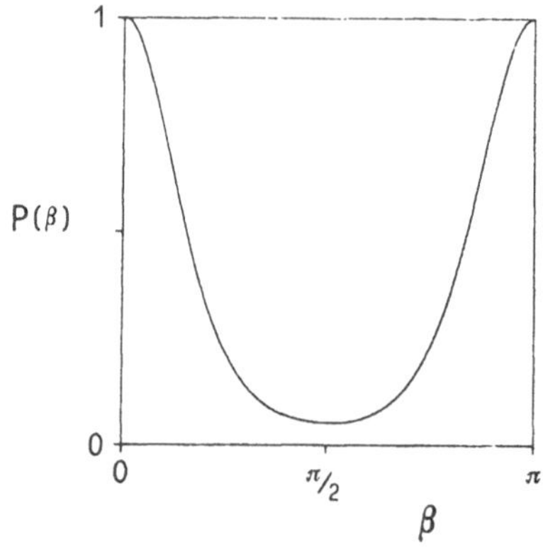

Figure 2. *An example of orientational distribution function P(β) corresponding to $<P_2>>0$ and $<P_4>>0$ (arbitrary units) plotted as a function of the angle β between molecule and mesophase symmetry axis.*

as before, and a negative $<P_4>$. This would suggest a different type of orientation distribution, for example a tilted one with a peak between 30.5 and 54.7 degrees. The qualitative physical significance of other combinations of order parameters can be obtained in a similar way. What we just said could be extended if we knew $<P_6>$ etc. Every higher order parameter restricts the bounds on $f(\cos\beta)$ and thus increases our knowledge on the system. It should be stressed that this does not mean that the expansion in Eq. (24) is so rapidly convergent that we only need the first few terms to reconstruct $f(\beta)$. Actually this will not be the case, at least in general. Any truncation of this kind would provide in fact a poor numeric approximation to $f(\beta)$ except for very low order parameters. The advantage of the orthogonal expansion in Legendre polynomials Eq. (24) or more generally in Wigner rotation matrices, Eq. (16) lies in fact in giving an expression which is at any level of approximation of well defined behaviour under rotation. We shall see when discussing the measurement of experimental properties that we do not need the whole distribution for computing second rank properties but just its second rank coefficient(s). Thus the fourth rank truncation is a poor approximation to $f(\beta)$ as such but in a number of cases the extra information is redundant anyway.

There may be cases, however, (e.g. in scattering techniques (11-12) where a knowledge of all $f(\beta)$ would be useful apart from being intellectually gratifying). The problem of finding the best approximation to the whole $f(\beta)$ starting from a knowledge of the first few order parameters or, equivalently, moments is solved using information theory (15-17). In this approach the most probable distribution is defined as that maximizing the entropy associated with the usual thermodynamic formula (1)

$$S = -k \int f(\omega) \ln f(\omega)\, d\omega. \tag{31}$$

It has been shown (16-17) that the best distribution in this respect has the form

$$f(\beta) = \exp\{ \sum_{L=0}^{L'} a_L P_L (\cos\beta)\} \tag{32}$$

where the coefficients a_L are obtained imposing the constraint that $\langle P_L \rangle$, L=0, ..., L' calculated from $f(\beta)$ have the known values.

Notice, however, that the information theory approach is in a way an "a posteriori" one. On the one hand it allows us to make optimum use of what we know. On the other hand there is no prediction on what the distribution should be at, say, different temperatures or what is its molecular origin. This brings us to the usefulness of order parameters in testing molecular theories. From this point of view a knowledge of $\langle P_4 \rangle$ as well as that of $\langle P_2 \rangle$ can be very useful to discriminate between various models of molecular organization inside the liquid crystal. To examine this important point we assume that every molecule is moving in an effective potential created by all the other molecules in the system $U(\cos\beta)$. This effective potential or pseudopotential or potential of mean torque (18) will obey the same symmetry restrictions introduced before for the singlet distribution. Thus it will be possible to approximate it as

$$U(\cos\beta) = \sum_L c_L P_L (\cos\beta). \tag{33}$$

where the coefficients c_L give the strength of the anisotropic potential. This effective potential can in turn be used to calculate order parameters etc. using the Boltzmann expression

$$f(\omega) = \exp\{-U(\omega)/kT\} \Big/ \int d\omega \exp\{-U(\omega)/kT\} \tag{34}$$

special case of Eq. (14). Thus, for example,

$$\langle P_L\rangle = \int_0^\pi d\beta \sin\beta P_L(\cos\beta)\exp\{-U(\cos\beta)/kT\} \Big/ \int_0^\pi d\beta \sin\beta \exp\{-U(\cos\beta)/kT\}. \tag{35}$$

The coefficients c_L give the strength of the anisotropic potential.

Consider as an example a truncation of the series Eq. (33) to the first symmetry allowed term, i.e.

$$U(x) = c_2 P_2(x), \text{ with } x = \cos\beta. \tag{36}$$

This truncation is consistent with a dispersion forces interaction (19) or more generally with any second rank type anisotropic interaction, whatever its physical origin (18); according to a simple molecular field theory of nematics, the coefficient c_2 is proportional to $\langle P_2\rangle$. Calculation of $\langle P_4\rangle$ versus $\langle P_2\rangle$ with the potential in Eq. (36) gives a universal curve without adjustable parameters where $\langle P_4\rangle$ is always positive.

It is therefore quite clear that if we wish to investigate the validity of using the distribution obtained from a molecular model giving Eq. (36) in describing ordering in a certain anisotropic liquid we should try to calculate not only $\langle P_2\rangle$ but $\langle P_4\rangle$ as well.

Similar arguments will hold for theoretical models for more complex molecules e.g. solutes deviating from cylindrical symmetry. There the predictions of the theory (20) should be tested by calculating the relevant order parameters in a range of thermodynamic conditions. In the next section we shall discuss which order parameters are needed for a molecule of a certain symmetry.

THE EFFECT OF SYMMETRY

We have seen in the previous section how the general expansion of the singlet orientational distribution $f(\omega)$ can be simplified when the mesophase and its constituent molecules are both cylindrically symmetric. In general, however, the exploitation of molecular and mesophase symmetry may be not quite so intuitive. One of the first problems in a systematic treatment will be listing the relevant order parameters for molecules and mesophases of various symmetry. We have shown in detail elsewhere (2) how a formal procedure for determining the independent order parameters can be introduced. We define the symmetry group of a phase as the group of transformations of the laboratory system that leave the singlet distribution $P^{(1)}(r,\omega)$ as well as the higher ones invariant. Similarly, we can define an effective symmetry for the molecule in terms of the group of molecular transformations leaving the singlet distribution unchanged. The same considerations hold, of course, for the purely orientational distribution $f(\omega)$ given by Eq. (16) which is of primary concern to us. In group theoretical language (21) we would say that $f(\omega)$ belongs to the totally symmetric representation of the group of the molecule and of the mesophase. Therefore one way of applying symmetry is to project the distribution onto the totally symmetric representation of these groups (21-23). Similar methods have been used to simplify the pair distribution and the intermolecular potential. Another way which we find convenient for the simplification of order parameters (2) relies on their definition as averages of Wigner rotation matrices

$$\langle D^L_{m,n}\rangle \doteq \int d\omega\ f(\omega)\ D^L_{m,n}(\omega). \tag{37}$$

A symmetry transformation $\underset{\sim}{O}$ does not alter $\int d\omega$ and also leaves $f(\omega)$ invariant, therefore the effect of $\underset{\sim}{O}$ on the order parameters just depends on how it transforms the Wigner matrix $D^L_{m,n}(\omega)$. This method becomes particularly convenient if restrictions on higher rank order parameters are to be found.

As a very simple example suppose that the chosen molecular z axis for a molecule is a k-fold symmetry axis. By applying the closure relation of the Wigner matrices (Eq. (A10)) the effect of a rotation of $2\pi/k$ around z is

$$\langle D^L_{m,n}\rangle = \sum_m \langle D^L_{m,q}\rangle\ D^L_{q,n}\ (2\pi/k,\ 0,\ 0)$$

$$= \langle D^L_{m,n}\rangle \exp\ (-i\ n\ 2\pi/k), \tag{38}$$

since $d^L_{ab}(0) = \delta_{ab}$. Eq. (38) can be satisfied when $n = 0, \pm k, \pm 2k, \ldots$ compatibly with $n \leq L$. Thus for a biaxial molecule a C_2 axis gives $n = 0, \pm 2, \ldots$ The restrictions induced by the

Table 2. *Effect of various molecular symmetry operations O_M on the Wigner rotation matrices (2). The notation for the symmetry operators is as follows. Here I stands for the inversion; $\sigma(xy)$ for a symmetry plane perpendicular to z; σ_ϕ for a plane making an angle ϕ with (zx). A rotation of $2\pi/k$ about z is written as $C_k(z)$ and a continuous rotation of γ about z as $C_\gamma(z)$. A rotation of π about an axis perpendicular to z and making an angle ϕ with x is indicated by $C_2(\phi)$. S_k stands for a k-fold rotoreflection axis.*

Operator O_M	$\langle Lm \vert O_M D \vert Ln \rangle$
$I = S_2$	$(-)^L D^L_{m,n}$
$\sigma(xy)$	$(-)^{L+n} D^L_{m,n}$
$\sigma(xz)$	$(-)^n D^L_{m,-n}$
$\sigma(yz)$	$D^L_{m,-n}$
σ_ϕ	$(-)^n \exp(-i2n\ \phi)\ D^L_{m,-n}$
$C_2(z)$	$(-)^n D^L_{m,n}$
$C_k(z)$	$\exp(-in2\pi/k) D^L_{m,n}$
$C_\gamma(z)$	$\exp(-in\gamma)\ D^L_{m,n}$
$C_2(x)$	$(-)^L D^L_{m,-n}$
$C_k(x)$	$(-)^{L-n} \exp(-in2\pi/k)\ D^L_{m,-n}$
$C_\gamma(x)$	$(-)^{L-n} \exp(-in\gamma)\ D^L_{m,-n}$
$C_2(y)$	$(-)^{L-n} D^L_{m,-n}$
$C_\beta(y)$	$\sum_q D^L_{m,q} d^L_{q,n}(\beta)$
$C_2(\phi)$	$(-)^L \exp(-i2n\phi)\ D^L_{m,-n}$
S_k	$(-)^{L-n} \exp(-in2\pi/k)\ D^L_{m,n}$

molecular point group can be worked out by studying first the effect of various molecular symmetry operations on the Wigner functions, which we summarize for convenience in Table 2. The effect of mesophase symmetries can also be easily obtained remembering that the order of operations applied in the laboratory and molecular frame is reversed so that for example, the first subscript, m, in $D^L_{m,n}$ is changed instead of n (cf. Appendix). In general an additional symmetry for the order parameters follows immediately from the reality of $f(\omega)$ and the relation $D^{L*}_{m,n} = (-)^{m-n} D^L_{-m,-n}$ for the Wigner functions which imply

$$\langle D^{L*}_{m,n} \rangle = (-)^{m-n} \langle D^{L*}_{-m,-n} \rangle \,. \tag{39}$$

The symmetry relations just determined allow us to find the relevant order parameters for molecules and phases of various symmetry. As an example we give in Table 3 the surviving order parameters of second and fourth rank for a molecule with a certain point group symmetry dissolved in a uniaxial phase (2). Finally, it is perhaps worth mentioning that the relations we have found tell us only about the order parameters that can be different from zero. Thus they indicate which order parameters we can try to measure but, of course, do not say anything about their magnitudes. Indeed it may well turn out that molecules of low symmetry exhibit order parameters consistent, within experimental error, with some higher symmetry. In particular we may find negligible deviations from cylindrical symmetry even if this is not guaranteed by the molecule point group. We talk in this case of effective cylindrical symmetry.

ORDERING MATRIX

An alternative definition for order parameters in uniaxial phases can be obtained by expanding the singlet orientational distribution $f(\omega)$ in terms of the direction cosines $\ell_\alpha = \ell_{z\alpha}$ (α = x,y,z) of the director with respect to a molecule fixed frame (24). Thus

$$8\pi^2 f(\omega) = 1 + 5 \sum_{\alpha,\beta} S_{\alpha\beta}\ell_\alpha\ell_\beta + 9 \sum_{\alpha,\beta,\gamma,\delta} S_{\alpha\beta\gamma\delta}\ell_\alpha\ell_\beta\ell_\gamma\ell_\delta + \dots \tag{40}$$

where the traceless second rank matrix

$$S_{\alpha\beta} = \langle 3\ell_\alpha\ell_\beta - \delta_{\alpha\beta}\rangle/2 \tag{41}$$

is called the Saupe ordering matrix (25) and

Table 3. *Independent orientational order parameters of second and fourth rank for uniaxial phases and molecules of various symmetry (2). n_2 and n_4 indicate the number of independent real quantities to be determined. Schonflies notation (21a) is used for the point group symbols.*

Molecular point group	n_2	$\langle D^2_{0,n}\rangle$	n_4	$\langle D^4_{0,n}\rangle$
C_1, C_i	5	$\langle D^2_{0,0}\rangle, \langle D^2_{0,1}\rangle, \langle D^2_{0,2}\rangle$	9	$\langle D^4_{0,0}\rangle, \langle D^4_{0,1}\rangle, \langle D^4_{0,2}\rangle, \langle D^4_{0,3}\rangle, \langle D^4_{0,4}\rangle$
C_s, C_2, C_{2h}	3	$\langle D^2_{0,0}\rangle, \langle D^2_{0,2}\rangle$	5	$\langle D^4_{0,0}\rangle, \langle D^4_{0,2}\rangle, \langle D^4_{0,4}\rangle$
C_{2v}, D_2, D_{2h}	2	$\langle D^2_{0,0}\rangle, \langle D^2_{0,2}\rangle = \langle D^{2*}_{0,2}\rangle$	3	$\langle D^4_{0,0}\rangle, \langle D^4_{0,2}\rangle = \langle D^{4*}_{0,2}\rangle, \langle D^4_{0,4}\rangle = \langle D^{4*}_{0,4}\rangle$
C_3, S_6	1	$\langle D^2_{0,0}\rangle$	3	$\langle D^4_{0,0}\rangle, \langle D^4_{0,3}\rangle$
C_4, C_{4h}, S_4	1	$\langle D^2_{0,0}\rangle$	3	$\langle D^4_{0,0}\rangle, \langle D^4_{0,4}\rangle$
C_{3v}, D_3, D_{3d}	1	$\langle D^2_{0,0}\rangle$	2	$\langle D^4_{0,0}\rangle, \langle D^4_{0,3}\rangle = \langle D^{4*}_{0,3}\rangle$
$C_{4v}, D_{2d}, D_{4h}, D_4$	1	$\langle D^2_{0,0}\rangle$	2	$\langle D^4_{0,0}\rangle, \langle D^4_{0,4}\rangle = \langle D^{4*}_{0,4}\rangle$
C_5, C_{5h}, C_{5v}	1	$\langle D^2_{0,0}\rangle$	1	$\langle D^4_{0,0}\rangle$
$D_{4d}, D_5, D_{5h}, D_{5d}$	1	$\langle D^2_{0,0}\rangle$	1	$\langle D^4_{0,0}\rangle$
$C_{3h}, C_6, C_{6h}, C_{6v}$	1	$\langle D^2_{0,0}\rangle$	1	$\langle D^4_{0,0}\rangle$
$D_{3h}, D_6, D_{6h}, D_{6d}$	1	$\langle D^2_{0,0}\rangle$	1	$\langle D^4_{0,0}\rangle$
$C_\infty, C_{\infty v}, C_{\infty h}, D_{\infty h}$	1	$\langle D^2_{0,0}\rangle$	1	$\langle D^4_{0,0}\rangle$

Table 4. *The components S_{ab}, with a, b = x, y, z of the Saupe ordering matrix in terms of the Wigner rotation matrix averages $\langle D^2_{m,n} \rangle$. Recall that $S_{ab} = S_{ba}$ and that $S_{xx} + S_{yy} + S_{zz} = 0$.*

$$S_{xx} = (\tfrac{3}{8})^{\frac{1}{2}}\{\langle D^2_{0,2}\rangle + \langle D^2_{0,-2}\rangle\} - \langle D^2_{0,0}\rangle/2$$

$$S_{xy} = i(\tfrac{3}{8})^{\frac{1}{2}}\{\langle D^2_{0,2}\rangle - \langle D^2_{0,-2}\rangle\}$$

$$S_{xz} = -(\tfrac{3}{8})^{\frac{1}{2}}\{\langle D^2_{0,-1}\rangle + \langle D^2_{0,1}\rangle\}$$

$$S_{yy} = -(\tfrac{3}{8})^{\frac{1}{2}}\{\langle D^2_{0,2}\rangle + \langle D^2_{0,-2}\rangle\} - \langle D^2_{0,0}\rangle/2$$

$$S_{yz} = i(\tfrac{3}{8})^{\frac{1}{2}}\{\langle D^2_{0,1}\rangle \; \langle D^2_{0,-1}\rangle\}$$

$$S_{zz} = \langle D^2_{0,0}\rangle$$

$$S_{\alpha\beta\gamma\delta} = \{35\langle \ell_\alpha \ell_\beta \ell_\gamma \ell_\delta\rangle - 5(\langle \ell_\alpha \ell_\beta\rangle\delta_{\gamma\delta} + \langle \ell_\alpha \ell_\gamma\rangle\delta_{\beta\delta} + \langle \ell_\alpha \ell_\delta\rangle\delta_{\beta\gamma} + \langle \ell_\beta \ell_\gamma\rangle\delta_{\alpha\delta} + \langle \ell_\beta \ell_\delta\rangle\delta_{\alpha\gamma} + \langle \ell_\gamma \ell_\delta\rangle\delta_{\alpha\beta}) + (\delta_{\alpha\beta}\delta_{\gamma\delta} + \delta_{\alpha\gamma}\delta_{\beta\delta} + \delta_{\alpha\delta}\delta_{\beta\gamma})\}/8. \quad (42)$$

The description of ordering in terms of Wigner matrices is normally convenient for theoretical manipulations in view of the simple transformation properties of these functions (cf. Appendix). However, the ordering matrix formalism is very frequently used in nuclear magnetic resonance (5), so we give in Table 4 the relation between $S_{\alpha\beta}$ and $\langle D^2_{0,n}\rangle$. Notice once more that the ordering matrix formalism as presented in this section and commonly employed applies to non uniaxial molecules in a uniaxial phase. A generalization to more complex phases will be given later on.

ORIENTATIONAL ORDER PARAMETERS FOR MORE COMPLEX PHASES

A classification of the possible symmetries of phases with long range order and of their defects has been put forward by Kleman and Michel using rather formal group theoretical arguments (26).

A theoretical classification of all the possible ordered phases has also been given by Goshen et al. (27). Here, however, we do not want to list all the possible order parameters for every phase since most of these are at the moment of purely academic interest. Instead we shall concentrate on purely orientational order parameters and say something about the smectic C phase. First we point out that phases possessing translational order should be given a space group like classification (27). Thus the symmetrization of the positional-orientational distribution functions and order parameters should be done by applying all the operations of the relevant space group. However, the purely orientational order parameters, which define the distribution $f(\omega)$, (Eq. (16)) can be classified simply according to the point group for the system. This is, of course, because the purely orientational order parameters are invariant under translation; they correspond to the k=0 mode in Eqs. (10) and (11). Thus to find out the orientational parameters for a phase of a given crystallographic point group we can use the arguments already outlined based on the Wigner matrices transformations. Consider an ideal smectic C for example. This phase is a tilted version of the smectic A phase, where the director or major optic axis makes an angle θ, known as the tilt angle, with Z parallel to the layer and Y parallel to the layer normal; here we use X,Y,Z for the laboratory frame. According to Table 2 the symmetry operations of the C_{2h} group give the restrictions

$$\langle D^L_{m,n}\rangle = (-)^m \langle D^L_{m,n}\rangle = (-)^{L+m}\langle D^L_{m,n}\rangle \, . \tag{43}$$

Thus L and m have to be even. The relevant second rank order parameters should then be

$$\langle D^2_{0,n}\rangle \; ; \; \langle D^2_{2,n}\rangle = (-)^n \langle D^{2*}_{-2,n}\rangle \, . \tag{44}$$

Notice that, even if the particles constituting the phase have cylindrical symmetry, we still have two independent order parameters, corresponding to the biaxiality of the phase. This is perhaps more transparent in cartesian coordinates. Let us define an ordering matrix Q

$$Q_{AB} = (3\langle \ell_A \ell_B\rangle - \delta_{AB})/2 \; ; \; A,B = X,Y,Z \tag{45}$$

formally identical to the Saupe ordering matrix. Here, however, it is the phase that is not cylindrically symmetric instead of the particles, and so we take $\ell_A \equiv \ell_{Az}$ to be the direction cosines for a molecule in the laboratory frame. Now, from the assumed symmetry of the smectic C it is clear that

$$Q = \begin{pmatrix} Q_{XX} & Q_{XY} & 0 \\ Q_{XY} & Q_{YY} & 0 \\ 0 & 0 & Q_{ZZ} \end{pmatrix} \tag{46}$$

This can be diagonalized by a rotation of θ about Z, provided

$$\tan 2\theta = 2Q_{XY}/(Q_{YY}-Q_{XX}).$$

The angle θ is just the tilt angle, which represents the orientation of the director in the (XY) plane. The eigenvalues of Q are

$$q_1 = -Q_{ZZ}/2 + p \tag{47a}$$

$$q_2 = -Q_{ZZ}/2 - p \tag{47b}$$

$$q_3 = Q_{ZZ} \tag{47c}$$

where $p = (\tfrac{1}{2})\{(Q_{XX}-Q_{YY})^2 + 4Q_{XY}^2\}^{\frac{1}{2}}$ is a biaxiality parameter.

If, on the other hand the constituent molecules of the biaxial phase are themselves biaxial the order parameters will be

$$\langle D^2_{0,0} \rangle \tag{48a}$$

$$\langle D^2_{0,2} \rangle = \langle D^2_{0,-2} \rangle \tag{48b}$$

$$\langle D^2_{2,0} \rangle = \langle D^2_{-2,0} \rangle \tag{48c}$$

$$\langle D^2_{2,2} \rangle = \langle D^2_{-2,2} \rangle = \langle D^2_{2,-2} \rangle = \langle D^2_{-2,-2} \rangle \tag{48d}$$

Other combinations of molecular and mesophase symmetry can be treated along the same lines as the need arises employing the results given in the previous section. The methods presented here apply also to the various types of discotic mesophases (28-29).

CONNECTION WITH EXPERIMENT

Irreducible Tensor Formulation

Having seen how orientational order parameters can be systematically introduced we now wish to examine how they can be experimentally determined.

Let us consider, as an example, the determination of a molecular property which behaves as a second rank tensor property. If $F^{(L,m)}$ are the irreducible spherical components of the tensor F, as defined in the Appendix and if F is symmetric then its only non-zero components will be those of rank L=0 and L=2. Let us consider the anisotropic, L=2, component. The tensor F is measured in a laboratory frame and its components can be related to the molecule fixed components $F^{(2,m)}_{MOL}$ by

$$F^{(2,n)}_{LAB} = \sum_m D^{2*}_{n,m}(\omega)\, F^{(2,m)}_{MOL} \, . \tag{49}$$

Taking an ensemble average we have, for a uniaxial mesophase,

$$\begin{aligned} \langle F^{(2,n)}_{LAB}\rangle &= \langle F^{(2,0)}_{LAB}\rangle \delta_{n0}, \\ &= \sum_m \langle D^{2*}_{0,m}\rangle\, F^{(2,m)}_{MOL} , \end{aligned} \tag{50}$$

where we have taken the laboratory Z axis to be parallel to the optical axis and neglected director fluctuations. We have moreover assumed F to be constant in the molecular frame. Substituting the definition of $F^{(2,0)}$ (cf. Table A1) we have for the average components of F parallel and perpendicular to the director

$$\langle F_{par}\rangle = f + (\tfrac{2}{3})^{\frac{1}{2}} \sum_m \langle D^{2*}_{0,m}\rangle\, F^{(2,m)}_{MOL} \tag{51}$$

$$\langle F_{per}\rangle = f - (\tfrac{1}{6})^{\frac{1}{2}} \sum_m \langle D^{2*}_{0,m}\rangle\, F^{(2,m)}_{MOL} \tag{52}$$

where f is trace of F. For a biaxial molecule the experimentally measurable anisotropy of $\langle F\rangle$ is

$$\langle F_{par}\rangle-\langle F_{per}\rangle=(\tfrac{3}{2})^{\frac{1}{2}}\{F^{(2,0)}_{MOL}\langle D^2_{0,0}\rangle+2\mathrm{Re}(F^{(2,2)}_{MOL}\langle D^{2*}_{0,2}\rangle)\} \tag{53}$$

where we have chosen the molecular frame as the principal frame of the ordering matrix.

Thus the measurement of at least two anisotropy values is required to determine both $\langle D^2_{0,0}\rangle$ and $\langle D^2_{0,2}\rangle$. Moreover the deviation from cylindrical symmetry parameter $\langle D^2_{0,2}\rangle$ becomes measurable only when the tensor F has an off axis component so that $F^{(2,2)} \neq 0$. If the molecule has effective cylindrical symmetry, in the sense that $\langle D^2_{0,m}\rangle = \langle D^2_{0,0}\rangle\delta_{m0}$, then we have

$$\langle P_2\rangle = \langle D^2_{0,0}\rangle$$

$$= \langle F^{(2,0)}_{LAB}\rangle / F^{(2,0)}_{MOL} . \tag{54}$$

Thus, the orientational order parameter $\langle P_2\rangle$ can be obtained from the measured anisotropy in $\langle F\rangle$. In practice various tensors F and thus various techniques have been employed. They range from NMR (5) to Electron Spin Resonance (31), Absorption Dichroism (32), Fluorescence Depolarization (33) and Diamagnetic susceptibility (34) measurements.

A similar analysis can obviously be performed for a tensor of higher rank. However, while it is relatively easy to determine $\langle P_2\rangle$ from the anisotropy of some second rank tensor F (e.g. the dipolar coupling between two nuclei) it is much more difficult to find a tensor quantity of higher rank which can be measured directly. One is then forced to resort to somewhat indirect methods. For example the fourth rank order parameter $\langle P_4\rangle$ can be extracted from measurements of the mean square value of a second rank quantity (33,35). In the limit of cylindrical symmetry, we have, in fact,

$$\langle (F^{(2,0)}_{LAB})^2\rangle = \{(\tfrac{1}{5}) + (\tfrac{2}{7})\langle P_2\rangle + (\tfrac{18}{35})\langle P_4\rangle\}(F^{(2,0)}_{MOL})^2. \tag{55}$$

In the Raman scattering technique described by Pershan (35), the quantity F_{MOL} is the differential polarizability tensor for a certain localized Raman mode. In Fluorescence Depolarization the average of the product of two different tensors i.e. the absorption and emission tensors is determined to obtain a fourth rank order parameter (33).

Non Uniaxial Phases

From a formal point of view the treatment of non uniaxial phases does not present any difficulties. Eq. (49) still holds but now more than one component $\langle F^{(2,n)}_{LAB}\rangle$ may be different from zero when averaging. This is perhaps more illuminating if we substitute once more for the cartesian components in the lab frame. Using Table A2 we have for a biaxial solute dissolved in a biaxial phase formed of biaxial particles and referring to principal frames

$$\langle F_{XX}\rangle = -F^{(0,0)}/3^{\frac{1}{2}} - (\langle D^2_{0,0}\rangle F^{(2,0)}_{MOL} + 2\mathrm{Re}\langle D^{2*}_{0,2}\rangle F^{(2,2)}_{MOL})/6^{\frac{1}{2}} \tag{56}$$

$$+ (\mathrm{Re}\langle D^{2*}_{2,0}\rangle F^{(2,0)}_{MOL} + 2\mathrm{Re}\langle D^{2*}_{2,2}\rangle F^{(2,2)}_{MOL})$$

$$\langle F_{YY}\rangle = -F^{(0,0)}/3^{\frac{1}{2}} - (\langle D^2_{0,0}\rangle F^{(2,0)}_{MOL} + 2\mathrm{Re}\langle D^{2*}_{0,2}\rangle F^{(2,2)}_{MOL})/6^{\frac{1}{2}} \tag{57}$$

$$- (\mathrm{Re}\langle D^{2*}_{2,0}\rangle F^{(2,0)}_{MOL} + 2\,\mathrm{Re}\langle D^{2*}_{2,2}\rangle F^{(2,2)}_{MOL})$$

$$\langle F_{ZZ}\rangle = -F^{(0,0)}/3^{\frac{1}{2}} + (\tfrac{2}{3})^{\frac{1}{2}}(\langle D^2_{0,0}\rangle F^{(2,0)}_{MOL} + 2\mathrm{Re}\langle D^{2*}_{0,2}\rangle F^{(2,2)}_{MOL}). \tag{58}$$

As we see the order parameters describing the phase biaxiality i.e. $\langle D^2_{2,0}\rangle$ and $\langle D^2_{2,2}\rangle$ cause $\langle F_{XX}\rangle$ to be different from $\langle F_{YY}\rangle$. This effect can in turn determine e.g. the splitting of "perpendicular" lines in NMR spectra as verified in experimental studies in Smectic C phase (5c,36). Notice that the observation of phase biaxiality does not require an off axis molecular tensor (i.e. it can be obtained even when we have $F^{(2,2)} = 0$). In a real situation observation or not of mesophase biaxiality will of course depend on the relative magnitude of the terms in Eqs. (56-59) and on the sensitivity of the experiment.

Cartesian Treatment

Let us consider again the measurement of a property which behaves as a second rank tensor $\langle F\rangle$. We imagine for completeness that all the nine components of the representative matrix can be measured in a given laboratory frame. The laboratory frame components will be connected to their molecular counterparts through a simple similarity transformation

$$\langle F_{AB}\rangle = \sum_{a,b} \langle R_{Aa}\, F_{ab}\, \tilde{R}_{bB}\rangle \quad , \quad A,B = X,Y,Z\ ;\ a,b = x,y,z \tag{59}$$

where capital and lower case letters are used for laboratory and molecule fixed components respectively and the upper tilde stands for the transpose. Here R_{Aa} is the rotation matrix connecting laboratory to molecule frame with components e.g. $R_{Aa} = \ell_{Aa}$, direction cosine of the a molecular axis. If the molecule is rigid or if the tensor F_{ab} does not fluctuate in the molecular frame (e.g. we have a dipolar coupling between two nuclei belonging to the same rigid fragment) then

$$\langle F_{AB}\rangle = \sum_{a,b} \langle R_{Aa} \tilde{R}_{bB}\rangle F_{ab} \tag{60}$$

We rewrite this equation by adding and substracting $f\delta_{AB}$ on the right hand side, and with $f = \mathrm{Tr}\, F$:

$$\langle F_{AB}\rangle = f\delta_{AB} + (\tfrac{2}{3}) \sum_{a,b} C_{AB,ab} F_{ab} \tag{61}$$

where the four indices matrix C, defined as

$$C_{ABab} \equiv \tfrac{3}{2}\langle R_{Aa} R_{Bb}\rangle - \delta_{AB}\, \delta_{ab}/2 \tag{62}$$

can be called a superordering matrix. It transforms in fact, as from Eq. (61), one second rank matrix into another, as a superoperator transforms one operator into another (37). It reduces to zero in an isotropic liquid in view of the properties

$$\mathrm{Tr}_A C \equiv \sum_{A} C_{AAab} = 0 \tag{63a}$$

$$\mathrm{Tr}_a C \equiv \sum_{a} C_{ABaa} = 0 \tag{63b}$$

$$\mathrm{Tr} C = \sum_{a,A} C_{AAaa} = 0 \,. \tag{63c}$$

We define in general a laboratory principal frame through the transformation which makes the C matrix block diagonal in the first two indices. In Monte Carlo simulations of non uniaxial phases (38) finding this frame might be a problem. Using this formalism a suitable tensor F can be defined in the molecular frame and all its average components $\langle F_{AB}\rangle$ determined in the lab frame. In the principal lab frame a tensor average as in Eq. (61) should be diagonal with

$$\langle F_{AA}\rangle = \langle F_{AB}\rangle\, \delta_{AB} = f + (\tfrac{2}{3}) \sum_{a,b} C_{AAab} F_{ab} \quad , \quad A = X,Y,Z. \tag{64}$$

In a uniaxial phase there are only two independent

quantities of the average tensor $\langle F \rangle$ e.g. the trace and the components parallel to the director i.e. $\langle F_{ZZ} \rangle$. Thus in turn only one block of the superordering matrix needs to be specified i.e.

$$S_{ab} = C_{ZZab} \tag{65}$$

In other words in a uniaxial phase the superordering matrix reduces to the usual Saupe matrix.

Another case mentioned earlier on is that of a biaxial phase formed of uniaxial particles. Here we only need

$$Q_{AB} = C_{ABzz}. \tag{66}$$

Thus in this case the superordering matrix can be reduced to the Q tensor defined by Eq. (45).

SUMMARY

Orientational order parameters can be introduced in general as expansion coefficients in the single particle distribution function. In this way we are naturally led to introduce Wigner rotation matrices averages as the order parameters for a rigid particle. For rigid molecules the number and type of non vanishing orientational order parameters can be determined by applying molecular and mesophase point group symmetry.

An equivalent cartesian representation especially convenient at second rank level in view of the many numerical algorithms dealing with matrix transformations can be introduced. This is the familiar Saupe ordering matrix for uniaxial phases and more generally a four indices superordering matrix.

ACKNOWLEDGMENT

Financial support from Ministero P.I. is gratefully acknowledged.

APPENDIX

IRREDUCIBLE TENSORS AND WIGNER ROTATION MATRICES

A tensor of rank n is a quantity that transforms under rotation as the nth direct power of a vector. The 3^n dimensional representation of the rotation group realized in this way can be decomposed into a set of irreducible representations $D^{(L)}$ each of dimension (2L+1). The matrix elements of the irreducible representation in a basis where the angular momentum operator J^2 and its projection J_Z are diagonal with eigenstates $|Lm\rangle$ can be written as

$$D^L_{m,n}(\alpha\beta\gamma) = \langle Lm|e^{-i\alpha J_Z} e^{-i\beta J_Y} e^{-i\gamma J_Z}|Ln\rangle, \qquad \text{(A1)}$$

where α, β, γ are Euler angles defined according to the convention of Rose (4). The transformation involved corresponds to a rotation of α around the z axis of a right handed laboratory frame, followed by a rotation of β around the new y axis then by a further rotation of γ about the current z axis. Anticlockwise rotations are taken to be positive. We also employ the notation $D^L_{m,n}$ (F' - F) for the rotation matrix carrying from F to F' when we wish to emphasize the frames F, F' involved. The matrix elements $D^L_{m,n}$ (αβγ) are called Wigner rotation matrices, Wigner functions or generalized spherical harmonics. Combinations of ordinary tensor components transforming according to the representation $D^{(L)}$ are called irreducible tensor components of rank L and denoted by $T^{(L,m)}$, e.g.

$$T^{(L,m)}_{MOL} = \sum_n D^L_{n,m}(\alpha\beta\gamma)\; T^{(L,n)}_{LAB} \;. \qquad \text{(A2a)}$$

$$T^{(L,m)}_{LAB} = \sum_n D^L_{m,n}(\alpha\beta\gamma)^* T^{(L,n)}_{MOL} \qquad \text{(A2b)}$$

where the LAB and MOL subscripts refer to the laboratory and rotated frame. The set of (2L+1) components of rank L is called an irreducible spherical tensor, $T^{(L)}$ of rank L.

Let us consider the case of a tensor T = a x b i.e. the direct product of two possibly different vectors. The explicit irreducible components of T are given in Table A1 in terms of their cartesian counterparts. Eq. (A2) illustrates the main reason for the usefulness of irreducible tensors in problems involving rotations i.e. that their transformation properties are very simple. It is therefore often convenient to rewrite cartesian tensor components in terms of the spherical ones. In

Table A1. *Irreducible spherical components $T^{(L,m)}$ of the tensor $T = a \times b$ in terms of the cartesian components of the vectors a, b.*

$$T^{(0,0)} = -(a_x b_x + a_y b_y + a_z b_z)/3^{\frac{1}{2}}$$

$$T^{(1,0)} = -i(a_y b_x - a_x b_y)/2^{\frac{1}{2}}$$

$$T^{(1,\pm 1)} = a_z b_x - a_x b_z \pm i(a_z b_y - a_y b_z)/2$$

$$T^{(2,0)} = (\tfrac{2}{3})^{\frac{1}{2}}\{a_z b_z - (a_x b_x + a_y b_y)/2\}$$

$$T^{(2,\pm 1)} = \mp\{a_x b_x + a_z b_x \pm i(a_y b_z + a_z b_y)\}/2$$

$$T^{(2,\pm 2)} = \{(a_x b_x - a_y b_y \pm i(a_x b_y + a_y b_x)\}/2$$

Table A2. *The cartesian components of a symmetric second rank cartesian tensor T_{ab} in terms of the spherical components $T^{L,m}$.*

$$T_{XX} = -T^{(0,0)}/3^{\frac{1}{2}} - T^{(2,0)}/6^{\frac{1}{2}} + (T^{(2,2)} + T^{(2,-2)})/2$$

$$T_{XY} = -i(T^{(2,2)} - T^{(2,-2)})/2$$

$$T_{XZ} = (T^{(2,-1)} - T^{(2,1)})/2$$

$$T_{YY} = -T^{(0,0)}/3^{\frac{1}{2}} - T^{(2,0)}/6^{\frac{1}{2}} - (T^{(2,2)} + T^{(2,-2)})/2$$

$$T_{YZ} = i(T^{(2,1)} + T^{(2,-1)})/2$$

$$T_{ZZ} = -T^{(0,0)}/3^{\frac{1}{2}} + (\tfrac{2}{3})^{\frac{1}{2}}T^{(2,0)}$$

Table A2 we show explicitly how this can be done for a symmetric second rank tensor. From Eq. (A1) it is apparent that we can express $D^L_{m,n}(\alpha\beta\gamma)$ as

$$D^L_{m,n}(\alpha\beta\gamma) = e^{-im\alpha}\, d^L_{m,n}(\beta)\, e^{-in\gamma}, \tag{A3}$$

The real quantities,

$$d^L_{m,n}(\beta) = \langle Lm|e^{-i\beta J_Y}|Ln\rangle, \tag{A4}$$

are called reduced or small Wigner matrices. Notice that for $\beta = 0$ Eq. (A4) gives $d^L_{m,n}(0) = \langle Lm|Ln\rangle = \delta_{m,n}$.

The functions $D^L_{m,n}(\alpha\beta\gamma)$ constitute a complete orthogonal set spanning the space of the angles α, β, γ. When one or two of the subscripts are zero the Wigner rotation matrices reduce respectively to spherical harmonics or Legendre polynomials.

$$D^L_{m,0}(\alpha\beta 0) = \{\, 4\pi/(2L+1)\,\}^{\frac{1}{2}} Y^*_{L,m}(\beta\alpha), \tag{A5}$$

$$D^L_{0,0}(0\beta 0) = d^L_{0,0}(\beta) = P_L(\cos\beta), \tag{A6}$$

where $Y_{L,m}$ is a spherical harmonic and P_L a Legendre polynomial. Some of their properties, which we frequently use, are as follows.

Orthogonality

$$\int_0^{2\pi} d\alpha \int_0^{\pi} d\beta\, \sin\beta \int_0^{2\pi} d\gamma\, D^{L^*}_{m,n}(\alpha\beta\gamma)\, D^{L'}_{m',n'}(\alpha\beta\gamma) \tag{A7}$$

$$= 8\pi^2 \delta_{m,m'}\delta_{n,n'}\delta_{L,L'}/(2L+1).$$

We also have the useful special cases concerning spherical harmonics

$$\int_0^{2\pi} d\alpha \int_0^{\pi} \sin d\beta Y_{Lm}(\alpha\beta)\, Y_{L'm'}(\alpha\beta)^* = \delta_{LL'}\delta_{mm'} \tag{A8}$$

and Legendre polynomials

$$\int_0^{\pi} d\beta \sin\beta \, P_L(\cos\beta) \, P_{L'}(\cos\beta) = 2\delta_{LL'}/(2L+1). \tag{A9}$$

Closure

This allows coupling two successive rotations $(\alpha_1\ \beta_1\ \gamma_1)$ and $(\alpha_2\ \beta_2\ \gamma_2)$ to give a total rotation of $(\alpha\beta\gamma)$ as

$$\sum_n D^L_{m,n}(\alpha_1\ \beta_1\ \gamma_1) D^L_{n,m'}(\alpha_2\ \beta_2\ \gamma_2) = D^L_{m,m'}(\alpha\beta\gamma), \tag{A10}$$

Symmetry

$$D^{L^*}_{m,n}(\alpha\beta\gamma) = (-)^{m-n} D^L_{-m,-n}(\alpha\beta\gamma) = D^L_{n,m}(-\gamma, -\beta, -\alpha). \tag{A11}$$

Products

The product of two Wigner rotations of rank L' and L'' with the same argument can be rewritten as a linear combination of Wigner rotation matrices of rank L according to the relation

$$D^{L'}_{m',n'} D^{L''}_{m'',n''} = \sum_{L=|L'-L''|}^{L'+L''} C(L'L''L;m'm'')C(L'L''L;n'n'')D^L_{m'+m'',n'+n''} \tag{A12}$$

where C(abc;de) is a Clebsch-Gordan coefficient (4).

Integral of Three Wigner Rotation Matrices

Coupling two Wigner rotations as in Eq. (A12) and using orthogonality relation Eq. (A7) the following useful integral can be obtained

$$\int_0^{2\pi} d\alpha \int_0^{\pi} \sin\beta d\beta \int_0^{2\pi} d\gamma \, D^{L''^*}_{m'',n''}(\alpha\beta\gamma) \, D^{L'}_{m',n'}(\alpha\beta\gamma) \, D^L_{m,n}(\alpha\beta\gamma) \tag{A13}$$

$$= 8\pi^2 \delta_{m+m',m''} \delta_{n+n',n''} \, C(LL'L'';mm') \, C(LL'L'';nn')/(2L''+1),$$

We now give the explicit expressions for the Wigner rotation matrices (and indirectly for the order parameters i.e. their orientational averages). From Eq. (A3) we see that what we really need are expressions for the small matrices $d^L_{m,n}(\beta)$. In

Table A3 we give explicit expressions for the most important cases L=0,2,4.

Table A3. *Explicit expressions for the small Wigner matrices* $d^L_{m,n}(\beta)$; *L=0,2,4.*

L = 0

$$d^0_{0,0} = 1$$

L = 2

$$d^2_{2,2} = d^2_{-2,-2} = (3 + 4\cos\beta + \cos 2\beta)/8$$

$$d^2_{2,1} = -d^2_{-2,-1} = -d^2_{1,2} = d^2_{-1,-2} = -(1 + \cos\beta)(\sin\beta)/2$$

$$d^2_{2,0} = d^2_{-2,0} = d^2_{0,2} = d^2_{0,-2} = 6^{\frac{1}{2}}(1 - \cos 2\beta)/8$$

$$d^2_{2,-1} = -d^2_{-2,1} = -d^2_{-1,2} = d^2_{1,-2} = (-2\sin\beta + \sin 2\beta)/4$$

$$d^2_{2,-2} = d^2_{-2,2} = (3 - 4\cos\beta + \cos 2\beta)/8$$

$$d^2_{1,1} = d^2_{-1,-1} = (\cos\beta + \cos 2\beta)/2$$

$$d^2_{1,0} = -d^2_{-1,0} = -d^2_{0,1} = d^2_{0,-1} = -6^{\frac{1}{2}}(\sin 2\beta)/4$$

$$d^2_{1,-1} = d^2_{-1,1} = (\cos\beta - \cos 2\beta)/2$$

$$d^2_{0,0} = (1 + 3\cos 2\beta)/4$$

L = 4

$$d^4_{4,4} = d^4_{-4,-4} = (35 + 56\cos\beta + 28\cos 2\beta + 8\cos 3\beta + \cos 4\beta)/128$$

$$d^4_{4,3}=-d^4_{-4,-3}=-d^4_{3,4}=d^4_{-3,-4}=2^{\frac{1}{2}}(-14\sin\beta-14\sin2\beta-6\sin3\beta-\sin4\beta)/64$$

$$d^4_{4,2}=d^4_{-4,-2}=d^4_{2,4}=d^4_{-2,-4}=7^{\frac{1}{2}}(4\cos\beta-4\cos2\beta-4\cos3\beta-\cos4\beta+5)/64$$

$$d^4_{4,1}=-d^4_{-4,-1}=-d^4_{1,4}=d^4_{-1,-4}=14^{\frac{1}{2}}(-6\sin\beta-2\sin2\beta+2\sin3\beta+\sin4\beta)/64$$

$$d^4_{4,0}=d^4_{-4,0}=d^4_{0,4}=d^4_{0,-4}=70^{\frac{1}{2}}(\sin\beta)^4/16$$

$$d^4_{4,-1}=-d^4_{-4,1}=-d^4_{-1,4}=d^4_{1,-4}=14^{\frac{1}{2}}(-6\sin\beta+2\sin2\beta+2\sin3\beta-\sin4\beta)/64$$

$$d^4_{4,-2}=d^4_{-4,2}=d^4_{-2,4}=d^4_{2,-4}=7^{\frac{1}{2}}(-4\cos\beta-4\cos2\beta+4\cos3\beta-\cos4\beta+5)/64$$

$$d^4_{4,-3}=-d^4_{-4,3}=d^4_{3,-4}=-d^4_{-3,4}=2^{\frac{1}{2}}(-14\sin\beta+14\sin2\beta-6\sin3\beta+\sin4\beta)/64$$

$$d^4_{4,-4}=d^4_{-4,4}=(35-56\cos\beta+28\cos2\beta-8\cos3\beta+\cos4\beta)/128$$

$$d^4_{3,3}=d^4_{-3,-3}=(7\cos\beta+14\cos2\beta+9\cos3\beta+2\cos4\beta)/32$$

$$d^4_{3,2}=-d^4_{-3,-2}=-d^4_{2,3}=d^4_{-2,-3}=14^{\frac{1}{2}}(\sin\beta-2\sin2\beta-3\sin3\beta-\sin4\beta)/32$$

$$d^4_{3,1}=d^4_{-3,-1}=d^4_{1,3}=d^4_{-1,-3}=7^{\frac{1}{2}}(3\cos\beta+2\cos2\beta-3\cos3\beta-2\cos4\beta)/32$$

$$d^4_{3,0}=-d^4_{-3,0}=-d^4_{0,3}=d^4_{0,-3}=35^{\frac{1}{2}}(-2\sin2\beta+\sin4\beta)/32$$

$$d^4_{3,-1}=d^4_{-3,1}=d^4_{-1,3}=d^4_{1,-3}=7^{\frac{1}{2}}(3\cos\beta-2\cos2\beta-3\cos3\beta+2\cos4\beta)/32$$

$$d^4_{3,-2}=-d^4_{-3,2}=-d^4_{-2,3}=d^4_{2,-3}=14^{\frac{1}{2}}(-\sin\beta-2\sin2\beta+3\sin3\beta-\sin4\beta)/32$$

$$d^4_{3,-3}=d^4_{-3,3}=(7\cos\beta-14\cos2\beta+9\cos3\beta-2\cos4\beta)/32$$

$$d^4_{2,2}=d^4_{-2,-2}=(5+2\cos\beta+4\cos2\beta+14\cos3\beta+7\cos4\beta)/32$$

$$d^4_{2,1}=-d^4_{-2,-1}=-d^4_{1,2}=d^4_{-1,-2}=2^{\frac{1}{2}}(-6\sin\beta+4\sin2\beta-14\sin3\beta-14\sin4\beta)/64$$

$$d^4_{2,0} = d^4_{-2,0} = d^4_{0,2} = d^4_{0,-2} = 10^{\frac{1}{2}}(3 + 4\cos 2\beta - 7\cos 4\beta)/64$$

$$d^4_{2,-1} = -d^4_{-2,1} = -d^4_{-1,2} = d^4_{1,-2} = 2^{\frac{1}{2}}(-3\sin\beta - 2\sin 2\beta - 7\sin 3\beta + 7\sin 4\beta)/32$$

$$d^4_{2,-2} = d^4_{-2,2} = (5 - 2\cos\beta + 4\cos 2\beta - 14\cos 3\beta + 7\cos 4\beta)/32$$

$$d^4_{1,1} = d^4_{-1,-1} = (9\cos\beta + 2\cos 2\beta + 7\cos 3\beta + 14\cos 4\beta)/32$$

$$d^4_{1,0} = -d^4_{-1,0} = -d^4_{0,1} = d^4_{0,-1} = -5^{\frac{1}{2}}(2\sin 2\beta + 7\sin 4\beta)/32$$

$$d^4_{1,-1} = d^4_{-1,1} = (9\cos\beta - 2\cos 2\beta + 7\cos 3\beta - 14\cos 4\beta)/32$$

$$d^4_{0,0} = (9 + 20\cos 2\beta + 35\cos 4\beta)/64$$

REFERENCES

1a. R. Balescu, *"Equilibrium and Non-Equilibrium Statistical Mechanics"*, Wiley, (1975).

1b. J.P. Hansen and I.R. McDonald, *"Theory of Simple Liquids"*, Academic Press, (1976).

1c. L. Landau and E.M. Lifshitz, *"Statistical Physics"*, Pergamon, (1969).

2. C. Zannoni, Chap. 3 in *"The Molecular Physics of Liquid Crystals"*, G.R. Luckhurst and G.W. Gray eds., Academic Press, (1979).

3. U. Fano, *Rev. Mod. Phys.* **29**, 74 (1957).

4. M.E. Rose, *"Elementary Theory of Angular Momentum"*, Wiley, (1957).

5a. J.W. Emsley and J.C. Lindon, *"N.M.R. Spectroscopy Using Liquid Crystal Solvents"*, Pergamon, (1965).

5b. P. Diehl and C.L. Khetrapal in *"NMR Basic Principles and Progress"*, P. Diehl, E. Fluck and R. Kosfeld eds., Vol. 1, Springer-Verlag (1969).

5c. J.W. Doane, in *"Magnetic Resonance of Phase Transitions"*, F.J. Owens, C.J. Poole Jr. and H.A. Farach, Academic Press (1979).

6. G.R. Luckhurst and G.W. Gray eds., *"The Molecular Physics of Liquid Crystals"*, Academic Press, (1979).

7. E.B. Priestley, P.J. Wojtowicz and P. Sheng eds., *"Introduction to Liquid Crystals"*, Plenum, (1975).

8. P.G. De Gennes, *"The Physics of Liquid Crystals"*, Oxford U.P. (1974).

9. S. Chandrasekhar, *"Liquid Crystals"*, Cambridge U.P. (1977).

10. C. Zannoni and M. Guerra, *Mol. Phys.*, **44**, 849 (1981).

11. A. Leadbetter, Chap. 13 in *"The Molecular Physics of Liquid Crystals"*, G.R. Luckhurst and G.W. Gray eds., Academic Press, (1979).

12. M. Kohli, K. Otnes, R. Pynn and T. Riste, *Z. Phys.* **B24**, 147 (1976).

13. M. Abramowitz and I.A. Stegun eds., *"Handbook of Mathematical Functions"*, Dover, (1964).

14. V. Zwetkoff, *Acta Physicoch. U.S.S.R.*, **10**, 557, (1939).

15a. H.E. Stanley, *"Introduction to Phase Transition and Critical Phenomena"*, Oxford U.P., (1971).

15b. A.B. Pippard, *"Elements of Classical Thermodynamics"*, Cambridge U.P., (1966).

16. E.T. Jaynes, *"The Maximum Entropy Formalism"*, MIT Press (1979).

17a. B.J. Berne and G.D. Harp, *Adv. Chem. Phys.*, **17**, 63, (1970).

17b. D.I. Bower, *J. Polym. Sci.*, **19**, 93 (1981).

18. G.R. Luckhurst, Chap. 4 in *"The Molecular Physics of Liquid Crystals"*, G.R. Luckhurst and G.W. Gray eds., Academic Press, (1979).

19. W. Maier and A. Saupe, *Z. Naturforsch.*, **13a**, 564, (1958); **14a**, 882 (1958); **14a**, 882 (1959); **15a**, 287 (1960).

20a. G.R. Luckhurst, C. Zannoni, P.L. Nordio and U. Segre, *Mol. Phys.*, **30**, 1345, (1975).

20b. D. Catalano, C. Forte, C.A. Veracini and C. Zannoni, *Israel J. Chem.*, **23**, 283 (1983) and refs. therein.

21a. L. Landau and E.M. Lifshitz, "*Quantum Mechanics*", Pergamon Press (1966).

21b. M. Lax, "*Symmetry Principles in Solid State and Molecular Physics*", Wiley (1974).

22. W.A. Steele, *J. Chem. Phys.* **39**, 3197 (1963).

23. L. Blum and A.J. Torruella, *J. Chem. Phys.* **56**, 303, (1972).

24. A.D. Buckingham, *Disc. Faraday Soc.*, **43**, 205 (1967).

25a. A. Saupe, *Z. Naturforsch*, **19a**, 161 (1964).

25b. A. Saupe, *Angew. Chem. (Int. Edn.)* **7**, 97 (1968).

26. M. Kleman and L. Michel, *Phys. Rev. Letts.*, **40**, 1387 (1978).

27. S. Goshen, D. Mukamel and S. Shtrikman, *Mol. Cryst. Liq. Cryst.*, **31**, 171 (1975).

28. S. Chandrasekhar, B.K. Sadashiva and K.A. Suresh, *Pramana*, **9**, 471 (1977).

29. J. Billard, J.C. Dubois, N.T. Nguyen and A. Zann, *Nouv. J. Chim.* **2**, 535, (1978).

30. J. Charvolin and B. Deloche, Chap. 15 in "*The Molecular Physics of Liquid Crystals*", G.R. Luckhurst and G.W. Gray eds., Academic Press, (1979).

31a. G.R. Luckhurst, in "*Electron Spin Relaxation in Liquids*", L.T. Muus and P.W. Atkins eds., Plenum, (1972).

31b. L.J. Berliner, "*Spin Labelling Theory and Applications*", Academic Press, (1976).

32. E.W. Thulstrup and J. Michl, *J. Phys. Chem.*, **84**, 82 (1980).

33a. C. Zannoni, *Mol. Phys.*, **38**, 1813 (1979); **42**, 1303 (1981).

33b. C. Zannoni, A. Arcioni and P. Cavatorta, *Chem. Phys. of Lipids*, **32**, 179 (1983).

34. A. Buka and W.H. de Jeu, *J. Phys.*, **43**, 361 (1982).

35a. P. Pershan, Chap. 17 in Ref. 6.

35b. K. Miyano, *J. Chem. Phys.* **69**, 4807 (1978).

35c. L.G.P. Dalmolen and W.H. de Jeu, *J. Chem. Phys.* **78**, 7353 (1983).

36. P.J. Bos, J. Pirs, P. Ukleja and J.W. Doane, *Mol. Cryst. Liq. Cryst.*, **40**, 59 (1977).

37. L.T. Muus, Chap. 1 in "*Electron Spin Relaxation in Liquids*", L.T. Muus and P.W. Atkins eds., Plenum (1972).

38a. G.R. Luckhurst and S. Romano, *Mol. Phys.*, **40**, 129 (1980).

38b. R. Hashim, University of Southampton, Chemistry Department Report (1981).

AN INTERNAL ORDER PARAMETER FORMALISM FOR NON-RIGID MOLECULES

Claudio Zannoni

Istituto di Chimica Fisica
Universita' di Bologna

INTRODUCTION

There has recently been much interest in describing orientational order in flexible molecules. The relevance of this problem is quite clear when considering that the molecules which form either thermotropic or lyotropic liquid crystals are in general not rigid. Typically we might have multi-ring molecules with some degree of internal rotation or molecules with floppy chain substituents. In addition to this, solutes with internal degrees of freedom dissolved in liquid crystals or membranes can be studied with various techniques (1). The assumption of rigidity (and quite often that of cylindrical symmetry) has usually been made in the past on the grounds of simplicity and of the inadequacy of experimental data in studying molecular structures at this level of sophistication. This latter point is no longer true; the quality and quantity of data becoming available, particularly from Nuclear Magnetic Resonance (2-4) techniques applied to isotopically substituted molecules now often demands going beyond the rigid molecule approximation.

The problem of a general formalism for describing orientational order in non-rigid particles is also becoming timely in the field of computer simulations (5). The first simulations of small alkyl chain systems and even of model bilayers are starting to appear (6). Once the attendant technical and computational problems are solved, these simulations can in fact give a huge amount of information. As an example, in a molecular dynamics simulation, all the positions and orientations of the submolecular fragments can be determined at every time step considered. Therefore, like in simulations of

J. W. Emsley (ed.), Nuclear Magnetic Resonance of Liquid Crystals, 35–52.

rigid particles the need arises of condensing this embarrassingly large amount of data in a set of quantities which on one hand contains the relevant information and on the other can hopefully be measured (7).

The description of orientational ordering in molecules with internal degrees of freedom has been considered by Emsley and Luckhurst (EL) using an equilibrium statistical mechanics approach (8) and by Burnell and de Lange (BL) using a time dependent picture (9). The two approaches are apparently quite different. EL have no time scale argument and express order parameters at Boltzmann averages in terms of a potential of mean torque depending on orientational as well as conformational degrees of freedom. They then partition this potential into an internal part $U_{int}(h)$ depending only on the internal state of the molecule and an external part depending both on conformation h and orientation ω. They are then able to assign an ordinary orientational ordering matrix to every conformation h. The orientational order parameter for the molecule is finally defined as an average over the singlet conformational distribution. In BL treatment the approach is kinetic. Rate equations for conformational changes are set up under assumptions of stochasticity and markovianicity and order parameters are recovered as time averages in the limit of very long times. We do not enter here into a detailed comparison of the two methods or in a discussion of their equivalence or not (10). We notice, however, that in both approaches the underlying philosophy is that of treating a non-rigid molecule as a mixture of conformers, considered in turn as normal rigid molecules.

We wish to briefly outline here an alternative approach, which is a generalization of that presented in the previous chapter and which takes into account from the start the differences between a rigid and a non-rigid molecule. For simplicity we assume a flexible molecule to have only rotational internal degrees of freedom, i.e. a multi-rotor molecule. Apart from these internal degrees of freedom the molecule is rigid in the sense that the assumption of fixed bond lengths and bond angles is made here. This is justified for most relatively large molecules since changes of "shape" are largely determined by torsional as opposed to, say, vibrational degrees of freedom.

To begin with let us consider the definition of the positional-orientational state of a molecule, assuming, as usual in the statistical mechanics of fluids (11), that it can be treated as a classical particle. The state of a rigid particle is specified as seen in the previous chapter by a position vector r and by three orientational parameters e.g. three Euler angles (12) or four quaternions linked by a normalization relation (7,13). This set of quantities defines the origin and the orientation of a coordinate frame fixed on the molecule (the molecular frame). We call the set of three parameters mentioned above the orientation ω of the particle. Typically we could choose the origin of the molecular frame on the centre of mass. We would then select three orthogonal axis in such a way as to put in evidence the symmetry of the molecule. For example, if the molecule contains a ternary or higher rotation axis, we might define this as the molecular z axis. Other symmetry elements could be exploited quite similarly using Group Theory (14). If, as it is often the case, the molecule does not have enough symmetry to unequivocally determine the particle fixed frame, we could typically choose the molecule frame as the one diagonalizing a certain molecular tensor of interest. For instance in treating dispersion interactions it might be convenient to adopt as a molecular system the principal axis system of the polarizability tensor. In Liquid Crystals applications it is often expedient to choose the molecular frame as that diagonalizing the second rank ordering tensor. It is clear that this type of operational definition is not unique and if symmetry is not high enough, different choices of tensor of interest could lead to different molecular frames. The arbitrariness in selecting a molecular system clearly represents no problem at all as long as each system can be precisely defined, as is the case for a rigid molecule. As an example in electron resonance it is common to work with at least two systems: a local one diagonalizing the magnetic interaction of interest (hyperfine, g,...) and a molecular system, diagonalizing the diffusion tensor. The same simple description does not apply to more complex molecules with mobile parts. In view of what we have said, defining an orientation requires the possibility of introducing a prescription for attaching a coordinate frame to the molecule. If the definition of molecular frame as that diagonalizing a certain tensor is adopted, it is apparent that an internal rotation will in general change the axis system even if overall tumbling has not taken place. In the most general case the state of a non rigid molecule can be described by a set of vectors defining the position in space of all the constituent atoms. Fortunately, however, this situation is more general than is normally needed, since a molecule contains collections of atoms whose relative positions are fixed. We call these rigid fragments. If we can attach a coordinate frame to every rigid fragment, then the state of a molecule can be given as the collection of orientations and positions of the rigid fragments in the laboratory frame r_1, ω_1, r_2, ω_2,..., where the subscripts label the various rigid parts. This description, although formally complete, has perhaps the disadvantage of not making immediately clear the relation between different parts of the same molecule.

We might also take the view that every conformation is like

a rigid molecule with its own molecular frame and its orientation with respect to the laboratory frame. However, we would still have the problem, when taking averages is required, of expressing every property of interest for the various conformations in the same coordinate frame, in order to sum the (suitably weighted) conformational contributions. This means in practice that a common molecular frame has to be chosen, and it has to be placed in a rigid part of the molecule. If it is not possible to assume that enough rigidly connected atoms exist to define such a frame, another, more general description will have to be adopted. It seems therefore that the possibility of defining something we can call the orientation of a non rigid molecule is linked to the possibility of defining in it a rigid part. Taking this view it is quite natural then to define the state of the flexible molecule by giving on one hand the position and orientation of its chosen rigid part with respect to the laboratory frame and on the other the orientations of the internally moving parts with respect to the molecular rigid frame.

In the next Section this approach will be examined in detail for the simplest case of a molecule formed by two rigid parts mobile one with respect to the other, and the concept of internal order parameter will be introduced. The relation to experimentally observable quantities will then be discussed and we shall look at the limiting situation where orientational and conformational degrees of freedom are decoupled. We shall then briefly make contact between the present formalism and that of Emsley and Luckhurst. The formalism will then be generalized to more complex molecules formed of various rotating units.

THE INTERNAL ORDER PARAMETER FORMALISM

To be specific let us consider a molecule formed of two rigid distinguishable parts M1 and M2 connected by a bond and rotating one with respect to the other. This could be for instance an unsymmetrically substituted biphenyl.

Let us take one of the fragments, M1, say, as the "rigid" part, where we place the molecular frame. The second part of the molecule can rotate through an angle ϕ around an axis r. We now imagine the molecule embedded in a uniform isotropic or anisotropic fluid phase. Thus the molecule we consider could be a mesogen in its nematic or isotropic phase or it could equally well be a solute molecule dissolved in such a phase.

The singlet distribution function (15) can be written as

$$f(r_1,\omega_1,r_2,\omega_2) = f_r(\omega_1,\omega_2)$$

$$= f(\omega_1, \phi), \tag{1}$$

where the dependence on the inter-fragment vector r is parametric as r is taken to be constant while internal or external reorientation takes place. We shall omit writing down explicitly this dependence on the internal axis orientation and distance parameter r from now on, when there is no risk of confusion. In the biphenyl type molecule mentioned above the internal rotation axis r is along the para axis. It should be noticed that by writing the distribution as a function of the two labelled orientations ω_1 and ω_2 we have made use of the physical distinguishability of the two groups i.e. that it is possible to devise an experiment that can tell which part we are looking at. If, on the contrary, there is permutation symmetry $P_{1,2}$ for the two parts we should consider just one-independent orientation ω and write Eq. (1) as $f(\omega, \phi)$. We now go back to Eq. (1) and rewrite the one particle distribution in a suitable composite basis set. In practice we expand the external orientation dependence in terms of Wigner rotation matrices and the internal rotation in Fourier series. We find

$$f(\omega,\phi) = \sum ((2L + 1)/16\pi^3)\ f^L_{mnq}\ D^{L*}_{mn}(\omega)\ \exp(-iq\phi), \tag{2}$$

where $q = 0, \pm1, \pm2, \ldots$ and as usual the sum is extended to all the coefficients not appearing on the left hand side. The angle ϕ, $0 \leqq \phi \leqq 2\pi$ is the dihedral rotation angle around r.

The orthogonality relation of the chosen basis immediately allows definition of the expansion coefficients as

$$f^L_{mnq} = \langle D^L_{m,n}(\omega)\ \exp(iq\phi)\rangle \tag{3}$$

where the angular brackets denote an orientational average over the distribution $f(\omega, \phi)$. As usual (15) the singlet distribution expansion coefficients are related to the order parameters for the system. We have here three types of order parameters, i.e.

(i) Purely Orientational Order Parameters

$$\langle D^L_{m,n}(\omega)\rangle = f^L_{mn0}. \tag{4}$$

This type of expansion coefficient is essentially an ordinary orientational order parameter for the molecular frame. It gives the average orientation of the chosen rigid part of the molecule with respect to the director frame. For a truly rigid molecule the ordering matrix is of course unique and the parameters determined give information on the total ordering matrix. Thus in the rigid molecule limit if we wish to employ symmetry to reduce the number of independent parameter it is the overall point symmetry of the molecule that should be employed and not the local one. If, instead, the rigid fragment is connected to the rest of the molecule via a single bond (i.e. if it is an internal rotor) then local symmetry operations may become feasible and be applied. As an example in the two ring type molecules already mentioned if the fragment where the molecular frame has been planted possesses D_2 symmetry, then the distribution in Eq. (2) can be simplified using this local symmetry independently of the second ring orientation. This amounts to saying that feasible local symmetry operations of this kind can be treated exactly as we have seen in the previous lecture. In the example given we shall have the restriction that L has to be even in Eq. (2). The possibility of having independent fragments implies, however, that in a flexible molecule the orientation of one sub-unit does not automatically determine the orientation of the other fragments.

We also notice that the expansion in Eq. (2) separates completely the laboratory frame operations from the internal motion. The rotation $D^L_{m,n}(\omega) \equiv D^L_{m,n}(M1\text{-}L)$ is that affected by mesophase symmetry operations. Thus, for example, we have for uniaxial mesophases

$$f^L_{mnq} = f^L_{onq}\, \delta_{mo} \tag{5}$$

whatever the conformation. Similarly if the phase has a plane of symmetry perpendicular to the director then L has to be even. Other operations can be applied as already seen in the previous chapter.

(ii) Purely Internal Order Parameters

We define purely internal order parameters as the averages

$$\langle \exp(iq\phi)\rangle = f^0_{00q} \; ; \quad q=0,\pm1,\pm2,... \tag{6}$$

These parameters describe the ordering of the second part of the molecule with respect to the first one irrespective of the overall orientation. They can be different from zero even in the isotropic phase if there is some preferential orientation of the second fragment around the internal axis. The internal order parameters for a rigid molecule with the second fragment at an angle ϕ_0 from the first one will just be $\exp(iq\phi_0)$. By suitable definition of axis system this becomes just one. A flexible molecule can be defined as a molecule with internal order parameters deviating from the theoretical rigid value whatever the frame. At the other extreme, we have $f^0_{00q} = 0$ when the distribution $f(\omega,\phi)$ does not depend on the internal angle ϕ. Apart from the trivial case of a cylindrical fragment this indicates a uniaxial distribution of the second sub-unit around the rotation axis. The coefficient $f^0_{00q} = 0$ thus indicates complete internal disorder around the rotation axis. Notice that no time argument is involved in going from a rigid to an internally disordered molecule. Internal order parameters are determined by the energy barrier to rotation. Thus if the energy barrier is $U(\phi)$ the internal order parameters are obtained from the Boltzmann expression

$$f^0_{00q} = \int d\phi \, \exp(iq\phi) \exp(-U(\phi)/kT) / \int d\phi \, \exp(-U(\phi)/kT), \tag{7}$$

where $U(\phi)$ is the internal potential of mean torque and k,T are respectively the Boltzmann constant and the absolute temperature. In an isotropic system and if the medium effect on the barrier can be neglected the $U(\phi)$ is the potential obtainable in principle from an ab-initio or semi-empirical theoretical calculation (16-19). In turn the determination of the parameters f^0_{00q} can provide information on the internal barrier.

(iii) Mixed Internal-External Order Parameters

The third type of coefficients in Eq. (2) are

$$\langle D^L_{m,n}(\omega) \exp(iq\phi)\rangle = f^L_{mnq} \,. \tag{8}$$

These terms describe coupling between internal and external degrees of freedom. They can be used to recover purely orientational order parameters for the second sub-unit from those of the first one. In fact, writing down explicitly the transformations between frames we have

$$\langle D^L_{m,n}(M2\text{-}L)\rangle = \sum_p \langle D^L_{m,p}(M1\text{-}L) \, D^L_{p,n}(M2\text{-}M1)\rangle$$

$$= \sum_p R^L_{p,n}(M2\text{-}M1) \, \langle D^L_{m,p}(M1\text{-}L) \exp(-in\phi)\rangle \tag{9a}$$

$$= \sum_p R^L_{p,n}(M2\text{-}M1) \, f^L_{mp-n} \tag{9b}$$

where we have introduced the notation

$$R^L_{a,b}(\alpha,\beta) \equiv D^L_{a,b}(\alpha,\beta,0)$$

$$\equiv \exp(-ia\alpha)\, d^L_{a,b}(\beta) \tag{10}$$

to indicate the transformation defining the rotation axis. Notice that here $-L \leq p,n \leq L$ so that only a subset of mixed order parameters is needed in obtaining the ordering matrix for the second ring. Eq. (9) constitutes a sort of sum rule for the mixed order parameters. For a rigid molecule Eq. (9) reduces to

$$\langle D^L_{m,n}(M2-L)\rangle = \sum_p \langle D^L_{m,p}(M1-L)\rangle\, D^L_{p,n}(M1-M2) \tag{11}$$

since the transformation linking the two frames is a time independent one. In this case the orientation $\omega = (M1-L)$ is sufficient to completely define the state of the particle. Eq. (11) simply gives the relation between the orientational order parameter when expressed in the two frames.

For a molecule where the internal rotation axis coincides with the molecular frame z axis, we have $R^L_{pn}(0,0) = \delta_{pn}$ and

$$\langle D^L_{mn}(M2-L)\rangle = \langle D^L_{mn}(M1-L)\exp(-in\phi)\rangle. \tag{12}$$

$$= f^L_{mn-n}$$

so that in this case the sum rule reduces to just one term. We have already mentioned when looking at purely orientational parameters that in the present treatment the different frame transformations employed to specify the orientational and internal state are clearly distinct, thus making it possible to apply symmetry simplifications referring to internal or external degreees of freedom. We have also said that mesophase symmetry does not put restrictions on the internal order parameters. Here, however, we have the possibility of implementing fragment symmetry. In simple cases this can be done by direct inspection. For example the order parameters for a molecule with a para (i.e. along the z axis) biaxial substituent are

$$f^L_{mnq},\ q = 0, \mp 2, \mp 4, \ldots$$

Similarly if we have a methyl substituent, or generally a substituent with C_3 symmetry around the internal rotation axis, then $q = 0, \mp 3, \ldots$. In more complex cases the methods developed within the group theory of non rigid molecules can be employed (20-22). Our purpose here is not that of giving a systematic treatment for various rotor symmetries but it is interesting that the formalism allows full exploitation of local symmetry if needed.

OBSERVABLES FOR A MOLECULE WITH ONE INTERNAL ROTOR

We now examine here the experimental determination of the various types of order parameters introduced. Let us consider the measurement of a tensor property F with irreducible components $F^{(2,m)}$. The observable averaged component of F along the director in the laboratory frame will be in a uniaxial phase,

$$\langle F^{(2,0)}\rangle = \sum_m \langle D^2_{0m}(M1-L)^* \, F^{(2,m)}_{M1}\rangle. \tag{13}$$

The property F can be a true overall molecular property, defined for a certain molecular conformation, e.g. polarizability or dipole moment. However, it can also be a sub-molecular, local property, such as the nuclear dipolar coupling $T^{(2,m)}_{ij}$ between two nuclei i and j of importance in NMR. In the latter case we may have various possibilities.

i) The two nuclei i and j belong to the fragment where the molecular frame has been planted. If this is the case the $T^{(2,m)}_{ij}$ are constant whatever the conformation and we have

$$\langle T^{(2,0)}_{ij}\rangle = \sum_m \langle D^2_{0m}(M1-L)^*\rangle \, T^{(2,m)}_{ij} \quad ; \ i,j \text{ on M1}. \tag{14}$$

Thus the experiment can give the ordering matrix of the rigid part, provided enough couplings are available. The situation here is just the same as that of a truly rigid molecule as we discussed earlier on.

ii) The two nuclei both belong to the second ("moving") part of the molecule. The molecular frame couplings $T^{(2,m)}_{ij}$ exhibit in this case a dependence on the internal rotation angle ϕ, but it is not difficult to show that this dependence can be removed. In fact

$$\langle T^{(2,0)}_{ij}\rangle = \sum_m \langle D^2_{0m}(M1-L)^* \, T^{(2,m)}_{ij}\rangle \ ; \ i,j \text{ on M2} \tag{15a}$$

$$= \sum_{m,n} \langle D^2_{0m}(M1-L)^* \, D^2_{mn}(M2-M1)^* \rangle \, T^{(2,n)}_{ij} \tag{15b}$$

$$= \sum_{m,n} (-)^m \, R^2_{mn}(M2-M1)^* \, f^2_{0-mn} \, T^{(2,n)}_{ij} \tag{15c}$$

where the dependence on internal motion has been removed by transforming $T^{(2,m)}_{ij}$ to a frame M2 fixed on the second fragment. Coupling the two Wigner rotation matrices with the closure relation, Eq. (15) becomes

$$= \sum_{n} \langle D^2_{0n}(M2-L)^* \rangle \, T^{(2,n)}_{ij} \quad ; \; i,j \text{ on M2} \tag{15d}$$

These type of couplings therefore give information about the alignment of the second fragment in the laboratory frame.

iii) Each of the two nuclei belongs to a different sub-unit. This is the most interesting situation from the point of view of getting information on intramolecular rotation, i.e. the distinguishing property of our flexible molecules. We have

$$T^{(2,0)}_{ij} = \sum_{m} D^2_{0m}(M1-L)^* \, T^{(2,m)}_{ij}(\phi) \quad ; \; i \text{ on M1, } j \text{ on M2} \tag{16}$$

where the coupling $T^{(2,m)}_{ij}$ is now a function of the internal rotation angle ϕ. By Fourier expansion of the tensor component $T^{(\,,m)}_{ij}(\phi)$ we can write

$$T^{(2,m)}_{ij}(\phi) = \sum_{q} (T^{(2,m)}_{ij})_q \exp(iq\phi). \tag{17}$$

The couplings $T^{(2,m)}_{ij}(\phi)$ and therefore the Fourier components $(T^{(2,m)}_{ij})_q$ can of course be computed when the geometry of the fragment is known. Substitution of Eq. (16) thus gives

$$\langle T^{(2,0)}_{ij} \rangle = \sum_{m} \sum_{q} \langle D^2_{0m}(M-L)^* \exp(iq\phi) \rangle \, (T^{(2,m)}_{ij})_q \tag{18a}$$

$$= \sum_{m} \sum_{q} (-)^m \, f^2_{0-m,q} (T^{(2,m)}_{ij})_q \tag{18b}$$

where $m = 0, \pm 1, \pm 2$; $q = 0, \pm 1, \pm 2, \ldots$. We see therefore that this type of coupling contains information on internal order parameters. Dipolar couplings between different pairs of nuclei i and j can be expressed in terms of the same set of order parameters $f^{0}_{0m,q}$. Notice, however, that the Fourier series in q is, strictly an infinite one and the number of parameters is not limited by an orthogonality selection rule as in the rigid solute case. On the other hand the convergence of the Fourier expansion of $T^{(2,m)}_{ij}$ will limit the number of relevant terms. If enough couplings can be collected a fitting procedure similar to that employed in the rigid molecule limit (3) can be used to extract a set of order parameters. Notice that if this can be pursued then a model for internal or overall reorientation does not have to be assumed. The determination of a set of order parameters would in turn allow methods mutuated from information theory (24-25) to be used for finding an approximate form of the single particle distribution as we shall briefly discuss later.

LIMITING CASES

A particularly simple limiting case arises when the probability of finding the molecule at a certain orientation ω and internal angle ϕ can be written as a product of orientational and conformational probabilities, i.e. when the two motions can be considered to be independent. In this case the mixed order parameters decouple:

$$f^{L}_{mnq} = \langle D^{L}_{m,n}(\omega)\ \exp(iq\phi)\rangle$$

$$= \langle D^{L}_{m,n}(\omega)\rangle\ \langle \exp(iq\phi)\rangle. \tag{19}$$

An order parameter for the second fragment becomes

$$\langle D^{L}_{m,n}(M2\text{-}L)\rangle = \sum_{p} \langle D^{L}_{m,p}(M1\text{-}L)\ D^{L}_{p,n}(M2\text{-}M1)\rangle$$

$$= \sum_{p} R^{L}_{p,n}(M2\text{-}M1)\ \langle D^{L}_{m,p}(M1\text{-}L)\rangle\ \langle \exp(-in\phi)\rangle. \tag{20}$$

Comparison with the rigid molecule expression Eq. (11) shows that Eq. (20) differs only by the substitution of exp(-inϕ) by ⟨exp(-inϕ)⟩ i.e. a rotation averaged over internal motions is substituted for the constant rotation. In this sense we can speak of an effective "averaged" molecule or pseudomolecule (8,23). We can see this also from the limiting expression for the coupling between nuclei on different frames, Eq. (16), which becomes

$$<T_{ij}^{(2,0)}> = \sum_m <D_{0,m}^{2*}(M1-L)> <T_{ij}^{(2,m)}>. \tag{21}$$

Notice that no argument about the relative speed of internal and overall rotation has been introduced.

RELATION TO EMSLEY-LUCKHURST FORMALISM

The EL formalism is treated in chapter 5. Here we consider the expression for $<A_{ZZ}>$ the component along the director of a partially averaged second rank tensor quantity A in the EL formalism, i.e.

$$<A_{ZZ}> = \sum_{a,b} \int dh\ P(h)\ A_{ab}(h)\ S_{ab}(h)\ ;\ a,b = X,Y,Z \tag{22}$$

where $A_{ab}(h)$ and $S_{ab}(h)$ are respectively the molecular property of interest and the ordering matrix calculated for the molecule in a conformation h, treated as a rigid particle. For a single rotor molecule the conformational label h reduces to the internal rotation angle ϕ. If we now expand the ϕ dependence of the ordering matrix and of the A tensor components in the molecular frame as

$$A_{ab}(\phi) = \sum (A_{ab})_n \exp(in\phi),$$

we find

$$<A_{ZZ}> = \sum_{n,n'} \sum_{a,b} (A_{ab})_n (S_{ab})_{n'} f^0_{00;n+n'} \tag{23}$$

where the parameters $f^0_{00;q}$, expansion coefficients of the internal distribution $P(\phi)$ correspond to our definition of purely internal order parameters. We see therefore that EL expressions can be written down in the present formalism. However, in the internal order parameters treatment the conformational information is disentangled from the rest. This shows the various types of different order parameters involved and their relation, if any.

MULTI-ROTOR MOLECULES

We consider a molecule with various rotating fragments as a system formed of N linked bodies of unspecified symmetry. Within the assumption of fixed bond length and valence angles we can think in terms of a molecular model where each sub-unit has one or more pins (bonds) r_i placed at a certain fixed orientation with respect to the local fragment frame Mi. The next body in the chain, if any, can be thought of as having a socket which fits in the previous pin (26). We can have various types of branched or linear systems according to the links realized between the bodies.

The methods required for molecules of different connectivity are quite similar. Here we shall therefore treat for convenience a simply connected system of N rotors (cf. Fig. (1)).

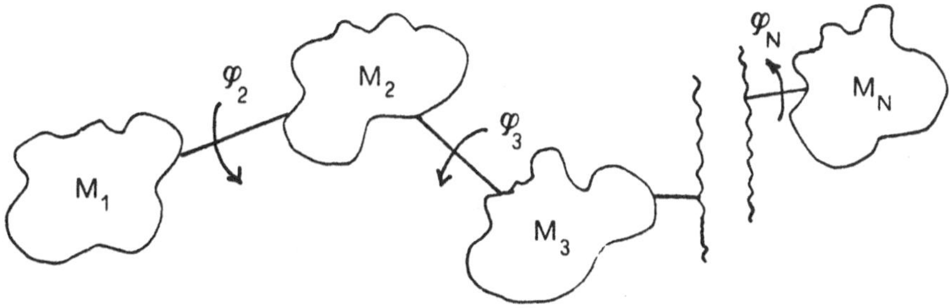

Figure 1. *A simply connected multi-rotor molecule.*

Here we imagine the molecular "rigid" frame to be fixed on the first (M1) fragment, although another sub-unit could equally well have been chosen, at least in general.

The system considered includes the case of a macromolecule with a chain attached or a nematogen with a methylene chain substituent. It also includes the case of polyphenyl type molecules.

The conformational-orientational state of a molecule of this kind can be given by specifying the rigid geometry (bond angles and bond lengths) once and for all and by giving the set of internal orientations $(\omega_1, \phi_2, \phi_3, \ldots, \phi_N)$. In the same sense the conformation alone is specified by $(\phi_2, \phi_3, \ldots, \phi_N)$. The singlet distribution function can now be written by generalizing Eq. (2) as

$$f(\omega, \phi_2, \phi_3, \ldots, \phi_N) = \sum\{(2L+1)/((8\pi^2)(2\pi)^{N-1})\} f^L_{mn,Q} D^{L*}_{mn}(\omega_1) \exp(-iQ.\Phi), \qquad (24)$$

where we have introduced the notation

$$Q = (q_2,\ q_3, \ldots,\ q_N) \tag{25a}$$

$$\Phi = (\phi_2,\ \phi_3, \ldots,\ \phi_N)\ . \tag{25b}$$

for the collection of labels q_i and of angles ϕ_i respectively. The coefficients in the expansion are explicitly

$$f^{L}_{mn,q_2q_3\ldots q_N} = \langle D^{L}_{mn}(\omega_1)\ \exp(i(q_2\phi_2+q_3\phi_3+\ldots+q_N\phi_N))\rangle \tag{26}$$

They represent generalized order parameters for the flexible molecule. A simple extension of the concepts introduced in the previous Sections allows interpreting these parameters as giving the correlation between the orientation of the various fragments.

Instead of keeping the treatment general let us now consider a specific example i.e. that of a molecule formed by three biaxial fragments M1, M2, M3 with a common rotation axis. This could be for example a suitably substituted terphenyl. The singlet distribution in a uniaxial phase will be

$$f(\omega_1,\phi_2,\phi_3) = \sum\{(2L+1)/32\pi^4\}f^{L}_{0n,q_2q_3}D^{L*}_{0n}(\omega_1)\exp(-iq_2\phi_2-iq_3\phi_3) \tag{27}$$

where L is even. We also have n, q_2 and q_3 even. If the distribution is even in ϕ_2 and ϕ_3 the order parameters should be invariant to a change in sign of q_2 and q_3. In order to consider the independent order parameters we can therefore limit ourselves to positive values of q_2, q_3 and of course of n.

Let us now consider again the experimentally determinable properties for the various rings. We consider dipolar couplings again even if other properties could of course be examined instead. We have:

i) First unit

$$\langle T^{(2,0)}_{ij}\rangle = \sum \langle D^{2}_{0n}(M1-L)\rangle\ T^{(2,n)*}_{ij}\ ;\ i,j \text{ on M1}$$

$$= \sum f^{2}_{0n,00}\ T^{(2,n)*}_{ij}\ ; \tag{28}$$

ii) Second unit

$$\langle T_{ij}^{(2,0)}\rangle = \sum \langle D^2_{0p}(M1-L)\ D^2_{pn}(M2-M1)\rangle\ T_{ij}^{(2,n)*};\ i,j \text{ on } M2$$

$$= \sum R^2_{pn}(M2-M1)\ \langle D^2_{0p}(M1-L)\ \exp(-in\phi_2)\rangle\ T_{ij}^{(2,n)*}$$

$$= \sum \langle D^2_{0n}(M1-L)\ \exp(-in\phi_2)\rangle\ T_{ij}^{(2,n)*}$$

$$= \sum f^2_{0n,-n0}\ T_{ij}^{2,n*}\ ; \tag{29}$$

iii) Third unit

$$\langle T_{ij}^{2,0}\rangle = \sum \langle D^2_{0p}(M1-L)\ D^2_{pq}(M2-M1)\ D^2_{qn}(M3-M2)\rangle T_{ij}^{(2,n)*};\ i,j \text{ on } M3$$

$$= \sum f^2_{0n,-n-n}\ T_{ij}^{(2,n)*} \tag{30}$$

Examining Eqs. (28-30) we see that all the data relative to the three rings depend on the generalized order parameters

$$f^2_{00,00}$$

$$f^2_{02,00}\ ,\ f^2_{02,20}\ ,\ f^2_{02,-2-2}\ .$$

Let us now consider an example of inter-fragment property, say the coupling $\langle T_{ij}^{(2,n)}\rangle$, where i is on M1 and j is on M2. We have, proceeding as usual

$$\langle T_{ij}^{(2,0)}\rangle = \sum \langle D^2_{0n}(M1-L)\ T_{ij}^{(2,n)*}(\phi_2)\rangle\ ;\ i \text{ on } M1,\ j \text{ on } M2$$

$$= \sum \langle D^2_{0n}(M1-L)\ \exp(iq_2\phi_2)\rangle\ (T_{ij}^{(2,n)*})_{q_2}$$

$$= \sum f^2_{0n,q_20}\ (T_{ij}^{(2,n)*})_{q_2}\ . \tag{31}$$

Here q_2 can be larger than two and order parameters like $f^2_{00,40}$ etc. may enter. Thus on the one hand more information on internal reorientation is contained, on the other hand, the number of couplings obtainable experimentally is limited and therefore complete information on internal motion will not be available. A recourse to some model for conformational equilibrium can be invoked as in other approaches (8,9). Another possibility is offered by information theory (24,25). This tells us that when the average value of a set of functions, here $f^2_{0n,Q} = D^2_{0n}(\omega_1)\exp(iQ.\Phi)$ is determined, the best distribution we can choose is

$$f(\omega_1,\Phi) = \exp\{\sum c_{20,nQ}\, D^2_{0n}(\omega_1)\, \exp(iQ.\Phi)\} \tag{32}$$

where the sum in the exponent extends to the functions whose average is measured. The coefficients $c_{20n,Q}$ are determined from the constraint that the available experimental data are reproduced. We see therefore that since this formalism identifies the order parameters for the system and their relation to experimentally available data it also offers a way to determining the best approximation to the orientational-conformational distribution available from NMR or other experimental techniques.

SUMMARY

We have briefly outlined a systematic approach to describing orientational and internal order in multi-rotor molecules. Contact with experiment has been made taking examples from NMR observables. The formalism can, however, be applied to other techniques and presents therefore a way of combining information on the system coming from the different approaches in a set of generalized order parameters. The quantities introduced should be particularly convenient to summarize the enormous quantity of configurational data obtained with simulation techniques in a relatively small set of parameters which can then be used to make contact with experiment. This systematic approach to the complete singlet distribution may be valuable when considering the amount of memory needed to store even a low resolution (large angular grid) multidimensional histogram even for a simple multi-rotor molecule.

ACKNOWLEDGMENT

Support from CNR and Min. P.I. (Rome) is gratefully acknowledged.

REFERENCES

1. G.R. Luckhurst and G.W. Gray eds., *"The Molecular Physics of Liquid Crystals"*, Academic Press, (1979).

2. J.W. Emsley, J.C. Lindon, *"NMR Spectroscopy Using Liquid Crystal Solvents"*, Pergamon, (1965).

3. P. Diehl and C.L. Khetrapal in *"NMR Basic Principles and Progress"*, P. Diehl, E. Fluck and R. Kosfeld eds., Vol. 1, Springer-Verlag (1969).

4. J.W. Doane, in *"Magnetic Resonance of Phase Transitions"*, F.J. Owens, C.J. Poole Jr. and H.A. Farach, Academic Press (1979).

5. C. Zannoni, Chap. 9 in *"The Molecular Physics of Liquid Crystals"*, G.R. Luckhurst and G.W. Gray eds., Academic Press (1979) and refs. therein.

6a. P. van der Ploeg and H.J.C. Berendsen, *J. Chem. Phys.* **76**, 3271 (1982).

6b. P. van der Ploeg and H.J.C. Berendsen, *Mol. Phys.* **49**, 233 (1983).

7. C. Zannoni and M. Guerra, *Mol. Phys.*, **44**, 849 (1981).

8a. J.W. Emsley and G.R. Luckhurst, *Mol. Phys.* **41**, 19 (1980).

8b. J.W. Emsley, G.R. Luckhurst and C.P. Stockley, *Proc. Roy. Soc.* **A381**, 117 (1982).

9. E.E. Burnell and C.A. de Lange, *J. Magn. Reson.* **39**, 461 (1980).

10. E.E. Burnell, C.A. de Lange and O.G. Mouritsen, *J. Magn. Reson.* **50**, 188 (1982).

11a. R. Balescu, *"Equilibrium and Non-Equilibrium Statistical Mechanics"*, Wiley, (1975).

11b. J.P. Hansen and I.R. McDonald, *"Theory of Simple Liquids"*, Academic Press, (1976).

11c. L. Landau and E.M. Lifshitz, *"Statistical Physics"*, Pergamon, (1969).

12. M.E. Rose, *"Elementary Theory of Angular Momentum"*, Wiley, (1957).

13. D.J. Evans, *Mol. Phys.* **34**, 317 (1977).

14a. L. Landau and E.M. Lifshitz, *"Quantum Mechanics"*, Pergamon Press (1966).

14b. M. Lax, "*Symmetry Principles in Solid State and Molecular Physics*", Wiley (1974).

15. C. Zannoni, Chap. 3 in "*The Molecular Physics of Liquid Crystals*", G.R. Luckhurst and G.W. Gray eds., Academic Press, (1979).

16. H. Perrin and J. Berges, *C.R. Acad. Sc. Paris*, **294**, Ser II, 1211 (1982).

17. F. Momicchioli, I. Baraldi and M.C. Bruni, *Chem. Phys.* **70**, 161 (1982).

18. W.J. Orville-Thomas (ed.), "*Internal Rotation in Molecules*", J. Wiley (1974).

19. P.J. Flory, "*Statistical Mechanics of Chain Molecules*", Wiley (1969).

20. S.L. Altmann, "*Induced Representations in Crystals and Molecules*", Academic Press (1977).

21. J. Maruani, Y.G. Smeyers and A. Hernandez Laguna, *J. Chem. Phys.* **76**, 3123 (1982).

22. G.D. Renkes, *Chem. Phys.* **57**, 261 (1981).

23. E.E. Burnell and C.A. de Lange, *Chem. Phys. Letts.* **76**, 268 (1980).

24. E.T. Jaynes, *Phys. Rev.* **106**, 620 (1957; **108**, 171 (1957).

25a. B.J. Berne and G.D. Harp, *Adv. Chem. Phys.*, **17**, 63 (1970).

25b. D.I. Bower, *J. Polym. Sci.*, **19**, 93 (1981).

25c. R.P.H. Kooyman, Y.K. Levine and B.W. van der Meer, *Chem. Phys.* **60**, 317 (1981).

26. M.R. Pear and J.H. Weiner, *J. Chem. Phys.* **71**, 212 (1979).

MOLECULAR FIELD THEORIES OF NEMATICS: SYSTEMS COMPOSED OF UNIAXIAL, BIAXIAL OR FLEXIBLE MOLECULES

G.R. Luckhurst

Department of Chemistry,
The University of Southampton, UK.

INTRODUCTION

A liquid crystal differs from a normal liquid by the existence of long range orientational order. An essential requirement for the formation of a liquid crystal is therefore some departure of the molecular shape from spherical symmetry. This deviation may take any form but the majority of liquid crystals are produced by rod-like molecules although in recent years liquid crystals composed of disc-like particles have been discovered. Indeed, their polymorphism has proved to be equally as rich as liquid crystals formed from rod-like molecules. Since most molecules are not spherically symmetric it is surprising that the occurrence of liquid crystals is not even more widespread than in fact it is. However molecular anisometry is a necessary but not sufficient condition for the formation of a liquid crystal phase. In addition we require the transition from the isotropic liquid to the liquid crystal to occur at a temperature above the freezing point. For most compounds the reverse is the case and the liquid freezes before the liquid crystal can be formed. Consequently the design of mesogenic compounds must be concerned with two quite different problems. Firstly the liquid crystal transition temperature should be as high as possible and this is achieved by making the molecular anisometry extremely large; for example the nematic - isotropic transition for p-quinquephenyl occurs at 423°C. In contrast the lower p-polyphenyls do not yield a liquid crystal phase because they crystallize first. The second problem therefore is to lower the melting point and this is readily achieved by the addition of one or two alkyl chains to the molecule because the chain flexibility necessarily increases the entropy of melting. The influence of the alkyl

J. W. Emsley (ed.), Nuclear Magnetic Resonance of Liquid Crystals, 53–83.

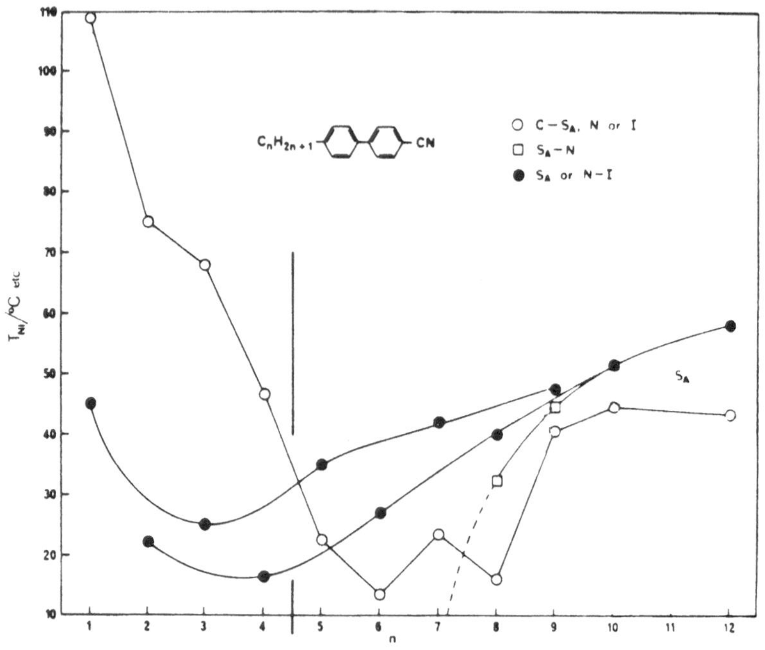

Figure 1. *The transition temperatures for the homologous series of 4-n-alkyl-4'-cyanobiphenyls.*

chain on the melting points is shown in Fig. (1) for the homologous series of the 4-n-alkyl-4'-cyanobiphenyls; there is clearly a dramatic reduction in the melting point as the length of the alkyl chain is increased, at least for the first six members of the series. As a consequence although the first four homologues do not give a stable liquid crystal phase all other members of the series can exist as a nematic, or smectic A phase or both. The length of the alkyl chain not only influences the melting point but also the liquid crystal transitions. For example the nematic - isotropic transition temperatures T_{NI} fall on one smooth curve for the members of the series containing an odd number of carbon atoms while the even members fall on a different curve; this is known as the odd - even effect.

The length of the alkyl chain influences other liquid crystal properties. The entropy of the nematic - isotropic transition also exhibits an odd - even effect as we can see from

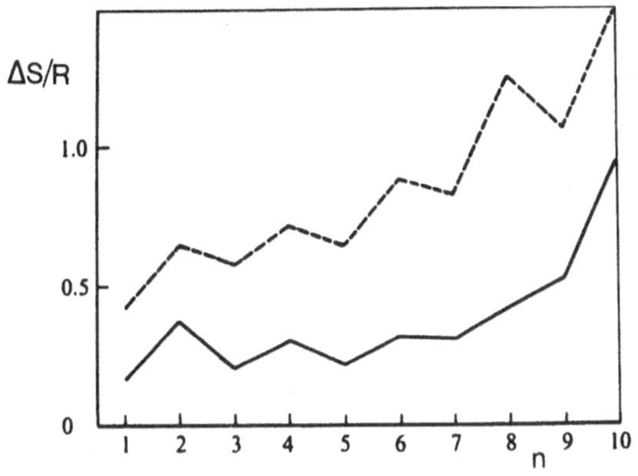

Figure 2. *The entropy of the nematic - isotropic transition $\Delta S/R$ for the 4,4'-di-n-alkoxyazoxybenzenes as a function of the number of carbon atoms in the alkyl chain. The dashed line shows the predictions of the Marcelja theory (1).*

the results for the 4,4'-di-n-alkoxyazoxybenzenes (1) which are shown in Fig. (2). As we might expect the behaviour of the system at the molecular level is also affected by the alkyl chain; thus the order parameter $\bar{P}_2^a$ for the rigid core changes with the chain length and some typical results, again for the 4,4'-di-n-alkoxyazoxybenzenes (2), are given in Fig. (3). It has often been argued both implicitly and explicitly that the alkyl chain in such mesogens exists in an all-trans configuration. However deuterium NMR studies have shown this assertion to be incorrect; in addition the order parameter for the methylene groups along the chain also exhibits an odd - even effect. The results given in Fig. (4) for the chain in 4-n-octyl-4'-cyanobiphenyl demonstrate this behaviour quite clearly (3).

Although experimentalists were aware of the profound influence of the alkyl chain on the properties of the liquid crystal, theoreticians have tended to ignore the inherent complications produced by the chain and have developed theories for mesophases formed of rigid particles. There is one notable exception which is the pioneering work by Marcelja who allowed explicitly for the alkyl chain in his theory (1). This was developed from the Maier-Saupe theory of nematics which is founded on a molecular field approximation applied to

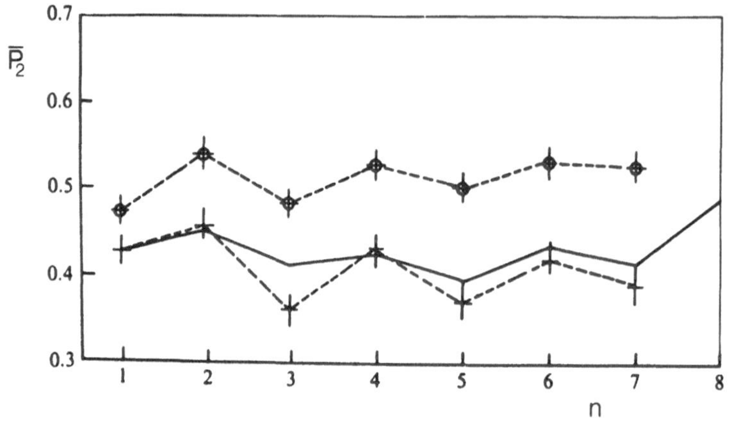

Figure 3. *The order parameters $\bar{P}_2^a$ for the 4,4'-di-n-alkoxyazoxybenzenes as a function of the number of carbon atoms in the alkyl chain at T_{NI} (+) and T_{NI} - 5°C (⊕) (2). The solid line shows the predictions of the Marcelja theory.*

anisotropic dispersion forces (4). In this discussion of theories for nematogens we find it helpful to begin with the Maier-Saupe theory of nematics and this is described briefly in the next section. We then go on to consider an extension of the Maier-Saupe theory to systems composed of non-cylindrically symmetric particles. This will be of value when we develop a theory of orientational ordering in nematics formed by flexible molecules which is used to understand results such as those shown in Fig. (4). In the final section we indicate how this theory can be extended to predict the influence of the chain length on the thermodynamic properties of the liquid crystal.

THE MAIER-SAUPE THEORY

This theory was developed to account for the behaviour of real nematogens but assumed, nonetheless, that the constituent molecules are both rigid and cylindrically symmetric. There have been several derivations of the Maier-Saupe theory each with its own degree of rigour (4). Here we give an essentially intuitive treatment which we shall find useful when we come to the more difficult problem of flexible molecules. Our starting point is the potential of mean torque U(β) which is defined in terms of the singlet orientational distribution function f(β) by

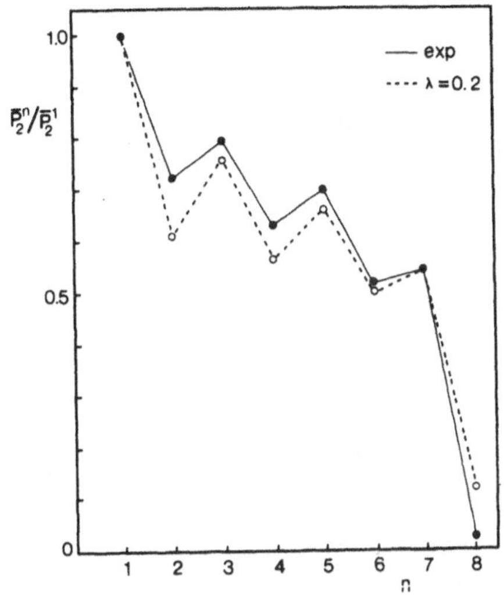

Figure 4. *The order parameters for the C-D directions in the n-octyl chain of 4-n-octyl-4'-cyanobiphenyl in the nematic phase as a function of the position of the methylene group in the chain. The values are normalized to the first segment (3).*

$$f(\beta) = \exp\{-U(\beta)/kT\}/Z \quad , \tag{1}$$

where the normalization Z is

$$Z = \int \exp\{-U(\beta)/kT\}\sin\beta d\beta \tag{2}$$

and β is the angle made by the molecular symmetry axis with the director. It is important not to confuse the potential of mean torque with the orientational potential of a single particle obtained by averaging the potential energy of the system over the configurations of its neighbours. As we shall see these two potentials are identical in certain theories based on the molecular field approximation but they are not in general equivalent.

The orientational order parameters which are of prime interest are those defined as the averages of the even Legendre polynomials (5) and these may be obtained from the singlet orientational distribution function,

$$\bar{P}_L = \int P_L(\cos\beta) f(\beta) \sin\beta d\beta \quad . \tag{3}$$

The potential of mean torque is not known and so we expand it in a complete basis set of Legendre polynomials

$$U(\beta) = -\sum_{L\neq 0} \varepsilon_L P_L(\cos\beta) \quad ; \tag{4}$$

the summation is restricted to even L because for nematics there is a plane of symmetry orthogonal to the director. The expansion coefficients result from the anisotropic interactions; they are determined by the orientational order in the mesophase and vanish in the isotropic liquid. To make the problem tractable it is important to reduce the number of unknown coefficients and so we truncate the series in Eq. (4) after the first term. This approximation appears drastic but is in accord with measurements of the singlet distribution function both for real systems (6) and for those studied in computer simulation experiments (7). An alternative way of testing this assumption is to determine the two order parameters $\bar{P}_2$ and $\bar{P}_4$ at a given temperature for there is a unique relationship between them provided the distribution function contains a single unknown. Although the second rank order parameter $\bar{P}_2$ is easily determined the measurement of $\bar{P}_4$ presents far greater difficulties. However values of $\bar{P}_2$ and $\bar{P}_4$ determined using electron resonance spin probe techniques (8) and the most recent Raman scattering studies (9) support this simple form for the potential of mean torque.

To proceed further and calculate the thermodynamic properties of the system requires additional assumptions. First we invoke the molecular field approximation which assumes that on average the orientation of one molecule is independent of the particular orientations of its neighbours. This can be expressed in a more precise way, it is equivalent to writing the N-body orientational distribution function as a product of N singlet orientational distribution functions (4)

$$f(\beta_1 \ldots . \beta_N) = \prod_{i=1}^{N} f(\beta_i) \quad . \tag{5}$$

Given the molecular field approximation the orientational contribution to the internal energy per particle is

$$U = \int \tilde{U}(\beta) f(\beta) \sin\beta d\beta \quad , \tag{6}$$

where the single particle orientational energy is distinguished from the potential of mean torque by the tilde; the entropy per particle is

$$S = -k\int \ln\{f(\beta)\}f(\beta)\sin\beta d\beta \quad . \tag{7}$$

The orientational Helmholtz free energy per particle is then

$$A = \int[\tilde{U}(\beta)f(\beta) + kT\ln\{f(\beta)\}f(\beta)]\sin\beta d\beta \tag{8}$$

and the nematic-isotropic transition occurs when the Helmholtz free energy of the nematic phase is equal to that of the isotropic liquid, provided the volume of the system is constant.

To evaluate these free energies we must make our second approximation which concerns the dependence of the expansion coefficients on the orientational order in the system. The single particle order is characterized by the averages of the even Legendre polynomials which form a complete set of order parameters but there are clearly too many for our requirements. For real nematics, at least near to T_{NI}, the largest order parameter is $\bar{P}_2$ and so it is reasonable and convenient to assume that ε_2 can be written as a polynomial in $\bar{P}_2$:

$$\varepsilon_2 = \sum_{n\neq 0} a_n \bar{P}_2^n \quad ; \tag{9}$$

the constant term is zero, because the interaction coefficient must vanish in the isotropic phase. The single particle orientational energy can now be obtained from the free energy because variations in $f(\beta)$ produced by changes in the order parameter should not change A (10). The angular dependence of the single particle energy is taken to have the same form $\{\equiv -\tilde{\varepsilon}_2 P_2(\cos\beta)$ as that for the potential of mean torque and so we must relate the expansion coefficients ε_2 and $\tilde{\varepsilon}_2$. This may be achieved by minimizing the Helmholtz free energy with respect to variations in the distribution function which are produced by changes in the order parameter $\bar{P}_2$. Thus

$$\int\delta f\{\tilde{\varepsilon}_2 P_2(\cos\beta) + \bar{P}_2(d\tilde{\varepsilon}_2/d\bar{P}_2)P_2(\cos\beta) + kT\ln ef(\beta) - \lambda\}\sin\beta d\beta = 0 \quad , \tag{10}$$

where λ is a Lagrange undetermined multiplier which appears because $f(\beta)$ is normalized. Since the quantity in curly brackets must vanish for all β the coefficient of $P_2(\cos\beta)$ obtained after writing $f(\beta)$ in terms of the potential of mean torque must also vanish, for this to occur we require

$$\varepsilon_2 = \tilde{\varepsilon}_2 + \bar{P}_2(d\tilde{\varepsilon}_2/d\bar{P}_2) \quad . \tag{11}$$

We see then that the potential of mean torque and the single particle energy are not in general identical.

In the Maier-Saupe theory the interaction coefficient ε_2 is assumed to be linear in the order parameter and so from Eq. (11) we find the same linear dependence of $\tilde{\varepsilon}_2$ on $\bar{P}_2$ but with

$$a_1 = 2\tilde{a}_1 \quad ; \tag{12}$$

the orientational Helmholtz free energy per particle is

$$A = \bar{u}_2\bar{P}_2^2/2 - kT \ln Z \quad , \tag{13}$$

where $\bar{u}_2$ has replaced a_1 to conform with the standard notation (4). The predicted behaviour of the system is obtained by evaluating the order parameter numerically from Eqs. (3) and (12). This is then used to determine the orientational Helmholtz free energy as a function of the scaled temperature $kT/\bar{u}_2$ and the nematic - isotropic transition is found when $kT_{NI}/\bar{u}_2$ is 0.2202. The gradient $(\partial A/\partial T)_V$ is discontinuous at T_{NI} and so the transition is first order; the entropy of transition $\Delta S/R$ is found to be 0.417 which is considerably larger than that observed experimentally. As the unknown parameter $\bar{u}_2$ is proportional to T_{NI} the order parameter is predicted to be a universal function of the reduced temperature T/T_{NI}, in reasonable but not complete agreement with experiment as we can see from the results shown in Fig. (5). The theory predicts therefore that the order parameter at the transition should be independent of the nematogen with a value of 0.429 but this prediction is not in agreement with the results in Fig. (3).

Some deviations from the predictions of the Maier-Saupe theory have been explained by assuming implicitly, a quadratic dependence for the interaction parameter ε_2 on the order parameter $\bar{P}_2$ (11). The distribution function now becomes

$$f(\beta) = \exp[\bar{u}_2\{\bar{P}_2 + \lambda\bar{P}_2^2\}P_2(\cos\beta)/kT]/Z \quad , \tag{14}$$

where λ is the ratio a_2/a_1. The effect of this additional parameter is either to strengthen ($\lambda>0$) or weaken ($\lambda<0$) the anisotropic molecular field which is responsible for the alignment of the molecule in the liquid crystal. As a consequence the transition temperature and order parameter at the transition may be either raised or lowered; some typical

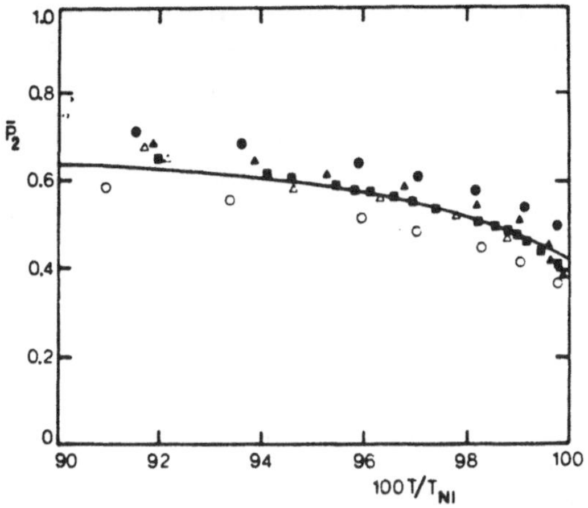

Figure 5. *The temperature dependence of $\bar{P}_2$ for 4,4'-dimethoxyazoxybenzene (○), 4,4'-diethoxyazoxybenzene (●), anisaldazine (■), 2,4-nonadienoic acid (▲) and 2,4-undecadienoic acid (△). The Maier-Saupe prediction is shown as the curve.*

results are given in Table 1. The marked variation of $\bar{P}_2^{NI}$ with λ appears to be in reasonable accord with the observed behaviour of different nematogens (4). However determination of the potential of mean torque from Monte Carlo simulation experiments does not support the introduction of a quadratic term in the interaction parameter ε_2, at least for the model nematogen studied (12).

Table 1. *The effect of the quadratic dependence of ε_2 on $\bar{P}_2$ on the properties of the liquid crystal.*

λ	$kT_{NI}/\bar{u}_2$	$\bar{P}_2^{NI}$
-0.2	0.2058	0.223
-0.1	0.2118	0.329
0	0.2202	0.429
0.1	0.2305	0.510
0.2	0.2426	0.571

In concluding this section we note that the Maier-Saupe theory of nematogens has been criticised for assuming that long range anisotropic dispersion forces are responsible for liquid crystal formation while ignoring the short range anisotropic repulsive forces. Such criticism would appear to have some validity especially as repulsive forces are known to be dominant in determining the structure of simple fluids. However although the starting point of the Maier-Saupe theory may be in doubt there is substantial evidence to suggest that the resultant potential of mean torque is essentially correct. As we have seen scattering experiments on real nematics have shown that the angular dependence of the potential of mean torque is well approximated by the second Legendre polynomial. In addition the linear dependence of the interaction coefficient ε_2 on the order parameter $\overline{P}_2$ has been demonstrated by computer simulation experiments on particles interacting via a Berne-Kushick-Pechukas potential which includes a short range anisotropic repulsive contribution (13). Theory, based on a van der Waals approach, also suggests that the potential of mean torque has the same form as that proposed by Maier and Saupe (14). In consequence the use of a Maier-Saupe like potential should not be taken necessarily to imply that long range dispersion forces dominate the anisotropic intermolecular potential since such a potential of mean torque is also consistent with short range anisotropic repulsive forces. The theories which are described in this chapter should therefore be generally applicable, especially those features dependent upon the potential of mean torque.

BIAXIAL MOLECULES

The molecules of all mesogens do not have the high symmetry which is assumed for them by the Maier-Saupe theory and so attempts have been made to extend this theory to biaxial molecules (15). When the molecule deviates from cylindrical symmetry the description of the orientational order becomes more complex; it may be described, for example, with the Saupe ordering matrix

$$S_{\alpha\beta} = (3\overline{\ell_\alpha \ell_\beta} - \delta_{\alpha\beta})/2 \quad , \tag{15}$$

where ℓ_α is the direction cosine of the molecular axis α with the director. There are five independent elements for S but in its principal axis system xyz this is reduced to just two, S_{zz} and $(S_{xx} - S_{yy})$. For cylindrically symmetric molecules $(S_{xx} - S_{yy})$ vanishes and we are left with S_{zz} which is equivalent to $\overline{P}_2$. The biaxial order parameter $(S_{xx} - S_{yy})$ is difficult to determine but has been measured for certain mesogens and found to be non-zero (16), consequently the extension of the Maier-Saupe theory to biaxial particles is of some importance.

Our starting point, as before, is the potential of mean torque but now we need two angles to define the orientation of a molecule; these are the spherical polar angles ($\beta\gamma$) which the director makes in a molecular coordinate system. A convenient set of functions with which to expand the potential of mean torque is the modified spherical harmonics (5) $C_{L,m}(\beta\gamma)$ and so we write

$$U(\beta\gamma) = -\sum_{L(\text{even})m} (-)^m \varepsilon_{L,m} C_{L,-m}(\beta\gamma) \quad . \tag{16}$$

The interaction coefficients ε_L which we encountered previously have been replaced by an interaction tensor which is written in irreducible form (5). The evaluation of the potential is again simplified by truncating the expansion after the second rank term. This still leaves five coefficients $\varepsilon_{2,m}$ because m may take values 2, 1, 0, −1, −2; these components are related to those of the cartesian tensor by

$$\begin{aligned} \varepsilon_{2,0} &= (2\varepsilon_{zz} - \varepsilon_{xx} - \varepsilon_{yy})/\sqrt{6} \quad , \\ \varepsilon_{2,\pm1} &= \mp(\varepsilon_{xz} \pm i\varepsilon_{yz}) \quad , \\ \varepsilon_{2,\pm2} &= (\varepsilon_{xx} - \varepsilon_{yy} \pm 2i\varepsilon_{xy})/2 \ . \end{aligned} \tag{17}$$

We can see that by working in the principal axis system of the interaction tensor the number of unknowns is reduced to two and the potential of mean torque becomes

$$U(\beta\gamma) = -\{\varepsilon_{2,0}P_2(\cos\beta) + \varepsilon_{2,2}(\tfrac{3}{2})^{\frac{1}{2}}\sin^2\beta\cos2\gamma\} \ . \tag{18}$$

The first term is equivalent to that for cylindrically symmetric particles and the second results entirely from the molecular biaxiality which is reflected in $\varepsilon_{2,2}$. A considerable advantage of this formulation is that the order parameters are given by the relatively simple results (15):

$$\overline{P}_2 = 2\pi Z^{-1}\int P_2(\cos\beta)I_0\{b(\tfrac{3}{8})^{\frac{1}{2}}\sin^2\beta\}\exp\{aP_2(\cos\beta)\}\sin\beta d\beta \tag{19}$$

and

$$\overline{\sin^2\beta\cos2\gamma} = 2\pi Z^{-1}\int\sin^2\beta I_1\{b(\tfrac{3}{8})^{\frac{1}{2}}\sin^2\beta\}\exp\{aP_2(\cos\beta)\sin\beta d\beta, \tag{20}$$

where $I_n(x)$ is a nth order modified Bessel function:

$$I_n(x) = \pi^{-1} \int_0^{\pi} \cos n\gamma \exp\{x\cos\gamma\}d\gamma \quad . \tag{21}$$

The parameters a and b are $\varepsilon_{2,0}/kT$ and $2\varepsilon_{2,2}/kT$ respectively and Z the orientational partition function is

$$Z = 2\pi \int I_0\{b(\tfrac{3}{8})^{\frac{1}{2}}\sin^2\beta\}\exp\{aP_2(\cos\beta)\}\sin\beta d\beta \quad . \tag{22}$$

The biaxial order parameter $\overline{\sin^2\beta\cos 2\gamma}$ is related to the elements of the Saupe ordering matrix and is just $(\tfrac{2}{3})(S_{xx} - S_{yy})$.

For a given scaled temperature $kT/\varepsilon_{2,0}$ the two independent order parameters depend on the ratio $\varepsilon_{2,2}/\varepsilon_{2,0}$ and so there is no unique relationship between S_{zz} and $(S_{xx} - S_{yy})$. However as both $\varepsilon_{2,0}$ and $\varepsilon_{2,2}$ are related to the orientational order we might expect their ratio λ to be independent of temperature and this expectation is supported by a more rigorous derivation of the potential of mean torque (15). We can therefore relate the biaxiality $(S_{xx} - S_{yy})$ to S_{zz} by calculating these order parameters for a given λ as a function of a. Some typical results are given in Fig. (6) and they show the anticipated increase in $(S_{xx} - S_{yy})$ with λ. They also show that the biaxiality passes through a maximum when S_{zz} is about 0.4. These predictions appear to be in accord with NMR investigations of pure liquid crystals and solutes dissolved in them (17). It would seem therefore that the form of the potential of mean torque is reasonably precise.

To proceed to a calculation of the thermodynamic properties is more difficult than in the previous problem because there are two second rank order parameters and their relation to the interaction tensor $\varepsilon_{2,m}$ requires a detailed calculation (15). However we may obtain an approximate relationship in the following way. The results in Fig. (6) for 4,4'-dimethoxyazoxybenzene reveal that near to the nematic isotropic transition where S_{zz} is about 0.4 the second order parameter $(S_{xx} - S_{yy})$ is much less than S_{zz}. We may assume therefore that the dependence of $\varepsilon_{2,m}$ on the biaxial order parameter, $\overline{\sin^2\beta\cos 2\gamma}$, can be ignored in comparison with $\overline{P}_2$ and so we have

$$\begin{aligned} \varepsilon_{2,0} &\simeq \overline{u}_{200}\,\overline{P}_2 \;, \\ \varepsilon_{2,2} &\simeq \overline{u}_{220}\,\overline{P}_2 \;; \end{aligned} \tag{23}$$

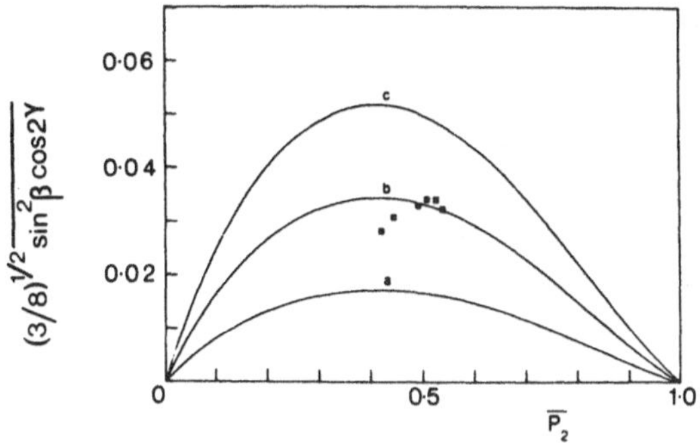

Figure 6. *The predicted dependence of* $(\frac{3}{8})^{\frac{1}{2}}\overline{\sin^2\beta\cos 2\gamma}$ *on* $\overline{P}_2$ *for* λ *equal to (a) 0.1, (b) 0.2 and (c) 0.3. The squares show the results observed for the nematogen 4,4'-dimethoxyazoxybenzene (15).*

this gives λ as $2\bar{u}_{220}/\bar{u}_{200}$ which should be independent of temperature as we had supposed. The orientational Helmholtz free energy per particle is obtained using the same procedure as before and we find

$$A = (kT/2)(a\overline{P}_2 + b(\tfrac{3}{8})^{\frac{1}{2}}\overline{\sin^2\beta\cos 2\gamma}) - kT\ln Z \ . \qquad (24)$$

We may determine the scaled transition temperature $kT_{NI}/\bar{u}_{200}$ for a given biaxiality λ by locating the temperature at which the free energies of the isotropic and nematic phases are identical. The results are listed in Table 2 together with the entropy of transition $\Delta S/R$ and the order parameters at the transition (15). The changes in these properties with λ are readily understood and are in reasonable agreement with experiment for some nematogens. It is important to note that although the biaxiality in the ordering matrix may be quite small the effect of the molecular biaxiality on other properties can be most marked.

FLEXIBLE MOLECULES: ORIENTATIONAL ORDER PARAMETERS

We shall now see how the theories which have been outlined can be extended to include the effects of chain flexibility. It is convenient to make such an extension in two stages. Here we shall deal with the form of the potential of mean torque and this will allow us to calculate the order parameters for the mesogen at a given temperature. In the following section we shall consider how the interaction tensors depend on the various order parameters and this will then enable us to obtain the thermodynamic properties of the nematogens.

Table 2. *The predicted dependence of the transition properties on the molecular biaxiality λ.*

λ	$kT_{NI}/\bar{u}_{200}$	S_{zz}^{NI}	$(S_{xx}^{NI} - S_{yy}^{NI})$	$\Delta S/R$
0	0.2202	0.429	0	0.417
0.1	0.2220	0.408	0.042	0.384
0.2	0.2280	0.341	0.086	0.275
0.3	0.2404	0.207	0.100	0.112

We begin with the conformational states available to the n-alkyl chain. These are most readily described by using the rotameric state model developed by Flory for polymers (18). In this model the bond lengths are taken to be fixed and the only angle which is allowed to vary is the bond rotation angle ϕ which is shown in Fig. (7) for part of an n-alkyl chain. Another view of this angle is given in Fig. (8) by looking along a C-C bond; there are clearly three special values of ϕ which correspond to the trans (t) and gauche ($g^{\pm}$) states of the groups R and R' with respect to one another. The internal energy of the chain varies markedly with the bond rotation angle ϕ as we can see from the results for n-butane given in Fig. (9). Indeed in the rotameric state model it is argued that the potential wells

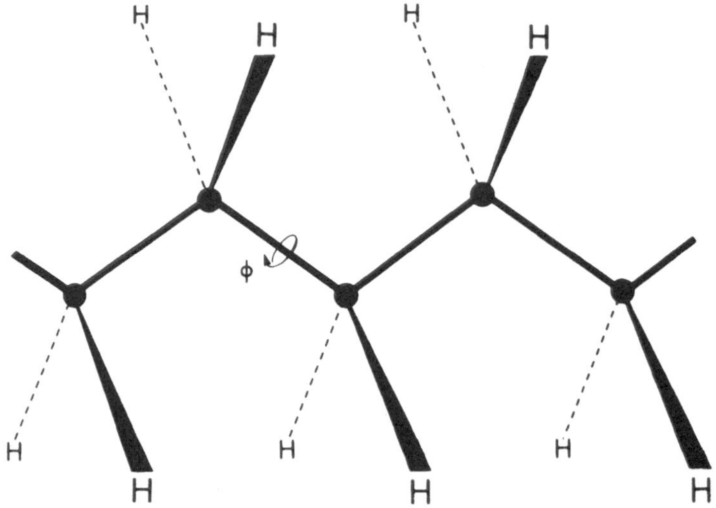

Figure 7. *The bond rotation angle ϕ for part of an n-alkyl chain in an all-trans conformation.*

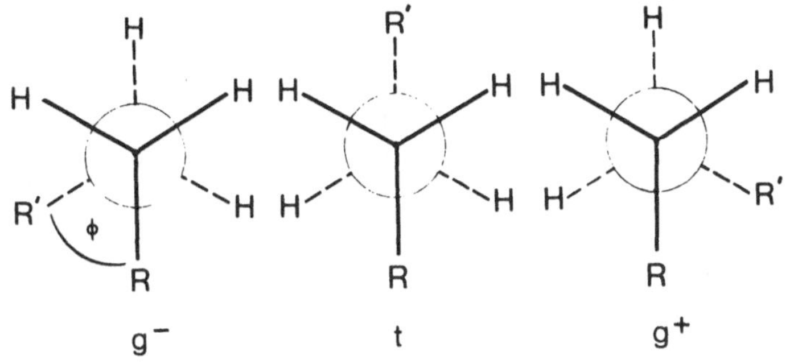

Figure 8. *The trans and the two symmetrical gauche states viewed along a C-C bond.*

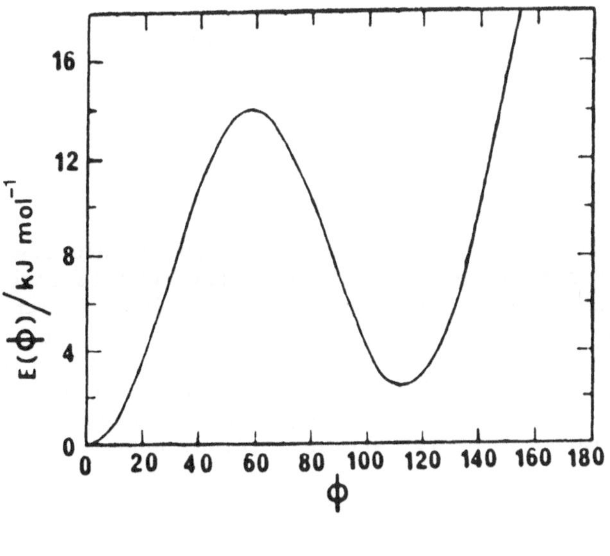

Figure 9. *The internal energy for n-butane as a function of the bond rotation angle for the central C-C bond (18).*

are so steep that the only values of ϕ with appreciable probability are those corresponding to the minima of the wells which are just the trans (ϕ = 0°) and gauche (ϕ = ±112°) states. This assumption therefore brings a dramatic reduction in the number of conformations available to the chain; the resultant discrete set of conformations can be designated by giving the rotational state of the groups about each C-C bond in the chain.

The conformation with the minimum energy is the all-trans (cf. Fig. (7)) and the energy of a conformer n with respect to this is obtained in the rotameric state model, from the number of gauche linkages N_g. The internal energy is then

$$U_{int}(n) = N_g E_{tg} , \tag{25}$$

where E_{tg} is the energy of a gauche state relative to a trans. In other words the model assumes that the rotational potential about a given bond is independent of the conformation of the remainder of the chain. Such an assumption is not strictly correct and it is possible to improve on Eq. (25) by the addition of a term $N_{g^\pm g^\mp} E_{g^\pm g^\mp}$ where $N_{g^\pm g^\mp}$ is the number of g^+g^- or g^-g^+ linkages since these linkages bring parts of the chain near to one another. For many purposes however the simpler result is quite adequate.

The normalized probability of finding a molecule in a particular conformation n in the isotropic phase is given by the usual Boltzmann factor

$$p^n = \exp\{-U_{int}(n)/kT\}/\sum_n \exp\{-U_{int}(n)/kT\} \quad . \tag{26}$$

This expression is not valid however in the liquid crystal phase because the anisotropic environment will tend to stabilize the more elongated conformers. To quantify this statement we must derive an expression for the potential energy of a molecule in a conformation n and a particular orientation ω with respect to the director. It is useful and natural to write the total potential $U(n,\omega)$ defined by an equation analogous to (1) as the sum of the internal energy and an external energy $U_{ext}(n,\omega)$ (19)

$$U(n,\omega) = U_{int}(n) + U_{ext}(n,\omega) \quad . \tag{27}$$

The internal energy of a conformer is assumed to be independent of its orientation. In contrast the external energy or potential of mean torque which results from the anisotropic interactions of a molecule with its neighbours depends on both the molecular conformation and orientation. The probability of finding a molecule with a particular orientation and conformation is then

$$f(n,\omega) = \exp[-\{U_{int}(n) + U_{ext}(n,\omega)\}/kT]/Z \quad , \tag{28}$$

where the total partition function is

$$Z = \sum_{n} \int \exp[-\{U_{int}(n) + U_{ext}(n,\omega)\}/kT]d\omega \quad . \tag{29}$$

At this stage we are only interested in the probability of a particular conformation independent of its orientation and so we must integrate the distribution function $f(n,\omega)$ over all orientations ω. This gives

$$P^n = \exp\{-U_{int}(n)/kT\}Q(n)/Z \quad , \tag{30}$$

where the orientational partition function for conformer n is

$$Q(n) = \int \exp\{-U_{ext}(n,\omega)/kT\}d\omega \quad ; \tag{31}$$

in the isotropic phase $U_{ext}(n,\omega)$ vanishes and we recover Eq. (26) for P^n. The term in $Q(n)$ therefore accounts for the difference in populations of conformer n between the isotropic and liquid crystal phases. To estimate this difference we require an expression for the external energy which we now develop.

By analogy with the theory for rigid biaxial particles we approximate the potential of mean torque for a particular conformation by

$$U_{ext}(n,\omega) = -\sum_{m} (-)^m \varepsilon^n_{2,m} C_{2,-m}(\omega) \quad . \tag{32}$$

Here $\varepsilon^n_{2,m}$ is the interaction tensor for one conformation and as there are many conformers (729 for the chain in 4-n-octyl-4'-cyanobiphenyl) we clearly need a method to relate the various $\varepsilon^n_{2,m}$. We are able to achieve this by assuming that each rigid sub-unit in the molecule makes its own contribution to the interaction tensor

$$\varepsilon^n_{2,m} = \sum_{j} \varepsilon^j_{2,m}(n) \quad . \tag{33}$$

The assumption of bond or group additivity is not new, indeed it has been used with some success to interpret the ordering matrices for rigid solutes dissolved in a liquid crystal (20). The group interaction tensors vary with conformation because their components are expressed in a common molecular frame. This dependence can be removed by transforming to a local frame and for segment j

$$\varepsilon^{j}_{2,m}(n) = \sum_{p} D^{2}_{m,p}(\Omega^{n})\varepsilon^{j}_{2,p} \quad , \tag{34}$$

where $D^{2}_{m,p}(\Omega^{n})$ is a Wigner rotation matrix (5) which transforms from the local to the molecular coordinate systems and Ω^{n} denotes the Euler angles connecting the frames for the nth conformer. We have assumed that the group parameters in their local frames are independent of conformation which should be a good approximation when the molecular segments are not conjugated. For molecules such as the 4-n-alkyl-4'-cyanobiphenyls the molecular frame is conveniently set in the aromatic core but this will not normally correspond with the principal axes for the interaction tensors $\varepsilon^{n}_{2,m}$. These axes are found therefore by diagonalizing the cartesian form of the interaction tensor. Given the principal components of the tensor it is then possible to calculate the principal elements of the ordering matrix from Eqs. (19) and (20).

The NMR experiment does not, in general, measure the ordering matrices of individual conformers because there is rapid exchange between the various conformations and so we observe a weighted average

$$\langle S_{\alpha\beta}\rangle = \sum_{n} p^{n} S^{n}_{\alpha\beta} \quad . \tag{35}$$

As we have seen we are able to calculate both the probability and the ordering matrix for a given conformation but we must now consider the variables contained in this calculation. The internal energy parameters E_{tg} and $E_{g\pm g\mp}$ are available from independent studies of n-alkanes although there is not complete agreement on the best values. The interaction tensors are not known and to reduce their number we suppose that, for a molecule such as a 4-n-alkyl-4'-cyanobiphenyl, there are just two kinds of segment. One is the aromatic core including both the cyano group and the first methylene group of the alkyl chain while the other is a carbon-carbon segment; the contribution of the C-H groups in the chain are taken to be negligible. In addition we assume that the interaction tensors for both kinds of segment are cylindrically symmetric, about the para axis for the core and the C-C axis for the carbon-carbon segments. We denote the unique components of the interaction tensors by X_a for the aromatic core and X_c for a chain segment. The total interaction tensor for a conformer is then

$$\varepsilon^{n}_{2,m} = X_a \, \delta_{0m} + X_c \sum_{j=2}^{N} C_{2,m}(\omega_j) \tag{36}$$

when the molecular frame has its z-axis along the para-axis of the core; the x axis lies in the plane formed by the first two C-C bonds of the alkyl chain. The chain contains N carbon atoms and the orientation of the jth C-C segment in the molecular frame is ω_j.

The theory (3) contains four variables E^*_{tg}, $E^*_{g\pm g\mp}$, X^*_a and X^*_c where the star indicates that the quantity has been divided by kT as required for the Boltzmann averages. Of these E^*_{tg}, X^*_a and λ_c, which is the ratio X^*_c / X^*_a, have the greatest influence on the orientational order. To illustrate the dependence of the order parameter profile for the nematogen 4-n-nonyl-4'-cyanobiphenyl we have calculated the order parameters S^i_{CD} for the C-D bond attached to the ith carbon atom in the chain using typical values for the three variables. These were X^*_a = 2.0, λ_c = 0.3 and E^*_{tg} = 1.0; the order parameters were then evaluated for a range of values of one variable while holding the other two constant. In Fig. (10) we show the order parameter profile calculated for λ_c = 0.3, E^*_{tg} = 1.0 and the core interaction parameter X^*_a equal to 1.5, 2.0 and 2.5. Increasing X^*_a corresponds either to lowering the temperature or increasing the anisotropic molecular interactions of the core. As expected therefore the extent of orientational ordering increases with X^*_a although the shape of the order parameter profile is essentially unchanged. This behaviour is in accord with that observed for the S^i_{CD} profile in a variety of mesogens (16).

We next consider the influence of λ_c on the variation of the order parameter S^i_{CD} along the nonyl chain. This variable determines the strength of the C-C chain segment interaction with the molecular field relative to that of the core. Accordingly we expect the orientational order to increase with increasing λ_c. This expected change does occur as we can see from Fig. (11) which shows calculations with λ_c equal to 0.2, 0.3 and 0.4, indeed increasing λ_c from 0.2 to 0.4 clearly has a major effect on S^i_{CD} presumably even though λ_c is relatively small because the chain contains a large number of segments. More importantly, perhaps, the shape of the order parameter profile also changes with λ_c unlike variations in X^*_a and so it should be possible to determine unique values for these variables from an observed profile.

Finally we consider the influence of E^*_{tg}, the energy of a gauche linkage relative to a trans on the order parameter profile. As E^*_{tg} is increased the all trans conformation will be

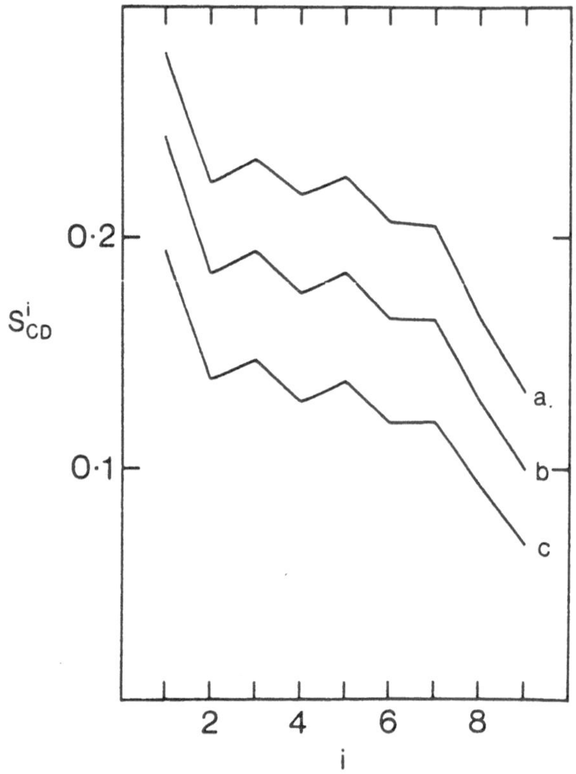

Figure 10. *The order parameter profile calculated for 4-n-nonyl-4'-cyanobiphenyl with* $\lambda_c = 0.3$, $E^*_{tg} = 1.0$ *and* $X^*_a = 2.5$ *(a), 2.0 (b) and 1.5 (c).*

favoured. This elongation of the chain will increase the overall orientational order and such behaviour is reflected by the S^i_{CD}, as we can see from the results shown in Fig. (12). These were obtained with $X^*_a = 2.0$, $\lambda_c = 0.3$ and E^*_{tg} equal to 0.8, 1.0 and 1.2. We also note that like X^*_a this variation in E^*_{tg} does not change the form of the order parameter profile. In principle this could complicate the analysis of experimental data to obtain the variables in the theory, however this problem is not severe because at least a range of acceptable values for E_{tg} is known.

We have tested the theory using experimental results for several cyanobiphenyls. We show in Fig. (4) the best fit to the order parameter for 4-n-octyl-4'-cyanobiphenyl; the optimum results for the variables were $E^*_{tg} = 1.0$ and $E^*_{g^\pm g^\mp} = 4.0$ which are in good agreement with other estimates of these conformational energies. The molecular field interaction parameters were $X^*_a = 2.0$ and $\lambda_c = 0.3$; as expected the core parameter is significantly larger than that for the C-C chain segment. The agreement between theory and experiment for the relative order parameters is quite good especially as the model contains only two completely free variables.

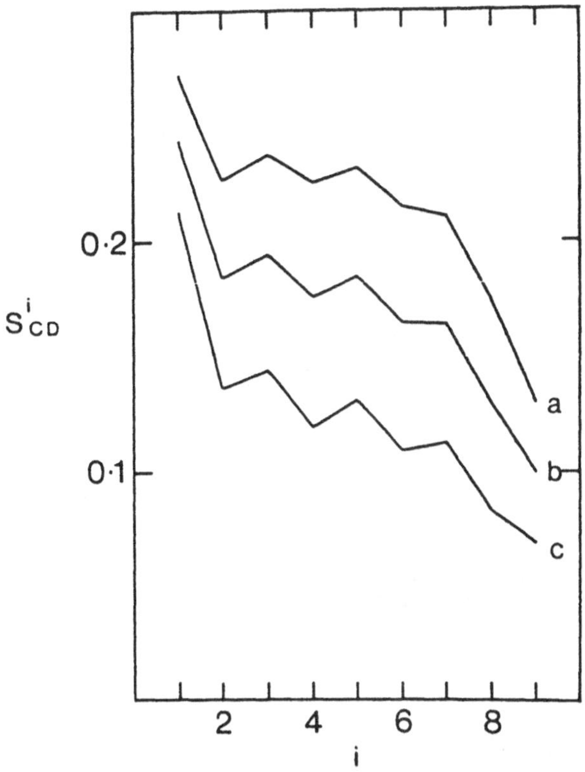

Figure 11. *The variation of S^i_{CD} along the alkyl chain in 4-n-nonyl-4'-cyanobiphenyl calculated with $X^*_a = 2.0$, $E^*_{tg} = 1.0$ and $\lambda_c = 0.4$ (a), 0.3 (b) and 0.2 (c).*

It is of considerable interest to see how the liquid crystal environment changes the probability distribution amongst the various conformers. Within the framework of our model this is a relatively easy task using typical values for the

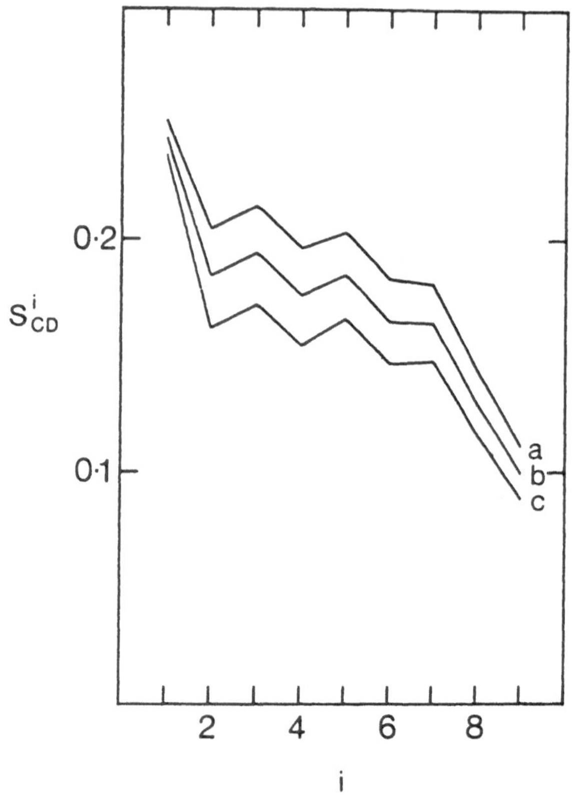

Figure 12. *The order parameter profile predicted for 4-n-nonyl-4'-cyanobiphenyl with* E^*_{tg} *= 1.2 (a), 1.0 (b) and 0.8 (c);* X^*_a *was set equal to 2.0 and* λ_c *to 0.3.*

parameters required in Eq. (30). Some of the results of such calculations are shown in Table 3 for fifteen of the most popular conformations of the alkyl chain for 4-n-octyl-4'-cyanobiphenyl. We see immediately that the all-trans conformer is stabilized by the nematic phase. Indeed some of these conformations with a single gauche linkage also have greater populations in the nematic phase. This contrasts with the assumptions of the kink model (21) which excludes all such conformers and replaces them with those containing $g^{\pm}tg^{\mp}$ linkages whereas our calculations show these to have a low probability (3).

Table 3. *The populations of the all-trans conformer and those containing a single gauche linkage in the isotropic and nematic phases of 4-n-octyl-4'-cyanobiphenyl. The parameters used in the calculations were* $E^*_{tg} = 1.0$, $E^*_{g^\pm g^\mp} = 4.0$, $X^*_a = 2.0$ *and* $X^*_c = 0.6$.

Conformation	Probability	
	Isotropic	Nematic
tttttt	5.54	10.80
$g^\pm$ttttt	2.04	1.14
t$g^\pm$tttt	2.04	3.35
tt$g^\pm$ttt	2.04	1.48
ttt$g^\pm$tt	2.04	3.17
tttt$g^\pm$t	2.04	2.19
ttttt$g^\pm$	2.04	3.35

The success of the theory (3) in predicting the orientational order in nematogens containing flexible chains provides some support for the form of the potential of mean torque which we have adopted. We proceed therefore in the final section to propose a particular dependence of the interaction parameters on the various orientational order parameters for the system. Such a development is necessary if we are to predict the thermodynamic properties of the nematic. In concluding this section we note that a similar potential has been developed by Marcelja (1), however in his calculations he restricts the orientation of the core to just three or five selected values. This quantization of the orientations can result in considerable errors and so in our calculations (3) all orientations are allowed.

FLEXIBLE MOLECULES: THE THERMODYNAMIC PROPERTIES

The problem of relating the interaction coefficients X_a and X_c to the orientational order in a nematic, composed of flexible molecules is a formidable one because there are so many different second rank order parameters for the system. For example, if we consider the nematic as a multicomponent mixture of the various conformers then there are two order parameters S_{zz} and $(S_{xx} - S_{yy})$ for each conformer. Even if we ignore the biaxiality in the ordering matrix which is small (3) there would still be a single order parameter for every conformation and there are many of these. However a given molecule sees an environment which is an average over the orientations and conformations of the neighbouring molecules. We might therefore use a weighted average of the order parameters for the conformers, as in Eq. (35),

$$\langle \bar{P}_2 \rangle = \sum_n p^n \bar{P}_2^n , \qquad (37)$$

where as before the conformational average is denoted by the angular brackets. The chain segments will interact with the molecular field to a different extent to the aromatic core and so we might assume

$$X_c = \varepsilon_c \langle \bar{P}_2 \rangle ,$$
$$X_a = \varepsilon_a \langle \bar{P}_2 \rangle . \qquad (38)$$

This approximation forces the ratio X_c/X_a to be temperature independent which would have the merit of simplifying the calculations considerably.

Although the relationships in Eq. (38) are attractively simple they do not reflect completely the molecular inhomogeneity resulting from the presence of the aromatic core and the alkyl chain. For example, in the molecular field theory of multicomponent mixtures of cylindrically symmetric particles (22) the interaction coefficient in the potential of mean torque for component i is given by

$$X_i = \sum_j \varepsilon_{ij} \phi_j v_j^{-1} \bar{P}_2^j , \qquad (39)$$

where ε_{ij} is determined by the strength of the anisotropic interaction between molecules i and j. When the components have different molecular volumes v the composition dependence of the X_i may be approximated by the volume fraction ϕ_j (23) instead of the mole fraction (22). If we consider the single component nematic to be a mixture of core and chain segments then, by analogy with the detailed derivation for real mixtures, we might expect the interaction parameters to be given by

$$X_c = \varepsilon_{cc} \phi_c v_c^{-1} \sum_j \langle \bar{P}_2^j \rangle + \varepsilon_{ac} \phi_a v_a^{-1} \langle \bar{P}_2^a \rangle \qquad (40)$$

and

$$X_a = \varepsilon_{ac}\phi_c v_c^{-1}\sum_j \langle \bar{P}^j \rangle + \varepsilon_{aa}\phi_a v_a^{-1}\langle \bar{P}_2^a \rangle \quad . \tag{41}$$

That is we take the aromatic core to be one component and the chain segment the other; thus v_c is the molecular volume for a C–C segment and v_a that of the core while ϕ denotes the corresponding volume fraction. Similarly ε_{cc} is a strength parameter caused by chain-chain interactions, ε_{aa} by core-core interactions and ε_{ac} by chain-core interactions. As the chain segments are identical the order parameters for the carbon-carbon segments of the chain are summed and we can define an average order parameter for the total chain as

$$\langle \bar{P}_2^c \rangle = \sum_j \langle \bar{P}_2^j \rangle \quad . \tag{42}$$

Similar expressions have been proposed by Marcelja (1) but by using less direct arguments.

The dependence of the interaction coefficients on the relative size of the core and chain via the volume fractions is an essential feature of the theory as the following argument demonstrates. In the limit that the chains do not interact with each other or with the aromatic core the system would consist of a set of rigid, cylindrically symmetric particles and so the equations should reduce to those of the Maier-Saupe theory. This follows directly from Eqs. (40) and (41), for X_c vanishes and X_a becomes

$$X_a = \varepsilon_{aa}\phi_a v_a^{-1}\langle \bar{P}_2^a \rangle \quad . \tag{43}$$

More importantly however we see, that after combining Eqs. (9) and (43), the nematic - isotropic transition temperature is

$$\begin{aligned} kT_{NI} &= 0.2202\varepsilon_{aa}\phi_a v_a^{-1} \quad , \\ &= 0.2202\varepsilon_{aa} v_m^{-1} \quad , \end{aligned} \tag{44}$$

where v_m is the molecular volume. Consequently as the length of the non-interacting chain increases, the molecular volume will also increase and so T_{NI} will decrease. This is qualitatively what we should expect, for the chains will tend to increase the average separation between the aromatic cores and so reduce their anisotropic interaction which in turn will lower the nematic - isotropic transition temperature. In fact all of the segmental interaction parameters in the theory are assumed to be scaled by the molecular volume

$$v_m = v_a + (N-1)v_c \tag{45}$$

for a chain containing N carbon-carbon segments, since the first methylene group is counted as part of the core.

The ratio of the interaction coefficients is

$$X_c/X_a = (\varepsilon_{ac}\langle\bar{P}_2^a\rangle + \varepsilon_{cc}\langle\bar{P}_2^c\rangle)/(\varepsilon_{aa}\langle\bar{P}_2^a\rangle + \varepsilon_{ac}\langle\bar{P}_2^c\rangle) \quad , \tag{46}$$

which will, in general, vary with temperature. Such a variation complicates the numerical evaluation of the equations in the theory considerably. We can avoid this difficulty by making the reasonable assumption that

$$\varepsilon_{ac} = (\varepsilon_{aa}\varepsilon_{cc})^{\frac{1}{2}} \tag{47}$$

which has proved to be successful in describing the anisotropic interaction between unlike species in binary mixtures of liquid crystals (22). With this geometric mean approximation the ratio of the interaction coefficients reduces to

$$X_c/X_a = (\varepsilon_{cc}/\varepsilon_{aa})^{\frac{1}{2}} \tag{48}$$

which is temperature independent. The approximation has the additional advantage of removing one of the variables from the theory.

To obtain the orientational Helmholtz free energy it is convenient to rewrite the potential of mean torque for a molecule in a given orientation and conformation by combining Eqs. (32) and (36). This gives, with the aid of the spherical harmonic addition theorem (5),

$$U_{ext}(n,\omega) = -\{X_a P_2(\cos\beta) + X_c \sum_j P_2(\cos\beta_j)\} \quad , \tag{49}$$

where the director makes an angle β with the aromatic core and β_j with the jth carbon-carbon segment of the chain. The Helmholtz free energy per molecule can now be obtained from the total single particle energy $\{U_{int}(n) + U_{ext}(n,\omega)\}$ by using the same procedure as we adopted in the Maier-Saupe theory. The only difference is that we must now average over the conformations of a molecule as well as its orientation; this gives

$$A = (X_a\langle\bar{P}_2^a\rangle + X_c\langle\bar{P}_2^c\rangle)/2 - kT\ln Z \quad , \tag{50}$$

where the total partition function is obtained from Eq. (29) as

$$Z = \sum_n Q(n)\exp\{-U_{int}(n)/kT\} \quad , \tag{51}$$

with the orientational partition function for conformation n as

$$Q(n) = \int\exp[\sum_m\{X_a^*\delta_{0m} + X_c^*\sum_j C_{2,m}(\omega_j)\}(-)^m C_{2-m}(\omega)]d\omega. \tag{52}$$

We are only interested in the difference in free energy between the nematic and the isotropic phases, for the transition from one to the other occurs when this is zero. We can obtain the Helmholtz free energy of the isotropic phase from Eqs. (50 - 52) by setting the orientational order and hence the interaction coefficients equal to zero. This gives

$$A_0 = -kT\ln Z_0 \quad , \tag{53}$$

where

$$Z_0 = 4\pi\sum_n \exp\{-U_{int}(n)/kT\} \quad ; \tag{54}$$

the factor of 4π is the orientational partition function for a molecule in the isotropic phase. The required difference in free energies per molecule is then

$$\Delta A = (X_a\langle\bar{P}_2^a\rangle + X_c\langle\bar{P}_2^c\rangle)/2-kT\ln(Z/Z_0) \quad . \tag{55}$$

At the transition ΔA is zero and so the entropy of transition is the change in internal energy divided by T_{NI} which gives

$$\Delta S/k = -(X_a^{NI}\langle\bar{P}_2^a\rangle_{NI} + X_c^{NI}\langle\bar{P}_2^c\rangle_{NI})/2kT_{NI}$$

$$+ (\langle\bar{U}_{int}\rangle_{NI} - \langle\bar{U}_{int}^0\rangle_{NI})/kT_{NI} \quad , \tag{56}$$

where the label NI indicates that the quantity is evaluated at the nematic - isotropic transition temperature.

This completes the derivation of the molecular field theory for nematics composed of non-rigid molecules and it is appropriate at this stage to consider some of the ways in which it differs from the model developed by Marcelja (1). According to his theory the order parameter for the aromatic core is, in our notation,

$$\langle\bar{P}_2^a\rangle = Z_a^{-1}\int P_2(\cos\beta)\exp\{X_a^* P_2(\cos\beta)\}\sin\beta d\beta \quad , \tag{57}$$

where the orientational partition function for the core is

$$Z_a = \int \exp\{X_a^* P_2(\cos\beta)\}\sin\beta d\beta \quad . \tag{58}$$

This seems to be a considerable oversimplification for the chain contributions to the molecular field experienced by the core via X_c are absent from the Boltzmann factor. As a consequence no conformational average is needed in the calculation of $\langle\bar{P}_2^a\rangle$, however the chain-core interactions are partially included because of the dependence of X_a on ε_{ac}. Such interactions are included in the new theory as we can see by combining Eqs. (19), (30), (35), and (36). The expression given by Marcelja for the order parameter of the chain seems to be flawed in a similar manner for it is

$$\langle\bar{P}_2^c\rangle = Z_c^{-1} \sum_n \sum_{\substack{\text{initial} \\ \text{orientations}}} \sum_j P_2(\cos\beta_j)\exp\{X_c^* \sum_j P_2(\cos\beta_j)\}. \tag{59}$$

(We have omitted a factor of 1.88/N which occurs as part of the Boltzmann factor of Eq. (10b) of (1) but which should, we believe, have been part of the pre-exponential factor). This expression differs from that in our theory for the core does not make any contribution to the molecular field experienced by the chain. It would appear that Marcelja has ignored the orientational coupling between the core and chain segments which is known to be of major importance (3). Eq. (59) also demonstrates another limitation of Marcelja's theory referred to previously for the integral over the segmental orientations has been replaced by a sum over a small number of orientations of the core with respect to the director.

We have used our new theory to predict the variation of the nematic - isotropic transition temperature along the homologous series of 4-n-alkyl-4'-cyanobiphenyls. In these calculations the conformational energy parameter E_{tg}^* was held fixed at 1.0; strictly the quantity E_{tg} is constant and so E_{tg}^* will vary with T_{NI} however such variation should be small for the cyanobiphenyls. The segmental volumes required in the calculations were taken from the tabulation by Bondi (24) and so were not used as adjustable parameters. The molecular interaction parameter ε_{aa} simply scales all of the transition temperatures and so the form of the T_{NI} profile is fitted by varying the single quantity λ_c which from Eq. (48) is $(\varepsilon_{cc}/\varepsilon_{aa})^{\frac{1}{2}}$. The best agreement is obtained with λ_c equal to 0.4 which is similar to but not identical with the value of 0.3 found by fitting the order parameter profiles for 4-n-pentyl and 4-n-octyl-4'-cyanobiphenyl (3). The nematic - isotropic transition temperatures predicted by the theory are compared in Fig. (13) with the experimental values for the alkyl

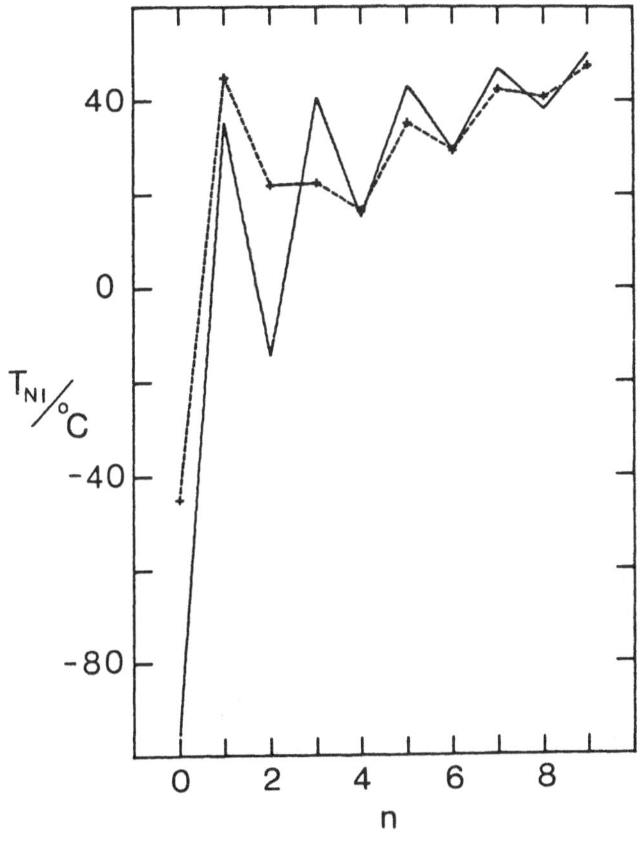

Figure 13. *The nematic - isotropic transition temperatures for the 4-n-alkyl-4'-cyanobiphenyls; the dashed line joins the observed values while the predicted T_{NI} are linked by the solid line*

cyanobiphenyls. The agreement is good for the higher members of the series but is not so impressive for the first few members. This discrepancy may well reflect the inadequacy of assuming that the cyanobiphenyl core may be treated as a cylindrically symmetric object.

REFERENCES

1. S. Marcelja, *J. Chem. Phys.*, **60**, 3599 (1974).

2. A. Pines, D.J. Ruben and S. Allison, *Phys. Rev. Letters*, **33**, 1002 (1974).

3. J.W. Emsley, G.R. Luckhurst and C.P. Stockley, *Proc. R. Soc., Lond.* **A381**, 117 (1982).

4. See, for example, G.R. Luckhurst in *"The Molecular Physics of Liquid Crystals"* Chapt. 4, editors G.R. Luckhurst and G.W. Gray, Academic Press, 1979.

5. M.E. Rose, *"Elementary Theory of Angular Momentum"* Wiley, 1957.

6. A.J. Leadbetter and E.K. Norris, *Molec. Phys.*, **38**, 669 (1979).

7. G.R. Luckhurst, P. Simpson and C. Zannoni, *Chem. Phys. Letters*, **78**, 429 (1981).

8. G.R. Luckhurst and R.N. Yeates, *J. Chem. Soc. Faraday Trans. II*, **72**, 996 (1976).

9. R. Seeliger, H. Haspeklo and F. Noack, *Molec. Phys.*, **49**, 1039 (1983).

10. P.G. de Gennes, *"The Physics of Liquid Crystals"* Oxford, 1974.

11. R.G. Horn and T.E. Faber, *Proc. R. Soc. Lond.* **A368**, 199 (1979); G.F. Kventzel and J. Katriel, *Molec. Cryst. Liq. Cryst.*, **84**, 93 (1982).

12. G.R. Luckhurst and P. Simpson, *Molec. Phys.*, **47**, 251 (1982).

13. A.L. Tsykalo and A.D. Bagmet, *Czech. J. Phys.*, **B28**, 1113 (1978); D.J. Adams and G.R. Luckhurst, unpublished results.

14. W.M. Gelbart and A. Gelbart, *Molec. Phys.*, **33**, 1387 (1977).

15. G.R. Luckhurst, C. Zannoni, P.L. Nordio and U. Segre, *Molec. Phys.*, **30**, 1345 (1975).

16. G.W. Gray, A. Mosley, J.W. Emsley and G.R. Luckhurst, *Molec. Phys.*, **35**, 1499 (1978); J.W. Emsley, G.R. Luckhurst and C.P. Stockley, *Molec. Phys.*, **44**, 565 (1981).

17. J.W. Emsley, R. Hashim, G.R. Luckhurst, G.N. Rumbles and F.R. Viloria, *Molec. Phys.*, **49**, 1321 (1983).

18. P.J. Flory, *"Statistical Mechanics of Chain Molecules"* Interscience 1969.

19. J.W. Emsley and G.R. Luckhurst, *Molec. Phys.*, **41**, 19 (1980).

20. J. Nehring and A. Saupe, *Molec. Cryst. Liq. Cryst.*, **8**, 403 (1969).

21. J. Seelig and W. Niederberger, *J. Amer. Chem. Soc.*, **96**, 2069 (1974).

22. R.L. Humphries, P.G. James and G.R. Luckhurst, *Symp. Faraday Soc.*, **5**, 107 (1971).

23. R.L. Humphries and G.R. Luckhurst, *Chem. Phys. Letters*, **23**, 567 (1973); D.E. Martire in *"The Molecular Physics of Liquid Crystals"* Chapt. 11, editors G.R. Luckhurst and G.W. Gray, Academic Press, 1979.

24. A. Bondi, *J. Phys. Chem.*, **68**, 441 (1964).

THE LANDAU-DE GENNES THEORY OF LIQUID CRYSTALS

G.R. Luckhurst

Department of Chemistry,
University of Southampton, U.K.

INTRODUCTION

The development of a molecular theory for liquid crystals requires a knowledge of the intermolecular potential and then the evaluation of the configurational partition function. Even for idealized systems the calculation of the partition function presents severe problems and we are forced to make approximations. However for a real mesogen such as a 4-n-alkyl-4'-cyanobiphenyl there is the added difficulty that we do not know the form of the pair potential and even though it might be approximated by a sum of atom-atom potentials the exact evaluation of the partition function would be an impossibly difficult task. As we have seen in chapter 3 the molecular field approximation allows us to make some progress in a logical manner for rigid particles while for real mesogenic molecules the effect of non-rigidity can be included only in a semi-intuitive manner.

An alternative approach to this general problem is to forsake any attempt at calculating the partition function and hence the configurational free energy from the intermolecular potential. Instead the free energy is written in terms of some variable which is of particular importance in characterizing the phase transition of interest, for example, the magnetization for ferromagnets. For the liquid-gas transition the relevant order parameter would be the difference in density of the two phases while for a real nematic - isotropic transition we might use the anisotropy in the diamagnetic susceptibility. Landau has argued that the free energy has a particularly simple dependence on the order parameter and its spatial derivatives at least near a

J. W. Emsley (ed.), Nuclear Magnetic Resonance of Liquid Crystals, 85–97.

second order phase transition (1). Here we describe the form of this free energy function and show de Gennes' extension to include the first order transition exhibited by nematogens (2). We shall then use this theory to account for the pretransitional behaviour exhibited by the isotropic phase as the transition to the nematic is approached. Finally we relate the phenomenological coefficients in the Landau-de Gennes theory to the molecular parameters which occur in the Maier-Saupe theory of nematics.

LANDAU THEORY OF SECOND ORDER PHASE TRANSITIONS

We begin by defining an order parameter Ψ which can be used to characterize the transition. Of course many order parameters may be needed to describe the phase but we shall assume that just one of these, Ψ, is dominant. In the low temperature ordered phase Ψ is non-zero while above the transition in the disordered phase it vanishes. As the transition is second order Ψ will become vanishingly small as we pass from the ordered to the disordered phase and so we might hope to write the free energy as a power series in Ψ. Strictly we should deal with the free energy density so that we may allow for spatial variations in the order parameter throughout the system. However for convenience we shall take the system to be homogeneous especially as such an assumption will not influence any of our results. The total free energy F of the system is then simply the free energy density f multiplied by its volume V. This free energy density will be a function of Ψ as well as pressure (or volume) and temperature T. We shall be concerned only with experiments for which the transition is driven by changes in the temperature and so the pressure (or volume) dependence of the expansion coefficients will be ignored.

We write therefore the free energy density as

$$f(\Psi,T) = f_0(T) + \alpha(T)\Psi + A(T)\Psi^2 + B(T)\Psi^3 + C(T)\Psi^4 + \ldots . \quad (1)$$

It is important to recognise the order parameter Ψ in this equation is not the equilibrium value but simply the value associated with some state or configuration of the system at the temperature T. The equilibrium value $\langle\Psi\rangle$ of the order parameter is the quantity which minimizes the free energy. Certain of the expansion coefficients vanish for systems exhibiting second order phase transitions. Consider, for example, the smectic A - smectic C transition where the tilt angle β passes continuously from zero in the A phase to a non-zero value in the C. Here the appropriate order parameter is the complex quantity

$$\Psi = \beta\exp(i\phi), \quad (2)$$

where ϕ is the azimuthal angle for the director projected on to the smectic plane (3). Since the free energy density is real, terms $\Psi\Psi^*$ and $(\Psi\Psi^*)^2$ may contribute but the linear and cubic terms clearly cannot, we have then

$$f(\Psi,T) = f_0(T) + A(T)\Psi^2 + C(T)\Psi^4 + \dots \quad . \qquad (3)$$

In addition the free energy is a symmetric function of the order parameter as required for such systems. For temperatures above the transition at T^* we expect the free energy to have a minimum at $\Psi = 0$ corresponding to a stable disordered phase. Conversely below T^* the minimum in the free energy should be at a non-zero value corresponding to a stable ordered phase. These forms of the free energy density are sketched in Fig. (1). To obtain such dependences we require A(T) to be positive above T^* but negative below the transition. Accordingly at the transition $A(T=T^*)$ must vanish and the simplest function to satisfy these requirements is

$$A(T) = a(T-T^*) \ . \qquad (4)$$

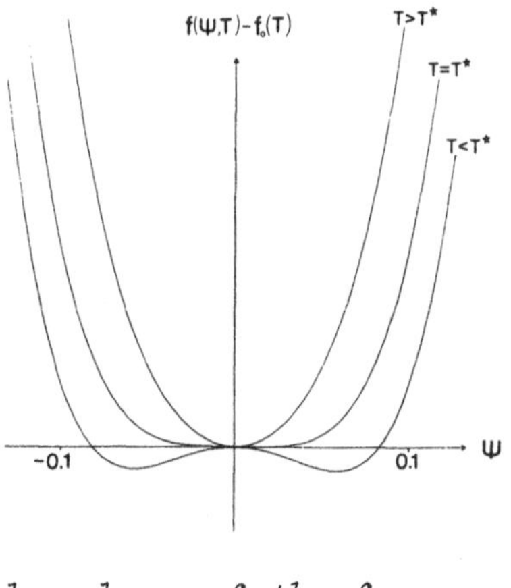

Figure 1. *The dependence of the free energy density $\{f(\Psi,T) - f_0(T)\}$ on the order parameter for a system exhibiting a second order phase transition. The curves were calculated from Eqs. (3) and (4) for temperatures above, at and below the transition at T^*.*

In addition the coefficient of the quartic term should be positive to obtain a stable, ordered phase below T^* and although C(T) may be temperature dependent it is assumed that this dependence is weak. Indeed we shall take the coefficient to be a constant equal to its value $C(T^*)$ at the transition. The free energy density $f_0(T)$ of the isotropic phase is also taken to be well behaved, at least in the region of the phase transition, and so the behaviour at the transition is determined by the temperature dependence of $A(T)$.

To find the equilibrium states we locate the minimum in the free energy density by differentiating it with respect to the order parameter and setting this equal to zero. The resultant cubic equation

$$2a(T-T^*)\Psi + 4C\Psi^3 = 0 \quad , \tag{5}$$

has solutions

$$\Psi = 0 \tag{6}$$

and

$$\Psi = \pm(a/2C)^{\frac{1}{2}}(T^*-T)^{\frac{1}{2}} \quad . \tag{7}$$

For temperatures above T^* the only real solution has $\Psi = 0$; it corresponds to the disordered state with its equilibrium value $\langle\Psi\rangle$ of zero. Below T^* $\Psi=0$ corresponds to the maximum in the free energy while the solutions $\pm(a/2C)^{\frac{1}{2}}(T^*-T)^{\frac{1}{2}}$ are symmetrically placed minima. The equilibrium order parameter in the low temperature phase therefore grows according to

$$\langle\Psi\rangle = (a/2C)^{\frac{1}{2}}(T^*-T)^{\frac{1}{2}} \quad . \tag{8}$$

and so the critical exponent for the tilt angle is predicted to be $\frac{1}{2}$. The tilt angle β in the smectic C phase of terephthalylidene-bis-(4-n-butylaniline) has been determined from the NMR spectrum of a probe molecule, methylene dichloride, dissolved in it (4). The temperature dependence of β is found to be in good agreement with that predicted by Eq. (8) as are measurements of the tilt angle in n-hexyl-4'- n-decyloxybiphenyl -4-carboxylate made by ESR spectroscopy and a paramagnetic probe (5). However other studies have produced exponents which are more in keeping with a prediction made by de Gennes (3) that the temperature dependence of the tilt angle should be of the form $(T^*-T)^{0.35}$.

The equilibrium entropy density of the system is available from the equilibrium free energy density since

$$s = -(\partial f/\partial T)_V \quad , \tag{9}$$

where we continue to indicate thermodynamic densities with lower case letters. Above the transition this gives

$$\begin{aligned} s &= -(\partial f_0(T)/\partial T)_V \ , \\ &= s_0 \end{aligned} \tag{10}$$

while for the ordered phase we find

$$s = s_0 + (a^2/2C)(T-T^*) \tag{11}$$

after substitution for the equilibrium order parameter from Eq. (8). We see therefore that the entropy density is continuous at the transition, in accord with its second order character. The equilibrium heat capacity density can be obtained from the equilibrium entropy density since

$$c_V = T(\partial s/\partial T)_V \quad . \tag{12}$$

In the disordered phase above T^*

$$\begin{aligned} c_V &= -T(\partial^2 f_0(T)/\partial T^2)_V \\ &= c_V^0 \end{aligned} \tag{13}$$

and in the ordered phase

$$c_V = c_V^0 + Ta^2/2C \quad . \tag{14}$$

The heat capacity is therefore discontinuous at the transition and the magnitude of the discontinuity is $a^2T^*/2C$. This discontinuity agrees with the second order, or continuous, nature of the phase transition.

THE LANDAU-DE GENNES THEORY OF NEMATICS

The nematic - isotropic transition is first order in that the entropy and the second rank orientational order parameter are discontinuous at the transition. However the entropy of transition is small ($\Delta S/R \sim 0.1$) and so we might expect to be able to extend the Landau theory for second order transitions to such weak, first order transitions and de Gennes has presented such an extension (2). The essence of this is the inclusion of a cubic term in the order parameter expansion of the free energy density. To see the origin of the cubic term we need to consider the nature of the order parameter used to characterize the nematic phase. In fact the quantity we require is a second rank tensor $\underset{\sim}{Q}$ whose components, in an arbitrary laboratory frame, are

$$Q_{\alpha\beta} = (3\overline{\ell_\alpha \ell_\beta} - \delta_{\alpha\beta})/2 \quad , \tag{15}$$

at least for cylindrically symmetric particles. For this idealised nematogen ℓ_α is the direction cosine between the laboratory α-axis and the molecular symmetry axis, $\delta_{\alpha\beta}$ is the Kroenecker delta and the bar indicates an average over a configuration of the system composed of N particles

$$\overline{\ell_\alpha \ell_\beta} = N^{-1} \sum_{i=1}^{N} \ell_\alpha^{(i)} \ell_\beta^{(i)} \quad , \tag{16}$$

where the direction cosines are for the ith particle. As we shall see the ordering tensor $\underset{\sim}{Q}$ is closely related to the familiar order parameter

$$S = (3\langle \ell_z^2 \rangle - 1)/2 \quad , \tag{17}$$

where ℓ_z is the direction cosine between the director and the molecular symmetry axis and now the angular brackets indicate an equilibrium average.

Since the free energy density is a scalar only scalar products of $\underset{\sim}{Q}$ or its trace can be included in f(Q,T). The trace of Q vanishes and so there can be no linear term at least in the absence of an external field; the quadratic term is just $\underset{\sim}{Q}.\underset{\sim}{Q}$, the quartic term involves both $\underset{\sim}{Q}.\underset{\sim}{Q}.\underset{\sim}{Q}.\underset{\sim}{Q}$ and $(\underset{\sim}{Q}.\underset{\sim}{Q})^2$, more importantly we can now include a cubic term $\underset{\sim}{Q}.\underset{\sim}{Q}.\underset{\sim}{Q}$. The form of these various products may be simplified because their gauge invariance means that we may choose any laboratory frame in which to evaluate them. Thus if we choose the z-axis to lie parallel to the director

$$\underset{\sim}{Q} = Q \begin{pmatrix} 1 & 0 & 0 \\ 0 & -\frac{1}{2} & 0 \\ 0 & 0 & -\frac{1}{2} \end{pmatrix} \tag{18}$$

since the nematic phase is uniaxial; the equilibrium value of the scalar Q would then be the order parameter S. Accordingly we obtain a simple expansion in powers of Q for the free energy density

$$f(Q,T) = f_0(T) + A(T)Q^2 + B(T)Q^3 + C(T)Q^4 + \ldots\ldots \quad . \tag{19}$$

As before we are guided in the form the expansion coefficients may take by considering the anticipated behaviour of the free energy density at various temperatures. For example, at high temperatures we expect f(Q,T) to have a minimum at Q=0 corresponding to the disordered or isotropic phase. At the nematic - isotropic transition T_{NI} the nematic and isotropic phases are in equilibrium. Consequently the free energy density must contain two equal minima; one of these is at S=0 while the other is at S_{NI} corresponding to the nematic phase at the transition. As the temperature is lowered further the minimum corresponding to the nematic phase must grow in depth although there will be a temperature range for which the minimum at Q=0 may be preserved. At very low temperatures we might expect to find a single minimum in the free energy density corresponding to the nematic phase. These various free energy densities are sketched in Fig. (2). Such curves may be obtained if the coefficients B(T) and C(T) are taken to be temperature independent while A(T) varies, as before, according to

$$A(T) = a(T-T^*) \quad ; \tag{20}$$

however as we shall see T^* will not now be the transition temperature.

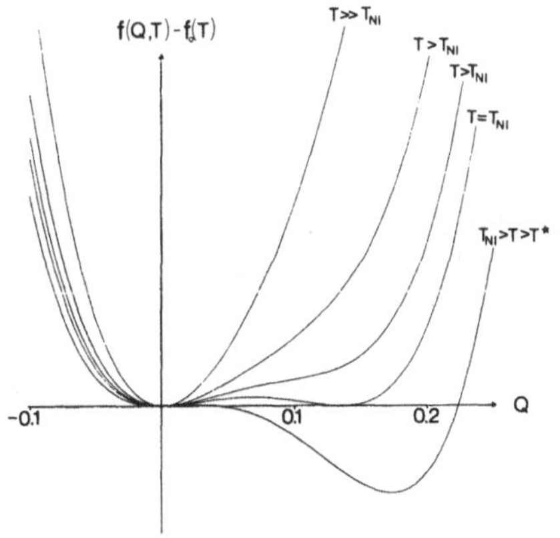

Figure 2. *The dependence of the free energy density* $\{f(Q,T) - f_0(T)\}$ *for a system of cylindrically symmetric particles exhibiting a nematic - isotropic transition. The curves were calculated from Eqs. (19) and (20) for temperatures above, at and below* T_{NI}.

To locate the nematic - isotropic transition we recall that the free energies of the two phases are the same at T_{NI} and so

$$a(T_{NI}-T^*)Q^2 + BQ^3 + CQ^4 = 0 \tag{21}$$

while these phases also correspond to minima in the free energy density; that is

$$2a(T_{NI}-T^*)Q + 3BQ^2 + 4CQ^3 = 0. \tag{22}$$

Solution of these equations gives the order parameter for the nematic phase at the transition as

$$S_{NI} = -B/2C \quad . \tag{23}$$

This clearly demonstrates the importance of the cubic term which is solely responsible for the first order character of the transition. The temperature at which the transition occurs may be obtained by substituting the equilibrium value of the order parameter S_{NI} into the expression for the free energy density and equating the values of f(Q,T) for the coexisting isotropic and nematic phases. This gives

$$T_{NI} = T^* + B^2/4aC \quad , \tag{24}$$

which shows that the transition temperature is higher by $B^2/4aC$ than the temperature T^* at which the coefficient of the quadratic term vanishes. Again we see that the cubic term produces the first order transition, for as B tends to zero so T_{NI} approaches the second order transition temperature T^*.

The importance of the cubic term is confirmed by evaluation of the entropy density of transition. The entropy density for the nematic phase is obtained from the free energy density via

$$s = -(\partial f(Q,T)/\partial T)_V \quad ,$$

$$= -(\partial f(Q,T)/\partial T)_{V,Q} - (\partial f(Q,T)/\partial Q)_{T,V}(\partial Q/\partial T)_V \quad . \tag{25}$$

However, as we have seen the condition for equilibrium is that the free energy density is a minimum with respect to the order parameter Q and so

$$s = -(\partial f(Q,T)/\partial T)_{V,Q} \tag{26}$$

which gives the equilibrium entropy density as

$$s = s_0 - a\, S_{NI}^2 \quad , \tag{27}$$

where s_0 is the entropy density $-(\partial f_0(T)/\partial T)_V$ for the isotropic phase. The entropy density of transition Δs is then

$$\Delta s = -aB^2/4C^2 \quad ; \tag{28}$$

thus confirming that the nematic - isotropic transition is first order. In the limit, as B tends to zero, the transition becomes second order.

It would be possible to fit the theory to experiment, however the number of adjustable parameters, a, B, C and T^* is large. In addition the assumptions concerning the temperature dependence of A and independence of B and C would cast doubt on the usefulness of such comparisons although they have been made (6). The Landau-de Gennes theory has been of considerable value, however, in understanding the pretransitional behaviour exhibited by nematogens and we now turn our attention to this aspect of the theory.

PRETRANSITIONAL BEHAVIOUR

One consequence of the weakness of the nematic - isotropic transition is that the isotropic phase acquires characteristics of the nematic to an extent which increases as the transition is approached. The major characteristic of the nematic phase is its long range orientational order and in the preceding isotropic phase the angular correlations grow as the temperature is lowered. This increase in angular correlation may be probed by a variety of experiments, for example the Cotton-Mouton effect, the Kerr effect, flow induced birefringence (7) and magnetic field induced quadrupolar splittings (8). In each of these experiments a minute orientational order is induced in the system by some external field and the magnitude of the effect observed is proportional to this field induced order. Here we show how the Landau-de Gennes theory may be used to predict the extent of the induced order and more importantly its temperature dependence.

It is convenient to take the applied field to be magnetic because this avoids the difficulties associated with internal field corrections when electric fields are used. For a given configuration of a system containing N cylindically symmetric particles the magnetic contribution to the free energy coming from the anisotropy in the diamagnetic susceptibility is

$$F_{mag} = -(\tfrac{1}{3})\Delta\chi B^2 \sum_i^N (3\ell_z^{(i)^2}-1)/2 \quad . \tag{29}$$

Here the magnetic flux density is denoted by B and should not be confused with the expansion coefficient in the free energy density, $\Delta\chi$ is the anisotropy $(\chi_{\|} - \chi_{\perp})$ where $\chi_{\|}$ is the component of the diamagnetic susceptibility parallel to the molecular symmetry axis and $\ell_z^{(i)}$ is the direction cosine of this axis for the ith molecule with the magnetic field. The order parameter Q for this configuration is

$$Q = N^{-1} \sum_{i}^{N} (3\ell_z^{(i)^2} - 1)/2 \tag{30}$$

and so the magnetic contribution to the free energy density is

$$f_{mag} = -(\tfrac{1}{3})\rho\Delta\chi B^2 Q \quad , \tag{31}$$

where ρ is the number density. We write the total free energy density as

$$f(Q,T,B) = f_0(B,T) + a(T-T^*)Q^2 + BQ^3 + CQ^4 - \chi Q \quad , \tag{32}$$

where χ is just $(\frac{1}{3})\rho\Delta\chi B^2$ and $f_0(B,T)$ contains the contribution from the scalar diamagnetic susceptibility. To find the equilibrium value of the field induced order parameter in the isotropic phase we seek the minimum in the free energy density from its derivative. This gives

$$2a(T-T^*)Q + 3BQ^2 + 4CQ^3 - \chi = 0 \quad , \tag{33}$$

and because the order parameter is small in the isotropic phase we may ignore the quadratic and cubic terms in Q. We then find the field induced order parameter S_B to be

$$S_B = \chi/2a(T-T^*) \quad . \tag{34}$$

This order parameter and hence the field induced birefringence or quadrupolar splittings are predicted to diverge at T^* and not the nematic - isotropic transition itself. The experimental results on many systems do indeed exhibit the temperature dependence given by Eq. (34) and have shown that $T_{NI}-T^*$ is approximately 1 K (7,8). Such behaviour is illustrated in Fig. (3) by the Cotton-Mouton constant, which is proportional to S_B, for 4-n-octyl-4'-cyanobiphenyl (9). Here the plot of the inverse Cotton-Mouton constant as a function of temperature is essentially linear; there is a small departure from linearity near the transition which may result from the neglect of quadratic and cubic terms in Eq. (33). The intercept T^* is slightly less than the nematic - isotropic transition. The Landau-de Gennes theory is however unable to make any prediction about this difference because the quantities occurring in the expansion of the free energy density are purely phenomenological. However, molecular field theory can be used to relate the expansion coefficients to molecular quantities and in the following section we show how this may be achieved.

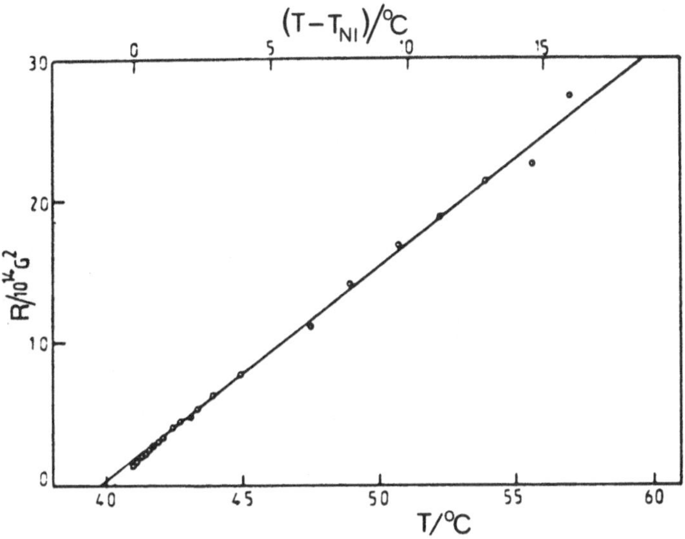

Figure 3. *The inverse of the Cotton-Mouton constant R for the isotropic phase of 4-n-octyl-4'-cyanobiphenyl plotted as a function of temperature (9).*

A MOLECULAR APPROACH

According to the molecular field approximation the orientational entropy S of a configuration is related to the singlet orientational distribution function f(β) by

$$S = -Nk \int_0^{\pi} f(\beta) \ln f(\beta) \sin\beta d\beta \quad , \tag{35}$$

where β is the angle between the molecular symmetry axis and the unique axis for the configuration, which is taken to be uniaxial. This simple result obtains because direct angular correlations are ignored in the molecular field approximation and so the N-body orientational distribution function is just the product of N singlet orientational distributions. In general the singlet distribution function may be expanded in a series of Legendre polynomials $P_L(\cos\beta)$

$$f(\beta) = \sum_L \tfrac{1}{2}\{2L+1\}\bar{P}_L P_L(\cos\beta) \quad , \tag{36}$$

where the $\bar{P}_L$ form a complete set of order parameters. When the orientational order is low, as it is in the isotropic phase subject to an external field, the distribution function is well approximated by the first two terms of the series

$$f(\beta) = \tfrac{1}{2} + (\tfrac{5}{2})Q(3\cos^2\beta-1)/2 \quad . \tag{37}$$

Substitution of this result into Eq. (35), expansion of the logarithm and evaluation of the integral gives the orientational entropy density as

$$s = -\rho k\,\{(\tfrac{5}{2})Q^2 - (25/21)Q^3 + (125/28)Q^4\}, \tag{38}$$

to fourth order in Q. The internal energy density, according to the Maier-Saupe theory, is simply

$$u = -(\rho/2)\bar{u}_2 Q^2 \quad , \tag{39}$$

where $\bar{u}_2$ is the Maier-Saupe interaction parameter (see chapter 3). The free energy density (u-Ts) for small Q in zero field is then

$$f(Q,T) = f_0(T) + (5\rho k/2)(T-\bar{u}_2/5k)Q^2 - (25/21)\rho kTQ^3 + (125/28)\rho kTQ^4 \quad . \tag{40}$$

Comparing coefficients in Eqs. (19), (20) and (40) gives a as $5\rho k/2$ and depends solely on the number density, a result which has also been obtained by more general arguments (10). In consequence **a** is only temperature independent at constant volume. More importantly it is clear that measurements of a will tell us little about the anisotropic interactions in a liquid crystal. This information is contained in the temperature T^* which is predicted to be $\bar{u}_2/5k$ whereas the nematic - isotropic transition temperature is given by the Maier-Saupe theory as $\bar{u}_2/4.54k$. The ratio T^*/T_{NI} is therefore predicted to be 0.908 which is in poor agreement with the experimental value of about 0.997; this failure of the theory results from the use of the molecular field approximation. The B coefficient is predicted by this molecular approach to be negative as required, however, we also see that it is temperature dependent although this dependence is relatively weak, especially in comparison to that of A in the vicinity of the phase transition. Similarly C is found to be positive, as required for the stability of the nematic phase and is also weakly temperature dependent. Both B and C are predicted to be independent of the anisotropic molecular interactions and to be determined solely by the number density. There do not appear to have been any measurements which would allow this prediction to be tested.

REFERENCES

1. See, for example, L.D. Landau and E.M. Lifshitz, *Statistical Physics* (Pergamon Press, Oxford 1969) Chapt. XIV.

2. P.G. de Gennes, *Liquid Crystals 3* eds. G.H. Brown and M.M. Labes (Gordon and Breach, London, New York, Paris, 1972) p. 21.

3. P.G. de Gennes, *The Physics of Liquid Crystals* (Clarendon Press, Oxford, 1974) p. 314.

4. Z. Luz and S. Meiboom, *J. Chem. Phys.*, **59**, 275 (1973).

5. G.R. Luckhurst and B.A. Timimi, *Phys. Lett.*, **75A**, 91 (1979).

6. E.B. Priestley, P.J. Wojtowicz and P. Sheng, *Introduction to Liquid Crystals* (Plenum Press, New York and London, 1974) Chapt. 10.

7. See, for example, S. Chandrasekhar, *Liquid Crystals*, (Cambridge University Press, Cambridge 1977) Chapt. 2.

8. G.S. Attard, P.A. Beckmann, J.W. Emsley, G.R. Luckhurst and D.L. Turner, *Molec. Phys.*, **45**, 1125 (1982); G.S. Attard, J.W. Emsley and G.R. Luckhurst, *ibid*, **48**, 639 (1983).

9. S.L. Zhang, Z.Y. Peng, J. Wu, T.H. Shen and N.Q. Wu, *Mol. Cryst. Liq. Cryst.*, **91**, 295 (1983).

10. R.L. Humphries and G.R. Luckhurst, *Proc. Roy. Soc.* **A382**, 307 (1982).

NMR SPECTRA IN LIQUID CRYSTALS THE PARTIALLY AVERAGED SPIN HAMILTONIAN

C.A. Veracini

Istituto di Chimica Fisica,
Università di Pisa, Italy.

INTRODUCTION

Nuclear Magnetic Resonance (NMR) spectroscopy is now a commonly used method to investigate both long range orientational order in liquid crystalline mesophases and molecular properties of partially oriented solutes (1). The NMR spectra of molecules in a liquid crystal environment are, in fact, dominated by second rank tensorial properties such as the shielding tensors $\tilde{\sigma}_i$, the dipolar couplings $\tilde{D}_{ij}$, the quadrupolar interactions $\tilde{q}_i$, partially averaged by the anisotropic molecular tumbling. Moreover, since the degree of the averaging processes depends on the orientational distribution, the equilibrium values of magnetic interactions, measurable from splittings and line positions in the NMR spectra, are functions of the orientational order.

The conditions of an effective averaging, i.e. that the frequency of molecular tumbling is much higher than anisotropic quantities expressed in frequency units, are usually met in most liquid crystals (1). Molecular translational diffusion, moreover, averages to zero the intermolecular interactions so that in the liquid crystal phase the intramolecular interactions can be detected without the complications of intermolecular ones which dominate the NMR spectra of solids (2).

Other spectroscopies (fluorescence (3), Raman (4), electron spin resonance (5), linear dichroism (6)) are used to investigate the orientational order of mesophases. These techniques, however, usually can measure only a single averaged component of a tensor $\tilde{T}_{\parallel}$ so that they do not give enough

J. W. Emsley (ed.), Nuclear Magnetic Resonance of Liquid Crystals, 99–121.

information for the study of orientational order. In this respect NMR has a number of advantages and in particular it may give access to many local submolecular properties. For example a suitable pair of protons allows the determination of an order parameter relative to the direction joining the two nuclei. Thus the presence of various pairs of protons gives a relatively large amount of experimental data. Moreover, if we assume the effect of isotopic substitution to be negligible, NMR of various active nuclei (and notably deuterium) can be performed to further increase the number of experimental data (1,7). The limitations of NMR should also be clearly put in evidence. To quote only a few:

a) measurements of partially averaged second rank tensors give access only to second rank order parameters, as will be shown later;

b) structural and orientational information are tightly connected. Quite often, because of the complexity of spectra or the scarce determination of spectral parameters, it is possible to obtain only partial information and a molecular structure must be assumed from other techniques;

c) large amplitude internal motions complicate the partial averaging process and it is not easy to separate the effects of overall molecular tumbling from the internal motions (8).

MACROSCOPIC ALIGNMENT

In the conditions normally used in NMR experiments the magnetic field *B* induces a uniform order in the mesophase so that the preferred orientation, identified by a unit vector *n* (the director), is the same for the whole sample. The magnetic fields of course are usually not strong enough to orient an isolated molecule and the interaction energy is due to a cooperative effect (9). The orientation dependent free energy of the interaction between the magnetic field and the liquid crystal can be expressed as (9):

$$G = -\Delta\chi B^2 \ (3\cos^2\alpha - 1)/6 \tag{1}$$

where $\Delta\chi = \chi_{\parallel} - \chi_{\perp}$ is the difference between the magnetic susceptibility parallel and perpendicular to the director, *B* is the magnetic field and α is the angle between the director and the field. The direction of minimal G, i.e. of the preferred alignment, will depend on the quantity $\Delta\chi$. For a large number of liquid crystals $\Delta\chi$ is positive and *n* will align along *B*; a few others (ZLI 1167 (10), some micellar lyotropics (11)) have negative $\Delta\chi$ and the perpendicular orientation, $n \perp B$, is preferred. Fluctuations in the alignment are always present so that a more or less peaked distribution around the direction of the preferred alignment is obtained depending also on the strength of B.

The response time for the alignment depends on the relative values of the magnetic torque density,

$$L_m = \tfrac{1}{2}\Delta\chi B^2 \sin 2\alpha,$$

and of the opposing viscous torque,

$$L_v = -\gamma_1 \frac{d\alpha}{dt},$$

where γ_1 is a viscosity coefficient. For most nematic liquid crystals it takes (at a field of some KGauss) only a few seconds to reorient (12). This is not the case for other systems such as amphiphylic or polybenzyl glutammate lyotropics where the alignment is relatively slow so that it can be easily followed by NMR spectroscopy; there is a change from a typical powder spectrum to the spectrum characteristic of a monodomain sample as shown in Fig. (1). Smectics and some cholesterics are very

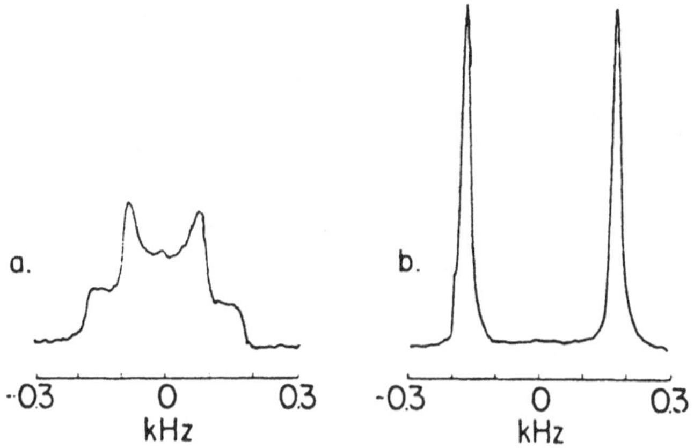

Figure 1. *Deuteron spectrum of* H_2O *in a nematic lyotropic phase a) immediately after being placed in the magnetic field, and b) after being in the magnetic field for several hours.*

viscous and do not align spontaneously, with the exception of some smectics in high fields (>4.6T).

SPIN HAMILTONIAN

An NMR spectrum contains two kinds of information: the position and intensity of the lines is related to the partially averaged interactions and is called "static" information; the dynamic information is contained in the lineshape or in the spin-lattice and spin-spin relaxation times. We will treat here only the static part: for a general treatment of the spin Hamiltonian we refer to the books of Abragam (14) and Slichter (15).

The positions and intensities of lines can be obtained by solving the Schrodinger equation with a spin hamiltonian containing interactions between nuclear magnetic moments and external fields, dipolar and indirect interactions between nuclear magnetic moments and eventually electrostatic interactions involving nuclear spins (1).

$$H = H_Z + H_D + H_J + H_Q \tag{2}$$

In this respect magnetic interactions for liquid crystal solutions are not different from those in solid state NMR. The difference arises mainly because of the motional averaging. In liquid crystal solutions molecules are free to tumble and diffuse and the spectra can be obtained by weighting the Hamiltonian with a proper orientational distribution function in order to reproduce the partially averaged interactions normally measured. Before treating in detail the various interactions, let us examine the effect of motional averaging on a generic tensor, T, of the kind involved in Eq. (2).

Motional Averaged Hamiltonian

The various terms of the Hamiltonian (2) can be expressed as scalar products U.T.V, where U and V are spin operators or the magnetic field (16). The magnetic field on the other hand constitutes a direction of quantization for the nuclear spins so that the operators U and V are conveniently expressed in a laboratory frame with the **z** axis along the field. The orientational order is usually described in a reference system with the z axis parallel to the director n. The molecular tensorial properties T, finally, are usually known with respect to an axis system placed on the molecule or a rigid fragment of the molecule. The problem therefore, of calculating averaged tensorial components resolved along the magnetic field $\tilde{T}_{zz}$, of second rank tensors consists in relating these different reference systems. For such purpose we can use tensors in a cartesian form or express them with their irreducible components on a spherical basis set.

Spherical Tensors

The terms of the hamiltonian operator (2) can be written as scalar products of two spherical tensors (16)

$$H_\lambda = \sum_{\ell} \sum_{m=-\ell}^{\ell} (-1)^m \, T_\lambda^{(\ell,m)} A_\lambda^{(\ell,-m)} \tag{3}$$

where T specifies the interaction and A is a spin operator constant in the laboratory frame; λ indicates the various interactions. For symmetric second rank tensors in uniaxial nonpolar mesophases only components with $\ell = 0$ and $\ell = 2$ do not vanish. We have, therefore, zero order contributions

$$H_\lambda^{(0)} = T_\lambda^{(0,0)} A_\lambda^{(0,0)} , \tag{4}$$

which is the only part surviving in isotropic phase spectra, and second rank components

$$H_\lambda^{(2)} = \sum_m (-1)^m \langle T^{(2,m)} \rangle A^{(2,-m)} , \tag{5}$$

where the brackets indicate a mean over both orientational and internal motions.

In the usual high field approximation only the part of the magnetic interactions commuting with the nuclear spin projection I_z is retained; in this case only secular terms (m = 0) are necessary:

$$H^{(2)} = \sum_\lambda \langle T^{(2,0)} \rangle_\lambda A^{(2,0)} . \tag{6}$$

In the case of uniaxial phases with $n \| B$ we do not need the high field approximation in order to delete nonsecular terms and m = 0 terms vanish because of the $D_{\infty h}$ symmetry of the phase.

Tables 1 and 2 give the relationships between $T^{(L,m)}$ and $T_{\alpha\beta}$, and $A^{(L,m)}$ and $A_{\alpha\beta}$.

In order to relate the observed $T^{(2,0)}$ to molecular constants and order parameters we write down the coupling $T_{zz}^{(2,0)}$ in the laboratory frame and then we transform it to a molecular frame. This requires the following coordinate systems:

B frame: this is our laboratory frame with z∥*B*,
d frame: a frame with z ∥ to the average director *d*
d' frame: a frame with z ∥ to the instantaneous director orientation *d'*.
m frame: a frame fixed on the molecule e.g. the frame that diagonalizes the S matrix for one of its rigid parts.

The relation then is the following:

$$\langle T^{(2,0)}\rangle = \sum_{\substack{mm'\\n}} D^{2*}_{0m}(d-B)\, D^{2*}_{mm'}(d-d')\, D^{2*}_{m'n}(m-d')\langle (T^{(2,n)})_{mol}\rangle \tag{7}$$

where the Wigner rotation matrix D^2_{pq} (F-F') connects frame F to frame F' and the angular brackets indicate an average over the

Table 1. *Relationship between $T^{(L,m)}$ the components of an irreducible tensor and $T_{\alpha\beta}$ ($\alpha,\beta = x,y,z$) the cartesian components.*

$$T^{(0,0)} = -1/\sqrt{3}\,(T_{xx} + T_{yy} + T_{zz}) = -1/\sqrt{3}\,\mathrm{Tr}(T)$$
$$T^{(1,0)} = -i/\sqrt{2}\,(T_{xy} - T_{yx})$$
$$T^{(1,\pm1)} = -\tfrac{1}{2}[T_{zx} - T_{xz} \pm i\,(T_{zy} - T_{yz})]$$
$$T^{(2,0)} = 1/\sqrt{6}[3T_{zz} - (T_{xx} + T_{yy} + T_{zz})]$$
$$T^{(2,\pm1)} = \mp \tfrac{1}{2}[T_{xz} + T_{zx} \pm i\,(T_{yz} + T_{zy})]$$
$$T^{(2,\pm2)} = \tfrac{1}{2}[T_{xx} - T_{yy} \pm i\,(T_{xy} + T_{yx})]$$

relevant motions. If we assume that the director fluctuations can be neglected $D^2_{mm'}(d-d') = \delta_{mm'}$. If furthermore the mesophase is uniaxial with respect to the director a δ_{m0} restriction in Eq. (7) follows and so the observed average tensor can be written

$$\langle T^{(2,0)}\rangle = D^{2*}_{00}(d-B) \sum_n \langle D^{2*}_{0n}\; T^{(2,n)}\rangle. \tag{8}$$

Table 2. *Relationship between irreducible and cartesian components of spin operators for magnetic interactions.*

Interaction	$A^{(0,0)}$	$A^{(1,0)}$	$A^{(1,\pm1)}$	$A^{(2,0)}$	$A^{(2,\pm1)}$	$A^{(2,\pm2)}$
nuclear Zeeman	$-\frac{1}{\sqrt{3}} I_z$	-	$-\frac{1}{2} I_\pm$	$\sqrt{(\frac{2}{3})}\, I_z$	$\mp \frac{1}{2} I_\pm$	-
dipole-dipole	-	-	-	$\frac{1}{\sqrt{6}} (3I_zS_z - I.S)$	$\mp\frac{1}{2}(I_zS_\pm + I_\pm S_z)$	$\frac{1}{2} I_\pm S_\pm$
J coupling	$-\frac{1}{\sqrt{3}} I.S.$	$-\frac{1}{2\sqrt{2}}(I_+S_- - I_-S_+)$	$\frac{1}{2}(I_zS_\pm - I_\pm S_z)$	$\frac{1}{\sqrt{6}}(3I_zS_z - I.S)$	$\mp\frac{1}{2}(I_zS_\pm + I_\pm S_z)$	$\frac{1}{2} I_\pm S_\pm$
quadrupole - electric field gradient	-	-	-	$\frac{1}{\sqrt{6}}(3I_z^2 - I(I+1))$	$\mp\frac{1}{2}(I_zI_\pm + I_\pm I_z)$	$\frac{1}{2} I_\pm I_\pm$

$I_\pm = I_x + i\, I_y$

Assuming, finally, the molecule to be a rigid one so that a unique order matrix exists, we have

$$\langle T^{(2,0)}\rangle = D^{2*}_{00}\,(d\text{-}B)\,\sum_n \langle D^{2*}_{0n}\rangle\, T^{(2,n)} \tag{9}$$

The quantities D^{2*}_{0n} are the set of order parameters; they are five, but the molecular symmetry can reduce this number as discussed by Zannoni in chapter 1. Since moreover in Eq. (9) we have a product of tensorial quantities and order parameters, it appears that absolute values of order parameters can be obtained only if the tensorial quantities are known with certainty.

In the case of biaxial phases the δ_{m0} restriction in Eq. (7) does not hold and the number of order parameters can be very high (1b, 16) so that the experiment cannot measure all of them in detail and simplifying assumptions (1b) are needed.

Magnetic Interactions in Cartesian Formalism (1d)

The component resolved along the field, T_{zz}, of a second rank tensor can be related to components of T in a molecular frame by

$$T_{zz} = \sum_{\alpha\beta} \cos\theta_{\alpha z}\, T_{\alpha\beta}\, \cos\theta_{\beta z} \tag{10}$$

where $\theta_{\alpha z}$ is the angle between z and α axes. By addition and subtraction of the trace of T (the sum of the diagonal elements),

$$T_{zz} = \tfrac{1}{3}(T_{aa}+T_{bb}+T_{cc}) + \tfrac{2}{3}\sum_{\alpha,\beta}^{abc} \tfrac{1}{2}(3\cos\theta_{\alpha z}\cos\theta_{\beta z}-\delta_{\alpha\beta})T_{\alpha\beta} \tag{11}$$

where $\delta_{\alpha\beta}$ is the kronecker delta. Taking equilibrium averages over the relevant motions in the liquid crystal phase (assumed here to be axially symmetric) we get

$$\langle T_{zz}\rangle = T^{iso} + \tfrac{2}{3}\sum_{\alpha,\beta}^{abc} \left\langle \frac{3\cos\theta_{\alpha z}\cos\theta_{\beta z}-\delta_{\alpha\beta}}{2}\, T_{\alpha\beta}\right\rangle \tag{12}$$

where T^{iso} is the trace of the tensor under consideration. In the case of separation of internal motions from reorientational ones

$$\langle T_{zz}\rangle = T^{iso} + \tfrac{2}{3}\sum_{\alpha\beta}^{abc} \left\langle \frac{3\cos\theta_{\alpha z}\cos\theta_{\beta z}-\delta_{\alpha\beta}}{2}\right\rangle T_{\alpha\beta} \tag{13}$$

where $\langle(3\cos\theta_{\alpha z}\cos\theta_{\beta z}-\delta_{\alpha\beta})/2\rangle = S_{\alpha\beta}$ are elements of the Saupe ordering matrix. The ordering matrix, because of the properties of direction cosines, is a symmetric traceless tensor which can be diagonalized so that for a rigid molecule, or for a rigid molecular fragment, only two $S_{\alpha\beta}$ elements are needed to describe the order of the principal axes (1). The S matrix has the transformation properties of a second rank tensor. The orientation along a particular axis (e.g. connecting i and j nuclei) can be expressed by elements in the x,y,z axis system:

$$S_{ij} = \sum_{p,q} \cos\theta_p^{ij}\, S_{pq}\, \cos\theta_q^{ij} \quad ; \; p,q = x,y,z \tag{14}$$

Only quantities $T^{iso} = \frac{1}{3}\,\mathrm{Tr}T$ are measurable in isotropic solutions. Since the direct dipolar couplings and quadrupolar interactions are both traceless, they give no observable splittings in isotropic phase spectra although they influence the linewidths and relaxation times. On the other hand the quantities

$$\tilde{T}_{zz}^{aniso} = \frac{2}{3} \sum_{\alpha,\beta}^{abc} S_{\alpha\beta} T_{\alpha\beta} \tag{15}$$

dominate the partially oriented spectra. In the case of an uniaxial phase oriented along the field any partially averaged tensorial quantity T is described by an axially symmetric tensor in diagonal form so that

$$\tilde{T}_{xx}^{aniso} = \tilde{T}_{yy}^{aniso} = -\frac{1}{2}\,\tilde{T}_{zz}^{aniso}$$

We can write down now the general spin hamiltonian (Eq. (2)) specifying the possible interactions, apart from higher rank ones. Within the usual high field approximation

$$H = \frac{1}{2\pi} \sum_i \gamma_i I_{iz}(1-\sigma_i^{iso} - \tilde{\sigma}_{zzi})B_z +$$

$$+ \sum_{i<j} \{J_{ij}^{iso}\, [I_{iz}I_{jz} + \tfrac{1}{2}\,(I_i^+I_j^- + I_i^-I_j^+)] +$$

$$+ [\tilde{J}_{zzij} + \tilde{D}'_{zzij}][I_{iz}I_{jz} - \tfrac{1}{4}(I_i^+I_j^- + I_i^-I_j^+)] +$$

$$+ \sum_i \frac{\tilde{q}_{zzi}}{4I_i(2I_i-1)}\, [3I_{iz}I_{iz} - I_i(I_i + 1)] \tag{16}$$

where

$\sigma_i^{iso} = \frac{1}{3}\,\mathrm{Tr}(\sigma)$ is the isotropic phase chemical shielding;

$J_{ij}^{iso} = \frac{1}{3}\,\mathrm{Tr}(J_{ij})$ is the isotropic indirect coupling;

$$\tilde{\sigma}_{zzi},\ \tilde{J}_{zzi},\ \tilde{D}'_{zzij} \text{ and } q_{zzi} = eQ_i\tilde{V}_{zzi}/h$$

are the partially averaged components of the shielding, indirect coupling, direct dipolar coupling and quadrupolar coupling tensors, respectively. $\tilde{V}_{zzi}$ is the component of the electric field gradient tensor.

By using a commonly adopted convention we call "experimental anisotropic coupling" the quantity $\frac{1}{2}\,(\tilde{D}'_{zzij} + \tilde{J}_{zzij})$, i.e.:

$$T_{ij}^{exp} = \tilde{D}_{zzij} + \frac{1}{2}\,\tilde{J}_{zzij} \qquad (17)$$

with $\tilde{D}_{zzij} = \frac{1}{2}\,\tilde{D}'_{zzij}$. We stress here that the quantities $\tilde{J}_{zzij}$ and $\tilde{D}_{zzij}$ always appear as a sum in the spin hamiltonian. Consequently they cannot be measured separately in the NMR spectra of oriented molecules. It is then very important to know experimentally or to evaluate theoretically the "pseudodipolar couplings" $\tilde{J}_{zzij}$ in order to get dipolar couplings $\tilde{D}_{zzij}$ which, as we shall see, are connected with the molecular geometry.

From Eq. (15) we realize that, if T^{iso} is known or zero, the $\tilde{T}_{zz}^{aniso}$ which reproduce the spectra with the hamiltonian of Eq. (16) are linearly related to the order matrix elements $S_{\alpha\beta}$ and tensorial components $T_{\alpha\beta}$. In the case, therefore, that the relevant tensors are available from other measurements (e.g. from solid state measurements) or can be assumed or evaluated, the relationship Eq. (15) can be used to study the molecular ordering (at least the second order elements in the expansion of the singlet orientational probability). On the other hand, if some of the measured interactions are utilized to determine the S matrix, Eq. (15) can be used to investigate molecular tensorial properties. This twofold use of Eq. (15) has been commonly adopted for dipolar and, although less extensively, for quadrupolar interactions. Since, however, the knowledge of the shielding and indirect coupling tensors is generally very poor, Eq. (15) is used to determine these quantities in favourable cases.

Let us now treat the various interactions in detail.

DIPOLAR COUPLING

The dipolar interaction in the hamiltonian of Eq. (16) can be derived by expanding the mutual energy of two magnetic dipoles μ_i and μ_j into the "dipolar alphabet" (15,2); by writing the hamiltonian in polar coordinates and using spin tensor operators:

$$H^D_{ij} = \frac{\gamma_i\gamma_j}{r^3_{ij}} \frac{h}{4\pi^2} \; [A + B + C + D + E + F] , \tag{18}$$

where

$$A = (1-3\cos^2\theta_{ij})I_{iz}I_{jz} \quad ; \quad B = -\tfrac{1}{4}(1-3\cos^2\theta_{ij})[I^+_i I^-_j + I^-_i I^+_j] \; ;$$

$$C = -\tfrac{3}{2}\sin\theta_{ij}\cos_{ij} e^{-i\psi}(I^+_i I_{jz} + I_{iz}I^+_j) \quad ; \quad D = C^* \; ;$$

$$E = -\tfrac{3}{4}\sin^2\theta_{ij} e^{-2i\psi} I^+_i I^+_j \; ; \qquad F = E^* \; .$$

In NMR experiments using large magnetic fields only terms A and B give first order contributions. In addition in mesophases having cylindrical symmetry and the director aligned along the field, the ensemble average of terms C - F is zero. In the hypothesis of uncorrelated motions we obtain

$$H^D_{ij} = \frac{\gamma_i\gamma_j \; h}{4\pi^2\langle r^3_{ij}\rangle} \; \langle 1-3\cos^2\theta_{ij}\rangle \; \{I_{iz}I_{jz} - \tfrac{1}{4}(I^+_i + I^-_i I^+_j)\} \tag{19}$$

where the brackets indicate the ensemble or time average and θ_{ij} is the angle between the internuclear vector r_{ij} and the magnetic field. The spatial part of the interaction,

$$\tilde{D}_{ijzz} = - \frac{\gamma_i\gamma_j}{4\pi^2} \left\langle\frac{1}{r^3}\right\rangle\left\langle\frac{3\cos^2\theta_{ij}-1}{2}\right\rangle = -K_{ij}\left\langle\frac{1}{r^3}\right\rangle S_{ij} \tag{20}$$

is nothing but the zz component of the dipolar tensor whose components in a molecular frame can be written as

$$D_{ij\alpha\beta} = -K_{ij} \; \frac{1}{r^3_{ij}} \quad (3\cos\theta_\alpha\cos\theta_\beta - \delta_{\alpha\beta}) \tag{21}$$

The relationship between the $\tilde{D}_{ijzz}$ and the tensorial elements $D_{ij\alpha\beta}$ can be derived from (13):

$$\tilde{D}_{ijzz} = -2K_{ij}\left\langle\frac{1}{r^3}\right\rangle\left[\frac{S_{aa}}{3}(3\cos^2\theta_{ija}-1) + \frac{S_{bb}}{3}(3\cos^2\theta_{ijb}-1) + \frac{S_{cc}}{3}(3\cos^2\theta_{ijc}-1) + 2S_{ab}\cos\theta_{ija}\cos\theta_{ijb} + 2S_{ac}\cos\theta_{ija}\cos\theta_{ijc} + 2S_{bc}\cos\theta_{ijb}\cos\theta_{ijc}\right]. \quad (22)$$

The splitting observable between two equivalent nuclei, remembering that H^D_{ij} of Eq. (18) does not commute with the transition operator I_x), and neglecting the pseudodipolar coupling is:

$$\Delta\tilde{\nu} = 3|D_{ijzz}|, \quad (23)$$

while for nuclei having different γ only the $I_{zi}I_{zj}$ part of Eq. (19) is first order and for the splitting we get:

$$\Delta\tilde{\nu} = |2D_{ijzz} + J_{ij}| \quad (24)$$

The dependence of dipolar splitting on ϕ, the angle between the director and the magnetic field, can be easily obtained from Eq. (9):

$$\Delta\nu(\phi) = \Delta\nu(0)(3\cos^2\phi - 1)/2 \quad (25)$$

We emphasize that the function $P^2(\cos\theta_{ij})$ is very steep for angles near 54.7° so that small uncertainties in θ_{ij} near this angle produce large errors in S_{ij} values derived by combining Eqs. (23) and (20) and assuming an r_{ij} value.

The large number of interproton dipolar couplings is responsible for the poorly resolved ^1H spectra of the mesophases (see Fig. (2)).

The information derived from such spectra is in fact not very selective. More details can be obtained, however, by deuteriating part of the molecules and eliminating the deuterium interactions either by double irradiation or spin echo techniques (18, 1c) as is shown in Fig. (3). With these procedures an accurate determination of the proton dipolar

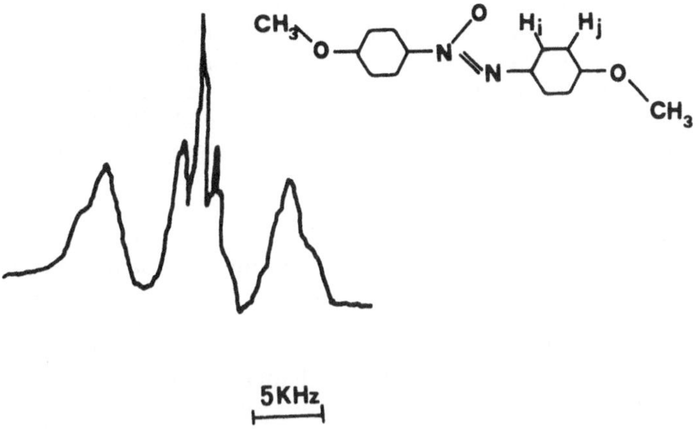

Figure 2. ^{1}H *NMR spectrum (60 MHz) of PAA at* T_{red} = *.98 (Visintainer et al. (17)).*

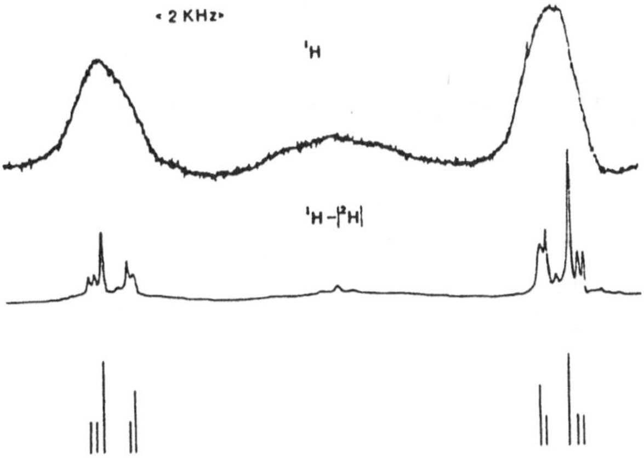

Figure 3. ^{1}H *NMR spectrum of 4-cyano-4'-n-pentyl-*d_{11}*-biphenyl-*d_4 *in the nematic phase at 31°C without and with* ^{2}H *decoupling (Emsley et al. (18)).*

couplings can allow a precise derivation of some of the order elements of non deuteriated molecular fragments (22).

The molecular organization, phase changes and angular dependence in various liquid crystals can be studied by adding to mesophases molecular probes (19) (often a symmetrical molecule with a low number of spins) and using the dipolar splittings of nuclei in the probe to monitor the mesophase behaviour. This procedure has, however, some drawbacks not only because the mesophase can be influenced by the host molecule but also because specific solute-solvent interactions can influence the probe ordering (20).

The determination of molecular structure in NMR spectroscopy using liquid crystal solvents is based on Eq. (21). In order to obtain D_{ij} from T_{ij}^{exp} the contribution of J_{ij}^{aniso} must be known, and for a more precise determination of geometry the experimental couplings have also to be corrected for the effect of molecular vibrations (1d) (see chapter 7).

NUCLEAR QUADRUPOLAR INTERACTION

For nuclei having a spin quantum number $I > \frac{1}{2}$ the interaction of the nuclear quadrupole moment with the electric field gradient can be expressed by the following hamiltonian (15),

$$H_Q=\sum_i \frac{e\,Q_i}{4I\,(2I-1)}[V_{x'x'}(3I_{x'}^2-I^2)+V_{y'y'}(3I_{y'}^2-I^2)+V_{z'z'}(3I_{z'}^2-I^2)]\ , \tag{25}$$

where (x'y'z') is the principal axis system of the electric field gradient tensor, $V_{\alpha\beta}$. By rearranging Eq. (25) and introducing the parameters $eq_{z'z'} = V_{z'z'}$ and $\eta = (V_{x'x'} - V_{y'y'})/V_{z'z'}$, and the ladder operators, then H_Q can be written as,

$$H_Q = \sum_i \frac{e^2qQ_i}{4I\,(2I-1)}\,[(3I_{z'i}^2-I_i^2) + \tfrac{1}{2}\eta(I_i^{+2}-I_i^{-2})]\ . \tag{26}$$

Many nuclei have $I > \frac{1}{2}$ and may give quadrupolar splittings on Zeeman energy levels. Nuclei such as Cl, Br, I and others, however, have large values of the quadrupolar moment (eQ_i), so that rapid quadrupolar relaxation gives very broad lines unless the electric field gradient is negligible.

Other nuclei like ^{10}B, ^{11}B, ^{14}N and in particular ^{2}H, have a relatively small value of (eQ) and therefore in principle they can give useful splittings in oriented media.

The low value of (eQ_i) for ^2H, in particular means that relatively sharp lines are observed in the NMR spectrum so that in oriented spin systems quadrupolar and sometime dipolar splittings between adjacent nuclei can be observed (1a, b, c).

Let us now consider for simplicity a single ^2H nucleus, without dipolar couplings, in a uniaxial liquid crystal uniformly oriented in a strong magnetic field with $n \parallel B$. In this case the quadrupolar hamiltonian acts as a perturbation on the Zeeman energy levels and we get from Eq. (26),

$$H = - \gamma B I_z/2\pi + \tilde{q}_{zz}[3I_z^2 - I(I + 1)]/4 \ , \tag{27}$$

where $\tilde{q}_{zz} = eQ\,\tilde{V}_{zz}/h$ and $\tilde{V}_{zz}$ is the component along the field of the partially averaged, axially symmetric, electric field gradient at the deuteron. Taking into account the selection rule $\Delta m = \pm 1$ we have two transitions with a splitting

$$\Delta\tilde{\nu} = 3\tilde{q}_{zz}/2. \tag{28}$$

The measured splitting can be related to the molecular quantities

$$q_{zz} = \tfrac{2}{3} \sum_{\alpha\beta} S_{\alpha\beta} q_{\alpha\beta} \tag{29}$$

and in a molecular frame where the $q_{\alpha\beta}$ tensor is diagonal then from Eqs. (28) and (29) we have,

$$\Delta\tilde{\nu} = \tfrac{3}{2}\, q_{aa}[S_{aa} + \tfrac{\eta}{3}(S_{bb} - S_{cc})] \ , \tag{30}$$

where

$$q_{aa} = \frac{e^2 q\, Q}{h} \qquad \text{and} \qquad \eta = \frac{q_{bb} - q_{cc}}{q_{aa}} \ .$$

Eq. (30) is the basis for the determination of q_{aa} and η in solutes from quadrupolar splittings. To this purpose not only S_{aa} and $S_{bb} - S_{cc}$ parameters must be known from another source (from dipolar splittings, for instance, which implies also an accurate determination of the molecular structure), but also measurements of $\Delta\tilde{\nu}$ values must be made on at least two samples with non correlated S values, and non distorting solvents must be used (21).

More widespread and interesting is the use of Eq. (24) for the study of orientational order. For this purpose, because normally η is not very large (its values fall in the range $-0.15 \leq \eta \leq 0.15$ (1a)) we neglect the second part of Eq. (30) so that,

$$\Delta\tilde{\nu} = \frac{3}{2} q_{aa} S_{aa} \quad . \tag{31}$$

For deuterium the principal axis a lies along the bond to the attached atom and q_{aa} for C-D bonds has typical values depending on hybridization:

$$\begin{aligned} q_{aa} &= 200 \pm 5 \ \text{kHz for sp bonds} \\ &= 185 \pm 5 \ \text{kHz for sp}^2 \text{ bonds} \\ &= 170 \pm 5 \ \text{kHz for sp}^3 \text{ bonds} \end{aligned}$$

It is possible therefore to obtain S^i_{CD} ($=S^i_{aa}$), the order parameters of the C-D bond orientations in molecules. The determination of order parameters of local C-D bonds and, as a consequence, of molecular fragments allows to study in a very selective way the orientational problem in liquid crystals (1a,b) and constitutes a test for the theories of molecular flexibility in mesophases (1c).

ANISOTROPY OF THE SHIELDING TENSOR

The theory of chemical shielding and computational methods for obtaining its components is widely illustrated in the literature (22,15). Measurements in liquid crystal solution are related to the quantity:

$$\sigma_{izz} = \sigma^{iso} + \frac{2}{3} \sum_{\alpha\beta} S_{\alpha\beta} \sigma_{i\alpha\beta}$$

Complications arise, however, since the experimental tensor is the sum of three different contributions all of them anisotropic (1d):

$$\sigma^{exp} = \sigma^{mol} + \sigma^{loc} + \sigma^{bulk} \ ;$$

σ^{mol} depends on the electronic distribution in the molecule and is the quantity of interest; σ^{loc} is a measure of the influence of neighbouring molecules and aggregates; σ^{bulk} depends on the shape and magnetic susceptibility of the sample. The importance of σ^{loc} and σ^{bulk} depends also on the method of measurement. Relationships for the evaluation of σ^{bulk}, given the geometry of the sample, have been derived (23). The contribution of σ^{loc} is less important for nuclei which have large shielding effects such as ^{13}C, ^{19}F and ^{31}P.

Referring the S values to abc, the principal axes of the shielding tensor, we obtain

$$\tilde{\sigma}_{izz} - \sigma_i^{iso} = \tfrac{2}{3} S_{aa}[\sigma_{aa} - \tfrac{1}{2}(\sigma_{bb} - \sigma_{cc})] +$$

$$+ \tfrac{1}{2} (S_{bb} - S_{cc})(\sigma_{bb} - \sigma_{cc})] \ . \tag{33}$$

Eq. (33) is a basis for the determination of orientational order if the shielding tensors are known by different measurements (e.g. by solid state NMR). Conversely, elements of the shielding tensor can be obtained if the S values can be determined accurately. For ^{13}C in natural abundance the "proton enhanced" ^{13}C technique of Pines et al. (24) has been applied to the study of orientational order as a function of temperature in a series of liquid crystals (25) (Fig. (4)). Although these measurements are in principle rich in information, in practice their interpretation is complicated by the lack of a detailed knowledge of the shielding tensor of the various ^{13}C so that it is not possible to obtain order parameters with good precision in this way.

We may write Eq. (32) in the form,

$$\delta_i^{LC}(\theta,T) - \delta_i^{iso}(T) = \tilde{\sigma}_{zzi} \ , \tag{34}$$

where δ_i^{LC} and δ^{iso} are the chemical shifts of nucleus i in the liquid crystal and isotropic solution respectively; θ is the angle between the director and the field and T is the temperature. The dependence on T arises from the $S_{\alpha\beta}$ elements, which for solutes can also be varied by changing the concentration of the solute. Various methods for studying the chemical shift tensors of solute molecules either by changing the temperature or orientation by means of an electric field or using smectic liquid crystals are reported in detail in the book of Emsley and Lindon (1d) and in a review by Lounila and Jokisaari (26) to which we refer also for an extensive literature on the subject. We would like, however, to mention here the method introduced by Khetrapal (27) of switching the orientation of the optic axis by 90° with respect to the applied magnetic field by mixing two liquid crystals of opposite diamagnetic anisotropies. The application of these techniques, which is useful for the determination of chemical shift anisotropy without the use of a reference compound, is illustrated for the ^{13}C of the cyano group of benzonitrile in Fig. (5) (28). In practice to investigate the $\sigma_{\alpha\beta}$ elements in a

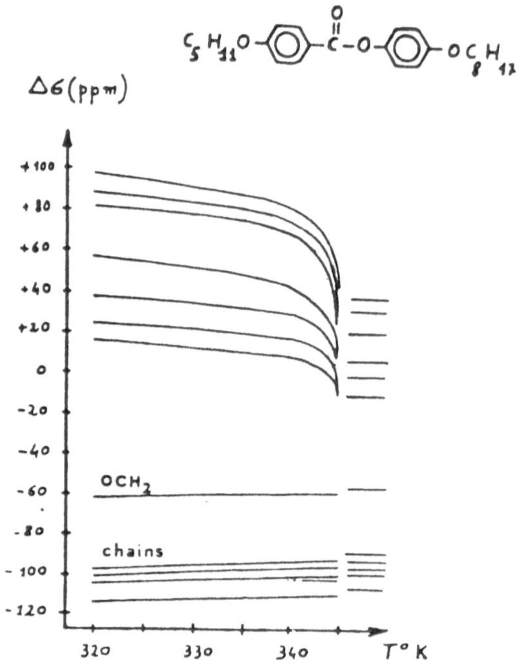

Figure 4. *Temperature dependence of the ^{13}C chemical shifts in the nematic phase of an alkoxybenzoate. (A. Losche (25)).*

molecule without symmetry a set of non-linearly dependent equations of the type of Eq. (34) is needed and, once again, precise values of $S_{\alpha\beta}$ must be determined. It is because of the difficulty of collecting a sufficient and meaningful set of data that the measurements have been concentrated on molecules of high symmetry (C_{3v} or higher) where only one S value is necessary so that only the chemical shift anisotropy

$$\Delta\sigma = [\sigma_{aa} - \tfrac{1}{2}(\sigma_{bb} + \sigma_{cc})]$$

is determined.

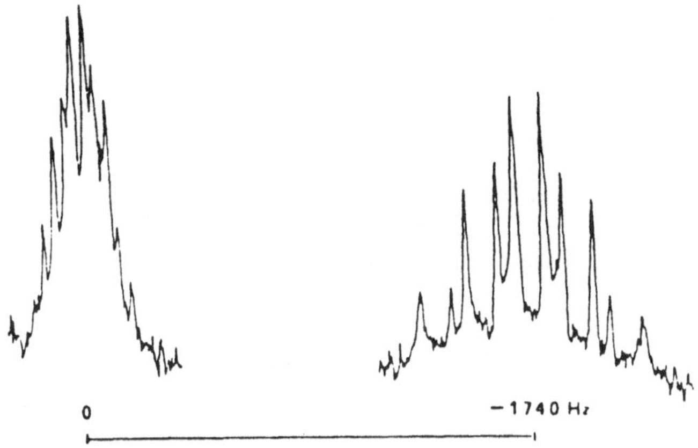

Figure 5. *^{13}C NMR spectra of benzonitrile a) partially oriented in ZLI 1167 ($\Delta\chi<0$, $n\perp B$), and b) partially oriented in a ZLI-PCH mixture ($\Delta\chi<0$, $n\parallel B$). The carbon of the $-C\equiv N$ group is 60% ^{13}C enriched.*

INDIRECT COUPLINGS

There are several detailed accounts of the theory of spin-spin indirect couplings (29,15,26). For an experimental determination of the anisotropy of the indirect coupling we rely on Eq. (15) which we rewrite as:

$$\tilde{T}_{ij}^{exp} = \tfrac{1}{3} \sum_{\alpha\beta} S_{\alpha\beta}(D_{ij\alpha\beta} + J_{ij\alpha\beta}) \ . \qquad (35)$$

The problem, analogous to that for the determination of other molecular quantities like q_{aa}, η and $\sigma_{\alpha\beta}$ is of knowing precise values of $S_{\alpha\beta}$ and of having a relatively small number of $J_{\alpha\beta}^{aniso}$ quantities to determine. One solution would be the use of many solvents, but these would all have to be non-distorting (26). Varying the temperature and other conditions does not help because the ratios of S values often remain constant. Undoubtedly it is very useful to know in advance that at least some of the coupling anisotropies are not very large.

The procedure of investigating molecules of well known geometry (determined by means of microwave spectroscopy or electron diffraction in the gas phase) and to compare the results with accurate, vibrationally corrected, geometrical structures from liquid crystal measurement, has pointed out that some indirect couplings like H-H, H-^{13}C, ^{15}N-H..... have small anisotropic contributions (26). On the other hand, other couplings like ^{13}C-^{13}C, ^{19}F-^{19}F, some H-^{19}F and for H-X couplings (where X is Hg, Cd, Si and Sn) a significant contribution from the anisotropic term was found (26). A straight forward procedure in order to prove J anisotropy is to consider that in molecules with high symmetry the ratios of purely dipolar couplings are determined geometrically. Thus, for hexagonal molecules like benzene and hexafluorobenzene we must have D(ortho): D(meta): D(para) = 1 : 0.1925 : 0.1250. When J^{aniso} is set to zero these ratios are found for the dipolar couplings derived from $\tilde{T}^{exp}$ for the protons in benzene (23), but not for the carbon-carbon couplings (31), nor for the couplings between fluorines in hexafluorobenzene (32).

Another way of testing the data for J anisotropy was found by MacLean and coworkers (33) in molecules containing parallel internuclear vectors like 1,1-difluoroethylene. In this case the ratio D_{HH}/D_{FF} in the case of negligible J anisotropy should be independent of the orientation. However, varying the concentration of the solute or the solvent this ratio changed from -0.129 to 0.875. Of course care should be taken in the analysis of data since other phenomena like site exchange (34) could influence the results. Another point to be stressed concerns a possible influence on data analysis of isotropic indirect couplings which often are simply assumed from isotropic spectra and not determined from anisotropic ones.

Returning to Eq. (35) we can conclude that measurements of J_{ij}^{aniso} elements are in practice only feasible by concentrating the attention on particular couplings (i.e. reducing the unknowns by assuming that some couplings have usually negligible anisotropy) and considering mainly symmetrical molecules. Even in that case an accurate determination of the orientation and therefore of the geometry, possibly in many solvents, is necessary.

REFERENCES

1.a) J. Charvolin and B. Deloche in *"The Molecular Physics of Liquid Crystals"* G.R. Luckhurst and G.W. Gray eds. Academic Press 1979.

b) J.W. Doane in *"Magnetic Resonance of Phase Transitions"* Academic Press 1979.

c) J.W. Emsley in *"Cristalli Liquidi"* GNCL editor, CLUT Torino 1982.

d) J.W. Emsley and J.C. Lindon *"NMR Spectroscopy Using Liquid Crystalline Solvents"*. Pergamon Press Oxford 1975.

2. M. Mehring *"Principles of High Resolution NMR in Solids"* Springer-Verlag, Berlin 1983.

3. C. Zannoni, *Mol. Phys.* **38**, 1813 (1979) (and references therein).

4. P.S. Pershan in *"The Molecular Physics of Liquid Crystals"*, G.R. Luckhurst and G.W. Gray eds., Academic Press, London 1979.

5. L.J. Berliner, *"Spin Labelling: Theory and Applications"* Academic Press, London 1976.

6. E. Sackmann and H. Mohwald, *J. Chem. Phys.* **58**, 5407 (1973).

7. J.C. Rowell, W.D. Phillips, L.R. Melby and R. Panar, *J. Chem. Phys.* **43**, 3443 (1965).

8.a) J.W. Emsley and G.R. Luckhurst *Mol. Phys.* **41**, 19 (1980).

b) E.E. Burnell and C.A. de Lange, *J. Magn. Reson.* **39**, 461 (1980).

9. P.G. de Gennes, *"The Physics of Liquid Crystals"*, Oxford 1975.

10. R. Eidenschink, D. Erdmann, A.J. Krause and L. Pohl, *Angew. Chem. Int. Ed. Engl.*, **17**, 2 (1978).

11. C.L. Khetrapal, A.C. Kunwar, A.S. Tracey and P. Diehl in *"NMR Basic Principles and Progress"* (P. Diehl, E. Fluck and R. Kosfeld, ed.) Vol. 9 Springer-Verlag Berlin 1975.

12. R.A. Wise, D.H. Smith and J.W. Doane, *Phys. Rev.* **A 7**, 1366 (1973).

13. P.J. Black, K.D. Lawson and T.J. Flautt, *Mol. Cryst. Liquid Cryst.* **7**, 201 (1969).

14. A. Abragam, *"The Principles of Nuclear Magnetism"*, Oxford 1961.

15. C.P. Slichter, *"Principles of Magnetic Resonance"*, Harper & Row, New York 1963.

16. P.L. Nordio and U. Segre in *"The Molecular Physics of Liquid Crystals"* G.R. Luckhurst and G.W. Gray eds., Academic Press, London 1979.

17. J.J. Visintainer, E. Bock, R.Y. Dong and E. Tomchuk, *Can. J. Phys.* **53**, 1483 (1975).

18. J.W. Emsley, G.R. Luckhurst, G.W. Gray and A. Mosley, *Mol. Phys.* **35**, 1499 (1978).

19. Z. Luz and S. Meiboom, *J. Chem. Phys.* **59**, 275 (1973).

20. D. Catalano, C. Forte, C.A. Veracini and C. Zannoni to be published in *Israel J. Chemistry.*

21. D. Catalano, C. Forte, C.A. Veracini - unpublished results.

22. N.F. Ramsey, *Phys. Rev.* **78**, 699 (1950).

23. A.D. Buckingham and E.E. Burnell, *J. Amer. Chem. Soc.* **89**, 3341 (1967).

24. A. Pines, D.J. Ruben and S. Allison, *Phys. Rev. Letters*, **35**, 1002 (1974).

25. A. Losche in *"Magnetic Resonance and Related Phenomena"* E. Kundla, E. Lippmaa and T. Saluvere Eds. Springer Verlag, Berlin (1979).

26. J. Lounila and J. Jokisaari in *"Progress in NMR Spectroscopy"*, vol. 15 (1982).

27. C.L. Khetrapal and A.C. Kunwar, *Chem. Phys. Letters*, **80**, 170 (1981).

28. C.A. Veracini unpublished results.

29. N.F. Ramsey, *Phys. Rev.* **91**, 303 (1953).

30. P. Diehl, H. Bösiger and H. Zimmerman, *J. Magn. Reson.* **33**, 113 (1979).

31. P. Diehl, H. Bösiger and J. Jokisaari, *Org. Magn. Reson.* **12**, 282 (1979).

32.a) L.C. Snyder and E.W. Anderson, *J. Chem. Phys.* **42**, 3336 (1965).

b) D. Catalano, L. Marcolini, C.A. Veracini, *Chem. Phys. Letters*, **88**, 342 (1982).

33.a) J. Gerritsen, G. Koopmans, H.S. Rollema and C. MacLean, *J. Magn. Reson.* **8**, 20 (1972).

b) J. Gerritsen and C. MacLean, *Mol. Cryst. and Liq. Cryst.* **12**, 97 (1971).

34. P. Diehl, S. Sykora, W. Niederberger and E.E. Burnell, *J. Magn. Reson.* **14**, 260 (1974).

STUDIES OF SOLUTES WITH INTERNAL ROTORS

C.A. Veracini

Istituto di Chimica Fisica dell'Università,
Pisa, Italy.

and

M. Longeri

Dipartimento di Chimica dell'Università
della Calabria,
Arcavacata di Rende (Cs), Italy.

INTRODUCTION

NMR using liquid crystal solvents (LCNMR) is a powerful tool for the investigation of molecular structure. The advantages and limitations of this technique have been extensively discussed (1). We review briefly the difficulties: the spectral complexity limiting the analyzable spin systems, the solute-solvent interactions and the difficult rationalization of the data in terms of separate structural and orientational parameters.

The presence of large, low frequency torsional motions (such as internal rotation and ring puckering) considerably complicates the interpretation of the NMR data compared to the case of "rigid" molecules, i.e. molecules showing only fast vibrational motions. On the other hand such difficulties are not peculiar to the LCNMR technique. Large intramolecular motions in "flexible" molecules are a basic source of problems for the investigation of molecular geometry. This is also the case for other experimental and theoretical methods (2). In fact the information available from the literature on this subject is, not surprisingly, lacking in results and often ambiguous.

J. W. Emsley (ed.), Nuclear Magnetic Resonance of Liquid Crystals, 123–146.

These aspects are well known and were a source of doubts and uncertainties in the studies of LCNMR of dissolved molecules (3). Meanwhile, and quite oddly indeed, in the studies by NMR of pure mesophases, which are commonly formed by molecules far richer in different kinds of internal motions, these problems were overlooked until recently (4), mesophase molecules being considered as rigid rods.

Without any doubt, as we shall see presently, when handling internal motions it is not possible to reach the same degree of precision on orientational and geometrical parameters which can be obtained, in some cases, for rigid molecules. However, interest in internal motion is still stimulating not only because some theoretical progress has been made recently, at least in properly posing the problem (5), but also because we need more knowledge about the many aspects of dealing with internal motions in small and medium sized molecules, in order to deal with the more challenging problem of treating the liquid crystal molecules themselves.

In the present chapter we will first present the theory of the conformational problem, discussing also the reliability of the various approximations that have to be made and, as a consequence, the range of validity of the results. We shall subsequently distinguish among different cases of internal motions studied so far, discussing their degree of confidence in selected examples.

THEORY

For a rigid molecule, or a rigid fragment of a molecule, ordered in a uniaxial liquid crystalline phase, the partially averaged component parallel to the director of the appropriate second rank tensor $A_{\|}$ is given by:

$$\tilde{A}_{\|} = A_0 + \frac{2}{3} \sum_{\alpha\beta}^{abc} S_{\alpha\beta} A_{\alpha\beta} \; ; \qquad (1)$$

a,b,c is a molecule fixed axis system where the tensor elements $A_{\alpha\beta}$ are defined, $S_{\alpha\beta}$ is an element of the Saupe order matrix and A_0 is the isotropic average of A. In the case of dipolar couplings, for instance, $A_{\alpha\beta} = D_{\alpha\beta}$ and A_0 is zero.

In the case of a flexible molecule we must deal with the effects of two simultaneous motional averagings: the internal molecular motion, which implies the existence of a number of different rotamers and their overall motions in the mean potential due to the anisotropic environment. Provided that the motions have frequencies greater than $(A_{aa} - \tilde{A}_{\parallel})$ where A_{aa} is the largest tensor component in the chosen molecular frame, the spectral analysis gives the statistically averaged quantities over the various motions. Problems arise about how to relate order matrices which can be extracted from the couplings of the various fragments and how to interpret inter-fragment couplings. For simplicity here we will adopt the statistical theory of Emsley and Luckhurst (5) (hereafter referred as E.L. Theory) and we will not repeat arguments based on relative times of reorientation and internal rotation (6).

According to the E.L. formalism the statistical averaged tensors are given by:

$$\tilde{A}_{\parallel} = \int d\chi d\Omega A_{\parallel}(\chi,\Omega)\exp\{-U(\chi,\Omega)/kT\}/Z \qquad (2)$$

where Z is the partition function,

$$Z = \int d\chi d\Omega \exp\{-U(\chi,\Omega)/kT\},$$

χ,Ω indicate the internal and orientational coordinates respectively, and $U(\chi,\Omega)$ the orientational potential energy which depends, as specified, on both conformation and orientation. From Eq. (2), writing $A_{\parallel}(\chi,\Omega)$ in a molecular frame we have:

$$\tilde{A}_{\parallel} = \sum_{\alpha\beta}^{abc} \int d\chi d\Omega A_{\alpha\beta}(\chi)\ell_{\alpha}\ell_{\beta}\exp\{-U(\chi,\Omega)/kT\}/Z \qquad (3)$$

where $\ell_{\alpha}\ell_{\beta} = \cos\theta_{\alpha z}\cos\theta_{\beta z}$, the z axis being along the director. The potential of mean torque can be written as a sum of two terms,

$$U(\chi,\Omega) = U_{int}(\chi) + U_{ext}(\chi,\Omega) \qquad (4)$$

where $U_{ext}(\chi,\Omega)$ represents the orientation energy of a particular conformation and $U_{int}(\chi)$ is assumed independent of Ω. We can perform the average over Ω:

$$\tilde{A}_{\parallel} = Z^{-1} \sum_{\alpha\beta}^{abc} \int d\chi A_{\alpha\beta}(\chi)\exp\{-U_{int}(\chi)/kT\}\overline{\ell_{\alpha}\ell_{\beta}}(\chi)Q(\chi) \qquad (5)$$

where

$$\overline{\ell_\alpha \ell_\beta}(\chi) = \int d\Omega \ell_\alpha \ell_\beta \exp\{-U_{ext}(\chi,\Omega)/kT\}/Q(\chi)$$

$$Q(\chi) = \int d\Omega \exp\{-U_{ext}(\chi,\Omega)/kT\}$$

If we define the function P(X) for the averaging over the internal motion we shall have

$$\tilde{A}_{\parallel} = \sum_{\alpha\beta}^{abc} \int \overline{\ell_\alpha \ell_\beta}(\chi) A_{\alpha\beta}(\chi) P(\chi) d\chi \tag{6}$$

which, by introducing the Saupe ordering matrix, can be written

$$\tilde{A}_{\parallel} = A_0 + \sum_{\alpha\beta}^{abc} \tfrac{2}{3} \int d\chi S_{\alpha\beta}(\chi) A_{\alpha\beta}(\chi) P(\chi) \tag{7}$$

and assuming a discrete number of conformations

$$\tilde{A}_{\parallel} = A_0 + \tfrac{2}{3} \sum_{\alpha\beta}^{abc} \sum_n p^n S^n_{\alpha\beta} A^n_{\alpha\beta} \tag{8}$$

where p^n is the statistical weight of the nth conformation.

In the limiting case in which U_{ext} does not depend on the conformational space, Eq. (8) can be written as

$$\tilde{A}_{\parallel} = A_0 + \tfrac{2}{3} \sum_{\alpha\beta}^{abc} S_{\alpha\beta} \langle A_{\alpha\beta} \rangle \tag{9}$$

where

$$\langle A_{\alpha\beta} \rangle = \sum_n p^n A^n_{\alpha\beta} \, .$$

The implications of Eq. (8) on conformational analysis are quite significant and frustrating. In fact the weights p^n and orientation parameters $S^n_{\alpha\beta}$ appear as products $p^n S^n_{\alpha\beta}$ and cannot be determined individually. Therefore, for a solution of Eq. (8) either the Saupe orientational matrices S^n (one for each conformer) should be known or we should know the conformer populations by means of different techniques or, in particular cases, by symmetry. In the absence of any reliable equation relating the $S^n_{\alpha\beta}$ elements of different conformers (again with the exception of symmetry in some particular cases), all that we can do is to test theoretical models of the products $p^n S^n$. Even so, the number of unknowns is usually very large (as the number of conformations can be) and a solution of Eq. (8) is not easy. The conformational analysis should, therefore, be restricted to a very limited number of flexible molecules or to the trivial case of molecules existing in solution as a single rigid conformer (n = 1).

The use of Eq. (9) is more attractive at first sight since, in this case, the conformational analysis can be carried out using only one averaged orientational matrix. This reduces the number of unknowns and provides an independent way of calculating the populations p^n of different conformers. On the other hand not only is experimental evidence in support of Eq. (9) lacking but many experiments are now available supporting just the opposite, i.e. the validity of Eq. (8).

APPROXIMATE METHODS

As shown in the previous section, we cannot proceed without some approximation, testing case by case. By analogy with the studies of rigid molecules, the comparison of the results with those from other techniques will tell us their degree of soundness.

In the case of an internal potential energy $V(\phi)$ having a limited number of steep minima (see Appendix (1)) the rotameric model can be assumed as a reasonable approximation. Also in this model, first of all, the groups are assumed to rotate rigidly, i.e. without geometrical relaxation, and the only values of angles having an appreciable probability are those corresponding to the minima of the potential wells ($p(\phi) \neq 0$ for $\phi_1, \phi_2 \ldots \phi_n$; ϕ is here some dihedral angle). Similarly, in the case of a ring puckering motion, sharp minima may exist for some values of an appropriate coordinate. The rotamers approximation is commonly used in chemistry; E.S.R., E.D., I.R. and microwave spectroscopy, because of their shorter time scale, are sometimes able to observe different spectra from different conformers.

Within this approximation we can use Eq. (8) where $p^n = p_\phi^n$ are now the populations of the conformers at $\phi_1, \phi_2, \ldots \phi_n$.

An attempt to improve this approximation can be made by introducing a population distribution centered around the minima of internal energy but assuming that there is no appreciable variation of the elements of the order matrices (1b) for slight deplacements from the equilibrium position.

Also within the simple rotamer approximation or with the introduction of weighting with potential wells, a solution of Eq. (8) in terms of p^n values and geometrical parameters is not always possible. As will be shown later, the ability to obtain a solution depends on the number of conformers, on their symmetry and on geometrical unknowns such as bond lengths and angles of the rotating "rigid" sub-units. Quite often, moreover, a correlation between geometry and some $S_{\alpha\beta}$ elements is found, producing uncertainties in both geometrical and orientational results (7,8).

As discussed so far, the indirect approach, i.e. to test at different levels of rigour the various possible models, from the simplest to the more complex, has been widely and usefully adopted. In this way it is very often possible to exclude quite rigorously and safely a number of situations even if a unique model cannot be definitely chosen.

Obviously, the greater the number of different models that can be tested, the greater is the confidence in the result that can be achieved. This fact, combined with the assumptions just made (high, sharp shaped torsional barrier and little dependence of the $S_{\alpha\beta}$ elements on torsional angle), reduces considerably the number and kind of molecules whose internal motion can be studied accurately by means of LCNMR.

For molecules with a very flat torsional potential the quantities $p^nS^n_{\alpha\beta}$ are not negligible in the regions around the potential minima, independently of how fast the $S^n_{\alpha\beta}$ elements change with the libration angle, because the p^n values remain relatively large in these regions. Also the shape of the rigid sub-units plays an important role. The presence, for instance, of medium sized highly unsymmetrical sub-units (such as vinylic, $-(C{=}O)X$, CH_2X groups) is such that many $S_{\alpha\beta}$ elements are quite large and have a consistent dependence on the torsional angle. Again in this case, the quantity $p^nS^n_{\alpha\beta}$, calculated away from the potential minima, is not negligible compared with its values for $\phi = \phi_1, \phi_2, \ldots \phi_n$.

In other words we can say that the form of $U_{ext}(\chi,\Omega)$ is a factor strongly influencing the reliability of the assumptions made so far and the stronger is the dependence of U_{ext} on χ the more critical is the possibility of investigating internal motions by LCNMR. Moreover, since the $S_{\alpha\beta}$ values are affected by the solvent, then the choice of the solvent can influence the degree of validity of the assumptions made so far.

APPLICATIONS

The previously described difficulties seem to have stimulated instead of discouraging researchers in this field. A relatively large number of molecules with internal rotations, in fact, have been studied. We will not attempt here to quote all the examples studied, but we will concentrate on those illustrating general principles. We shall distinguish, perhaps arbitrarily, the following situations:

A) molecules with only one conformer (or symmetry related conformers);
B) molecules with two or many conformers;
C) methyl rotations;
D) structures involving ring puckering motions:

A1) Molecules with a single locked conformer

A good example of a molecule that could exist in principle either in the cis or trans conformation is 1,3-butadiene (9) (see Fig. (1)).

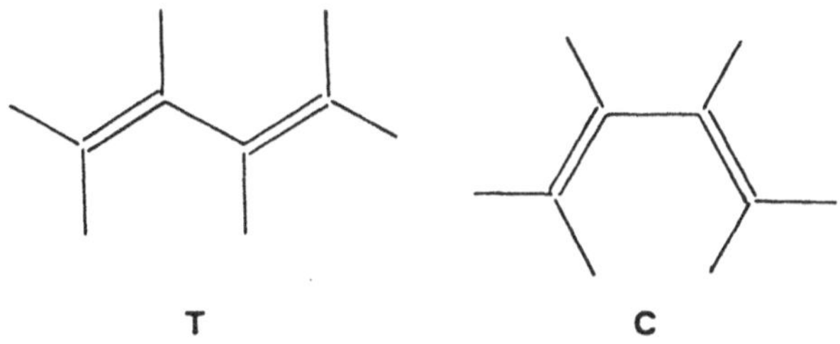

Figure 1. *The trans (T) and cis (C) forms of butadiene.*

Electron diffraction (ED) studies (10) indicate that the molecule exists as a trans isomer in the gas phase. A torsionally averaged structure, however, with a best-fit angle of 10° was not excluded on the basis of experimental results (11). The LCNMR data have been fitted with only the trans planar structure which has a C_{2h} group symmetry and, therefore, with a single order matrix with three nonvanishing elements. The NMR data were found not to improve on assuming a twist angle. However, since vibrational averaging was not performed, a contribution of libration is still not safely excluded. This is confirmed by the differences found for the C-H bond lengths with respect to the ED data (1.2%) and by a longer total shape of the molecule. In summary there is general agreement with the theoretical calculations (12) (see Fig. (2)) and other experimental results (10,11) but the details of $V(\phi)$ near 0° have still to be investigated.

Other molecules were found locked in a single position, or nearly so, like pyridin-2,6-dicarbaldehyde (13), selenophen-2-

carbaldehyde (14), salicylaldehyde (15) and 2,2'-bipyridyl (16) to quote only a few.

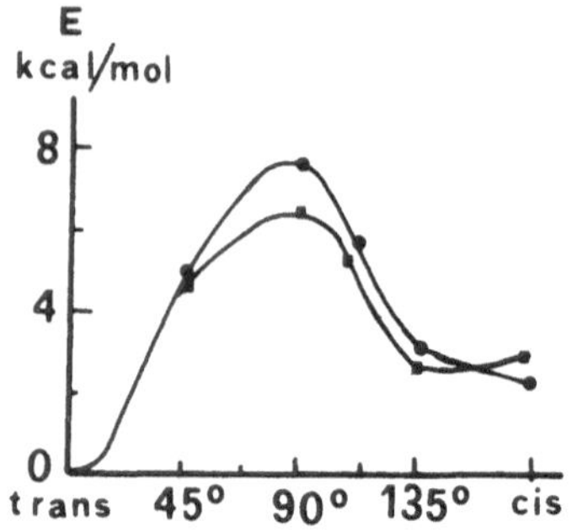

Figure 2. *Ground state potential curve for the internal rotation of 1,3-butadiene (● SCF-, ■ CI treatment).*

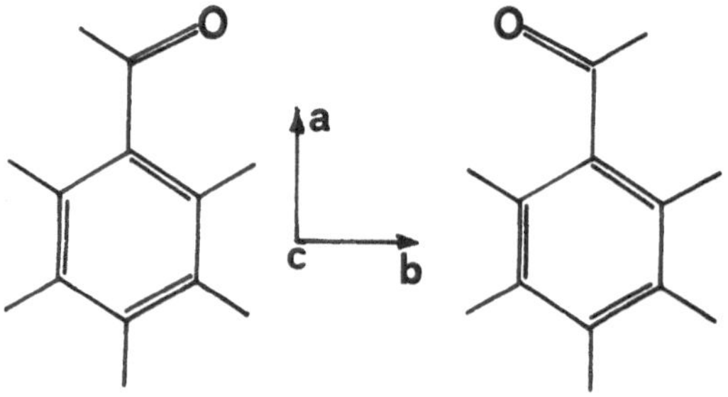

Figure 3. *The two conformations of benzaldehyde generated by rotation about axis a through 180°. The abc axes are fixed in the plane of the rings.*

A2) Molecules with two planar equivalent conformers

This is one of the most studied cases. Benzaldehyde is a good example which, as is well known also from other techniques (17), has the two forms (a) and (b) in equilibrium with a potential barrier of 21 kJ/mole in the gas phase and 33 kJ/mole in the liquid state.

With respect to the axis system in the figure, the two conformers (1) and (2) have symmetry related order matrices (18), i.e. the same diagonal elements and $S^1_{ab}=-S^2_{ab}$. However, as pointed out by Burnell et al. (19), we can describe this case in a suitable axis system on the formyl group with a single order matrix, since the flipping of the benzene ring between the two planar forms does not alter the orientation.

Indeed, the necessity of S_{ab} elements for fitting the dipolar couplings in this and many other similar cases (20) (at least with the axis system of Fig. (3)) was a clear proof that the conditions for application of Eq. (9) are not valid since in that case only diagonal elements of the mean matrix would have been necessary.

In a detailed study of benzaldehyde (21), including the analysis of ^{13}C sub-spectra, the dipolar couplings were used, after vibrational corrections, not only to obtain an accurate structure of the benzene ring but also to investigate the potential energy of internal rotation. A potential of the form

$$V(\phi) = \frac{V_2}{2}(1 - \cos 2\phi)$$

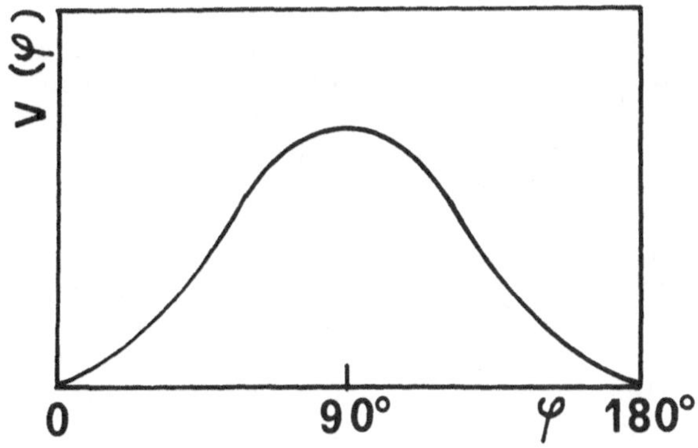

Figure 4. *A typical potential curve for the internal rotation in molecules having two planar equivalent conformers.*

(see Fig. (4)) was introduced in order to calculate the dipolar couplings of the aldehydic proton, averaging on a classical distribution of non planar forms corresponding to torsional motion. The dependence of $S_{\alpha\beta}$ on the torsional angle was not considered. Unfortunately, notwithstanding the fact that eight dipolar couplings were available, either the depth of the potential was calculated to be too large or, introducing only a small correlation (1°) of the rotation to the in-plane vibration of the ortho proton, large variations resulted for the potential and the C-H aldehydic bond distance, giving model-dependent solutions of the problem. Nevertheless, the necessity of introducing some contribution from non planar forms was clearly established.

The importance of taking the torsional motion into account emerged also in a proton study of styrene (22) where this motion was introduced directly in the vibrational corrections. In fact, by taking into account a distribution around potential minima, a significant improvement of the results was obtained even if the slightly anomalous values of the C-H bond lengths and angles for the vinyl group leads to the suspicion that at least off-diagonal $S_{\alpha\beta}$ elements depend significantly on the torsional angle. Many other examples of molecules with a similar potential which have been studied include pyridine-4-carbaldehyde (18), benzoylfluoride (20), pentafluorobenzaldehyde (23), acetophenone (24), anisole (25), p-NO_2 anisole ^{13}C (26).

A3) Molecules with non planar equivalent conformers

In many molecules having a biphenyl-like structure a competition between non-bonded interactions and conjugation produces a potential for the internal rotation as depicted in Fig. (5). Here again, from the motional point of view we can speak of an unique order matrix in a suitable axis system or of symmetry related order matrices. In 3-phenylthiophene (27), for instance, the planar model does not fit the dipolar couplings. The molecule in a twisted position is neither planar nor symmetric so that each conformer requires five S elements. By symmetry, however, the diagonal elements are the same while the off diagonal ones have their signs related by symmetry.

The molecules in table 1 were found by LCNMR to conform to this potential, with the values of the twist angle, ϕ, given in the table. It is perhaps worth noticing that many of these molecules were found to be planar in the solid state and twisted in the gas phase (28).

In the case of 4,4'-bypyridil (29) a weighting of the dipolar couplings with a potential function of the type depicted in Fig. (5) has been used. In this case the symmetry limits the Fourier expansion to the terms:

$$V(\phi) = \frac{V_2}{2}(1-\cos 2\phi) + \frac{V_4}{2}(1+\cos 4\phi) + \frac{V_8}{2}(1+\cos 8\phi) \quad (10)$$

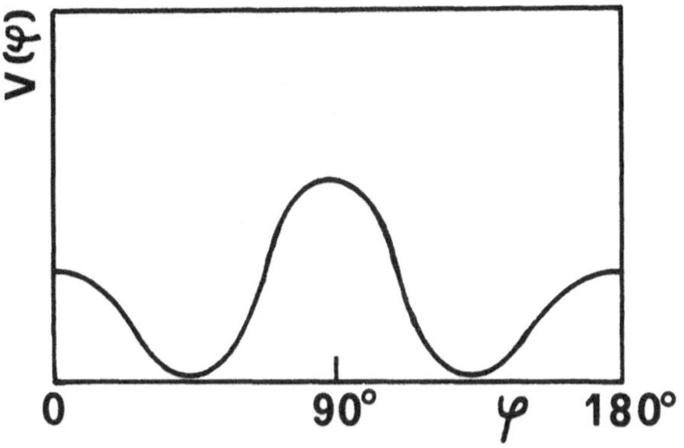

Figure 5. *A typical potential curve for the internal rotation in molecules having a biphenyl-like structure (ϕ = 0° refers to the planar conformation).*

Table 1. *Values of the dihedral angle ϕ found by LCNMR in various biphenyl-like molecules. The angle is measured from planar conformation. The value of ϕ observed for biphenyl in the gaseous state (electron diffraction) is 42°.*

Compound name and structure	Remark
3,3',5,5'-tetrachlorobiphenyl 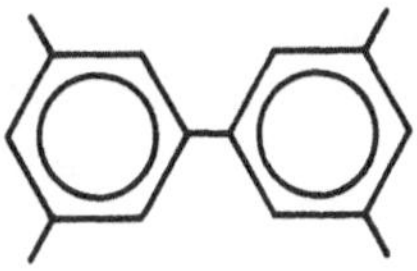	The molecule cannot exist in a single planar or orthogonal conformer. A twisted conformer with ϕ=34° fits the only direct dipolar coupling sensitive to the intramolecular motion (28).

Compound name and structure	Remark
4,4'-dichlorobiphenyl 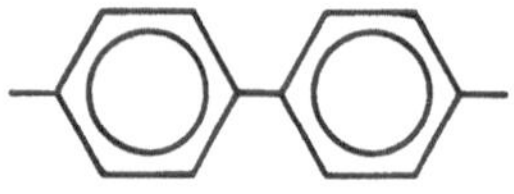	A twisting angle of $\phi=42°$ was obtained using a potential barrier which takes into account the influence of the conjugation potentials (3.3 kcal/mole) and nonbonding interaction (2.0 kcal/mole). The effect of the liquid crystalline solvent was also introduced (estimated value 4.5±1 kcal/mole) (58).
3,4,4',5-tetrabromobiphenyl 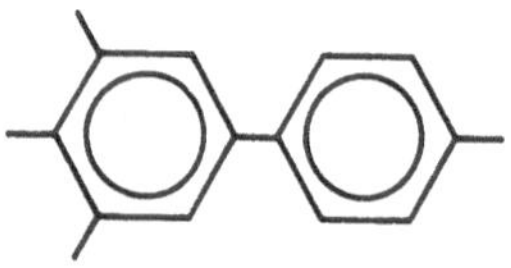	Three dipolar couplings are available. The molecule does not exist in the planar form in the nematic phase. Using a classical model it was shown that the twisting angle must be $\phi=33°-34°$ compared with $\phi=42°$ obtained for unsubstituted biphenyl in gaseous phase. Only one nondiagonal $S_{\alpha\beta}$ was used (59).
4,4'-bipyridyl 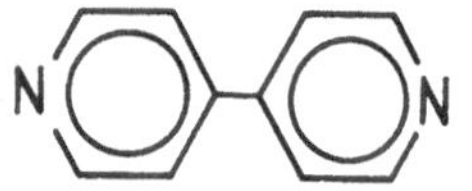	A twisting angle of $\phi=29.58°$ was found. By weighting the dipole couplings with an internal potential $2V(\phi) = V_2(1-\cos 2\phi)+V_4(1+\cos 4\phi)$ a potential barrier with $V(90°) > V(0°)$ was obtained (29).
Variously substituted 4'-bromodiphenyl 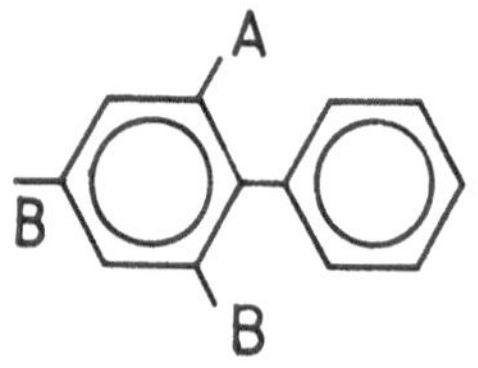 1 A=Cl, B=Cl 2 A=Cl, B=H 3 A=Br, B=Br 4 A=Br, B=H 5 A=I, B=H	Using a model based on an equilibrium among classical, twisted, symmetry related, conformers, ϕ was found to be 68° for 1 and 2, 70° for 3 and 4, 77° for 5 compared with 35° when positions 2,6 are not substituted. Only one off-diagonal $S_{\alpha\beta}$ was used (60).

Compound structure and name	Remark
3-Phenyl-1,2,5,oxa, thia and selena-diazole 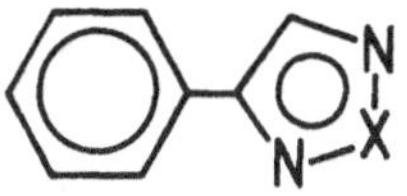	The three inter ring dipolar couplings were fitted assuming the molecule existing as twisted conformers with ϕ ranging from 18° to 21° through the series. Planar and orthogonal conformers were excluded. Only one off diagonal $S_{\alpha\beta}$ was used (61).
Phenylisoxazoles (1) 3-Phenylisoxazole 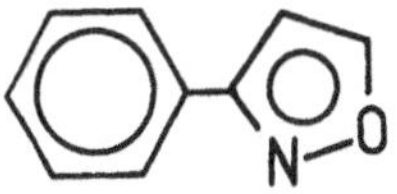 (2) 4-Phenylisoxazole (3) 5-Phenylisoxazole	A twisting angle of 18° for 1 and 3 and 20° for 2 was found using a model based on an equilibrium among classical conformers. Angles and bond distances were best fitted when five $S_{\alpha\beta}$ elements were used (62).
2,2'-bipyrimidine 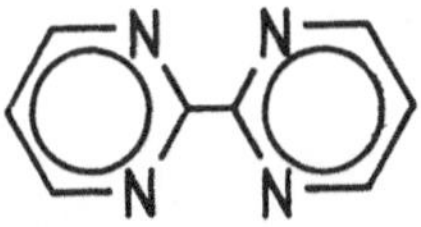	The calculated dipolar couplings are in agreement with a model assuming an equilibrium among twisted, symmetry related conformers. The twisting angle (ϕ=41°) is in fairly good agreement with ab initio and H.E. computations (ϕ=40°) (63,64).
3-Phenylthiophene 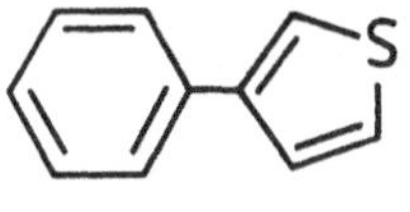	Various conformational models (planar, orthogonal, equilibrium between planar and orthogonal conformers) were discussed and discarded. The best fit was obtained for a twisted conformer model with ϕ=24° against an ab initio value of 30°. The use of a potential energy function gives a better agreement with the unsubstituted thiophenic coordinates (27).

A least squares fit of the data was obtained with a negligible value of V_8 and a deep minimum of $V(\phi)$ for $\phi = 29.58°$ was indicated. In this case the barriers at 0° and 90° are determined only by V_2 and V_4: $V(0°) = V_4$ and $V(90°) = V_2 + V_4$. Also, when V_8 is negligible then ϕ_{min} is determined only by the ratio V_2/V_4. In fact, by differentiating $V(\phi)$ we get

$$\phi_{min} = \cos^{-1}(0.5 + V_2/8V_4)^{\frac{1}{2}} , \qquad (11)$$

which for $\phi_{min} = 29.58°$ gives $V_2/V_4 = 2.05$. We can see, therefore, that although the barrier to the internal rotation could not be determined with certainty, the qualitative conclusion that $V(90°) > V(0°)$ was reached.

The treatment leading to Eq. (11) in the case of negligible V_8 is general and could be applied to other biphenyls.

The introduction of a distribution function, as in the previous case, was sometimes attempted and, as is shown for 3-phenylthiophene, was useful in obtaining a better fit of the dipolar data and a more consistent geometry.

The p-Br- and p-NO_2-aniline (8) also belong to this group of molecules with non-planar equivalent conformers, although with a totally different potential curve. For these molecules (^{15}N enriched samples, were used) the LCNMR was successful in demonstrating not only the non-planarity of the NH_2 but also a substituent effect on the out of plane angle of this group.

B) Molecules with two or more conformers

The presence of two non symmetry-related conformers in equilibrium introduces severe complications. Let us consider in detail the case of 2,2'-biselenophene (30). This molecule may have either an intramolecular energy curve, as in Fig. (6a), as is likely for the similar molecules of 2,2'-bifuryl (31) and 2-2'-bithienyl (32), or, as suggested by semiempirical M.O. calculation, a curve such as the one in Fig. (6b). Assuming for the selenophene rings the undistorted selenophene geometry from microwave measurements (33) and considering the inter-ring distance d and the θ bond angle (see Fig. (7)) as adjustable parameters, the fifteen dipolar couplings obtained from H-H and Se-H dipolar interactions were used to test the following models:

a) the planar Se-Se cis structure (C_{2v} point symmetry);
b) the planar Se-Se trans structure (C_{2h} symmetry);
c) a non-planar conformation (C_2 symmetry);
d) a mixture of cis and trans planar structures;

e) a mixture of cisoid and transoid twisted conformations.

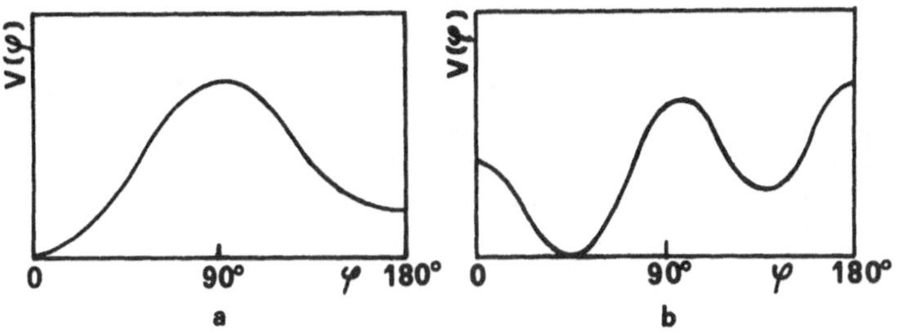

Figure 6. *Two possible potential curves for the internal rotation in molecules having non symmetry-related conformations of minimal energy.*

The three situations a) b) and c) can be easily excluded since, on the basis of the assumed geometry, they do not fit the dipolar couplings. Large r.m.s. deviations for all values of d and θ were obtained with these models (see table 2). (The model c) is equivalent to model e) with W_T = 100%).

Table 2.

Model	a)	b)	d)	e)*
R.M.S. (Hz) deviation	407	88	4(D_{13}>22Hz)	0.7
Geometrical parameters				d = 1.472Å; θ = 127.1° Φ_C = 22.4° Φ_T = 30.6° W_T = 77.6%

*with the assumption $S_{yyC} = S_{yyT}$

The models d) and e) can be tested by deriving the products $p^n S^n_{\alpha\beta}$ from the assumed geometry and the measured dipolar couplings. The $S^n_{\alpha\beta}$ can then be displayed as a function of p^n. Only model e) gave a satisfactory fit. For each rotamer twisted by Φ another with a -Φ is possible so that two cisoid and two transoid rotamers may exist in equilibrium (the twist angles Φ_C of the cisoid rotamers may be different from those of transoid ones Φ_T). By symmetry the Φ and $-\Phi$ rotamers have order matrices with equal diagonal elements and off diagonal ones of opposite sign. In summary this model requires six order elements $S_{\alpha\beta}$ to define the orientation. The fitting of all the dipolar couplings within experimental errors was possible with values of $\Phi_C \cong \Phi_T = 25° \pm 5°$ and reasonable values of d and θ. In Fig. (7) the S_{yy} and S_{zz} values are displayed as a function of W_T, the weight of the transoid conformer, for a case having $\Phi_C = \Phi_T = 25°$, d = 1.47 Å and $\theta = 127°$.

The physical limits of order parameters $1 \geqq S_{\alpha\beta} \geqq -0.5$ restrict the possible values of $W_T (0.9 \geqq W_T \geqq 0.3)$. We have to consider moreover that a normal S_{zz} value for the nematic mesophase with an added solute is of the order of 0.6~0.5 so that, quite probably, since the solute has usually lower S values than the solvent, a further limitation of W_T $(0.85 > W_T > 0.5)$ is very likely. In table 2 the geometrical and orientational results are given which were obtained using the model e) with the additional assumption of $S_{yyC} = S_{yyT}$. The uniqueness and usefulness of information given by LCNMR in this case is apparent, even at a level of rigour undoubtedly lower than in the previous cases.

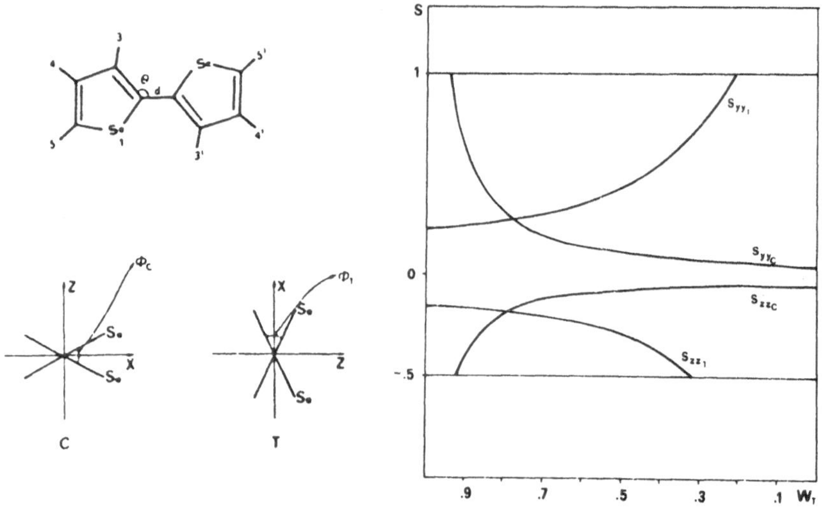

Figure 7. *2,2'-biselenophene structure and choice of axis system for cisoid (C) and transoid (T) rotamers. Diagonal elements of the order matrix versus the weight of the transoid conformer* W_T.

Similar methods were used in the multinuclear (^1H, ^2H, ^{13}C) study of thioanisole and substituted thioanisoles (34). For these molecules, by means of LCNMR it was possible to exclude rigorously a series of models, such as planar structure (see Fig. (4)). A discrimination, however, between either twisted conformers (see Fig. (5)) or a mixture of planar and orthogonal forms (see Fig. (8)) cannot be done.

Many other cases involving two or many conformers were also studied even within the rather crude approximation of Eq. (9) (i.e. neglecting the dependence of $U(\chi,\Omega)$ on χ). It is relevant that, although based on wrong hypotheses, these results are in agreement with information from other techniques (1-d).

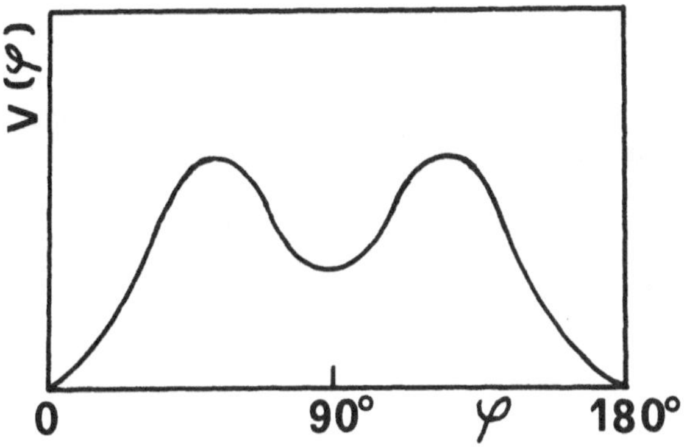

Figure 8. *Possible potential function for the internal rotation in thioanisoles, implying a mixture of planar and orthogonal forms as stable conformers.*

C) Molecules with methyl rotation

When, as in the case of toluene (35), the barrier is very low (60 J mol^{-1}) with a six-fold axis, it is likely that $U_{ext}(\chi,\Omega)$ has a small dependence on χ. A model which assumes the same barrier as in the gas phase and a unique order matrix was found to be consistent with the measured dipolar couplings. The results were not, however, very sensitive to the height of the hindering potential.

A better sensitivity to the barrier height for the internal rotation was found in two studies of o-halotoluenes where the potential barrier to the methyl group rotation was estimated as 5.01, 1.07 and 2.16 KJ/mole for chloro, bromo and iodo compounds, respectively (36,37).

In o-xylene the internal rotation is more complicated since the possibility of a geared rotation exists. An accurate study of this molecule concluded, however, that the methyl rotation is not coupled (38).

Other work involving methyl rotation, studied by LCNMR, includes 2,6-dichlorotoluene and 3,5-dichlorotoluene (39), p-chlorotoluene (40), γ-picoline (41), p-xylene (42), and mesitylene (43).

D) Molecules with ring puckering motions

The cases of nonrigid molecules having ring puckering motions investigated by LCNMR mainly concern examples of equilibrium between two puckered structures. In these cases the NMR of oriented molecules proved to be a powerful tool in deciding whether the planar conformation or non-planar ones were the most stable. Thus, while the molecules of tropone (44,45) and cyclopentadiene (46,47) were found to be essentially planar, an interesting case of nonplanarity was found in cyclobutane. In a first study by Meiboom and Snyder (48) an equilibrium between only two bent structures was assumed for this molecule; the best-fit bending angle was of 27°, to be compared with values of 35°, 33.3° ± 0.5°, 27° measured in the gas phase by Electron Diffraction, Infrared and Microwave spectroscopies, respectively. Cole and Gilson (49) reinvestigated the problem with the introduction of a potential function taking into account the ring puckering and the methylene rocking vibrations. Within this model they obtained for the best-fit angle a value of 29°, but only a slight improvement in the fitting of the data.

In a study of ring puckering of 1,4-dioxene (50) as well as in a similar deuterium investigation of 2,4,6-cycloheptatriene (51), it was possible to demonstrate once more the inadequacy of Eq. (8) in describing the internal motion. In most of these studies the dependence of $S_{\alpha\beta}$ on the internal motions was disregarded. In small globular molecules, however, with low potential barriers and large internal motion as in the case of cyclopentene (52), a strong dependence of $S_{\alpha\beta}$ on internal motion is possible, so that the problem is difficult to handle (53).

Other molecules of this type studied by LCNMR include 2,5-dihydrofuran (54), trimethylene oxide and sulfide (55,56) and cyclobutanone (57).

CONCLUSIONS AND PERSPECTIVES

Looking at the results as a whole, LCNMR seems to operate with success in conformational studies principally in excluding or confirming models. In such work LCNMR furnishes precious information which sometimes cannot be obtained otherwise. On the other hand, the limitations pointed out so far do not allow straightforward solutions of the conformational problems and, in addition, the quantitative determination of energy barriers appears at the moment not to give meaningful results.

There is however a continuous effort in progress in this field even if step by step. The improvements from an experimental point of view can be envisaged in obtaining more and better experimental data, either by analysing the spectra of the compounds under investigation over wider ranges of temperature and concentration, in various solvents, or by using ^{13}C satellites and multinuclear NMR.

Theoretical progress, on the other hand, with statistical studies of order parameters could help a great deal. A better understanding of how the molecular orientation depends on the structure starts to give good results, at least as far as liquid crystal molecules are concerned, and there is some hope that it will also help for solute molecules.

APPENDIX

Quantum mechanical rigid rotor

For a rigid rotor the time-independent Schrödinger equation is:

$$H_r \psi_r = E_r \psi_r \tag{1}$$

where

$$H_r = -\frac{\hbar^2}{2I_r}\frac{\partial^2}{\partial\phi^2} + V(\phi)$$

and I_r is the reduced moment of inertia (2). The potential $V(\phi)$ can be expressed as a Fourier series

$$V(\phi) = \tfrac{1}{2}\sum_n V_n\,(1 - \cos n\phi) \tag{2}$$

where ϕ is the rotational angle. The rotational problem is then easily solved first calculating the elements of the hamiltonian within the harmonic series basis set and then diagonalizing it. The eigenfunctions can be expressed as

$$\psi_r(\phi) = e^{i\sigma\psi}\sum_{-\infty}^{+\infty} A_{kr}e^{ink\phi}$$

where σ is an integer with values $-\frac{n}{2}$, $-\frac{n}{2}+1,\ldots\frac{n}{2}$ for n odd and $\frac{n}{2}$, $\frac{n}{2}-1,\ldots 0$ for n even. The A_{kr} coefficients are the elements of the R matrix which diagonalizes the hamiltonian. The probability $P(\phi)$ is then given by

$$P(\phi) = \frac{\sum_r p_r(\phi)e^{-E_r/kT}}{\sum_r e^{-E_r/kT}} \quad (4)$$

where $p_r(\phi)$, the probability that the molecular conformation is described by an angle ϕ in the rotational state with energy E_r, is given by

$$p_r(\phi) = \psi_r^*(\phi)\psi_r(\phi) \ . \quad (5)$$

The summation in Eq. (4) converges very quickly at room temperature in many cases.

To give an example, in the case of a two-well potential of form

$$V(\phi) = V_1(1-\cos\phi)/2 + V_2(1-\cos 2\phi)/2 + V_3(1-\cos 3\phi)/2 \ , \quad (6)$$

the probability distribution weighted over the first 40 rotational states is displayed in Fig. (9).

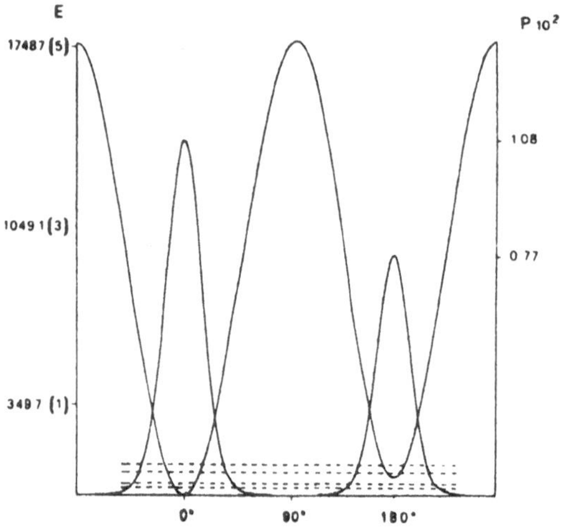

Figure 9. *Energy levels and angular probability density $P(\phi)$ in the case of a two-well potential. The energy is in cm^{-1}. In parentheses the corresponding kcal/mole are given.*

REFERENCES

1.(a P. Diehl, C.L. Khetrapal in *"NMR Principles and Progress"* Vol. 1 Springer-Verlag Berlin 1969.

(b J.W. Emsley, J.C. Lindon *"NMR Spectroscopy Using Liquid Crystal Solvents"* Pergamon Press Oxford 1975.

(c J. Bulthuis, C.W. Hilbers, C. MacLean, *MTP Int. Review of Science, Physical Chemistry*, series 1, Vol. 4, Butterworths London 1972.

(c L. Lunazzi in *"Determination of Organic Structures by Physical Methods"* (F.C. Nachod, J.J. Zuckerman and E.W. Randall eds.) Vol. 6, Academic Press New York 1976.

2. W.J. Orville-Thomas in *"Internal Rotation in Molecules"* J. Wiley London 1974.

3.(a P. Diehl, *Pure Appl. Chem.* **32**, 111 (1972).

(b P. Diehl and W. Niederberger in *"Specialist Periodical Report NMR"* Vol. 3 Chemical Society London 1974.

4. J.W. Emsley, G.R. Luckhurst, C.P. Stockley, *Proc. R. Soc. London*, **A 381**, 117 (1982).

5. J.W. Emsley, G.R. Luckhurst, *Mol. Phys.* **41**, 19 (1980).

6. E.E. Burnell, C.A. De Lange, O.G. Mouritsen, *J. Magn. Reson.* **50**, 188 (1982).

7. T.C. Wong, *Mol. Phys.* **34**, 921 (1977).

8. G. Fronza, R. Mondelli, F. Lelj, E.W. Randall, C.A. Veracini, *J. Magn. Reson.*, **37**, 275 (1980).

9. A.L. Segre, S. Castellano, *J. Magn. Reson.* **7**, 5 (1972).

10. K. Kuchitsu, T. Fukujama, Y. Morino, *J. Mol. Struct.* **1**, 463 (1967) and refs. therein.

11.(a E.B. Reznikova, V.I. Tulin, V.M. Tatevskii, *Opt. Spektronsk.*, **13**, 364 (1962).

(b A.L. Segre, L. Zetta, A. Di Corato, *J. Mol. Spectr.*, **32**, 296 (1969).

12. B. Dumbacher, *Theoret. Chim. Acta (Berl.)* **23**, 346 (1972) and refs. therein.

13. P.L. Barili, M. Longeri, C.A. Veracini, *Mol. Phys.* **28**, 1101 (1974).

14. P. Bucci, G. Chidichimo, F. Lelj, M. Longeri, N. Russo, *J. Chem. Soc. Perkin II*, 109 (1979).

15. P. Diehl, P.M. Enrichs, *J. Magn. Reson.* **5**, 134 (1971).

16. J.W. Emsley, J.G. Garnett, M.A. Long, L. Lunazzi, G. Spunta, C.A. Veracini, A. Zandanel, *J. Chem. Soc. Perkin II*, 853 (1979).

17. M. Hansen, H.J. Jacobsen, unpublished results; M. Hansen, Thesis, University of Aarhus, Denmark 1971.

18. K.G. Orrell, V. Sik, *J. Chem. Soc. Faraday II*, **72**, 941 (1976).

19. E.E. Burnell, C.A. De Lange, *J. Magn. Reson.* **39**, 461 (1980).

20. T.C. Wong, E.E. Burnell, *J. Magn. Reson.* **22**, 227 (1976).

21. P. Diehl, J. Jokisaari, J. Amrein, *Org. Magn. Reson.* **13**, 451 (1980).

22. J.W. Emsley, M. Longeri, *Mol. Phys.* **42**, 315 (1981).

23. J.W. Emsley, J.C. Lindon, D.S. Stephenson, *J. Chem. Soc. Perkin II*, 1508 (1975).

24. J.W. Emsley, J.C. Lindon, J.M. Street, G.E. Hawkes, *J. Chem. Soc. Faraday Trans. II*, **72**, 1365 (1976).

25. P. Diehl, H. Huber, A.C. Kunwar, M. Reinhold, *Org. Magn. Reson.* **9**, 374 (1977).

26. D. Catalano, M. Longeri, C.A. Veracini, unpublished results.

27. G. Chidichimo, A. Liguori, M. Longeri, C.A. Veracini, *J. Magn. Reson.* **51**, 438 (1983).

28. A.D'Annibale, L. Lunazzi, A.C. Boicelli, D. Macciantelli, *J. Chem. Soc. Perkin II*, 1396 (1973) and refs. therein.

29. J.W. Emsley, D.S. Stephenson, J.C. Lindon, L. Lunazzi, S. Pulga, *J. Chem. Soc. Perkin II*, 1541 (1975).

30. G. Chidichimo, F. Lelj, M. Longeri, N. Russo, C.A. Veracini, *Chem. Phys. Letters*, **67**, 384 (1979).

31. L. Bellitto, C. Petrongolo, C.A. Veracini, M. Bambagiotti, *J. Chem. Soc. Perkin II*, 314 (1977).

32.(a P. Bucci, M. Longeri, C.A. Veracini, L. Lunazzi, *J. Am. Chem. Soc.* **96**, 1305 (1974).

(b C.L. Khetrapal, A.C. Kunwar, *Mol. Phys.* **28**, 441 (1974).

33. N.M. Pozdeev, O.B. Akulinin, A.A. Shapkin, N.M. Magdesieva, *Dokl. Acad. Nauk SSSR* **185**, 384 (1969).

34. J.W. Emsley, M. Longeri, C.A. Veracini, D. Catalano, G.F. Pedulli, *J. Chem. Soc. Perkin II*, 1289 (1982).

35. P. Diehl, H.P. Kellerhals, W. Niederberger, *J. Magn. Reson.* **4**, 352 (1971).

36. P. Diehl, P.M. Henrichs, W. Niederberger, *Mol. Phys.* **20**, 139 (1971).

37. P. Diehl, P.M. Henrichs, W. Niederberger, J. Vogt, *Mol. Phys.* **21**, 377 (1971).

38. E.E. Burnell, P. Diehl, *Mol. Phys.* **24**, 489 (1972).

39.(a P. Diehl, C.L. Khetrapal, W. Niederberger, P. Partington, *J. Magn. Reson.* **2**, 181 (1970).

39.(b P. Diehl, H.P. Kellerhals, W. Niederberger, *J. Magn. Reson.* **3**, 230 (1970).

40. P. Diehl, F. Moia, *Org. Magn. Reson.*, **15**, 326 (1981).

41. C.L. Khetrapal, A. Saupe, *J. Magn. Reson.*, **9**, 275 (1973).

42. D. Canet, J. Barriol, *Mol. Phys.* **27**, 1705 (1974).

43. D. Canet, Ph.D. Thesis, University of Nancy, France (1973).

44. J.W. Emsley, J.C. Lindon, *Mol. Phys.* **25**, 641 (1973).

45. C.A. Veracini, F. Pietra, *J. Chem. Soc. Chem. Comm.*, 1262 (1972).

46. J.W. Emsley, J.C. Lindon, D.S. Stephenson, M.C. McIvor, *Mol. Phys.* **28**, 93 (1974).

47. C.A. Veracini, M. Guidi, M. Longeri, A.M. Serra, *Chem. Phys. Letters* **24**, 99 (1974).

48. S. Meiboom, L.C. Snyder, *J. Chem. Phys.* **52**, 3857 (1970).

49. K.C. Cole, D.F.R. Gilson, *J. Chem. Phys.* **60**, 1191 (1974).

50. C.A. De Lange, K.J. Peverelli, *J. Magn. Reson.* **16**, 159 (1974).

51. C.A. Veracini, unpublished results.

52. D.S. Stephenson, G. Binsch, *Mol. Phys.* **43**, 697 (1981).

53. C.R. Counsell, J.W. Emsley, G.R. Luckhurst, *Mol. Phys.* **43**, 711 (1981).

54. K.C. Cole, D.F.R. Gilson, *Can. J. Chem.* **52**, 281 (1974).

55. K.C. Cole, D.F.R. Gilson, *J. Chem. Phys.* **56**, 4362 (1972).

56. A. D'Annibale, L. Lunazzi, G. Fronza, R. Mondelli, S. Bradamante, *J. Chem. Soc. Perkin II*, 1908 (1973).

57. C.L. Khetrapal, A.C. Kunwar, A. Saupe, *Mol. Phys.* **25**, 1405 (1973).

58. W. Niederberger, P. Diehl, L. Lunazzi, *Mol. Phys.* **26**, 571 (1973).

59. L. Lunazzi, D. Macciantelli, *Gazz. Chim. Ital.* **105**, 657 (1975).

60. L.D. Field, S. Sternhell, A.S. Tracey, *J. Amer. Chem. Soc.* **99**, 5249 (1977).

61. C.A. Veracini, A. De Munno, V. Bertini, M. Longeri, G. Chidichimo, *J. Chem. Soc. Perkin II* 561 (1977).

62. C.A. Veracini, A. De Munno, M. Longeri, G. Chidichimo, V. Bertini, *J. Chem. Soc. Perkin II* 572 (1979).

63. J. Courtieu, Y. Gounnelli, C. Duret, P. Gonord, S.K. Kan, *Org. Magn. Reson.* **6**, 622 (1974).

64. F. Lelj, N. Russo, G. Chidichimo, *Chem. Phys. Letters* **69**, 530 (1980).

MOLECULAR STRUCTURE FROM DIPOLAR COUPLING

P. Diehl

The University of Basel,
Klingelbergstr. 82,
CH-4056 Basel.

THE BASIC EQUATIONS

The determination of molecular structure from NMR spectra of partially oriented molecules, which was first suggested in 1963 by Saupe and Englert (1), is based on the last term of the following effective Hamiltonian:

$$H = -(2\pi)^{-1}\sum_i \gamma_i\, I_{zi}\, (1-\sigma_i^{iso}-\sigma_i^{aniso})\, B_z$$

$$+ \sum_{i<j} \{T_{ij}^{iso}\, [I_{zi} I_{zj} + \tfrac{1}{2}\, (I_i^+ I_j^- + I_i^- I_j^+)]$$

$$+ T_{ij}^{aniso}\, [I_{zi}\, I_{zj} - \tfrac{1}{4}\, (I_i^+ I_j^- + I_i^- I_j^+)]\} \tag{1}$$

where $T_{ij}^{iso} = J_{ij}^{iso} = J_{ij}$ is the well known indirect coupling constant and $T_{ij}^{aniso} = J_{ij}^{aniso} + 2D_{ij}$ is the sum of the pseudo dipolar coupling, J_{ij}^{aniso}, and the dipolar or direct coupling D_{ij}. The dipolar coupling is defined as

$$D_{ij} = -\frac{\mu_0 \hbar \gamma_i \gamma_j}{8\pi^2}\, \tfrac{1}{2} < \frac{3\cos^2\theta_z - 1}{r_{ij}^{\;3}} > \tag{2}$$

Here μ_0 is the permeability in vacuo, r_{ij} is the internuclear distance, θ_z is the angle between the magnetic field and the vector **r** connecting the nuclei i and j. If r_{ij} and θ_z are

J. W. Emsley (ed.), Nuclear Magnetic Resonance of Liquid Crystals, 147–180.

independent of each other, D_{ij} may be simplified as follows:

$$D_{ij} = -\frac{\mu_0 \hbar \gamma_i \gamma_j}{8\pi^2} S_{ij} \left\langle \frac{1}{r_{ij}^3} \right\rangle \tag{3}$$

with $S_{ij} = \frac{1}{2} \langle 3 \cos^2\theta_z - 1\rangle$, the order parameter of the axis passing through i and j, with respect to the applied magnetic field in the z-direction.

The order parameter in turn can be expressed in terms of the order parameters of the molecular axis system (1,2,3) and the angle α between the magnetic field and the direction of the optic axis:

$$S_{ij} = \tfrac{1}{2} (3 \cos^2\alpha - 1) [\tfrac{1}{2} S_{33}(3 \cos^2\theta_3 - 1)$$

$$+ \tfrac{1}{2} (S_{11} - S_{22})(\cos^2\theta_1 - \cos^2\theta_2) + 2S_{12} \cos\theta_1 \cos\theta_2 \tag{4}$$

$$+ 2S_{13}\cos\theta_1\cos\theta_3 + 2S_{23}\cos\theta_2 \cos\theta_3]$$

In the spectral analysis we determine the so called "direct coupling" D_{ij}^{exp} which actually is equal to $\frac{1}{2}T_{ij}^{aniso}$ and consequently consists of $D_{ij} + \frac{1}{2}J_{ij}^{aniso}$, i.e. contains a contribution due to the anisotropy of the indirect coupling. The value of T_{ij}^{aniso} depends upon molecular orientation as follows:

$$T_{ij}^{aniso} = \tfrac{1}{2}(3 \cos^2\alpha - 1)\cdot\tfrac{2}{3}\{S_{33}[T_{33} - \tfrac{1}{2}(T_{11}+T_{22})] + \tfrac{1}{2}(S_{11}-S_{22})(T_{11}-T_{22})$$

$$+ S_{12}(T_{12}+T_{21}) + S_{13}(T_{13}+T_{31}) + S_{23}(T_{23}+T_{32})\} \tag{6}$$

THE PROCEDURE OF STRUCTURE DETERMINATION

The structural analysis begins with the selection of the solute molecule (about 20 mole%) and the solvent liquid crystal. The solvent is heated to the isotropic phase and the solute carefully dissolved by shaking the sample which has been degassed and sealed under vacuum. The spectrum is recorded with particular attention to temperature homogeneity because direct couplings are quite sensitive to variations in temperature resulting in line broadening. Usually double wall tubes are used with D_2O as a lock-substance in the interwall space. The water also averages out thermal gradients by rapid convection.

Preferably liquid crystals are selected which may be rotated in the magnet without changing the alignment of the director with the field. The signal to noise ratio (S/N) is made as good as possible because it affects the D_{ij}^{exp}, and hence the structure precision via the error of the line positions Δ

$$\Delta = [\nu_{\frac{1}{2}}^2/(S/N) + \nu_p^2/12]^{\frac{1}{2}} \tag{7}$$

where ν_p is the point resolution.

After a good spectrum has been obtained (line width ($\nu_{\frac{1}{2}}$) of the order of one hertz for protons), the next step is the spectral analysis by computer. It proceeds in a way similar to the isotropic case. Approximate starting values of parameters, (of which there is now one more type, the dipolar coupling), are taken from the isotropic spectra, (δ and J), and are calculated from an approximate molecular structure and guessed order parameters (D). Or, if a good guess is impossible, then automatic programs are applied. Alternatively analysis programs may be used which try to reproduce the spectrum i.e. the dipolar couplings by keeping an approximate molecular structure constant and varying the values of $S_{\alpha\beta}$ exclusively during the iterative fit. Such programs find the approximate solution much more quickly and from quite wrong starting values, because the number of variables is very small, usually two or three. Of course the final solution must still be obtained by fitting all the dipolar couplings independently.

The resulting values of D_{ij}^{exp} are fed into a further computer program (which will be discussed later) and used to determine the order as well as the structure parameters.

DIFFERENCES BETWEEN SPECTRA OF ISOTROPIC AND OF ORIENTED MOLECULES

As pointed out above, the spectra of the oriented molecules are more complex because they contain more parameters. Furthermore the dipolar coupling is, contrary to the indirect one, also observable between chemically equivalent nuclei. The protons of a methyl group, for example, produce a triplet with intensity ratio 1:2:1 and a splitting of 3D(HH). The protons of benzene display a spectrum of approximately 70 transitions which depend upon the three D(HH) but also on the J(HH). Weak interaction which may simplify the spectra of isotropic liquids usually exists exclusively between different types of nuclei (e.g. ^{13}C and H) in the oriented case.

On the other hand there is a simplification of spectra

observable which behaves very much like "weak coupling". As an example we may cite a three spin system in which one direct coupling is much stronger than the others ($D_{AB} > D_{AC}, D_{BC}$). In the 3x3 submatrix of the hamiltonian of this system all the off-diagonal elements connecting the diagonal elements containing D_{AB} with the other diagonal element may be neglected. As a result the matrix decays exactly like in the transition from an ABC to an ABX spectrum of isotropic systems. Consequently we observe ab type doublets with splitting $3D_{AB}$ and centres of gravity at:

$$\frac{1}{2}(\nu_A + \nu_B) \pm \frac{1}{2}\left[\frac{1}{2}(J_{AC} + J_{BC}) + (D_{AC} + D_{BC})\right]$$

The repeated spacing is x:

$$x = \frac{1}{2}(J_{AC} + J_{BC}) + (D_{AC} + D_{BC}) \qquad (8)$$

The parameters which may directly be read from the spectrum are D_{AB}, $\nu_A+\nu_B$, ν_C and x. Further information is lost, and particularly D_{AC} and D_{BC} cannot be separated from each other.

Finally for spin ≥ 1 the spectra of the oriented molecules are further complicated by the quadrupolar interaction.

Like the temperature gradients the presence of concentration gradients seriously affects the spectra by broadening particularly the lines in the wings. A temperature increase of 1% as well as a solute concentration increase by 5% approximately reduce the degree of order by 1%, and consequently gradients of this order of magnitude lead to line broadening of 10 Hz in the wings of spectra with dipolar couplings of 1000 Hz.

RELATIONS BETWEEN DIRECT COUPLINGS AND MOLECULAR STRUCTURE

The simplest molecule that can be studied for structure determination is a linear system of three spins. Taking, for example, three protons A,B, and C we observe the three couplings:

$$D_{AB} = -(1/r_{AB}^3)\, S_{AB}\, K_{HH}$$

$$D_{AC} = -(1/r_{AC}^3)\, S_{AC}\, K_{HH}$$

$$D_{BC} = -(1/r_{BC}^3)\, S_{BC}\, K_{HH}$$

with $K_{HH} = \frac{\mu_0 \hbar \gamma_H \gamma_H}{8\pi^2}$ and $S_{AB} = S_{AC} = S_{BC}$

From two out of the three observed couplings we can derive the internuclear distance ratio, thus,

$$r_{AB}/r_{BC} = (D_{BC}/D_{AB})^{\frac{1}{3}} \tag{10}$$

This simple case illustrates a basic property of all direct couplings. They never define an absolute length, but only relative distances, that is the shape of molecules. In order to determine absolute order parameters we always have to assume an absolute internuclear distance.

Internuclear angles can be measured absolutely. In a methyl group the proton-proton and proton-carbon dipolar couplings determine α, the HCH angle, as follows:

$$r_{HH}/r_{CH} = \sqrt{3}\sin\beta = 2\sin(\alpha/2) \tag{11}$$

where β is the angle between the C-H bond and the C_3 symmetry axis. The value of r_{HH}/r_{CH} is derived from D(CH)/D(HH):

$$D(CH)/D(HH) = \left(\frac{\gamma_H}{\gamma_C}\right)\left(\frac{r_{HH}}{r_{CH}}\right)^3\left[\left(\frac{r_{HH}}{r_{CH}}\right)^2 - 2\right].$$

Looking at a system of four spins at the corner of a planar rectangle, we can see that there are three different dipolar couplings: D_{12}, D_{13} and D_{23}. The system has effective C_2 symmetry and must therefore be characterized by two order parameters S_{11} and S_{22}. The molecular shape is fully determined by one parameter, the distance ratio (r_{12}/r_{23}). Obviously we have a case of three measured direct couplings for the three unknown parameters, and hence the system is exactly defined.

For this rectangular spin system we can still find an expression which relates the molecular shape to the three couplings:

$$(D_{12}/D_{23})(r_{12}/r_{23})^5 - (D_{13}/D_{23})\{1 + (r_{12}/r_{23})^2\}^{\frac{5}{2}} + 1 = 0. \tag{12}$$

There is no longer an analytical solution to this equation, so that already for this simple system, as of course is the case for more complex ones, the solution must be found iteratively by computer. But it is possible to discuss the uniqueness of solutions. Instead of interpreting Eq. (12) as a fifth order equation for the unknown molecular shape (r_{12}/r_{23}), we discuss it as an ensemble of equations for the linear relation between (D_{12}/D_{23}) and (D_{13}/D_{23}), one (at least) for each value of r_{12}/r_{23}. We then see immediately that these linear relations have a common envelope which is defined as:

$$(D_{13}/D_{23}) = |(D_{12}/D_{23})|/[(D_{12}/D_{23})^{\frac{2}{3}} + 1]^{\frac{3}{2}} . \tag{13}$$

It also becomes clear that a measured set of dipolar couplings (D_{12}/D_{23}) and (D_{13}/D_{23}) usually leads to two possible solutions because there are two tangents of the envelope (see Fig. (1)).

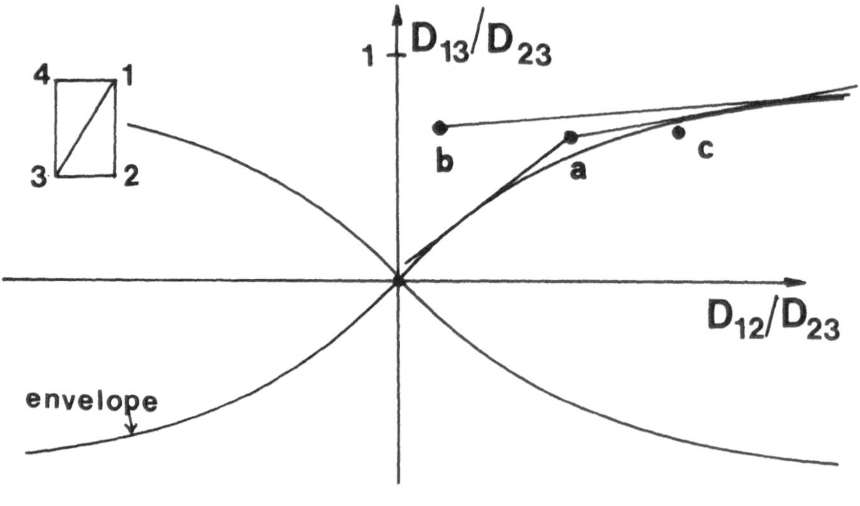

Figure 1. *Solutions for the shapes of rectangular spin systems. Depending upon the position of the measured point (•), there are two (a), one (b), or zero (c) tangents to the envelope, i.e. solutions.*

Actually we can find the following conditions for the existence of two solutions:

$$|D_{12}| > |D_{13}| < |D_{23}|$$

and for only one solution:

$$|D_{12}| < |D_{13}| < |D_{23}|$$

The situation becomes particularly difficult if the measured ratios of couplings correspond to a point near to or on the envelope, because then the two tangents are quite similar and we have two almost identical geometrical solutions for the molecular shape, and there is no way to identify the correct one except to run a new spectrum with varied temperature or solvent, or (see later) to include ^{13}C-satellite information. The critical region corresponds to equal order parameters for the directions parallel to the two sides of the rectangle.

This discussion illustrates that the problem of uniqueness and precision of a geometrical solution can be very complex. It is usually difficult if not impossible to predict whether a set of measured D_{ij} is ideal or not, that is whether it leads to a well defined unique solution with small errors or to a result with extreme errors. It is consequently recommended also from this point of view to record several spectra for the same molecule in various solvents.

For this simple four-spin system and indeed for spin systems in general one should make sure that there are enough measurable dipolar couplings for a determination of the unknown structure, as well as orientation, parameters. The counting is done in the following way. The number of measurable different D-couplings in systems without symmetry is equal to n(n-1)/2 for n nuclei. These must define the five order parameters as well as the 3(n-2) coordinates of the nuclei with respect to a basis which is formed by the distance between two nuclei. Consequently the following condition can be derived:

$$n(n-1)/2 - 5 \; -(n-2)3 \geqq 0$$

$$\text{i.e.} \quad n \geqq 7. \tag{16}$$

For planar systems we find

$$n(n-1)/2 - 3 \; -(n-2)2 \geqq 0$$

$$\text{i.e.} \quad n \geqq 5, \tag{17}$$

and, as we have seen already for planar systems with C_2-symmetry, the relation is : $n \geqq 4$. We can see, that for unsymmetrical molecules the system must be quite large (seven nuclei) in order to be defined.

As pointed out above, a system of three spins in a molecule with C_2-symmetry is in general underdetermined. If, however, we assume, that there are no solvent effects, we may combine the data from two spectra recorded in different solvents. We then measure six dipolar couplings from which the two sets of two order parameters for the two solvents and the two unknown coordinates of the third nucleus with respect to a fixed basis can be derived.

For such a measurement it is particularly important to choose solvents which induce a substantially different ratio of order parameters S_x/S_y in the two cases. This necessity can be demonstrated analytically by first fixing nuclei 1 and 3 along the y-axis at a distance a. Nucleus 2 has the coordinates (x,y); (see Fig. (2)).

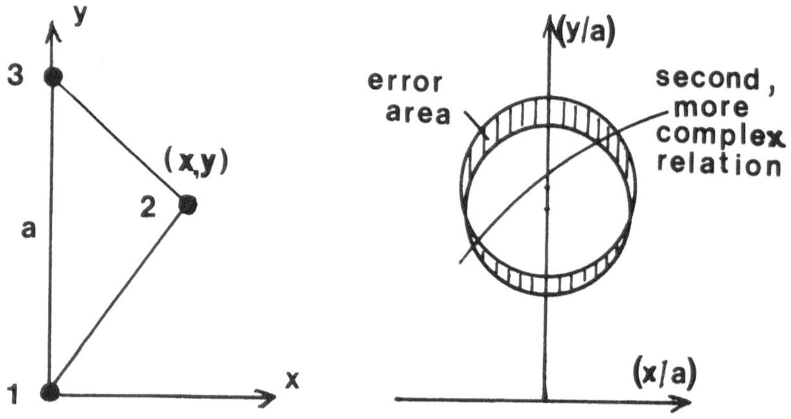

Figure 2. *Errors for structure determined in two different solvents.*

The ratios of direct couplings are:

$$D_{12}/D_{13} = (S_x/S_y)g_1 + g_2 ,$$

$$D_{23}/D_{13} = (S_x/S_y)g_3 + g_4 ,$$

with

$$g_1 = (x/a)^2/[(x/a)^2 + (y/a)^2]^{\frac{5}{2}}$$

$$g_2 = (y/a)^2/[(x/a)^2 + (y/a)^2]^{\frac{5}{2}}$$

$$g_3 = (x/a)^2/\{(x/a)^2 + [1-(y/a)]^2\}^{\frac{5}{2}}$$

$$g_4 = [1-(y/a)]^2/\{(x/a)^2 + [1-(y/a)]^2\}^{\frac{5}{2}} \qquad (18)$$

In the second liquid crystal solvent we have equal values of g_1-g_4 but $S_x{}'$ and $S_y{}'$ instead of S_x and S_y. The measured couplings are D'_{12}, D'_{13} and D'_{23}. We can form the following difference:

$$(D_{12}/D_{13}) - (D'_{12}/D'_{13}) = g_1[(S_x/S_y) - (S'_x/S'_y)] \qquad (19)$$

and derive the relation:

$$\frac{(D_{12}/D_{13})-(D'_{12}/D'_{13})}{(D_{23}/D_{13})-(D'_{23}/D'_{13})} = \frac{g_1}{g_3} = P = \frac{\{(x/a)^2+[1-(y/a)]^2\}^{\frac{5}{2}}}{\{(x/a)^2+(y/a)^2\}^{\frac{5}{2}}} \qquad (20)$$

or

$$(x/a)^2 + [(y/a) - 1/(1-P^{\frac{2}{5}})]^2 = P^{\frac{2}{5}}/(1-P^{\frac{2}{5}})^2 \qquad (21)$$

This equation represents a circle with centre at $(x/a) = 0$; $(y/a) = 1/(1-P^{\frac{2}{5}})$ with a radius of

$$R = P^{\frac{1}{5}}/(1-P^{\frac{2}{5}}) \qquad (22)$$

The solution for (x/a) and (h/a) must be on this circle.

Obviously P is now very sensitive to errors if $[(D_{12}/D_{13})-(D'_{12}/D'_{13})]$ is small i.e. if the ratio (S_x/S_y) is close to (S'_x/S'_y).

As a practical example we derive that an error of 1% in D_{12}/D_{13} can easily lead to an error of 10% in $[(D_{12}/D_{13})-(D'_{12}{}'/D'_{13}{}')]$ and to 20% in P. Finally the centre of the circle has an error of 8% and the radius of 9%. The error area (see Fig. (2)) is the difference between the two circles with different centres and different radii. Consequently even if the further relation between (x/a) and (y/a) which can be obtained from equations (18) is exact, the final error of the parameters depends quite critically and in a complex way on the intersection of the curve representing this relation and the mentioned error area.

THE COMPUTER PROGRAMS FOR STRUCTURE DETERMINATION

The direct couplings which have been derived in the spectral analysis together with the variance - covariance error matrix form the input to the computer program, which iteratively derives the molecular structure as well as the order parameters and their variance - covariance matrix. One such program is SHAPE (2) which also allows the determination of absolute angles with their errors; intramolecular motion can also be included. The program which considers molecular symmetry can also fit simultaneously several sets of values of dipolar couplings for the same molecule in various solvents.

There also exists a PASCAL version of the program which allows the use of a spectrometer dedicated computer such as an ASPECT 2000.

In the program SHAPE all the D_{ij}^{exp} are assumed independent from each other. The fitting process is performed by minimising the function

$$\sum_i (D_i^{exp} - D_i^{calc})/\sigma_i^2 \tag{23}$$

where the index i distinguishes between the different dipolar couplings and σ_i^2 is the variance of D_i^{exp}. For determining the errors of the molecular geometry the entire variance-covariance error-matrix of the D_{ij}^{exp} is transformed in order to determine the errors in the molecular geometry, that is here also the covariance is considered. In fact, for strongly interdependent D_i^{exp}, the fitting process should be modified to minimize the function,

$$\sum_{ij} A_{ij} (D_i^{exp} - D_i^{calc}) (D_j^{exp} - D_j^{calc}) \quad , \tag{24}$$

where A_{ij} are elements of the inverse of the variance-covariance matrix.

An equivalent approach is to construct new parameters I_i which are linear combinations of the D_i^{exp} so that the variance-covariance matrix of the I_i is diagonal. These transformation coefficients are directly available from various spectral analysis programs. They are called error-vectors and are listed with the corresponding standard errors σ_i'. The fitting process is then performed by minimizing the function (3),

$$\sum_i (I_i^{exp} - I_i^{calc})^2/\sigma_i'^2 \ . \tag{25}$$

In general the resulting geometrical parameters will be slightly different and their errors will usually be smaller than the ones derived by formula (23).

TYPES OF MOLECULES STUDIED SO FAR

The method of NMR of oriented molecules for structure determination has been applied to a large number of molecules. All together there are now approximately 500 structures determined. Of these there are roughly 30% with four spins, 20% with five spins, 30% with six spins, 10% with seven spins and 10% with eight and more spins. The resulting structural data are collected in books and reviews (4). Quite clearly the large spin systems producing complex spectra of 1000 or more transitions are still avoided. Proton NMR form the predominant part of these spectra, but there are also some papers on ^{2}H-, ^{19}F- and ^{31}P-resonance. Even less used are ^{11}B, ^{13}C, ^{14}N, ^{15}N, ^{17}O, ^{111}Cd, ^{113}Cd, ^{117}Sn, ^{119}Sn and ^{199}Hg.

Spectral simplifications by substitution of ^{2}H with ^{1}H-{^{2}H} decoupling as well as multiple quantum transitions have so far been used rarely. The method of using satellites from rare nuclei in the ^{1}H-spectra has been increasingly applied and will be discussed later.

BASIC THEORY FOR VIBRATION CORRECTIONS

As we have seen above, NMR of oriented molecules allows the measurement of ratios of internuclear distances as well as absolute angles. In fact, the presence of molecular vibrations at IR-frequencies makes these parameters time dependent. Actually, if we consider the mean amplitudes of vibrations for bonded atomic pairs (C-H: 0.078 Å; C-F: 0.045 Å; C-C: 0.050Å; Si-H: 0.089 Å) and for nonbonded pairs (H-C-H: 0.135 Å; F-C-F: 0.056 Å; H-Si-H: 0.166 Å) we see that the definition of an interatomic distance is quite problematic. The angles between internuclear and molecular axes may vary simultaneously with the variation of interatomic distances and consequently the degree of order of these axes is also time dependent. Obviously, in order to be comparable with results from different spectroscopies, the data derived from NMR of oriented molecules must be corrected for vibration.

Ideally we are interested in the equilibrium structures r_e which minimize the electronic potential within the Born-Oppenheimer approximation. In practice we assume that nuclei vibrate around this minimum and the actual motion may be expanded in terms of the anharmonic and the harmonic vibration. Of these the second can be computed relatively easily from the known harmonic force constants. The resulting molecular structure which is corrected for harmonic effects is called the r_α-structure, where r_α is the distance between average nuclear positions at thermal equilibrium. The difference between r_e and r_α is due to the anharmonicity of the vibrational potential; r_α has the advantage of being internally consistent, that is there are no shrinkage effects (on a straight line the distance $\overline{AC}$ is equal to the sum of $\overline{AB} + \overline{BC}$), and the result can be compared with different spectroscopies.

In electron diffraction, for example, the measured quantity for a benzene C-H bond is

$$\langle r_{ED} \rangle = r_e + \langle \Delta z \rangle + (\langle \Delta x^2 \rangle + \langle \Delta y^2 \rangle)/2r_e \ , \tag{26}$$

and in microwave spectroscopy

$$\langle r_{MW} \rangle = r_e + \langle \Delta z \rangle - (3\langle \Delta z^2 \rangle - \langle \Delta x^2 \rangle - \langle \Delta y^2 \rangle)/2r_e \ . \tag{27}$$

The NMR result contains the variation of the degree of order of the internuclear axis, and is derived as follows:

$$D = D^e + \sum_\alpha \left(\frac{\partial D}{\partial \alpha}\right)_e \langle \Delta\alpha \rangle + \frac{1}{2} \sum_\alpha \left(\frac{\partial^2 D}{\partial \alpha^2}\right)_e \langle \Delta\alpha^2 \rangle \tag{28}$$

with $\alpha = x, y, z$

$$D = -K \left\langle\left\langle \frac{S}{r^3} \right\rangle_m\right\rangle_v \tag{29}$$

where the averages are over the molecular tumbling $\langle\ \rangle_m$ and over the vibrational motion $\langle\ \rangle_v$. With the assumption that the vibrational motion does not affect the molecular orientation, D can be written as:

$$D = -K \sum_{i,j} S_{ij} \left\langle \frac{\cos\alpha_i \cos\alpha_j}{r^3} \right\rangle \tag{30}$$

with α_i as the angle between the molecular axis system at equilibrium and the instantaneous internuclear axis r.

Then, for a benzene C-H bond we have the instantaneous distance

$$r^2 = (r_e + \Delta z)^2 + \Delta x^2 + \Delta y^2$$

and the instantaneous degree of order

$$S = S_{zz} \; [(\Delta x^2 - 2\,\Delta y^2 + (r_e + \Delta z)^2]/r^2$$

$$D = -\,(KS_{zz}/r_e{}^3)\;[1-3\,\langle\Delta z\rangle/r_e + 6\,\langle\Delta z^2\rangle/\,r_e{}^2$$

$$-3\,\langle\Delta x^2\rangle/\,2r_e{}^2\; -9\langle\Delta y^2\rangle\;/\;2r_e{}^2] \qquad (31)$$

Consequently the direct coupling which is measured:

$$D = -KS_{zz}/r_d{}^3$$

defines the effective internuclear distance r_d as follows:

$$r_d = r_e + \langle\Delta z\rangle + [\langle\Delta x^2\rangle + 3\langle\Delta y^2\rangle - 4\langle\Delta z^2\rangle]/2r_e \qquad (32)$$

The general theory for vibration corrections can now be formulated (5).

We start from the relation

$$D = -\,K \sum_{\alpha,\beta} S_{\alpha\beta}\,\phi_{\alpha\beta} \qquad (33)$$

with

$$\phi_{\alpha\beta} = \langle r_\alpha \cdot r_\beta\,/r^5\rangle$$

Again assuming that the $S_{\alpha\beta}$ do not vary as a function of the vibration, we expand $\phi_{\alpha\beta}$ for small amplitudes of vibration. In the resulting equation

$$\phi_{\alpha\beta} = \phi^e_{\alpha\beta} + \phi^a_{\alpha\beta} + \phi^h_{\alpha\beta} + \ldots \qquad (34)$$

The indices e,a and h correspond to the equilibrium geometry, the anharmonic correction and the harmonic correction. These depend upon the angles between the internuclear axis and the molecular coordinate system ($\zeta_\alpha = \cos\phi_\alpha$, and $\alpha,\beta,\gamma,\delta = x,y,z$) as well as on the covariance matrices $C_{\alpha\beta}$ of the vibrational amplitudes Δ, ($r = R+\Delta$).

$$C_{\alpha\beta} = \langle\Delta\alpha \cdot \Delta\beta\rangle_v .$$

We find

$$\phi^e_{\alpha\beta} = \zeta_\alpha\zeta_\beta/R^3 \tag{35}$$

$$\phi^a_{\alpha\beta} = -\sum_\gamma (5\zeta_\alpha\zeta_\beta\zeta_\gamma - \delta_{\alpha\gamma}\zeta_\beta - \delta_{\beta\gamma}\zeta_\alpha)\langle\Delta_\gamma\rangle_v/R^4$$

$$\phi^h_{\alpha\beta} = [C_{\alpha\beta}-5\sum_\gamma \zeta_\gamma(C_{\alpha\gamma}\zeta_\beta+C_{\beta\gamma}\zeta_\alpha) + \tfrac{5}{2}\zeta_\alpha\zeta_\beta \sum_{\gamma\delta} C_{\gamma\delta}(7\zeta_\gamma\zeta_\delta-\delta_{\gamma\delta})]/R^5$$

The dipolar coupling D is defined as:

$$D = D^e + d^a + d^h \tag{36}$$

where

$$D^e = -K\sum_{\alpha\beta} S_{\alpha\beta}\,\phi^e_{\alpha\beta}$$

$$d^a = -K\sum_{\alpha\beta} S_{\alpha\beta}\,\phi^a_{\alpha\beta}$$

$$d^h = -K\sum_{\alpha\beta} S_{\alpha\beta}\,\phi^h_{\alpha\beta}$$

The corrected D-value is:

$$D^d = D - d^h = D^e + d^a \tag{37}$$

In order to be able to compute ϕ^h (and d^h), the quantities $C_{\alpha\beta}$ must be known. They can be derived from a complete solution of the vibrational problem in terms of vibrational frequencies and normal coordinates, the molecular potential being approximated by a quadratic function.

The advantage of this approach is that the quantities of interest, the ϕ's, can be calculated separately and are stored as a characteristic matrix for a molecule which easily allows the computing of the corrections for various S-tensors. Furthermore the ϕ's immediately display general properties of the vibration corrections. First, d^h decreases with R^{-5}, so that the relative corrections d^h/D^e decreases with R^{-2}. Consequently large corrections will be observed for directly bonded nuclei. Similarly d^a/D^e decreases generally more slowly with R^{-1}. Second the relations for the ϕ's allow simple analytical expressions to be derived for corrections in molecules with symmetry.

For two nuclei on the z-axis with $S_{zz} = -2S_{xx} = -2S_{yy}$ we find

$$d^h = 6\,(C_{zz} - C_{xx})\,D^e/R^2 \tag{38}$$

and $d^a = -3\,\langle\Delta z\rangle_v\,D^e/R$

For two nuclei on a z-axis with $S_{xx} \neq S_{yy}$ we obtain

$$d^h = (D^e/R^2)[\tfrac{1}{2}(12C_{zz}-7C_{xx}-5C_{yy}) + (C_{yy}-C_{xx})(S_{yy}/S_{zz})] \tag{39}$$

Here the relative correction can become very large if S_{zz} approaches zero. Actually d^h is still finite for $S_{zz} = 0$, as can be seen from the equivalent relation

$$d^h = -\,(K/R^5)[(C_{yy}-C_{xx})S_{yy} + \tfrac{1}{2}(12C_{zz}-7C_{xx}-5C_{yy})S_{zz}] \tag{40}$$

If the y-axis coincides with an at least three-fold symmetry axis, we have $S_{xx} = S_{zz} = -\tfrac{1}{2}\,S_{yy}$ and

$$d^h = \tfrac{1}{2}\,(D^e/R^2)(12C_{zz}-3C_{xx}-9C_{yy}) \tag{41}$$

As an example we find for benzene ^{13}C-H bonds with $C_{xx} = 1.41\times10^{-2}\,\text{Å}^2$, $C_{yy} = 2.26\cdot10^{-2}\,\text{Å}^2$ and $C_{zz} = 0.59\cdot10^{-2}\,\text{Å}^2$ that d^h is approximately 8% of D^e. If such a correction is neglected, the measured distance is wrong by roughly one third of 8% i.e. by 2.7%.

The vibration correction as described above has been programmed for the computer and is available as the program VIBR (5). It calculates atomic coordinates from a versatile structure code, converts valence, or central force field constants into Cartesian force field constants, calculates normal coordinates and vibrational frequencies, evaluates the covariance matrices and corresponding terms ϕ^h and, finally, the d^h. It also allows the calculation of the corrections for the various individual vibrational modes. By use of this facility it can be seen, for example, that for the benzene molecule the corrections to the H-H direct couplings, which are of the order of 0.5 to 1.5%, stem predominantly from the C-H bend (B_{2g}, B_{2u}, E_{1g}, E_{2u}) and the ring deformation (E_{2u}) modes.

We have pointed out above that for internuclear directions

with a zero order parameter the vibration correction is particularly problematic. In the next section we therefore derive the continuum of directions in molecules which have $S_{ij} = 0$, the so called S = 0 cone.

THE CONTINUUM OF DIRECTIONS WITH $S_{aa} = 0$ IN ORIENTED MOLECULES; THE $S_{aa} = 0$ CONE

The degree of order of an arbitrary axis *a* fixed in a molecule can be related to ordering matrix elements referred to another molecular frame by,

$$S_{aa} = \sum_{p,q} \cos\alpha_p \cos\alpha_q S_{pq} \tag{42}$$

where p,q = x, y, z, and α_p, α_q are the angles between the axis *a* and the molecular x,y,z-axes.

If we write $\cos\alpha_p = \frac{p}{\ell}$ we find that for all the axes which have $S_{aa} = 0$ there exists a general equation:

$$x^2 S_{xx} + y^2 S_{yy} + z^2 S_{zz} + 2xzS_{xz} + 2xyS_{xy} + 2yzS_{yz} = 0 \tag{43}$$

This equation describes the surface of an elliptic cone in three dimensions which can be transformed by rotation of the principal axes which makes the product terms equal to zero. The resulting equation is

$$u^2 S_{uu} + v^2 S_{vv} + w^2 S_{ww} = 0$$

or $$u^2 S_{uu} + v^2 S_{vv} - w^2 (S_{uu} + S_{vv}) = 0 \tag{44}$$

and with $S_{uu} = \frac{1}{c^2}$ and $S_{vv} = \frac{1}{d^2}$ we find

$$\frac{u^2}{c^2} + \frac{v^2}{d^2} - \frac{w^2}{[c^2 d^2/(c^2+d^2)]} = 0 \tag{45}$$

This cone includes the point u = v = w = 0. It intersects the planes $w = \pm cd/(c^2+d^2)^{\frac{1}{2}}$ in ellipses which are given by $u^2/c^2 + v^2/d^2 = 1$. The two angles defining one half of the aperture of the cone γ and δ are derived as:

$$\tan\gamma = (1 + d^2/c^2)^{\frac{1}{2}} \text{ and } \tan\delta = (1 + c^2/d^2)^{\frac{1}{2}}$$

so that

$$\gamma = \tan^{-1} (1 + S_{uu}/S_{vv})^{\frac{1}{2}}$$

$$\delta = \tan^{-1} (1 + S_{vv}/S_{uu})^{\frac{1}{2}}$$

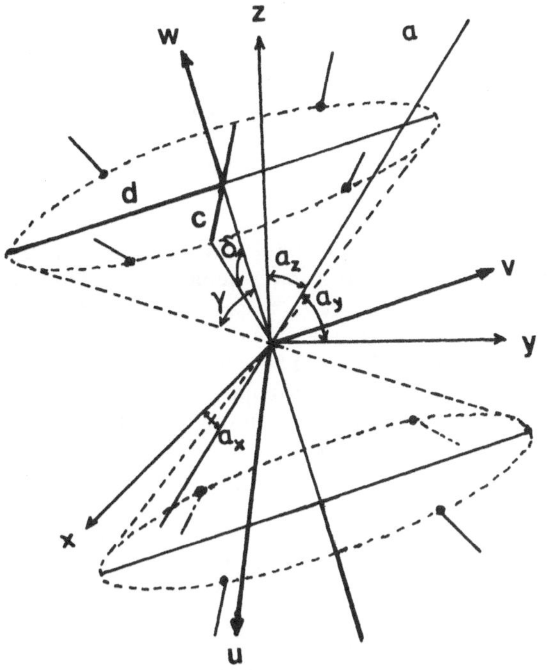

Figure 3. *The* $S_{aa} = 0$ *cone.*

The existence of such a cone implies, that for molecules which are three-dimensional there will quite probably always be internuclear directions of small or zero degrees of order. For planar molecules on the other hand directions with $S_{aa} = 0$ can only arise if the two perpendicular molecular axes in the plane have opposite degrees of order.

Finally it may be mentioned that the diagonals $|u|=|v|=|w|$ are always axes of $S_{aa} = 0$ (magic directions!).

THE USE OF SATELLITES FOR INCREASED SENSITIVITY

For the determination of the structure of organic molecules it is of course necessary to gain information on the proton as well as on the carbon coordinates. The carbon positions may be derived from the dipolar coupling constants $D(^{13}CH)$ which can be observed either in the ^{13}C-spectra of the oriented molecule or as the ^{13}C-satellites in the proton spectra. The second method has a considerable advantage in sensitivity because for an equal number of nuclear pairs ^{1}H and ^{13}C the ^{1}H-satellite signal is proportional to the ^{1}H sensitivity, whereas the ^{13}C-signal is proportional to the ^{13}C-sensitivity. Consequently the gain in signal to noise for an equal sample volume is of the order of sixty. This argument may be used for all the non proton nuclei and the gain in signal to noise may be quite considerable (e.g. 935 for ^{15}N and 29600 for ^{57}Fe).

The satellite method also has its drawbacks. One of these is the difficulty of detecting the weak satellites which often appear very close to the 100 times (case of ^{13}C) stronger transitions which arise in molecules with ^{12}C. This difficulty, however, can be avoided by difference spectroscopy. In this method (AISEFT = Abundant Isotope Elimination by Fourier Transform) (6) the free induction decays of the proton spectrum with the non proton nucleus decoupled are subtracted from the free induction decays of the same spectrum without decoupling. In the resulting difference the positive signals are exclusively due to the satellites of nuclei coupled to the non-proton (e.g. ^{13}C or ^{15}N) and at the positions of the original strong lines there appear negative signals with intensities equal to the satellites.

A second difficulty, which cannot be eliminated, is the complexity of the spectra. For each possible different position of the non proton nucleus there will be a corresponding satellite spectrum, with all the spectra superimposed on each other. For example in o-chlorotoluene we observe a 7-proton spectrum with seven 8-spin satellite spectra superimposed. Obviously, for molecules with low symmetry, we very quickly reach an upper limit of complexity. In general the samples must be very pure because peaks from a one percent impurity could easily be mistaken as satellite lines. This problem can be eliminated by difference spectroscopy. The couplings derived from the satellites are usually substantially less precise than the proton-proton couplings because the signal to noise ratio in the satellites is much smaller.

THE USE OF SATELLITES IN UNDERDETERMINED CASES

In the section on relations between dipolar coupling and molecular structure we have seen that small spin systems may be underdetermined, that is, the number of observed dipolar coupling constants is not sufficient for a determination of both the unknown nuclear coordinates and the order parameters.

As an example we can look at proton systems with C_2-symmetry. These are exactly determined for four nuclei. Consequently, for example, in polysubstituted chlorobenzenes only the ortho- and para-disubstituted cases provide structure information. However, if we include the ^{13}C-satellite information, we drastically increase the number of observed coupling constants. In the 4-proton system we now have twelve new couplings in the ortho- and eight in the para-disubstituted case. The number of unknowns has increased by only six and four coordinates respectively so that these systems are now highly overdetermined. But also the tri- and tetra-substituted molecules are now tractable. For the tetra-substituted cases it is necessary to use two different liquid crystal solvents in order to increase the number of measured direct couplings so that it exceeds the number of unknowns. For the penta-substituted benzene we can see that in each liquid crystal we observe four dipolar couplings and introduce two new unknown degrees of order. Consequently, we must take spectra in three different solvents to give twelve dipolar couplings and with the six unknown order parameters as well as the five coordinates there are eleven unknowns.

POSSIBLE STRUCTURE PRECISION

At first sight the method of oriented molecules seems to be an extremely precise method for structure determination. Dipolar couplings of the order of 10^3 Hz between protons can be measured with an error of 10^{-1} Hz. The resulting relative precision of 10^{-4} in the ratios of dipolar couplings corresponds to a relative precision of roughly 10^{-5} in the distance ratios. It consequently appears possible to distinguish distances which differ by less than 10^{-4} Å with respect to a fixed basis.

In general this extreme precision will be meaningless. There are various reasons for this limitation which will be discussed later, such as the anisotropy of the indirect coupling, correlation of molecular shape with angular orientation and complex formation. But also the vibration correction will in general not be extremely precise; it depends upon the force field used as well as upon the approximations which are made. Its relative precision is perhaps 5% so that a vibration correction of 2.5% in a C-H bond length introduces an uncertainty of roughly 0.1%, so that vibration corrections already limit the precision to approximately 10^{-3} Å. As we will see later, in practice, the precision is of the order of 10^{-2} Å in bond lengths and 10^{-1} degrees in angles.

LIMITATION OF STRUCTURE PRECISION DUE TO THE ANISOTROPY OF THE INDIRECT COUPLING CONSTANTS

As pointed out earlier the total anisotropic coupling measured in the spectra of oriented molecules contains contributions from the anisotropy of the indirect coupling (8). From Eq. (6) we can derive that D_{ij}^{exp}, the experimental "dipolar coupling" is defined in the principal axis system of the order matrix as:

$$D_{ij}^{exp} = D_{ij} + \tfrac{1}{2} J_{ij}^{aniso} = \tfrac{1}{2}(3\cos^2\alpha - 1)\{S_{33}[-\frac{K}{r_{ij}^3} \cdot \tfrac{1}{2}(3\cos^2\theta_3 - 1) + \tfrac{1}{3}\Delta J]$$

$$+ (S_{11}-S_{22})[-\frac{K}{r_{ij}^3}\, \tfrac{1}{2}\,(\cos^2\theta_1 - \cos^2\theta_2) + \tfrac{1}{6}\,(J_{11}-J_{22})]\} \qquad (47)$$

$$\text{with } \Delta J = J_{33} - \tfrac{1}{2}\,(J_{11} + J_{22}) \;;\; K = \frac{\mu_0 \hbar \gamma_i \gamma_j}{8\pi^2}$$

Actually, this mixture of dipolar, or direct, with indirect contributions is not surprising, if we remember that the indirect coupling may be considered as the screening of the dipolar coupling between nuclear magnetic moments which arises from the electrons in the internuclear space. This screening in a way resembles the chemical shift and because of the missing rotational symmetry of the electronic structure, it is a tensor with non zero trace. Of course, this "screening" of the dipolar coupling physically contains contributions, as for example the contact interaction, which cannot be interpreted simply as currents.

On inspection of Eq. (47) we see that the dipolar and the indirect coupling contributions have the same dependence on the order parameters, but they are not zero under the same conditions. For instance D_{ij} is zero if

$$S_{33}(3\cos^2\theta_3 - 1) = -\,(S_{11}-S_{22})\,(\cos^2\theta_1 - \cos^2\theta_2)\;. \qquad (48)$$

Then J_{ij}^{aniso} has the value

$$J_{ij}^{aniso} = \frac{1}{2}\,(3\cos^2\alpha-1)(S_{11}-S_{22})\cdot\frac{1}{3}[(J_{11}-J_{22})-2\Delta J\,\frac{(\cos^2\theta_1-\cos^2\theta_2)}{(3\cos^2\theta_3-1)}]. \tag{49}$$

In a plane ($\cos\theta_3 = 0$), D_{ij} disappears for

$$S_{11}/S_{22} = -\tan^2\theta_1\ . \tag{50}$$

Similarly, we can see that under certain circumstances J_{ij}^{aniso} is equal to zero. Actually the condition is,

$$S_{11}/S_{22} = -\,(J_{33}-J_{22})/(J_{33}-J_{11})\ . \tag{51}$$

From Eq. (47) it may also be seen that $J_{ij}^{aniso}/2D_{ij}$ is a meaningful quantity for the description of the relative importance of the J-anisotropy contribution only for cases where $S_{11} = S_{22}$, that is for rotational symmetry. In general, for $S_{11} \neq S_{22}$ the ratio may become zero or infinity.

If we want to determine a molecular structure we must of course try to make the necessary correction for the coupling anisotropy, that is we must measure J_{ij}^{aniso}. Unfortunately this proves to be rather difficult. The first problem which arises is the one of determinacy.

If a molecular structure is defined by *G* geometrical parameters and its symmetry requires the determination of *O* order parameters, then the *C* measured coupling constants should normally, that is without coupling anisotropy, fulfil the following inequality,

$$C > O + G\ . \tag{52}$$

If we include the $C\cdot O$ terms due to the coupling anisotropy, then we must have

$$C > G + O + C\cdot O\ . \tag{53}$$

This condition is generally not fulfilled and only a measurement in several (n) liquid crystals provides

$$n\cdot C > G + n\cdot O + C\cdot O$$

$$\text{i.e. } n(C-O) > G + C\cdot O\ , \tag{54}$$

which will be true for large enough n, if $C > 0$. On the other hand measurements in several different liquid crystals are only helpful if the molecular structure is not solvent dependent.

Further problems arise from the vibration corrections, which are performed on the D_{ij}^{exp} but should be applied to the D_{ij} exclusively. Also the elements J_{ij} must be assumed independent of the solvent. Finally, any correlation effects, particularly stemming from complex formation (see later) seriously affect the determination of the small J_{ij} contributions.

In conclusion we can say that the measurement of the J-tensor elements is generally very difficult but is the only basis for the correction of the D_{ij}^{exp}. The methods of determination are either the use of several liquid crystal solvents, or in principle the application of Eq. (49), although for $D_{ij} = 0$ there will be additional difficulties from vibration corrections.

The existence of anisotropic J-contributions to the D_{ij}^{exp} can be demonstrated in molecules with high symmetry, as for example hexafluorobenzene in that the measured couplings do not fulfil the required geometrical relations as for example that D(ortho)/D(para)=8 in a regular hexagon. Furthermore, large apparent solvent effects on the molecular structure can in principle be assigned to J-anisotropy. In molecules, finally, where the position of a nucleus is defined by more than one direct coupling, the large residual of couplings in the iterative process of geometry determination may be identified as anisotropy of the J.

The anisotropy of spin-spin coupling constants has been studied in approximately fifty different molecules of which forty contained the nuclear pairs F-F. The following orders of magnitude (J^{aniso}/D) have been derived either experimentally or theoretically (8)

H - H	$\sim 10^{-4}$	F - C	$\sim 10^{-2}$
C - H	$\sim 10^{-4}$ to 10^{-3}	F - F	$\sim 10^{-2}$ to 10^{-1}
N - H	$\sim 10^{-2}$	C - C	$\sim 10^{-2}$ to 10^{-1}
F - H	$\sim 10^{-3}$ to 10^{-2}	C - N	$\sim 10^{-2}$ to 10^{-1}

LIMITATION OF STRUCTURE PRECISION DUE TO CORRELATION

We have mentioned earlier that the separation of the average $<(3\cos^2\theta_2-1)\cdot r_{ij}^{-3}>$ into an inverse third power of the internuclear distance and a degree of order is an allowed operation only if there is no connection between r_{ij} and θ_{ij}. Quite generally, if the molecular shape depends upon the angles between the molecular axis system and the liquid crystal director, that is if we have "correlation", then the calculation of the average is possible only if we introduce this interdependence analytically. Obviously, this will in general be a very complex problem.

In order to study the various consequences of correlation we can formulate a particularly simple model in which we consider two nuclei on an axis at an angle θ with respect to the optic axis of the liquid crystal, which is also the magnetic field direction. The internuclear distance is assumed to depend upon θ as follows,

$$r = r_0(1 + f(\theta))$$

with

$$f(\theta) = \alpha\,[\cos(2\theta)-1] \tag{55}$$

so that for

$$\theta = 0°,\ 180°,\ r \text{ has the value } r_0$$

and for

$$\theta = 90°,\ \text{it is } r_0(1-2\alpha).$$

Then we find

$$\begin{aligned} r^{-3} &= r_0^{-3}(1-3f(\theta)) \\ D' &= -Kr_0^{-3}\langle P_2(\cos\theta)(1-3f(\theta))\rangle \\ &= -Kr_0^{-3}\{S-3\langle P_2(\cos\theta)\cdot\alpha[\cos(2\theta)-1]\rangle\} \end{aligned} \tag{56}$$

where $P_n(\theta)$ are the Legendre functions.

The resulting D' is

$$D' = -Kr_0^{-3}[(1+\tfrac{20}{7}\alpha)\ S- \tfrac{4}{5}\alpha - \tfrac{72}{35}\alpha\langle P_4(\cos\theta)\rangle + \text{higher terms}] \tag{57}$$

This "dipolar coupling" contains the normal dipolar coupling

($-Kr_0^{-3} \cdot S$) plus several terms (correlation terms) which are proportional to α, the amplitude of the variation of molecular shape. There is a mixed contribution proportional to $\alpha \cdot S$ and a terms proportional to α exclusively. This term is particularly interesting because it does not depend upon the orientation $S = S_2$ or $S_4 = \langle P_4(\cos\theta) \rangle$. It seems reasonable to assume that it should disappear with $S = 0$; i.e. with zero orientation. However, if the $S = 0$ originates in an average of positive and negative degrees of order because of a complex formation between solute and solvent (see later) then α remains finite. The factor is certainly small, being the relative change of bond length or bond angle as a function of the molecular orientation. On the other hand it may become important for small values of S. If for instance we assume $\alpha = 10^{-3}$ and $S = 10^{-2}$ then the relative contribution of the correlation term to the observed direct coupling is

$$-K\, r_0^{-3} \cdot \alpha / - K\, r_0^{-3} \cdot S = \alpha / S = 10^{-1} \; .$$

Neglecting correlation we make an error of 10% in a ratio of dipolar couplings, that is of 3% in a distance ratio.

The term proportional to $\langle P_4(\cos\theta) \rangle$ will of course depend upon the relative importance of this higher degree of order. A numerical study of this question reveals that if we describe the distribution of the oriented molecules with the following probability density

$$\omega(\theta) = A\, e^{-B \cos^2\theta} \tag{58}$$

then (S_4) is approximately equal to 1 to 10% of (S_2) and (S_6) is smaller than 1% of (S_2) if B is between 1 and 3. Correlation undoubtedly exists as can be inferred from the observed direct couplings in tetrahedral molecules. This phenomenon is discussed later.

VIBRATION - ROTATION COUPLING

In order to explain observed direct couplings in tetrahedral molecules such as methane, which for symmetry reasons should have all the D_{ij} equal to zero, the following potential of second order tensorial form describing the interaction between the liquid crystal solvent and the solute has been suggested (9)

$$V = -\tfrac{1}{2} F_{ij}\, \beta_{k\ell}\, (Q_m)\, \phi_{ik}(\alpha,\beta,\gamma)\, \phi_{j\ell}(\alpha,\beta,\gamma) \; . \tag{59}$$

Here F_{ij} is some tensorial property of the liquid crystal referred to the laboratory axes and $\beta_{k\ell}$ is a tensorial property of the solute molecules referred to body fixed axes, and which depends upon vibrational normal coordinates Q_m. The ϕ_{ik} are direction cosines between laboratory fixed i-axes and body fixed k-axes; they depend upon the angles which determine the orientation of the molecule. In Eq. (59) the summation convention over repeated indices is applied.

Only the anisotropic part of F_{ij} is considered since it is this which leads to deformation effects. Its average, $\overline{F}$, contains the non-zero elements $\overline{F}_{xx} = \overline{F}_{yy} = -\frac{1}{2}\overline{F}_{zz}$.

$$\overline{F} = -\tfrac{1}{3} G = -\tfrac{1}{3}(F_{\parallel} - F_{\perp}) \ . \tag{60}$$

$\beta_{ij}(Q_m)$ is expanded as,

$$\beta_{ij}(Q_m) = \beta_{ij}^{0} + \sum_m \beta_{ij}^{m} Q_m + \ldots \text{ with } \beta_{ij}^{m} = \frac{\partial \beta_{ij}}{\partial Q_m}\Big|_{Q_m = 0} \ . \tag{61}$$

Because V now contains both vibrational (Q_m) as well as rotational (α, β, γ) coordinates, then we certainly have correlations between these two motions. The authors conclude that "given a particular deformation, some orientations will be more likely than others" and give the name "vibration-rotation coupling" to this phenomenon.

For a more complete description a rotation-vibration wave function is introduced and the coupling is taken into account by first order perturbation theory. The resulting wave function is used to derive the direct couplings as follows:

$$D_{pq} = \frac{G}{10} \sum_{n'} \frac{1}{\omega'^{2}_{n}} \left(\beta^{n'}_{\alpha\beta} d^{n'}_{pq\alpha\beta} - \tfrac{1}{3}\, \beta^{n'}_{\alpha\alpha}\, d^{n'}_{pq\beta\beta}\right) \tag{62}$$

with ω'_n = angular frequency of the normal mode n'

$$d^{n'}_{pq\alpha\beta} = \frac{\partial}{\partial Q_m} \left(- \frac{\mu_0}{8\pi^2} \frac{\hbar \gamma_p \gamma_q}{r^3_{pq}} C^{pq}_{\alpha}\, C^{pq}_{\beta} \right) \Big|_{Q_m = 0} \tag{63}$$

C^{pq}_{α} = cosines of angle between molecule fixed axis α and internuclear axis pq.

Finally, the $\beta^{n'}_{ij}$ can be interpreted or identified as the derivatives of the electric polarisability tensor with respect to the normal mode vibration of the solute molecule.

By fitting the $\beta_{ij}^{n'}$ for the three contributing normal modes of CH_4 it is possible to reproduce the observed dipolar couplings.

MEAN-FIELD DEFORMATION

There is an alternative theoretical approach to correlation (10) which for CH_4 leads to the same result as the vibration-rotation interaction, but which is more transparent as far as the interpretation of its physical meaning is concerned.

The following hamiltonian of a vibrating molecule which is subject to an external liquid crystal mean field can be formulated in normal coordinates,

$$H = \tfrac{1}{2} \sum_k \dot{Q}_k^{\,2} + \tfrac{1}{2} \sum_k \omega_k^{\,2} Q_k^{\,2} + U(Q_k) \tag{64}$$

$U(Q_k)$ can be expanded

$$U(Q_k) = \sum_k U_k Q_k \text{ with } U_k = \frac{\partial U(Q_k)}{\partial Q_k} \Big|_{Q_k = 0} \tag{65}$$

so that

$$H_k = \tfrac{1}{2} \dot{Q}_k^{\,2} + \tfrac{1}{2} \omega_k^{\,2} Q_k^{\,2} + U_k Q_k \tag{66}$$

The last term can be identified as the dependence of internal upon external molecular coordinates or more clearly as the dependence of molecular shape upon its orientation (i.e. correlation).

The hamiltonian can be transformed to new variables

$$\xi_k = Q_k - \bar{Q}_k = Q_k + \frac{U_k}{\omega^2_k} \tag{67}$$

with ξ_k measuring the displacement of Q_k from the new equilibrium $\bar{Q}_k$.

We then have

$$D_{ij} = - P_2(\cos\alpha) \frac{\mu_0 \hbar \gamma_i \gamma_i}{8\pi^2} \sum_{\alpha\beta} \langle \phi_{\alpha\beta}^{ij} \, s_{\alpha\beta} \rangle \tag{68}$$

where

$$\phi_{\alpha\beta} = (\cos\theta_{ij\alpha} \cdot \cos\theta_{ij\beta})/r_{ij}^3$$

and $$s_{\alpha\beta} = \tfrac{1}{2}\,(3\cos\theta_{z'\alpha}\cos\theta_{z'\beta} - \delta_{\alpha\beta})$$

$\phi_{\alpha\beta}$ is the molecular geometry dependent term and $s_{\alpha\beta}$ is the instantaneous molecular orientation.

Expansion of $\phi^{ij}_{\alpha\beta}$ leads to:

$$\phi^{ij}_{\alpha\beta} = \phi^{eij}_{\alpha\beta} + \sum_k \phi^{ij}_{\alpha\beta,k}\,(\xi_k + \bar{Q}_k)$$

$$+ \tfrac{1}{2} \sum_{k,\ell} \phi^{ij}_{\alpha\beta,k\ell}(\xi_k\xi_\ell + \bar{Q}_\ell\xi_k + \bar{Q}_k\xi_\ell + \bar{Q}_k\bar{Q}_\ell) + \ldots \quad (69)$$

The quantum mechanical expectation value of $\phi^{ij}_{\alpha\beta}$ is:

$$< \psi\, |\phi^{ij}_{\alpha\beta}|\, \psi> = \phi^{eij}_{\alpha\beta} + \sum_k \phi^{ij}_{\alpha\beta,k}\bar{Q}_k + \tfrac{1}{2} \sum_{k\ell} \phi^{ij}_{\alpha\beta,k\ell}\bar{Q}_k\,\bar{Q}_\ell$$

$$+ \tfrac{1}{2} \sum_{k\ell} \phi_{\alpha\beta,k\ell} < \psi|\xi_k\xi_\ell|\,\psi> + \ldots \quad (70)$$

The first term is the equilibrium value, the last one is the well known harmonic vibration correction. The second and third terms are the new deformational (correlation) correction terms. From these we obtain

$$d^d_{ij} = P_2(\cos\alpha)\, \frac{\mu_0\hbar\gamma_i\gamma_j}{8\pi^2} \sum_{\alpha,\beta} \sum_k \frac{1}{\omega_k^2}\, \phi^{ij}_{\alpha\beta,k} <U_k s_{\alpha\beta}> \quad (71)$$

If the orienting potential is

$$U = -A \sum_{\gamma\delta} \alpha_{\gamma\delta}\, s_{\gamma\delta} \quad \text{, we have} \quad (72)$$

$$U_k = -A \sum_{\gamma\delta} \alpha_{\gamma\delta,k}\, s_{\gamma\delta} \quad ; \quad \alpha_{\gamma\delta,k} = \frac{\partial\alpha_{\gamma\delta}}{\partial Q_k}\Big|_{Q_k} = 0$$

and

$$d_{ij}^{d} = -P_2(\cos\alpha)\,\frac{\mu_0\hbar\gamma_i\gamma_j}{8\pi^2}\,A\sum_{\alpha\beta,\gamma\delta}\sum_{k}\frac{1}{\omega_k^{2}}\,\phi^{ij}_{\alpha\beta,k}\cdot\alpha_{\gamma\delta,k}\langle s_{\alpha\beta}s_{\gamma\delta}\rangle \tag{73}$$

Finally

$$d_{ij}^{d} = -P_2(\cos\alpha)\,\frac{\mu_0\hbar\gamma_i\gamma_j}{8\pi^2}\,A\sum_{k}\frac{1}{\omega_k^{2}}\,\{\tfrac{3}{10}\overline{(\phi_k^{ij}\alpha_k)} - \tfrac{1}{10}\,\bar{\phi}_k^{ij}\bar{\alpha}_k \tag{74}$$

$$+ \tfrac{2}{7}\sum_{\alpha\beta}\,[3(\phi_k^{ij}\alpha_k)_{\alpha\beta} - \phi^{ij}_{\alpha\beta,k}\,\bar{\alpha}_k - \bar{\phi}_k^{ij}\alpha_{\alpha\beta,k}]\;S_{\alpha\beta}\}$$

here $\bar{\phi}_k^{ij} = \mathrm{Tr}(\phi^{ij}_{\alpha\beta,k}) = \sum_{\gamma}\phi^{ij}_{\gamma\gamma,k}$ etc.

$$\text{and}\quad (\phi_k^{ij}\alpha_k)_{\alpha\beta} = \sum_{\gamma}\phi^{ij}_{\alpha\gamma,k}\,\alpha_{\gamma\beta,k} \tag{75}$$

For $S_{\alpha\beta} = 0$ we have the same formula (62) as derived for vibration-rotation coupling.

Application of relation (74) to a particularly simple case, H_2, shows that the molecule is elongated (for positive $\Delta\alpha_1$) when its axis is parallel with the liquid crystal director and shortened by one half of the elongation amplitude if it is perpendicular.

The angle dependence of the elongation is:

$$\varepsilon = \frac{1}{\sqrt{2m}}\,\frac{A}{\omega_1^{2}}\,\Delta\alpha_1\,\tfrac{1}{2}\,(3\cos^2\theta_{z'3} - 1) \tag{76}$$

with $\Delta\alpha_1 = \alpha_{\parallel}^{1} - \alpha_{\perp}^{1} = \frac{\partial}{\partial Q_1}(\alpha_{33}-\alpha_{11})_{Q_1=0} = \frac{\partial\Delta\alpha}{\partial Q_1}\Big|_{Q_1=0}$

The correlation terms is:

$$d^{d} = D^{e}[-\,\frac{6}{\sqrt{2m}}\cdot\frac{1}{\omega_1^{2}}\,\frac{\Delta\alpha_1}{\Delta\alpha}\cdot\frac{1}{r_e}\,kT\,(1 + \tfrac{5}{7}\,S + \ldots)] \tag{77}$$

$$(\Delta\alpha = \alpha_{\parallel} - \alpha_{\perp})$$

One can see that in this case, as in fact quite in general the deformation (correlation)-corrections are proportional to $1/r_e$ so that they decay with distance more slowly than the vibration corrections ($1/r_e^2$).

LIMITATION OF STRUCTURE PRECISION DUE TO COMPLEX FORMATION; THE TWO-SITE APPROXIMATION

A further possible reason for correlation between molecular orientation and shape is a specific interaction, such as complex formation. Choosing again a particularly simple model of a linear molecule in which we observe two direct couplings D_A and D_B we assume that there are two sites (e.g. free molecule (index 1) and complex (index 2)) for the solute with different degrees of order S_1 and S_2. Furthermore, if the geometry ($g = -k/r^3$) of the molecule is slightly site dependent

$$g_{A_2} = g_{A_1} (1 + \beta_A) \; ; \; \beta_A << 1$$

$$g_{B_2} = g_{B_1} (1 + \beta_B) \; ; \; \beta_B << 1$$

we can see that:

$$D_A = \rho S_1 g_{A_1} + (1-\rho) S_2 g_{A_2} = g_{A_1} [\rho S_1 + (1-\rho) S_2 (1 + \beta_A)] \quad (78)$$

$$D_B = \rho S_1 g_{B_1} + (1-\rho) S_2 g_{B_2} = g_{B_1} [\rho S_1 + (1-\rho) S_2 (1 + \beta_B)] \; .$$

Here ρ is the relative concentration at site 1 and $(1-\rho)$ at site 2. The molecular geometry is determined by the ratio D_B/D_A which, for $\beta_A = \beta_B$ or $\beta_A = \beta_B = 0$ is equal to g_{B1}/g_{A1}. For finite and different β's, however, the ratio may deviate considerably from g_{B1}/g_{A1}. Actually it is equal to zero for $\rho S_1 = -(1-\rho) S_2 (1+\beta_B)$ or to infinity for $\rho S_1 = -(1-\rho) S_2 (1+\beta_A)$. This peculiar behaviour arises because for $D_B = 0$, then D_A still has a residual value of $g_{A1} (1-\rho) S_2 (\beta_A - \beta_B)$, and for $D_A = 0$, then D_B is equal to $g_{B1} (1-\rho) S_2 (\beta_B - \beta_A)$; D_A or D_B can be zero only if S_1 and S_2 have opposite signs. If we want to find out how "far" from the points of singularity (0 or ∞) the apparent molecular geometry still deviates from the true geometry by α%, we can solve the equation

$$\frac{D_B(\alpha)}{D_A(\alpha)} = \frac{g_{B_1} [\rho S_1 + (1-\rho) S_2 (1+\beta_A)]}{g_{A_1} [\rho S_1 + (1-\rho) S_2 (1+\beta_B)]} = \frac{g_{B_1}}{g_{A_1}} [1 + \frac{\alpha}{100}] \; . \quad (79)$$

The result is that

$$\frac{D_B(\alpha)}{D_B(D_A=0)} = 1 + \frac{100}{\alpha} \text{ ; respectively } \frac{D_A(\alpha)}{D_A(D_B=0)} = - \frac{100}{\alpha} , \qquad (80)$$

and

$$D_B(\alpha) = g_{B_1} S_2 (1-\rho)(\beta_B-\beta_A)(1 + \frac{100}{\alpha})$$

$$D_A(\alpha) = g_{A_1} S_2 (1-\rho)(\beta_B-\beta_A)(\frac{100}{\alpha}) . \qquad (81)$$

The ratio of dipolar couplings has an error of α% (due to correlation) if the D_{ij} are approximately equal to $(100/\alpha)$ times the residuals ΔD_{ij}. With typical ΔD_{ij} values of several Hz for protons, we see that an error of 3% in a ratio of dipolar couplings, i.e. of 1% in a distance ratio, may still persist for values of D_{ij} of the order of 100 Hz. It is also possible to define an amplification factor M which indicates to what extent the true molecular deformation between the sites is amplified in the apparent deformation. The result is

$$M = \frac{\alpha}{100(\beta_B-\beta_A)} = \frac{S}{1 + S(1+\beta_A)} , \qquad (82)$$

with

$$S = \frac{(1-\rho)S_2}{\rho S_1} . \qquad (83)$$

For $S = -1$ the amplification M is equal to the inverse of the true deformation

$$M(S = -1) = 1/\beta_A , \qquad (84)$$

consequently the amplification factor can easily reach values of the order of 10^2.

A corresponding two site theory can also be developed for molecules with two order parameters. The results are quite similar to the ones given above.

A practical example of complex formation and corresponding apparent structure deformation has been observed for acetylene (11). It displayed an apparent variation of its structure by 30%

as a function of temperature. This behaviour could be explained with a true deformation of 1% and two sites of opposite signs for the order parameter.

A similar phenomenon was reproduced artificially for the molecule methylalcohol (12), which orients with order parameters of opposite signs in two different liquid crystals. In the mixture of these two solvents the methyl angle varied apparently by several degrees which could be attributed to a true variation of 0.3 degrees.

Finally in the molecule cyclobutadiene iron tricarbonyl the observed variation of the ratios of dipolar couplings in cyclobutadiene with the liquid crystal solvent (13) was attributed to complex formation. Here, the molecule was assumed to bond to a site which induces order parameters of opposite sign in the molecular plane which rotates and deforms slightly with the rotation. The deformation necessary to explain the observed variation of the ratio of D_{ij} from $2 \cdot \sqrt{2} = 2.828$ to 2.547 is of the order of 0.1% with an order parameter ratio of -1.2.

SUMMARY

NMR of oriented molecules is a very fast and efficient method of structure determination for small molecules (ten magnetic nuclei or less) or small parts of larger spin-systems. The precision which may easily be reached is of the order of 10^{-2} Å and 10^{-1} angular degrees. In many cases a higher precision is possible but not meaningful, as long as the solvent effects on the structure are not fully understood. As an example I cite thiophene (14) in mixtures of two liquid crystal solvents. (See Fig. 4).

In fact, the described NMR method, in contrast with other spectroscopies, deals with molecules which are not free but interact strongly with the liquid crystal solvent. At present it is not clear whether this interaction can in general be successfully described by a mean field or whether in some cases it must be looked at as a specific interaction with the "surface" of the solvent so that the picture of sites is more realistic. It is possible that the solute molecule spends most of its time in collisions with the solvent molecules from which it picks up orientation but is at the same time considerably distorted. In other words we are probably seeing the distorted molecules with their correlation between orientation and molecular shape.

On the other hand the correlation phenomenon seems to be a

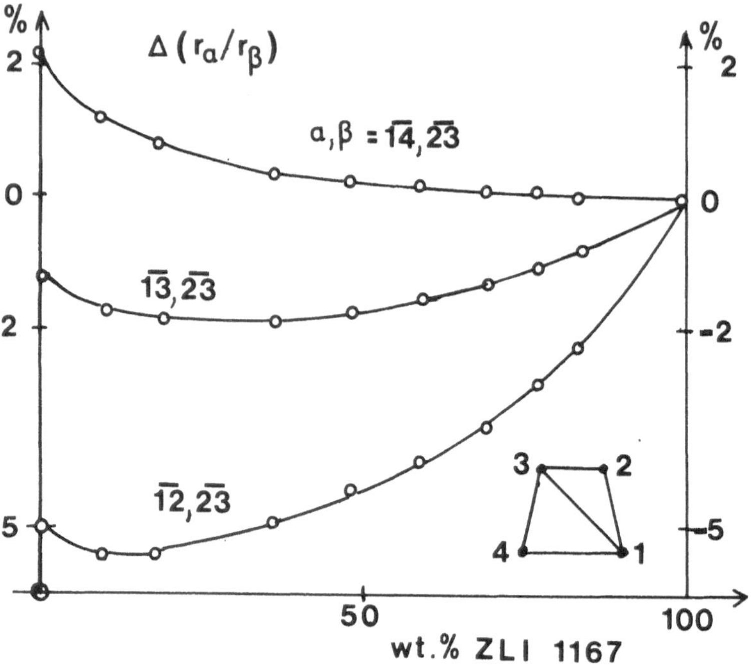

Figure 4. *Solvent effects on the structure of thiophene in the mixture (ZLI 1167, Phase IV).*

particularly interesting and rewarding subject for further detailed studies, and which will be difficult to analyze by any other spectroscopy.

REFERENCES

1. A. Saupe and G. Englert, *Phys. Rev. Lett.* **11**, 462 (1963). G. Englert and A. Saupe, *Z. Naturforsch.* **19a**, 172 (1964).

2. P. Diehl, P.M. Henrichs and W. Niederberger, *Mol. Phys.* **20**, 139 (1971).

3. J. Lounila and T. Väänänen, *Mol. Phys.* **49**, 859 (1983).

4. P. Diehl and C.L. Khetrapal in *"NMR Basic Principles and Progress"*, P. Diehl, E. Fluck, R. Kosfeld, Eds. Springer-Verlag, Heidelberg, (1969), Vol. 1.

- P. Diehl and P.M. Henrichs in *"NMR Specialist Periodical Reports"*; R.K. Harris, Ed. Chem. Soc. London (1972), Vol. 1.

- J. Bulthuis, C.W. Hilbers and C. MacLean, *MTP Int. Rev. Sci. Phys. Chem. Ser. 1*, 4, 201 (1972).

- P. Diehl and W. Niederberger in *"NMR Specialist Periodical Reports"*; R.K. Harris, Ed. Chem. Soc. London (1974), Vol. 3.

- L.C. Snyder and S. Meiboom in *"Critical Evaluation of Chemical and Physical Structural Information"*, D.R. Lide and M.A. Paul, Eds. National Academy of Sciences, Washington, D.C. (1974).

- J.W. Emsley and J.C. Lindon in *"NMR Spectroscopy Using Liquid Crystal Solvents"*, Pergamon Press, Oxford (1975).

- C.L. Khetrapal, A.C. Kunwar, A.S. Tracey and P. Diehl in *"NMR Basic Principles and Progress"*; P. Diehl, E. Fluck, R. Kosfeld, Eds. Springer-Verlag, Heidelberg (1975), Vol. 9.

- L. Lunazzi in *"Determination of Organic Structures by Physical Methods"*; F.C. Nachod, J.J. Zuckermann, E.W. Randall, Eds. Academic Press, New York, (1976), Vol. 6, Chapter 6, pp. 335-413.

- P. Diehl in *"NMR Specialist Periodical Reports"*; R.K. Harris, Ed. Chem. Soc. London (1976), Vol. 5.

- C.L. Khetrapal and A.C. Kunwar in *"Advances in Magnetic Resonance"*; J.S. Waugh, Ed. Academic Press, London, (1977), Vol. 9.

- C.L. Khetrapal and A.C. Kunwar in *"NMR Specialist Periodical Reports"*; R.J. Abraham, Ed. Chem. Soc. London, (1979), Vol. 8.

- C. Schumann in *"Handbook of Liquid Crystals"*; H. Kelker and R. Hatz, Verlag Chemie, Heidelberg, (1980), Chapter 10, pp. 426-509.

5. S. Sýkora, J. Vogt, H. Bösiger and P. Diehl, *J. Magn. Reson.* **36**, 53 (1979).

6. D. Canet, J.P. Marchal and J.P. Sarteaux, *C.R. Acad. Sci, Paris* **71**, 279 (1974).

7. P. Diehl, H. Bösiger and H. Zimmermann, *J. Magn. Reson.* **33**, 113 (1979).

8. J. Lounila and J. Jokisaari in *"Prog. N.M.R. Spectr."*; J.W. Emsley, J. Feeney, L.H. Sutcliffe, Eds. Pergamon Press,

Oxford, (1982), Vol. 15.

9. J.G. Snijders, C.A. de Lange and E.E. Burnell, *J. Chem. Phys.* **77**, 5386 (1982).

10. J. Lounila and P. Diehl, *J. Magn. Reson.* **56**, 254 (1984).

11. P. Diehl, S. Sýkora, W. Niederberger and E.E. Burnell, *J. Magn. Reson.* **14**, 260 (1974).

12. P. Diehl, M. Reinhold, A.S. Tracey and E. Wullschleger, *Mol. Phys.* **30**, 1781 (1975).

13. P. Diehl, F. Moia, H. Bösiger and J. Wirz, *J. Mol. Struct.* **98**, 297 (1983).

14. P. Diehl, J. Jokisaari, S. Müller and T. Väänänen, *Org. Magn. Reson.* **21**, 143 (1983).

ON THE ORIENTATION OF SMALL MOLECULES IN ANISOTROPIC SOLVENTS

C.A. de Lange*, J.G. Snijders* and E.E. Burnell**

* Free University,
Departments of Physical and Theoretical Chemistry,
De Boelelaan 1083,
1081 HV Amsterdam,
The Netherlands.

** University of British Columbia,
Department of Chemistry,
2036 Main Mall,
Vancouver (B.C.),
Canada V6T 1Y6.

INTRODUCTION

Although the description of the orientation of rigid molecules in an anisotropic environment is a straightforward problem, it has not been quite so easy to account for the effects of internal motion completely satisfactorily. For solutes with large-amplitude intramolecular motions the problem of how to treat their average orientation has given rise to some controversy. A first attempt to phrase the problem in terms of a kinetic approach in which the ratio of the relevant time scales for internal motion and reorientation plays an important role (1,2) was followed by a seemingly alternative description using equilibrium statistical mechanics (3). In a later paper it was shown that a more general theory encompassed both these approaches as special cases, hence demonstrating their essential equivalence (4). Such theories have led to at least a qualitative understanding of how to account for intramolecular motions. Despite these efforts, a detailed description of even small molecules showing small anisotropic couplings and with only small-amplitude internal vibrations is still lacking. A case in point are molecules which in their equilibrium conformation possess tetrahedral or higher symmetry. Of course,

J. W. Emsley (ed.), Nuclear Magnetic Resonance of Liquid Crystals, 181–205.

under the assumption of rigidity such molecules, when dissolved in an isotropic phase, should show no dipolar or quadrupolar splittings in their NMR spectra. However, in most studies on molecules in this category anisotropic splittings have been observed. They were interpreted correctly from the beginning as arising from some sort of interplay between molecular vibrations and solute reorientation (5,6).

To approach the problem of vibration-rotation interaction in a more detailed and quantitative fashion, a study of a number of simple molecules whose electronic structure and vibrational force fields are well known, seems indicated. For this purpose the molecules hydrogen, methane and their deuteriated analogues would appear eminently suitable. The physical properties of these compounds are extremely well characterized and they are expected to be chemically virtually inert towards most liquid-crystal solvents.

In the following we shall describe a general theory for the orientation of small molecules dissolved in liquid-crystal solvents. Recent experimental results for the hydrogens and methanes in a number of thermotropic phases will be discussed in terms of this theory. Finally some remarks will be made about the actual physical mechanisms which are responsible for the interaction between solute and solvent.

In this chapter we shall make no attempt to discuss all the finer points of the theory in great detail, but we shall concentrate instead on the highlights of a research programme which has been going on for a number of years now. For a thorough discussion of various more subtle intricacies of the theoretical and experimental aspects the interested reader is referred back to the original papers. A full account of our work on the hydrogens is given in (7). An extensive experimental study of methane and its deuteriated analogues dissolved in various nematic phases (8) has been followed by a theoretical treatment of the dipolar (9) and quadrupolar (10,11) anisotropic couplings in these molecules. In (12) a number of interesting conclusions regarding the orienting mechanisms of small molecules in liquid-crystal solvents have been reached. Finally, a review article in which the theory is summarized, extended and applied to all the available experimental results has appeared in the literature at the end of 1983 (13).

THEORY

General Treatment

In order to describe orientational effects of small solute

molecules in liquid-crystal environments in a quantitative fashion, a model is needed for the interaction between solute and solvent. The simple model we shall consider here pictures the liquid-crystal environment as providing an average second order mean field tensor, F_{ij}, that interacts with some second order tensorial property, β_{ij}, of the solute molecules. The tensor β_{ij} is supposed to be determined by the electronic structure and hence the geometry of the molecule and therefore depends on its vibrational (normal) coordinates Q_m or, equivalently, its symmetry coordinates S_k which are related to the normal coordinates through

$$S_k = \sum_m L_{km} Q_m. \tag{1}$$

Many actual interactions are of this kind, (e.g. electric field-polarizability, electric field gradient-quadrupole moment), but much of what follows will not depend on the actual physical basis of the interaction, so for the sake of generality we will not specify the interaction further at this point, but will come back to it in a later section. Let us assume for simplicity that the liquid crystal, in NMR experiments, has cylindrical symmetry around the external magnetic field z-axis. The tensor F_{ij} should then reflect this symmetry. Hence, the interaction can be written, ignoring the isotropic part of F_{ij}, as

$$V = -\tfrac{1}{2} F_{ij}\beta_{ij}(S_m) = -\tfrac{1}{3} G\, \beta_{k\ell}(S_m) S_{k\ell}(\Omega), \tag{2}$$

with $G = F_{\parallel}-F_{\perp}$. (Einstein convention of summation over repeated tensor indices). Here we have transformed the tensor from laboratory fixed i,j axes to molecule fixed k,ℓ axes. The orientational operator $S_{k\ell}$ depending on the orientation Ω, is defined as:

$$S_{k\ell}(\Omega) = \tfrac{3}{2}\, \Phi_{kz}(\Omega)\Phi_{\ell z}(\Omega) - \tfrac{1}{2}\delta_{k\ell} \tag{3}$$

where the Φ_{kz} are the direction cosines between the molecule fixed k axes and the laboratory fixed z-axis.

Note that the interaction V depends on both the orientation (i.e. rotational coordinates) and geometry (i.e. vibrational coordinates) of the solute molecule. Hence, it will induce a correlation between these two modes of motion. In the following we shall treat V as a perturbation on the zeroth order problem, for which we will take the harmonic approximation for the vibrational problem, and the rigid rotor for the rotational degree of freedom. The zeroth order wavefunction will then be a simple product of harmonic oscillator wavefunctions for each normal mode and a rigid rotor wavefunction.

The observables that we shall be concerned with here, such as dipolar and quadrupolar couplings which can be measured by NMR techniques, have the following general form:

$$A(S_m,\Omega) = a_{k\ell}(S_m)S_{k\ell}(\Omega) \tag{4}$$

i.e. they depend on both the rotational and vibrational coordinates and have the above multiplicative structure. In particular we shall be interested in the thermodynamic expectation value of the operator A:

$$A(T) = \sum_n P_n(T)\langle n|A(S_m,\Omega)|n\rangle \tag{5}$$

$$P_n(T) = e^{-E_n/kT}/\sum_j e^{-E_j/kT}$$

where n labels the rotational-vibrational quantum states and P_n is the familiar Boltzmann-factor. If we neglect vibrational excitation at temperature T, then without the perturbation of Eq. (2) we find

$$A^{(0)}(T) = \langle 0|a_{k\ell}(S_m)|0\rangle \sum_j P_j^{(0)}(T)\langle J|S_{k\ell}(\Omega)|J\rangle \tag{6}$$

where $|0\rangle$ is the vibrational groundstate and J is a shorthand notation for all rotational quantum numbers which label the unperturbed rotational states. It can easily be shown that the sum in Eq. (6) vanishes for a rigid rotor so that for a freely rotating and vibrating molecule (or for one interacting with its environment in an isotropic way) observables of form given by Eq. (4) will have a vanishing expectation value. In order to get a finite contribution one has to take into account the change in the rotational-vibrational wavefunction induced by the anisotropic interaction of Eq. (2). By standard perturbation theory we have to first order:

$$|0J\rangle^{(0+1)} = |0J\rangle^{(0)} + \sum_{\substack{n\neq 0 \\ J'\neq J}} \frac{|nJ'\rangle\langle nJ'|V|0J\rangle}{E_{0J}^{(0)}-E_{nJ'}^{(0)}} \tag{7}$$

where the sum is over all vibrational and rotational excitations.

We now develop the $\beta_{k\ell}(S_m)$ tensor into a Taylor series around the equilibrium ($S_m = 0$) geometry, and assume it is justified to truncate after the linear term:

$$\beta_{k\ell}(S_m) = \beta_{k\ell}^{eq} + \sum_m \beta_{k\ell}^{m} S_m + \ldots \tag{8}$$

with

$$\beta_{k\ell}^{eq} = \beta_{k\ell}(S_m = 0) \quad \text{and} \quad \beta_{k\ell}^{m} \equiv \frac{\partial \beta_{k\ell}}{\partial S_m}\bigg|_{S_m = 0}$$

Similarly we develop $a_{ij}(S_m)$ into a Taylor series around the equilibrium geometry and retain only the first two terms:

$$a_{ij}(S_m) = a_{ij}^{eq} + \sum_m a_{ij}^{m} S_m + \ldots \tag{9}$$

with a notation as in Eq. (8). Using these Taylor expansions and the well-known properties of the harmonic oscillator wavefunctions we obtain after some algebra:

$$A(T) = A^{rig}(T) + A^{nrig}(T) \tag{10a}$$

$$A^{rig}(T) = \sum_J P_J^{(1)}(T)\, a_{ij}^{eq} \langle J|S_{ij}|J\rangle - \tfrac{2}{3} G\, a_{ij}^{eq} \beta_{k\ell}^{eq} \sum_{\substack{J,J' \\ J' \neq J}} P_J^{(0)}(T) \frac{\langle J|S_{ij}|J'\rangle\langle J'|S_{k\ell}|J\rangle}{E_J^{(0)} - E_{J'}^{(0)}} \tag{10b}$$

$$A^{nrig}(T) = -\tfrac{2}{3} G \sum_{\substack{m \\ J,J'}} P_J^{(0)}(T) \frac{\hbar}{2\omega_m} a_{ij}^{m} \beta_{k\ell}^{m} \frac{\langle J|S_{ij}|J'\rangle\langle J'|S_{k\ell}|J\rangle}{E_J^{(0)} - E_{J'}^{(0)} - \hbar\omega_m} . \tag{10c}$$

We have separated A(T) into two contributions, the first of which would already be present in a perfectly rigid rotor, while the second only arises because the molecule can vibrate.

The matrix elements which occur in Eq. (10) can be easily calculated in the spherical and symmetrical top case. Often, however, a classical treatment of the rotational degree of freedom is justified and leads to considerable simplification:

$$A^{rig}(T) = \frac{G}{10kT} \left(- \tfrac{1}{3}\, a_{ii}^{eq} \beta_{kk}^{eq} + a_{ij}^{eq} \beta_{ij}^{eq}\right) \tag{11}$$

Similarly, if we neglect the rotational energy differences in the non-rigid part (10c) we can perform the J' summation using the closure relation for the rotational wavefunctions and obtain

$$A^{nrig}(T) = \frac{1}{10}\, G \sum_{k,\ell} F_{k\ell}^{-1} \left(\frac{\partial a_{ij}}{\partial S_k} \frac{\partial \beta_{ij}}{\partial S_\ell} - \tfrac{1}{3} \frac{\partial a_{ii}}{\partial S_k} \frac{\partial \beta_{jj}}{\partial S_\ell}\right) \tag{12}$$

where $F_{k\ell}$ is the isotopically invariant harmonic force matrix related to the vibrational frequencies ω_m by

$$\sum_{k\ell} L_{km}\, F_{k\ell}\, L_{\ell m} = \omega_m^{\ 2}. \tag{13}$$

The derivatives with respect to symmetry coordinates in Eq. (12) are all taken at the equilibrium geometry. Note that while the rigid contribution of Eq. (11) depends explicitly on temperature, the non-rigid contribution given in Eq. (12) does not. Of course both contributions depend on temperature through the liquid-crystal field anisotropy G, which is determined by the motion of the liquid-crystal molecules themselves.

In the following the isotope dependence of the quantities $A^{rig}(T)$ and $A^{nrig}(T)$ will be important. For $A^{nrig}(T)$ its isotope dependence hinges on that of $\partial a/\partial S$ and $\partial \beta/\partial S$, where a and β are referred to body fixed axes. Even though the relation between the coordinates S and internal coordinates can be taken to be isotope invariant, nevertheless, $\partial a/\partial S$ and $\partial \beta/\partial S$ can vary with isotope substitution. This is because the definition of body fixed axes depends on the Eckart conditions of zero angular momentum which in turn are related to the isotope dependent moments of inertia. However, the $\partial a/\partial S$ and $\partial \beta/\partial S$ for different isotopically substituted molecules differ at most by a rotation of the equilibrium tensors a_{ij}^{eq} and β_{ij}^{eq}. Therefore if one of these tensors is isotropic, the corresponding derivatives are isotope invariant.

Dipolar Couplings

The dipolar couplings which can be measured by NMR studies of molecules dissolved in liquid crystals can be expressed as the temperature and quantum averages of an operator D_{pq}, where p and q label the two nuclei which are coupled:

$$D_{pq}(S_m,\Omega) = -\frac{\hbar}{2\pi}\frac{\gamma_p\gamma_q}{r_{pq}^3(S_m)}\Phi_{pq,k}(S_m)\Phi_{pq,\ell}(S_m)\,S_{k\ell}(\Omega). \tag{14}$$

Here γ_p,γ_q are the gyromagnetic ratios of the nuclei, r_{pq} their internuclear distance and $\Phi_{pq,k}$ the cosine of the angle between the internuclear p-q direction and the molecule fixed k-axis. $S_{k\ell}$ is the orientation operator. Since the operator D_{pq} has the general form of Eq. (4), the theory of the previous section applies with

$$a_{k\ell}(S_m) = -\frac{\hbar}{2\pi}\frac{\gamma_p\gamma_q}{r_{pq}^3}\Phi_{pq,k}\Phi_{pq,\ell} \tag{15}$$

The required parameters $a_{k\ell}^{eq}$ and $\partial a_{k\ell}/\partial S_m$ can easily be calculated from a knowledge of the geometry of the molecule under consideration. Note that the tensor $a_{k\ell}^{eq}$ is not isotropic, so that $\partial a/\partial S$ will be isotope dependent due to the Eckart conditions.

Quadrupolar Couplings

Molecules which contain nuclei with $I > \frac{1}{2}$ when dissolved in liquid crystals show spectra with quadrupolar couplings. These couplings can be expressed as temperature and quantum averages of an operator B_μ where μ labels the nuclei in the molecule (14):

$$B_\mu(S_m,\Omega) = \tfrac{1}{2}\frac{eQ_\mu}{h}V_{k\ell}^{\mu}(S_m)S_{k\ell}(\Omega) \tag{16}$$

Here eQ_μ stands for the nuclear quadrupole moment of nucleus μ and $V_{k\ell}^{\mu}$ is minus the intramolecular electric field gradient tensor at the position of nucleus μ. Again this operator has the general form of Eq. (4) and the present theory is applicable with the geometry-dependent factor

$$b_{k\ell}(S_m) = \tfrac{1}{2}\frac{eQ_\mu}{h}V_{k\ell}^{\mu}(S_m). \tag{17}$$

The parameters $b_{k\ell}^{eq}$ and $\partial b_{k\ell}/\partial S_m$ cannot be obtained simply from the positions of the nuclei in the molecule alone. The electrons contribute significantly to the intramolecular electric field gradients and a proper determination of these quantities requires an electronic structure calculation. We note again that the tensor $b_{k\ell}^{eq}$ is not isotropic, and therefore the quantities $\partial b/\partial S$ will be isotope dependent.

THE HYDROGEN MOLECULE

Because the hydrogen molecule has axial symmetry around the internuclear z' axis, the β tensor possesses only two independent elements $\beta_{\parallel}(=\beta_{z'z'})$ and $\beta_{\perp}(=\beta_{x'x'}=\beta_{y'y'})$. Assuming a rigid structure then the geometrical part of the dipolar coupling operator of Eq. (15) takes the simple form

$$a^{eq}_{z'z'} = -\frac{\hbar}{2\pi}\frac{\gamma_p\gamma_q}{r_{eq}^{3}} \; ; \; a^{eq}_{ij} = 0 \text{ for } i,j \neq z'. \tag{18}$$

Using these assumptions we obtain for the observed dipolar couplings from Eq. (10b):

$$D^{rig}_{pq}(T) = a^{eq}_{z'z'} \langle S_{z'z'}\rangle_T \; , \tag{19}$$

with

$$\langle S_{z'z'}\rangle_T = \frac{G(\beta_{\parallel}-\beta_{\perp})}{3kT} \sum_J P_J^{(0)}(T)\langle J|S_{z'z'}|J\rangle^2$$

$$-\tfrac{2}{3}\, G\,(\beta_{\parallel}-\beta_{\perp}) \sum_{\substack{J,J' \\ J\neq J'}} P_J^{(0)}(T)\, \frac{\langle J|S_{z'z'}|J'\rangle\langle J'|S_{z'z'}|J\rangle}{E_J^{(0)} - E_{J'}^{(0)}} \, . \tag{20}$$

Since the summations over J and J' can be evaluated from the well-known unperturbed rigid rotor wavefunctions and energies, the only remaining quantity on the right-hand side of (20) is $G(\beta_{\parallel}-\beta_{\perp})$. This quantity which has the dimension of energy, is unknown a priori because it depends on the specification of the precise ordering mechanism. In the following we shall treat it as an adjustable parameter to be obtained from a fit to experiment.

At this point it is important to realize the consequences of the Pauli principle in homonuclear diatomics such as H_2 whose nuclei are fermions and D_2 whose nuclei are bosons. The Pauli principle places restrictions on the total wavefunction and therefore H_2 occurs in two different modifications; one with total nuclear spin I = 1 which combines only with odd-numbered rotational states J (ortho-H_2), the other with total spin I = 0 (and hence no NMR spectrum) in combination with even-numbered J states (para-H_2). For D_2 a similar distinction exists between ortho-D_2 (total nuclear spin I = 0 or 2, with even J states) and para-D_2 (total nuclear spin I = 1, with odd J states). For HD there is no similar restriction on the spin and rotational states of the molecule. As a consequence, the J summations in

Eq. (20) have to be carried out over the proper set of J values, depending on the molecule under consideration. Moreover, the dependence of $E_J^{(0)}$ on the moments of inertia of the species means that the observed dipolar couplings for the members of the hydrogen series are expected to show an isotope dependence beyond the trivial effect of gyromagnetic ratios. When the quantum mechanical description of the rotational levels is replaced by a classical averaging over the rotational degree of freedom, then Eq. (11) reduces to:

$$D_{pq}^{c\ell}(T) = \frac{G(\beta_{\parallel}-\beta_{\perp})}{15\ kT}\, a_{z'z'}^{eq} \ . \qquad (21)$$

The equilibrium bond lengths and electronic structures of the hydrogens are identical and therefore this classical average is isotope independent and hence predicts identical dipolar couplings for the different species (apart from the effect of the γ's).

From the experimental dipolar couplings (scaled to H_2 values) a remarkable isotope effect is evident. In all liquid-crystal solvents used the major effect of substituting a hydrogen in our probe molecules by a deuterium is to increase the observed (scaled) dipolar coupling by about 7%. Apparently the classical rotational averaging of Eq. (21) cannot describe the experimental results, and a quantum mechanical averaging over the rotational levels as in Eq. (20) is required. According to Eq. (19) and Eq. (20), one parameter $G(\beta_{\parallel}-\beta_{\perp})$ should suffice to describe the dipolar couplings in all the hydrogens. This single parameter $G(\beta_{\parallel}-\beta_{\perp})$ can be obtained at different temperatures and in various liquid crystals from a least squares fitting procedure. The orientation parameters $\langle S_{z'z'}\rangle_T$ can thus be calculated from Eq. (20) and are presented graphically in Fig. (1).

From this figure it is clear that the experimental results can be reproduced satisfactorily. Apparently the isotope effect on the dipolar couplings in the hydrogens in liquid-crystal solvents is a purely quantum mechanical effect which persists at room temperature. For the hydrogens, which are unusual molecules in the sense that their rotational levels are so widely spaced, classical averaging over the rotational degree of freedom (see Fig. (1)) is clearly too crude an approximation. A point worth stressing is that the parameter $G(\beta_{\parallel}-\beta_{\perp})$ does not show the same sign in different liquid-crystal solvents. An understanding of sign and magnitude of $G(\beta_{\parallel}-\beta_{\perp})$ would require a physical model for the solute-solvent interaction (see later section) which may differ in different solvents, as well as a detailed description of the solvents themselves.

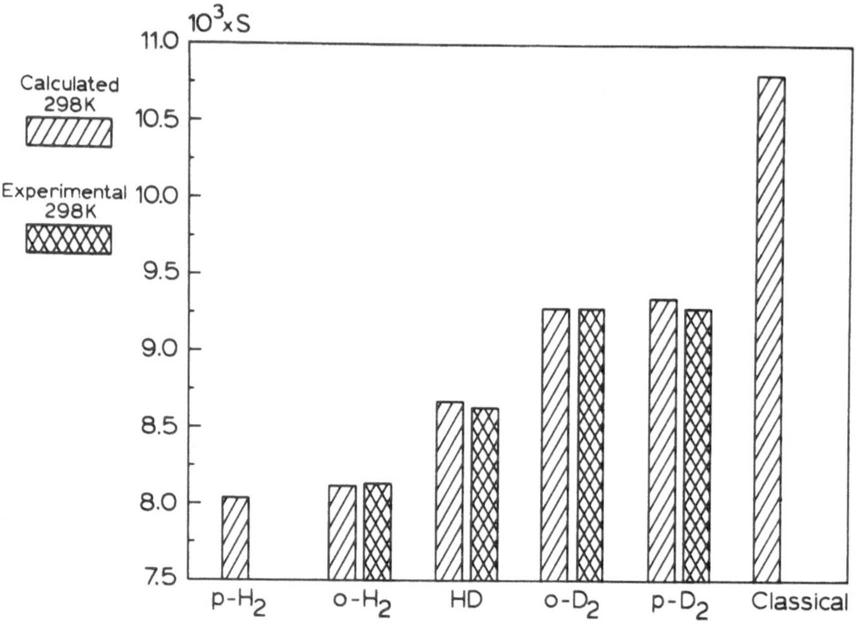

Figure 1. *Orientation parameters of the isotopically substituted hydrogen molecules dissolved in the liquid crystal solvent 1132.*

Deuterium quadrupolar couplings have also been observed for the molecules HD and ortho- and para-D_2 dissolved in various liquid crystals. Within experimental error the quadrupolar couplings measured for ortho- and para-D_2 have been found to be identical in the temperature range studied. The assumption of rigidity and the axial symmetry of the hydrogens result in one independent component $V_{\parallel}^{eq} = V_{z'z'}^{D}$ $(= -2V_{x'x'}^{D} = -2V_{y'y'}^{D})$. Under these conditions Eq. (16) simplifies to:

$$B_D^{rig} = 3eV_{\parallel}^{eq}\, Q_D \langle S_{z'z'} \rangle_T / h \; . \tag{22}$$

Since $\langle S_{z'z'} \rangle_T$ is already known from the dipolar analysis, the only additional parameter needed in Eq. (22) is $V_{\parallel}^{eq}$. This quantity $V_{\parallel}^{eq}$ has been determined for hydrogen using virtually exact wavefunctions (15). Combining the results of these highly correlated calculations with quadrupolar couplings from molecular beam magnetic resonance experiments (16) an accurate value of $Q_D = 2.860 \times 10^{-27}$ cm^2 is obtained (15), leading to

$$eV_{\|}^{eq}\ Q_D/h = 226.000 \text{ kHz}\ . \qquad (23)$$

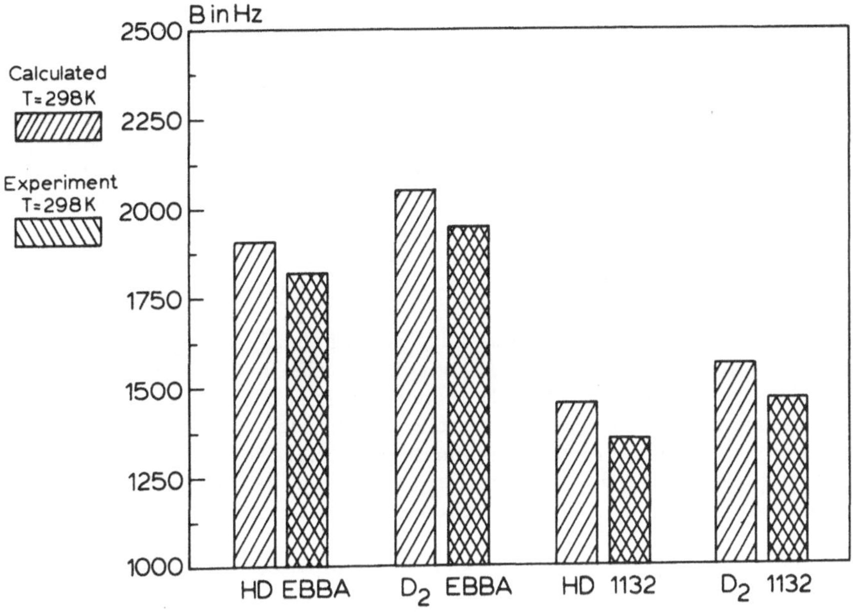

Figure 2. *Observed and calculated absolute values of the quadrupolar couplings in HD and D_2.*

Using this result the quadrupolar couplings B_D can be calculated directly from Eq. (22). In Fig. (2) the experimental quadrupolar couplings and those calculated as described above are compared graphically. It appears that considerable discrepancies far beyond experimental error show up. It is also noteworthy that for HD and D_2 dissolved in one liquid-crystal solvent at one temperature these deviations are approximately constant. This important realization will provide the clue to an explanation of these discrepancies. We shall come back to this point later.

THE METHANE MOLECULE

When methane and its deuteriated analogues are dissolved in liquid-crystal solvents, the NMR spectra of the solutes show dipolar couplings despite the fact that all these molecules possess an *equilibrium* tetrahedral configuration. Let us first consider the symmetrically substituted species CH_4 and CD_4 which have an *average* tetrahedral structure as well. In purely tetrahedral systems the tensor β_{ij}^{eq} is isotropic by symmetry, leading to a zero rigid contribution from Eq. (10b). Therefore in these two species the observed dipolar couplings arise from non-rigid effects alone, through a coupling of molecular vibrations and rotations.

A fully tetrahedral molecule possesses four normal vibrational modes, viz. a symmetric breathing mode of A_1 symmetry, a bending mode of E symmetry, and two modes of F_2 symmetry, one predominantly bending, one stretching. The A_1 mode preserves tetrahedral symmetry and does not play a role in a vibration-rotation coupling mechanism, as is evident from Eq. (10c). The remaining three modes are expected to participate in the non-rigid terms given in Eq. (10c). Unlike the case of the hydrogens, the methanes have rotational levels which are relatively close together. Although in the hydrogens classical averaging over the rotational degree of freedom turned out to be an unwarranted approximation, the classical approach for the methanes is much more reasonable. Therefore, the dipolar couplings in CH_4 and CD_4 can be expressed as in Eq. (12) when the definition of $a_{k\ell}(S_m)$ is taken from Eq. (15).

The quantities $\partial a_{ij}/\partial S$ and $\partial \beta_{ij}/\partial S_\ell$ merit closer attention. The former can be calculated from the known methane equilibrium geometry and are isotope dependent. The quantities $\partial \beta_{ij}/\beta S_\ell$ are derivatives of the methane β tensor elements with respect to the symmetry modes. It is natural to assume that the β tensor is entirely determined by the electronic structure of the molecule at the geometry for which we need its value. Hence, it is independent of the masses of the various nuclei in the molecule, and it should therefore be identical for different isotopically substituted molecules at the same geometry when referred to a bond-axis fixed coordinate system. Moreover, for a tetrahedral structure the β^{eq} tensor is isotropic, leading to isotope independent $\partial \beta_{ij}/\partial S_\ell$ even in a (moment of inertia dependent) body fixed axis system. It can be shown from group theory that there exists only one independent $\partial \beta_{ij}/\partial S_\ell$ element for each symmetry mode. The A_1 mode cannot contribute to the mechanism which causes the dipolar splittings and therefore the dipolar couplings which arise from the non-rigid effect are completely determined by the molecular geometry and force field and three independent parameters $G\beta_2$, $G\beta_3$ and $G\beta_4$, one for each relevant symmetry mode.

We now focus our attention on the remaining molecules of the methane series, viz. CH_3D, CH_2D_2 and CHD_3. Since the quantities $\partial\beta/\partial S$ are isotope independent, the non-rigid effect in these unsymmetrically substituted methanes depends on precisely the same parameters $G\beta_2$, $G\beta_3$ and $G\beta_4$ as found for CH_4 and CD_4. However, the effect of vibrational anharmonicity means that these molecules do not possess a fully tetrahedral average structure, and rigid contributions to the dipolar couplings should be considered as well. In the non-tetrahedral average configuration the β-tensor is no longer isotropic. The deviations from isotropy can be expressed as:

$$\beta_{ij}^{av} = \beta_{ij}^{eq} + \sum_m \frac{\partial\beta_{ij}}{\partial Q_m}\bigg|_{Q_m = 0} \langle Q_m \rangle \quad +\ldots \tag{24}$$

Here, the Q_m signify the normal modes of the unsymmetrically substituted methane. The derivatives $\partial\beta_{ij}/\partial Q_m$ are again directly related to our three independent parameters $G\beta_2$, $G\beta_3$ and $G\beta_4$ through the L-matrix transforming normal modes of the unsymmetrically substituted methane into symmetry modes of the fully tetrahedral species. The $\langle Q_m \rangle$ are the average values of the normal coordinates in the vibrational ground state. In view of the small deviations from tetrahedral symmetry, truncation of the series after the second term should be justified. In a purely harmonic potential, all the $\langle Q_m \rangle$ would of course vanish, leaving only the isotropic β_{ij}^{eq} part and hence no orientation due to a rigid effect. In an anharmonic potential, however, this is no longer the case, and those Q_m which belong to a totally symmetric irreducible representation of the point group of the substituted molecule can acquire a non-zero value. Since for methane the cubic anharmonic force field is known in detail (17), the quantities $\langle Q_m \rangle$ can be calculated. Thus, for the non-symmetrically substituted methanes in addition to the non-rigid effect (Eq. (12)), a rigid contribution (compare Eq. (11))

$$D_{pq}^{rig}(T) = \frac{G}{10kT}\left(a_{ij}^{eq}\,\beta_{ij}^{av} - \tfrac{1}{3}\,a_{ii}^{eq}\,\beta_{kk}^{av}\right) \tag{25}$$

is obtained. The label *eq* again indicates that the geometrical factors are to be calculated at the equilibrium geometry.

In summary, the dipolar couplings in methane and its deuteriated analogues can be calculated from a knowledge of three parameters $G\beta_2$, $G\beta_3$ and $G\beta_4$, together with the methane geometry and harmonic and cubic anharmonic force field. In the symmetrically substituted methanes only contributions due to the non-rigid effect are obtained, whereas in the non-symmetrically substituted compounds a balance between rigid and non-rigid contributions is expected.

The compounds $^{13}CH_4$, $^{13}CH_3D$, CH_2D_2, CHD_3 and CD_4 have been studied thoroughly in a number of liquid-crystal solvents and different dipolar couplings in molecules of the series have been determined in each case. For each liquid crystal the theoretical expressions for the dipolar couplings have been fitted to the available experimental values by least squares adjustment of the three parameters $G\beta_2$, $G\beta_3$ and $G\beta_4$. The agreement between theoretical and experimental dipolar couplings is excellent. As an illustration the dipolar HH, HD and DD couplings (scaled to HH with the appropriate γ-ratio) are represented graphically in Fig. (3) (in EBBA) and Fig. (4) (in 1132). The rigid and non-rigid contributions to the total dipolar couplings are given as well.

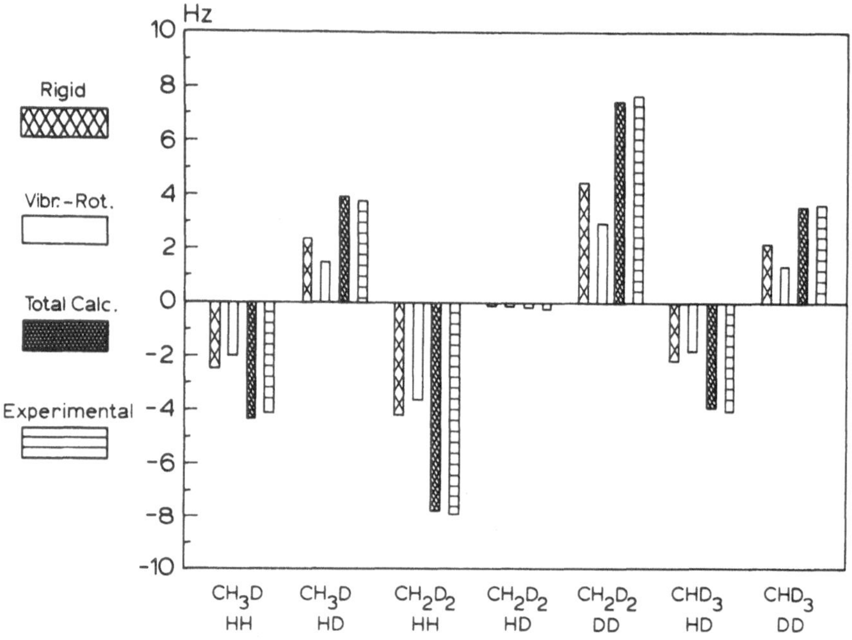

Figure 3. *Observed and calculated scaled dipolar couplings for isotopically substituted methanes dissolved in the liquid crystalline solvent 1132.*

The ^2H NMR spectra of the deuteriated methanes dissolved in liquid crystals 1132 and EBBA are dominated by deuterium quadrupolar splittings as shown in Fig. (5). As in the case of the dipolar couplings, the quadrupolar couplings are in general also composed of a rigid and a non-rigid part, according to equations (25), (12) and (17). Clearly the electric field gradient tensor $-V^{\mu}_{k\ell}$ plays an important role.

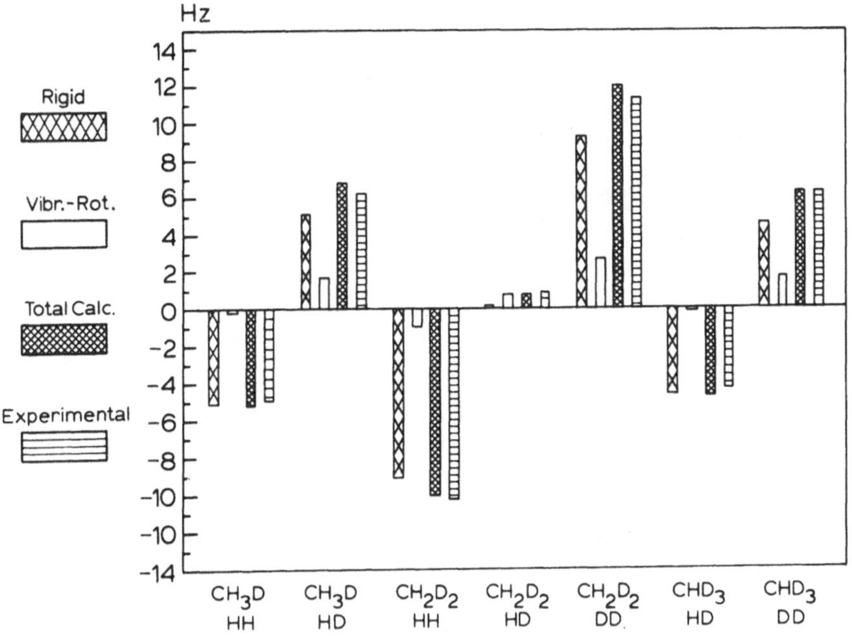

Figure 4. *Observed and calculated scaled dipolar couplings for isotopically substituted methanes dissolved in the liquid crystalline solvent EBBA.*

In the tetrahedral equilibrium geometry the local C_{3v} symmetry means that $V^{\mu}_{k\ell}$ is diagonal and cylindrical in any coordinate system that has one of its axes (the z' axis) along the C-D_μ bond and can therefore written as

$$V^{\mu eq}_{k\ell} = V^{eq}_{\parallel} \left(\tfrac{3}{2}\Phi^{eq}_{z'k}\Phi^{eq}_{z'\ell} - \tfrac{1}{2}\delta_{k\ell}\right) , \tag{26}$$

with $V^{eq}_{\parallel} = V^{eq}_{z'z'}$. The $\Phi^{eq}_{z'k}$ are the cosines of the angles between the C-D_μ bond axis and the molecule fixed k axis, at the equilibrium geometry. At non-equilibrium geometries the field gradient tensor will in general no longer have this form, since the local symmetry at nucleus μ will be lower than C_{3v}. However, ab initio electronic structure calculations (11) show that, to an excellent approximation, the field gradient tensor nevertheless remains diagonal along the bond axis and retains

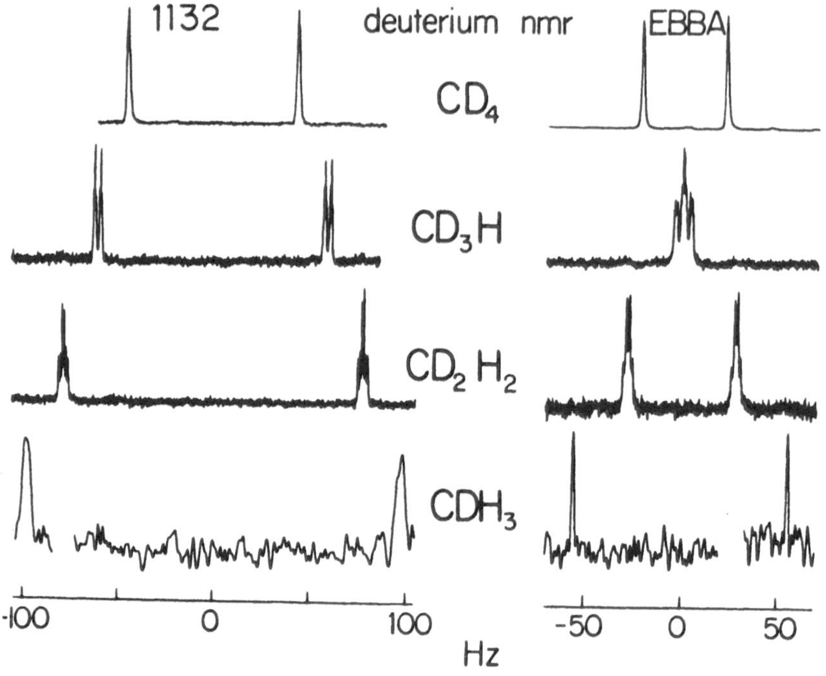

Figure 5. *Deuteron spectra of the isotopically substituted methanes dissolved in liquid crystalline solvents.*

its cylindrical symmetry around this axis, even at moderate distortions from equilibrium geometries. Inspection of equations (25), (17) and (26) shows that the rigid contribution is proportional to the quantity $V_{\parallel}^{eq}$. Moreover, the rigid effect is clearly isotope dependent since it is non-zero for the unsymmetrically deuteriated methanes and vanishes for CH_4 and CD_4.

For the non-rigid contribution we note that only the first term in Eq. (12) survives in the quadrupolar case because V_{ij}^{μ} is traceless. The non-rigid contribution is therefore proportional to $\partial V_{k\ell}^{\mu}/\partial Sm$, and the properties of this quantity completely determine the isotope dependence of the non-rigid effect. Of course $V_{\parallel}$ and $\phi_{z'k}$ do vary with geometry. For the derivatives of the field gradient with respect to symmetry modes we can therefore write:

$$\partial V^{\mu}_{k\ell}/\partial S_m = \frac{3}{2} V^{eq}_{\parallel} \left(\frac{\partial \Phi_{z'k}}{\partial S_m} \Phi^{eq}_{z'\ell} + \Phi^{eq}_{z'k} \frac{\partial \Phi_{z'\ell}}{\partial S_m}\right) +$$

$$+ \frac{\partial V_{\parallel}}{\partial S_m} \left(\frac{3}{2} \Phi^{eq}_{z'k} \Phi^{eq}_{z'\ell} - \frac{1}{2} \delta_{k\ell}\right) \tag{27}$$

where all derivatives are taken at the equilibrium geometry. In this equation the second term is isotope invariant, since it depends only on the equilibrium geometry and on the $\partial V_{\parallel}/\partial S_m$, which are also isotope invariant because they are electronic properties referred to a bond axis, not to a body fixed (moment of inertia dependent) axis system. The first term, however, is not isotope invariant since, apart from the isotope invariant equilibrium field gradient $V^{eq}_{\parallel}$, it depends on the direction cosine derivatives $\partial \Phi_{z'k}/\partial S_m$, which do vary with isotope substitution because of their dependence on the molecule fixed axis system.

In summary, the quadrupolar couplings can be described as the sum of two types of terms. One type is isotope dependent and is proportional to $V^{eq}_{\parallel}$, whereas the other type is isotope independent, and proportional to $\partial V_{\parallel}/\partial S_k$ evaluated at the equilibrium geometry.

The quadrupolar couplings can now be calculated if the quantities $V^{eq}_{\parallel}$ and $\partial V_{\parallel}/\partial S_k$ are known, since all other ingredients are available from the methane force field and geometry, or from the analysis of the dipolar couplings ($G\beta_i$). Moreover, electronic structure calculations (11) show that $\partial V_{\parallel}/\partial S_k$ is completely negligible for the E and F_2 bend modes ($V_{\parallel}$ is almost completely determined by the C-D bond length, irrespective of where the other hydrogens are). The A_1 stretch mode does of course not contribute. We are therefore left with only two parameters, $V^{eq}_{\parallel}$ and $\partial V_{\parallel}/\partial S(F_s)$, the derivative of the electric field gradient with respect to the F_2 stretch symmetry mode.

These two parameters $V^{eq}_{\parallel}$ and $\partial V_{\parallel}/\partial S(F_s)$ can now be fitted by a least squares procedure to the experimental quadrupolar couplings measured for CH_3D, CH_2D_2, CHD_3 and CD_4 dissolved in a variety of liquid crystals. The quadrupolar couplings are reproduced quite accurately when this procedure is followed (Fig. (6)). Ideally, the $V^{eq}_{\parallel}$ and $\partial V_{\parallel}/\partial S(F_s)$ obtained in this manner should not depend on the liquid crystal, since these parameters are supposed to be intrinsic properties of the methane solute. From Fig. (7) it is apparent that this condition of liquid crystal independence is well satisfied for the equilibrium field gradient $-V^{eq}_{\parallel}$. These field gradients of ~185

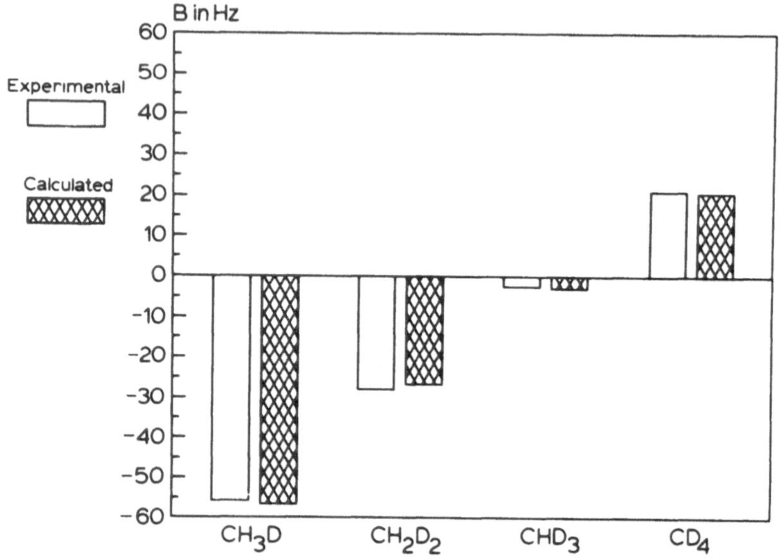

Figure 6. *Deuterium quadrupolar couplings for isotopically substituted methanes dissolved in the liquid crystalline solvent EBBA.*

kHz compare very well with those obtained from electronic structure calculations on methane, and are typical of the values measured in other, non-tetrahedral, hydrocarbons for a deuterium nucleus bonded to a carbon atom. On the other hand, the fit parameters $\partial V_{\|}/\partial S(F_s)$ are far from liquid-crystal independent. In fact, variations of more than an order of magnitude in this parameter occur throughout the series of solvents used (See Fig. (8)). This state of affairs is clearly unacceptable and shows that in the above theory some essential ingredient is missing. Somehow there must be an additional source of quadrupolar coupling that depends on the liquid crystal. Inclusion of such an extra contribution to the quadrupolar couplings should remove the inconsistencies in the values obtained for $\partial V_{\|}/\partial S(F_s)$ without at the same time sacrificing the excellent results for $V_{\|}^{eq}$. Since the quantity $V_{\|}^{eq}$ is contained in an isotope dependent term while $\partial V_{\|}/\partial S(F_s)$ is incorporated in a contribution which is isotope invariant, it is to be expected that an additional mechanism which would rectify the present unsatisfactory situation would have to be isotope invariant. Inclusion of such a mechanism would leave the fit values of $V_{\parallel}^{eq}$ unaffected while the electric field gradient derivatives $\partial V_{\parallel}/\partial S(F_s)$ would be changed. A mechanism with precisely these properties will be proposed as an additional source of quadrupolar coupling in the next section.

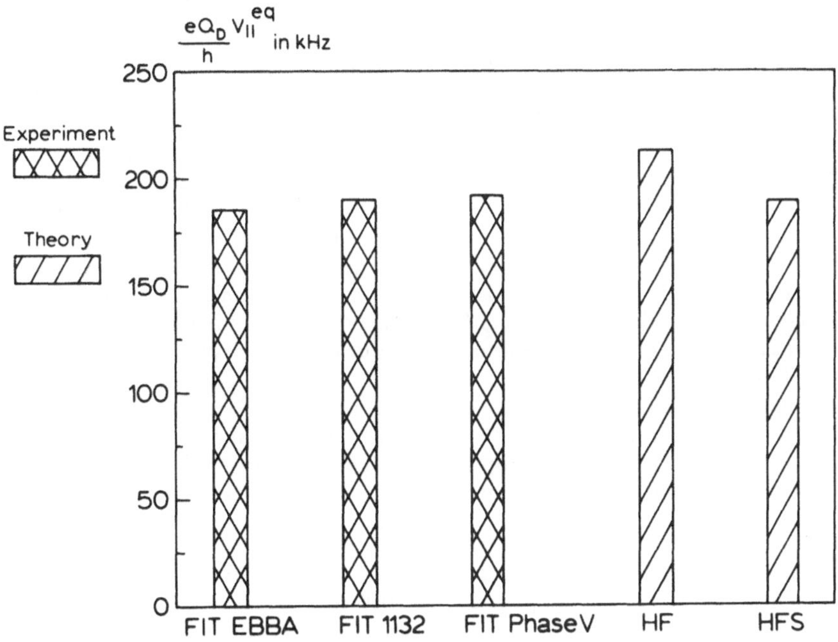

Figure 7. *Electric field gradients in methanes.*

EXTERNAL FIELD GRADIENTS

In the previous section we have established the need for an extra isotope independent mechanism to account for the quadrupolar couplings in the methanes. We also recall that in the hydrogen case we found a discrepancy between calculated and experimental quadrupolar couplings, which was approximately isotope invariant as well. It is therefore tempting to assume that both deviations from the theory presented above have a common cause. The mechanism we propose is the interaction of the nuclear quadrupole moments with an average external field gradient $-\bar{V}_{ij}^{ext}$ which is due to the liquid crystal environment. The assumed cylindrical symmetry of the liquid crystal around the axis of the magnetic field (z-axis) means that the only independent component of (minus) the external electric field gradient is $\bar{V}_{zz}^{ext}(=-2\ \bar{V}_{xx}^{ext} =-2\ \bar{V}_{yy}^{ext})$. The interaction of the nuclear quadrupole moment with the external field gradient gives rise to an additional contribution to the quadrupolar coupling given by:

$$B_D^{ext} = 3e\bar{V}_{zz}^{ext}Q_D/4h \ . \tag{28}$$

Note that this interaction is independent of the orientation of the solute and therefore also isotope independent; in fact it depends only on the liquid crystal and not on the solute. Moreover, if we assume that the methanes experience the same average field gradient as the hydrogens when dissolved in a particular liquid crystal at equal temperatures, then the additional contribution to the quadrupolar coupling B_D^{ext} should be identical whether a deuterium nucleus is bonded in hydrogen or in methane. Therefore, although we cannot calculate B_D^{ext} since $\bar{V}_{zz}^{ext}$ is a priori unknown, we can transfer the B_D^{ext} found for the hydrogens as the difference between theory and experiment, to the quadrupolar couplings in the methanes. We can then recalculate the fit to the methane quadrupole couplings, taking this extra contribution B_D^{ext} into account. As noted earlier, due to the isotope invariance of B_D^{ext}, the quality of the fit and the optimal negative internal equilibrium field gradient $V_{\|}^{eq}$ remain unchanged, while new values are now obtained for the internal field gradient derivatives $\partial V_{\|}/\partial S(F_s)$.

In Fig. (8) the values for $\partial V_{\|}/\partial S(F_s)$ obtained from the least squares fit are compared without and with inclusion of the external field gradient mechanism. It is seen that through adding this extra mechanism the quantity $\partial V_{\|}/\partial S(F_s)$ now varies much less with liquid crystal as can be expected of an intrinsic molecular property. For the liquid crystals EBBA and Phase V the parameters are rather close, while for 1132 a somewhat larger value is obtained. It is noteworthy that the original deviation of more than an order of magnitude for 1132 has been substantially reduced. The remaining discrepancies could be caused by a slightly different average external field gradient experienced by the methanes as compared to the hydrogens, for example through a different sampling of the intermolecular solvent regions by the two solutes. In Fig. (8) we also give the field gradient derivatives as calculated by two different electronic structure calculations (Hartree-Fock and Hartree-Fock-Slater). Although complete agreement between these calculations and the present fit to experiment is not obtained, it is nevertheless pleasing that these theoretical values for $\partial V_{\|}/\partial S(F_s)$ are of the same order of magnitude as the field gradient derivatives obtained from experiment, thus confirming the plausibility of the proposed additional quadrupolar coupling mechanism.

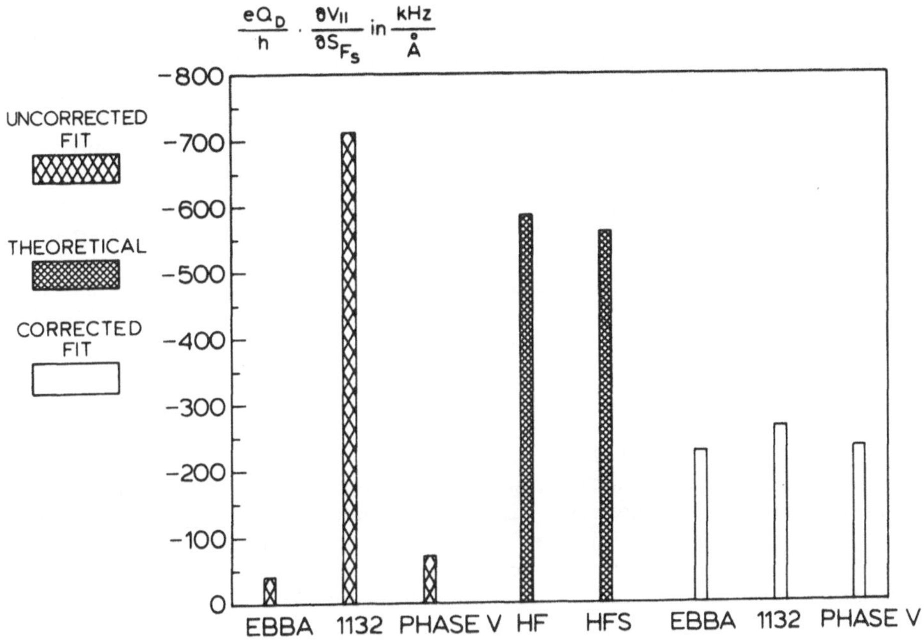

Figure 8. *Methane electric field gradients in the absence (uncorrected) and presence (corrected) of external electric field gradients.*

When the difference between experimental and theoretical quadrupolar couplings in the hydrogens is taken for B_D^{ext} and when $Q_D = 2.860 \times 10^{-27}$ cm^2 is used, values for $\bar{V}_{zz}^{ext}$ in different liquid-crystal solvents and at different temperatures can be obtained from Eq. (28).

SOLUTE-SOLVENT INTERACTION MECHANISMS

From the foregoing it has become clear that the molecular theory of orientation which assumes a second-order tensorial interaction mechanism between solute and solvent has been very successful in explaining the experimental results obtained for the hydrogens and methanes. Although the theory does not require anything more specific about the mechanism than that it takes a second-order tensor form, a more detailed understanding of the solute-solvent interaction is still a primary goal. When the interaction is assumed to be electric in nature, the following qualify as possible mechanisms: (i) the coupling between the anisotropy in the solute polarizability and the anisotropy in the mean squared electric field in the solvent, and (ii) the coupling between the molecular quadrupole moment and the anisotropy in the solvent electric field gradient. Interactions via the solute electric dipole moment and hyperpolarizabilities do not have the required second-order tensorial form. Of course, the present molecular theory hinges on the implicit assumption that the entire solute molecule experiences the same solvent "field". However, the success of a general second-order tensorial mechanism in describing the experimental results obtained until now seems well established. In an attempt to be more specific about the actual orienting mechanism we shall consider the available evidence.

When the experimental quadrupolar splittings in the hydrogens and the methanes are considered independently, discrepancies appear to be present in either case. As shown above, these discrepancies, which are seemingly independent, can all be removed at the same time by introducing an electric field gradient anisotropy in the liquid-crystal solvent, thus doing away with the need for ad hoc rationalizations. However, when a sizable electric field gradient anisotropy is present, its coupling with the solute molecular quadrupole moment must be considered as an orienting mechanism. For the hydrogens, where the rigid effect dominates, the importance of this orienting mechanism can now be estimated. Defining the molecular electric quadrupole moment (18) as

$$Q_{k\ell} = \tfrac{1}{2} \sum_i e_i \, (3r_{ki} \, r_{\ell i} - r_i^2 \, \delta_{k\ell}) \tag{29}$$

we obtain for a cylindrically symmetric solvent environment and an axially symmetric solute molecule the following interaction potential:

$$V = \tfrac{1}{2} \, \bar{V}_{zz}^{ext} \, Q_{z'z'} \, S_{z'z'} \, . \tag{30}$$

A method to estimate $\bar{V}_{zz}^{ext}$ has been discussed in the previous section, and the H_2 molecular quadrupole moment is known: $Q_{z'z'} = 0.662 \times 10^{-26}$ esu.cm^2 (19). In the case of the rigid hydrogens Eq. (2) of the general theory reduces to

$$V = -\tfrac{1}{3} \, G \, (\beta_{\parallel} - \beta_{\perp}) S_{z'z'} \tag{31}$$

In Fig. (9) the values of $-\frac{3}{2} \, \bar{V}_{zz}^{ext} \, Q_{z'z'}$ and those of $G(\beta_{\parallel} - \beta_{\perp})$ obtained from the least squares fits to the dipolar couplings of the hydrogens in a number of liquid-crystal solvents are compared. Apparently, in all liquid-crystal solvents used the molecular quadrupole moment mechanism for the orientation of the hydrogens can be held responsible for an appreciable contribution to the overall orientation.

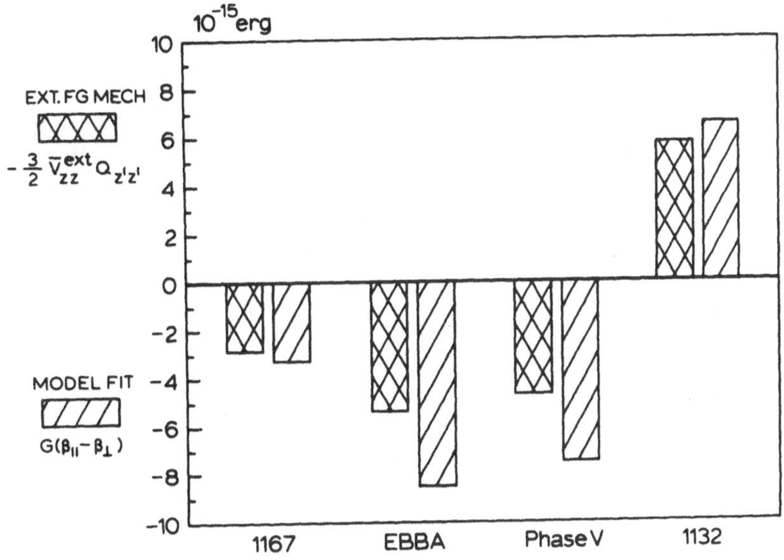

Figure 9. *Interaction between an external electric field gradient and the molecular quadrupole moment as an orienting mechanism for the hydrogen molecule dissolved in liquid crystalline solvents.*

Although there are indications that the interaction between external electric field gradient and molecular quadrupole moment may also be important in other molecules, the situation is less clear-cut than in the case of the hydrogens. It may be expected that in the majority of cases not a single orientation mechanism will dominate but that, on the contrary, various interaction terms will contribute to some extent. The problem is a tantalizing one, but before the matter can be settled the scarce experimental evidence available to date will have to be amplified. What is clearly needed is an extension of the previous experiments to a series of molecules whose physical properties (e.g. electric dipole moments, quadrupole moments, polarizabilities) are accurately known. A systematic study of such systems and an interpretation of the experimental results in terms of the general molecular theory of orientation could lead to a better understanding of the solute-solvent interactions which are operative. Until that time the discussion on orienting mechanisms is doomed to remain somewhat speculative.

ACKNOWLEDGEMENTS

The authors are grateful to Drs. E.J. Baerends and G.N. Patey for stimulating discussions, and to the North Atlantic Treaty organization for financial support (CAdL, EEB, grant nr. 1954).

REFERENCES

1. E.E. Burnell and C.A. de Lange, *J. Magn. Res.* **39**, 461 (1980).

2. E.E. Burnell and C.A. de Lange, *Chem. Phys. Letters* **76**, 268 (1980).

3. J.W. Emsley and G.R. Luckhurst, *Mol. Phys.* **41**, 19 (1980).

4. E.E. Burnell, C.A. de Lange and O.G. Mouritsen, *J. Magn. Res.* **50**, 188 (1982).

5. L.C. Snyder and S. Meiboom, *J. Chem. Phys.* **44**, 4057 (1966).

6. D. Bailey, A.D. Buckingham, F. Fujiwara and L.W. Reeves, *J. Magn. Res.* **18**, 344 (1975).

7. E.E. Burnell, C.A. de Lange and J.G. Snijders, *Phys. Rev.* A **25**, 2339 (1982).

8. E.E. Burnell and C.A. de Lange, *J. Chem. Phys.* **76**, 3474 (1982).

9. J.G. Snijders, C.A. de Lange and E.E. Burnell, *J. Chem. Phys.* **77**, 5386 (1982).

10. J.G. Snijders, C.A. de Lange and E.E. Burnell, *J. Chem. Phys.* **79**, 2964 (1983).

11. J.G. Snijders, W. van der Meer, E.J. Baerends and C.A. de Lange, *J. Chem. Phys.* **79**, 2970 (1983).

12. G.N. Patey, E.E. Burnell, J.G. Snijders and C.A. de Lange, *Chem. Phys. Letters* **99**, 271 (1983).

13. J.G. Snijders, C.A. de Lange and E.E. Burnell, *Israel J. of Chem.* **23**, 269 (1983).

14. P. Diehl and C.L. Khetrapal, *NMR Basic Principles and Progress*, Vol. 1, Springer, New York, 1969; C.L. Khetrapal, A.C. Kunwar, A.S. Tracey and P. Diehl, *NMR Basic Principles and Progress*, Vol. 9, Springer, New York, 1975.

15. R.V. Reid and M.L. Vaida, *Phys. Rev.* A **7**, 1841 (1973); R.V. Reid and M.L. Vaida, *Phys. Rev. Lett.* **34**, 1064 E (1975); D.M. Bishop and L.M. Cheung, *Phys. Rev.* A **20**, 381 (1979).

16. R.F. Code and N.F. Ramsey, *Phys. Rev.* A **4**, 1945 (1971).

17. D.L. Gray and A.G. Robiette, *Mol. Phys.* **37**, 1901 (1979).

18. A.D. Buckingham, *Quarterly Reviews* **8**, 183 (1959).

19. D.E. Stogryn and A.P. Stogryn, *Mol. Phys.* **11**, 371 (1966).

DIFFUSIVE AND COLLECTIVE MOTIONS IN LIQUID CRYSTAL PHASES

Giorgio Moro, Ulderico Segre and Pier Luigi Nordio

Institute of Physical Chemistry,
University of Padua, Italy.

MOLECULAR MOTIONS IN LIQUID CRYSTALS

In a liquid crystal phase several kinds of motion take place simultaneously, some of them occurring essentially at the molecular level (rotations, translations, conformational changes) some others involving large numbers of molecules (collective order fluctuations). These last motions are typical of the liquid crystalline states of the matter, but can manifest themselves also in the isotropic phase near to the isotropic-liquid crystal phase transition, giving rise to characteristic pretransitional phenomena.

All motions reflect the shape of the instantaneous potential fields determined on each molecules by the surroundings, and a knowledge of the details of the dynamics provides information on the dominant molecular interactions, the intensity and durations of forces and torques, and the persistence of inertial effects. From this point of view, however, collective motions are less informative than the molecular motions, because the macroscopic elastic constants and viscosities, in terms of which collective fluctuations are described, are very indirectly related to molecular properties. On the other hand, models for the roto-translational diffusion, which have been well tested for isotropic liquids, can be extended to anisotropic phases, although with difficulties which increase with the structural complexity of the phases under investigation; the detailed description of these motions is very sensitive to the nature of the anisotropic forces responsible for the organization of the liquid crystal structures.

J. W. Emsley (ed.), Nuclear Magnetic Resonance of Liquid Crystals, 207–230.

The aim of this chapter is therefore to present the theoretical models which describe the dynamical processes in anisotropic liquid phases, in order to provide a tool capable of analysing the appropriate experiments and which gives answers to the following questions:

i) interpretation of the molecular dynamics in terms of brownian diffusion (sequences of small displacements), jump diffusion (random jumps of arbitrary size) or inertial motions;

ii) the relative importance of collective effects;

iii) an estimate of the couplings between different motions;

iv) information on anisotropic potentials from the dynamical processes;

v) characterization of the dominant processes at the observation frequency characteristic of NMR experiments.

MOTIONAL PROCESSES AND SPIN RELAXATION

Molecular motions have the effect of modulating the anisotropic magnetic interactions, producing a random time-dependent perturbation which ultimately induces spin relaxation (1). This effect is generally described by the so-called "stochastic Liouville equation" (2). Let X denote the stochastic variables, i.e. the variables which are randomly modulated by the dynamical processes (e.g., the molecular orientations), and which are characterized by a probability distribution P(X). We then introduce a spin-density operator $\sigma(X,t)$ in terms of which the macroscopic magnetic observables are defined as:

$$M_\alpha(t) = N \gamma_e \hbar \, \mathrm{Tr}\, S_\alpha \int dX \, \sigma(X,t) \tag{1}$$

S_α being the quantum-mechanical spin operator associated with the physical property M_α. The spin-density operator obeys the Liouville equation:

$$(\partial/\partial t)\, \sigma(X,t) = [\, -\, iH(X)^\times + R_X]\, \sigma(X,t) \tag{2}$$

where $H^\times$ denotes commutation with the Hamiltonian of the spin system, and R_X is an operator governing the time evolution of the stochastic variables. The determination of the actual form of R_X is the objective of this chapter, but at the moment we shall consider it to be known. The initial condition for $\sigma(X,t)$ is given by the Maxwell-Boltzmann distribution:

$$\sigma(0) = P(X) \exp [- H(0) / kT] / Z , \qquad (3)$$

where H(0) is the Hamiltonian for the system at the initial time t = 0.

The formal solution for the Liouville equation is

$$\sigma(X,t) = \exp\{[- i H(X)^{\times} + R_X]t\} \sigma(0) \qquad (4)$$

or, after taking the Fourier-Laplace transform:

$$\sigma(X, \omega) = \{i[\omega - H(X)^{\times}] + R_X\}^{-1} \sigma(0) \qquad (5)$$

In most NMR experiments, the characteristic relaxation times for the spin states are much longer than those for the "lattice" variables X. This situation is referred to as the fast motional limit. Under these conditions the right-hand side of Eq. (5) can be expanded but with the expansion truncated at the second-order term. In this way, the usual Redfield expressions (1, 3) for the spin relaxation times are obtained. Since the spin Hamiltonian can always be expressed in terms of spherical tensors, in the form (1a):

$$H = \sum_{\mu} \sum_{j,m} T_{\mu}^{(j,m)^*} S_{\mu}^{(j,m)} \qquad (6)$$

where $S_{\mu}^{(j,m)}$ are spin operators in the laboratory frame and $T_{\mu}^{(j,m)}$ are related to specific magnetic interactions (chemical shielding tensor, dipolar or quadrupolar interactions), the Redfield solutions of the spin relaxation problem are usually given in terms of the spectral density functions $J_m^{j\mu\nu}(\omega)$, Fourier-Laplace transforms of the correlation functions of the perturbing Hamiltonian:

$$J_m^{j\mu\nu}(\omega) = \mathrm{Re} \int_0^{\infty} dt\, e^{-i\omega t}\, \overline{T_{\mu}^{(j,m)^*}(t)\, T_{\nu}^{(j,m)}} \qquad (7)$$

CORRELATION FUNCTIONS

In the interpretation of spectroscopic lineshapes or the relaxation of macroscopic observables, one almost invariably makes use of the concept of a correlation function. Let the functions $f_p(X)$ and $f_q(X)$ be elements of a complete, orthonormal set. The correlation function for a random stationary process X = X(t) is defined as:

$$g_{pq}(t) = \overline{f_p(t)^* \, f_q(0)}$$

$$= \iint dXdX_0 \; f_p(X)^* \; f_q(X_0) \; P(X_0) \; P(X_0;\, X,t). \tag{8}$$

In this expression, $P(X_0)$ is the equilibrium distribution and $P(X_0;\, X,t)$ the conditional probability, which is assumed to obey a kinetic equation of the type

$$(\partial/\partial t) \; P(X_0;X,t) = RP(X_0;X,t), \tag{9}$$

subjected to the initial conditions

$$P(X_0;X,0) = \delta(X - X_0), \tag{10}$$

with R an evolution operator acting on the random variable X, such that RP(X)=0. A formal solution of Eq. (9) is:

$$P(X_0;X,t) = e^{Rt} \; \delta(X - X_0). \tag{11}$$

Substitution into Eq. (8) gives:

$$g_{pq}(t) = \int dX_0 \; (e^{R^\dagger t} \; f_p(X_0))^* \; P(X_0) \; f_q(X_0), \tag{12}$$

or, in a bra-ket notation:

$$g_{pq}(t) = \langle f_p \mid e^{Rt} \; P \mid f_q \rangle \tag{12'}$$

Comparison of the Eqs. (8) and (12) shows that $f_p(t)$ obeys the equation of motion:

$$f_p(t) = e^{R^\dagger t} \; f_p(0) \tag{13}$$

with $R^\dagger$ the adjoint of R. This result is analogous to that found in quantum-mechanics, showing that the time-dependence of distribution functions (or density matrices) and dynamical variables are governed by evolution operators which are the adjoints one of the other.

We may also derive a kinetic equation for the correlation function itself:

$$(\partial/\partial t)\ g_{pq}(t) = \langle f_p \mid R\ e^{Rt}\ P \mid f_q \rangle$$

$$= \sum_k \langle f_p \mid R \mid f_k \rangle \langle f_k \mid e^{Rt}\ P \mid f_q \rangle$$

$$= \sum_k R_{pk}\ g_{kq}\ (t) \tag{14}$$

or, in terms of the correlation tensor $\underset{\sim}{g}(t)$ of elements $g_{pq}(t)$:

$$(\partial/\partial t)\ \underset{\sim}{g}(t) = \underset{\sim}{R}\cdot\underset{\sim}{g}(t). \tag{14'}$$

From the above definitions, it follows

$$R^{\dagger} = P^{-1}\ R\ P. \tag{15}$$

PROJECTION OPERATOR TECHNIQUES

In the following, we shall often make use of "reduction" or "projection" procedures, which allow us to concentrate our attention on some relevant properties of a system, taking into account only in an averaged way all the interacting surroundings, or "bath". Examples of this procedure are commonly encountered when calculating static properties by using single particle distribution functions; here, however, we wish to derive dynamical equations for a "test" particle, in a liquid crystal phase, the effects of all other interacting particles being treated statistically. In somewhat different contexts, we are also interested in measurable properties of a system, where some variables change so quickly that they have been averaged away in the lapse of time characteristic of the measure, or simply ignored for lack of knowledge. In these cases, we might speak of the "relevant" part of the information, as opposed to its "irrelevant" complement. The mathematical treatment of the problem requires the concept of projection operators, as introduced in molecular dynamics by Mori and Zwanzig (4, 5), and which we shall outline here following the formulation of Hynes and Deutch (6).

Let the dynamical properties of a system be completely defined in terms of functions F(t), in the multidimensional space $\{X^N\} \equiv \{X_0,\ X_1 \ldots X_N\}$, where X_0 is the "system" or "relevant" variable, and the other are generically attributed to the "bath".

The functions F obey the equation of motion

$$(\partial/\partial t)\ F(t) = i\ L\ F(t) \tag{16a}$$

$$F(t) = \exp\ (i\ L\ t)\ F(0) \tag{16b}$$

with L the appropriate time-evolution operator which for a classical system is given by the Poisson bracket with the Hamiltonian. The complete structure of F may not be accessible to our knowledge, or we may not be interested in it; we then introduce a projection operator P, that we leave unspecified at this stage but which should enable us to extract the dependence of F $\{X^N\}$ only upon the relevant variables X_0.

The only property required for P is to be idempotent, i.e. $P^2 = P$. (This follows from the fact that when acting on a function of the reduced variable X_0, it must leave it unaltered).

We have, after defining $Q = 1 - P$:

$$(\partial/\partial t)\ PF(t) = (PiLP)\ PF(t) + (PiLQ)\ \ Q\ F(t) \tag{17a}$$

$$(\partial/\partial t)\ QF(t) = (QiLQ\ )\ PF(t) + (QiLP)\ \ Q\ F(t) \tag{17b}$$

After solving Eq. (17b) for QF, substitution into (17a) gives the following result for $f(t) = PF(t)$:

$$(\partial/\partial t)f(t) = \{PiLP + \int_0^t ds\ PiLQ\ \exp[QiLQ\ (t-s)]QiLP\}f(s) \tag{18}$$

This equation is exact because it is the result of formal manipulations of the equation of motion given in Eq. (16); it has been simply rewritten in a way which is hopefully more suitable for perturbation expansions. As it stands, Eq. (18) is in effect intractable; note that the exponential in the integral is actually an operator which acts on everything to its right. At this stage, an important approximation whose physical significance will be appreciated in specific contexts, is commonly performed on Eq. (18). It is based on the consideration that the evolution operator can always be written as $L = L_0 + L_b$, with L_b acting upon the bath coordinates only (while L_0 depends in general upon the bath variables too), to obtain

$$QLQ = L_b + QL_0\ Q \cong L_b\ . \tag{19}$$

The first equality derives from the definition of Q, which projects any function onto the subspace of the bath variables. The approximation can be considered the zero-order term of a resolvent expansion (7); it consists essentially in the adoption of a "time scale separation", according to which the bath variables vary faster than the system variables (6). The important consequence of the approximation is that the truncated exponential operator no longer acts on f(t), and so Eq. (18) becomes:

$$(\partial/\partial t)\ f(t) = i\ \Omega\ f(t) - \int_0^t ds\ K(t-s)\ f(s) \tag{20}$$

with the two operators Ω and K expressed as

$$\Omega = \mathcal{P}\ \mathcal{L}\ \mathcal{P} \tag{21a}$$

$$K(t) = \mathcal{P}\ \mathcal{L}\ (1-\mathcal{P})\ \exp\ (i\ \mathcal{L}_b\ t)\ (1 - \mathcal{P})\ \mathcal{L}\ \mathcal{P}. \tag{21b}$$

It is often useful to work with the Fourier-Laplace transform of Eq. (20)

$$\tilde{f}(\omega) = \int_0^\infty dt\ e^{-i\omega t}\ f(t)$$

$$= [i(\omega - \Omega) + K(\omega)]^{-1}\ f(0) \tag{22}$$

CHOICE OF THE PROJECTION OPERATOR

The average value of any function $F\ \{X^N\}$ is defined as

$$\bar{F} = \int \{d\ X^N\}\ P\{X^N\}\ F\ \{X^N\}, \tag{23}$$

where $P\ \{X^N\}$ is the distribution function

$$P\{X^N\} = Z^{-1} e^{-H\ \{X^N\}\ /\ kT} \tag{24}$$

with Z the partition function. Let us rewrite Eq. (23) in the form

$$\bar{F} = \int dX_0\ \{dX_1{}^N\}\ P\{X^N\}\ F\{X^N\}$$

$$= \int dX_0\ P(X_0)\ \langle F\rangle_b\ (X_0), \tag{25}$$

where $P(X_0)$ is now the singlet distribution function:

$$P(X_0) = \int d\{X_1{}^N\}\ P\{X^N\}. \tag{26}$$

and $\langle F\rangle_b\ (X_0)$ is the bath-average of the multiparticle function $F\{X^N\}$. A judicious choice for P is therefore:

$$\mathcal{P}F\ \{X^N\} = \langle F\rangle_b\ (X_0) \tag{27}$$

or

$$\mathcal{P}F\ \{X^N\} = P(X_0)^{-1} \int \{dX_1{}^N\}\ P\{X^N\}\ F\{X^N\}\ . \tag{28}$$

Note that the factor $P(X_0)^{-1}$ is dictated from the condition for $\mathcal{P}$ to be idempotent:

$$\mathcal{P}^2F\{X^N\} = \mathcal{P}\ \langle F\rangle_b\ (X_0) = \langle F\rangle_b\ (X_0). \tag{29}$$

It should be noted that this procedure leads to a form of the evolution operator which is suitable for dynamical variables, i.e. the adjoint of the operator for distribution functions. This latter operator must be obtained with a different choice for the projection operator, based on the property that for multivariate distributions:

$$P(X_0,t) = \int \{d\ X_1{}^N\}\ P(\{X^N\},\ t). \tag{30}$$

Therefore a convenient choice for $\mathcal{P}$ is one which allows us to project $P(X_0,t)$ out from the total probability density. Again from the condition for $\mathcal{P}$ to be idempotent, we find:

$$\mathcal{P}\ (\cdots) = [P\{X^N\}\ /\ P(X_0)] \int \{dX_1{}^N\}\ (\cdots). \tag{31}$$

In the following, we always write the evolution operators in the form suitable for distribution functions.

THE FORMAL DERIVATION OF THE DIFFUSION EQUATION

The diffusion equation is derived from the classical many-particle equations of motion by means of a repeated sequence of contraction (or projection) onto subspaces defined by a lower number of variables. The first of these operations reduces the many-body problem to an equation for the probability distribution of coordinates and momenta of a "test" particle only, called the "generalized Fokker Planck equation".

The total Hamiltonian for a system composed of one test particle and N bath particles is:

$$H = H_0 + H_b \tag{32a}$$

$$H_0 = \frac{\underset{\sim}{p}_0^2}{2m} + \tfrac{1}{2}\,\underset{\sim}{L}_0 \cdot \underset{\sim}{I}_0^{-1} \cdot \underset{\sim}{L}_0 + V_0,$$

$$V_0 = \sum_{j=1}^{N} U(\underset{\sim}{r}_0 - \underset{\sim}{r}_j,\ \Omega_0,\ \Omega_j) \tag{32b}$$

$$H_b = \sum_{n=1}^{N} \left(\frac{\underset{\sim}{p}_n^2}{2m_n} + \tfrac{1}{2}\,\underset{\sim}{L}_n \cdot \underset{\sim}{I}_n^{-1} \cdot \underset{\sim}{L}_n + V_n\right),$$

$$V_n = \sum_{j>n}^{N} V(\underset{\sim}{r}_n - \underset{\sim}{r}_j,\ \Omega_n,\ \Omega_j) \tag{32c}$$

where $\underset{\sim}{p}$ and $\underset{\sim}{L}$ denote linear and angular momentum respectively, $\underset{\sim}{I}$ is the inertia tensor, and $\Omega \equiv (\alpha,\beta,\gamma)$ is the set of Eulerian angles defining the orientation of a particle relative to a reference frame.

The time evolution of the system is governed by the Liouville operator, that for classical systems is obtained in conformity with the Hamilton equation as (8):

$$\frac{d}{dt} = i\,\mathcal{L} \equiv \sum_k \left(\frac{\partial H}{\partial P_k}\frac{\partial}{\partial Q_k} - \frac{\partial H}{\partial Q_k}\frac{\partial}{\partial P_k}\right) \tag{33}$$

$(Q_k,\ P_k)$ being a pair of conjugate variables (coordinate and momentum) of the system. One then obtains:

$$i\mathcal{L} = i\mathcal{L}_0 + i\mathcal{L}_b \tag{34a}$$

$$i\mathcal{L}_0 = \frac{\underset{\sim}{p}_0}{m}\cdot\frac{\partial}{\partial \underset{\sim}{r}_0} + \underset{\sim}{F}_0 \cdot \frac{\partial}{\partial \underset{\sim}{p}_0} + i\,\underset{\sim}{L}_0 \cdot \underset{\sim}{I}^{-1} \cdot \underset{\sim}{M}_0 + \underset{\sim}{N}_0 \cdot \frac{\partial}{\partial \underset{\sim}{L}_0} + \underset{\sim}{J}_0 \cdot \frac{\partial}{\partial \underset{\sim}{L}_0} \tag{34b}$$

$$i\,\mathcal{L}_b = \sum_n \left[\frac{\tilde{p}_n}{m_n}\cdot\frac{\partial}{\partial \tilde{r}_n} + \tilde{F}_n\cdot\frac{\partial}{\partial \tilde{p}_n} + i\,\tilde{L}_n\cdot\tilde{I}_n^{-1}\cdot\tilde{M}_n + \right.$$
$$\left. + \tilde{N}_n\cdot\frac{\partial}{\partial \tilde{L}_n} + \tilde{J}_n\cdot\frac{\partial}{\partial \tilde{L}_n}\right] \tag{34c}$$

where $\tilde{F}_k = -(\partial V/\partial \tilde{r}_k)$ and $\tilde{N}_k = -i(\tilde{M}_k V)$ are the force and torque acting on the k-th particle, $\tilde{M}_k$ is the generator of infinitesimal rotations (9) about three orthogonal axes fixed in the k-th particle, and which is mathematically equivalent to the quantum-mechanical angular momentum operator, and whose explicit expressions can be found in Barchewitz (10), and $\tilde{J}_k$ is a precessional term with components:

$$J_\alpha = L_\beta L_\gamma (I_\gamma^{-1} - I_\beta^{-1})\,\varepsilon_{\alpha\beta\gamma} \tag{35}$$

$\varepsilon_{\alpha\beta\gamma}$ being the complete antisymmetric Levi-Civita tensor defining the cyclic permutations (8).

If the entire set of the test and bath variables is again denoted by $\{X^N\}$, with $X \equiv (r, p, \Omega, L)$, then Eq. (16) with $-i$ replaced for i can be used to describe the time evolution of the multi-particle distribution function.

The resultant system of first-order differential equations can be solved by computer simulation of the molecular dynamics, for a (hopefully) sufficiently large sample of bath molecules. Here we shall treat the problem statistically, by averaging in a suitable way the interactions between a single particle and the rest of the fluid. Since the complete one-particle distribution function $P(X_0,t)$, in the coordinate and momentum space, can be obtained by integration on the bath variables, the choice for the projector P given in Eq. (31) seems to be the most appropriate. However, we prefer to think of $\mathcal{L}$ as acting on a general function $F(\{X^N\},t)$ and to use the projector given in Eq. (27), which avoids carrying along the multi-particle distribution function $P\{X^N\}$. One should remember that this procedure leads to the adjoint forms of the operators Ω and K.

From the definitions given in the Eqs. (21) and (22) one obtains

$$P\,\mathcal{L}\,P = P\mathcal{L}_0 P = \langle \mathcal{L}_0 \rangle_b \tag{36}$$

$$(1 - P)\, iL_0 \; P = i(L_0 - \langle L_0 \rangle_b)$$
$$= (\underset{\sim}{F}_0 - \langle \underset{\sim}{F}_0 \rangle_b) \frac{\partial}{\partial \underset{\sim}{p}_0} + (\underset{\sim}{N}_0 - \langle \underset{\sim}{N}_0 \rangle_b) \frac{\partial}{\partial \underset{\sim}{L}_0} \tag{37}$$

The final result reads:

$$\Omega = - \langle L_0 \rangle_b \tag{38a}$$

$$-K(\omega) = \frac{\partial}{\partial \underset{\sim}{p}_0} \cdot \underset{\sim}{\beta}^{TT}(\omega)\, P \cdot \frac{\partial}{\partial \underset{\sim}{p}_0} P^{-1} + \frac{\partial}{\partial \underset{\sim}{p}_0} \cdot \underset{\sim}{\beta}^{TR}(\omega)\, P \cdot \frac{\partial}{\partial \underset{\sim}{L}_0} P^{-1} +$$
$$+ \frac{\partial}{\partial \underset{\sim}{L}_0} \cdot \underset{\sim}{\beta}^{RT}(\omega)\, P \cdot \frac{\partial}{\partial \underset{\sim}{p}_0} P^{-1} + \frac{\partial}{\partial \underset{\sim}{L}_0} \cdot \underset{\sim}{\beta}^{RR}(\omega)\, P \cdot \frac{\partial}{\partial \underset{\sim}{L}_0} P^{-1} \tag{38b}$$

where $\underset{\sim}{\beta}^{TT}$, $\underset{\sim}{\beta}^{TR}$ and $\underset{\sim}{\beta}^{RR}$ have the physical meaning of translational, roto-translational and rotational friction tensors, and are expressed in microscopic terms as

$$\underset{\sim}{\beta}^{AB}(\omega) = P(X_0)^{-1} \int_0^\infty dt\; e^{-i\omega t} \int \{d\, X_1^N\}\; \underset{\sim}{\Delta}^A \exp(iL_b t)\; P\{X^N\}\; \underset{\sim}{\Delta}^B \tag{39}$$

with

$$\underset{\sim}{\Delta}^T = \underset{\sim}{F}_0 - \langle \underset{\sim}{F}_0 \rangle_b \; , \; \underset{\sim}{\Delta}^R = \underset{\sim}{N}_0 - \langle \underset{\sim}{N}_0 \rangle_b \tag{40}$$

This result is a form of the fluctuation-dissipation theorem, according to which the hydrodynamical friction tensors are related to Fourier transforms of force (or torque) correlation functions. Note that the friction tensors have an implicit dependence on X_0, but we shall ignore this point here (11). If torques and forces relax rapidly compared with the test particle variables, then the frequency dependence of the frictional coefficients β can be neglected, at least when working in the frequency range characteristic of the fluctuations of X_0. The one-particle evolution operator obtained in this way shall be denoted by Γ, and the corresponding equation for the distribution function is known (7,12) as the stochastic Fokker-Planck equation. In full, omitting for simplicity the subscript used to denote the tagged particle, and the brackets indicating bath averages (because now all quantities refer to single particles), one has:

$$\Gamma = \Gamma^T + \Gamma^R + \Gamma^{TR} + \Gamma^{RT} \tag{41}$$

$$\Gamma^T = -\frac{\underset{\sim}{p}}{m} \cdot \frac{\partial}{\partial \underset{\sim}{r}} - \underset{\sim}{F} \cdot \frac{\partial}{\partial \underset{\sim}{p}} + \frac{\partial}{\partial \underset{\sim}{p}} \cdot \underset{\sim}{\beta}^{TT} P \cdot \frac{\partial}{\partial \underset{\sim}{p}} P^{-1} \tag{42a}$$

$$\Gamma^R = -i \underset{\sim}{L} \cdot \underset{\sim}{I}^{-1} \cdot \underset{\sim}{M} - (\underset{\sim}{N}+\underset{\sim}{J}) \frac{\partial}{\partial \underset{\sim}{L}} + \frac{\partial}{\partial \underset{\sim}{L}} \cdot \underset{\sim}{\beta}^{RR} P \cdot \frac{\partial}{\partial \underset{\sim}{L}} P^{-1} \tag{42b}$$

$$\Gamma^{TR} + \Gamma^{RT} = \frac{\partial}{\partial \underset{\sim}{p}} \cdot \underset{\sim}{\beta}^{TR} P \cdot \frac{\partial}{\partial \underset{\sim}{L}} P^{-1} + \frac{\partial}{\partial \underset{\sim}{L}} \cdot \underset{\sim}{\beta}^{RT} P \cdot \frac{\partial}{\partial \underset{\sim}{p}} P^{-1} \tag{42c}$$

The Smoluchowski equation is a contraction of the Fokker-Planck equation on positional coordinates only. This is physically correct when linear and angular momenta are effectively randomized in the mean period of the molecular displacements. This is essentially true for ordinary liquids except when composed of very small molecules; in liquid crystal phases, the strong local potentials are expected to quench efficiently the particle momenta.

The route we shall follow to derive the Smoluchowski equation is to define a projector which averages linear and angular momenta over a Maxwell-Boltzmann distribution. According to Eq. (28):

$$P(\ldots) = (2\pi kT)^{-3} (m^3 I_x I_y I_z)^{-\frac{1}{2}} \int d\underset{\sim}{p}\, d\underset{\sim}{L} \exp\left[-\frac{1}{2kT}(\underset{\sim}{L} \cdot \underset{\sim}{I}^{-1} \cdot \underset{\sim}{L} + \underset{\sim}{p}^2/m)\right] (\ldots) \tag{43}$$

Again, using the equations (21) with $\Gamma = \Gamma_0 + \Gamma_b$ substituted for $i\,L$, and Γ_b now including all terms with differential operators on the $(\underset{\sim}{p},\underset{\sim}{L})$ variables, we have:

$$i\,\Omega = P\,\Gamma_0\,P = 0 \tag{44}$$

because the thermal averages $\bar{p}$ and $\bar{L}$ are zero;

$$(1-P)\,\Gamma_0\,P = -\frac{\underset{\sim}{p}}{m} \cdot \frac{\partial}{\partial \underset{\sim}{r}} - i\,\underset{\sim}{L} \cdot \underset{\sim}{I}^{-1} \cdot \underset{\sim}{M} \tag{45}$$

Proceeding as before, one obtains for the reduced stochastic operator, that we shall denote now by R:

$$R = R^T + R^R + R^{TR} + R^{RT} \tag{46}$$

$$R^{T} = \frac{\partial}{\partial \underset{\sim}{r}} \cdot \underset{\sim}{D}^{T} P \cdot \frac{\partial}{\partial \underset{\sim}{r}} P^{-1} \tag{47a}$$

$$R^{R} = -\underset{\sim}{M} \cdot \underset{\sim}{D}^{R} P \cdot \underset{\sim}{M} P^{-1} \tag{47b}$$

$$R^{TR} + R^{RT} = i\left(\frac{\partial}{\partial \underset{\sim}{r}} \cdot \underset{\sim}{D}^{TR} P \cdot \underset{\sim}{M} + \underset{\sim}{M} \cdot \underset{\sim}{D}^{RT} P \cdot \frac{\partial}{\partial \underset{\sim}{r}}\right) P^{-1} \tag{47c}$$

where the elements $\underset{\sim}{D}^{AB}$ have now the physical meaning of diffusion tensors. The matrixes $\underset{\sim}{D}$ of diffusion tensors and $\underset{\sim}{\beta}$ of friction tensors are related by the generalized Einstein relation (12, 13):

$$\underset{\sim}{D} = \begin{pmatrix} \underset{\sim}{D}^{T} & \underset{\sim}{D}^{TR} \\ \underset{\sim}{D}^{RT} & \underset{\sim}{D}^{R} \end{pmatrix} = (kT)^{2} \beta^{-1} \tag{48}$$

$$\underset{\sim}{\beta} = \begin{pmatrix} \underset{\sim}{\beta}^{TT} & \underset{\sim}{\beta}^{TR} \\ \underset{\sim}{\beta}^{RT} & \underset{\sim}{\beta}^{RR} \end{pmatrix} \tag{49}$$

The roto-translational couplings are zero for axially symmetric molecules (14) but should in principle be taken into account in the cases of lower symmetry.

EXPLICIT FORM OF THE DIFFUSION EQUATION FOR CYLINDRICAL MOLECULES

For axially symmetric molecules, the diffusion equation can be written as:

$$(\partial/\partial t)\, P(\underset{\sim}{r}, \Omega, t) = \{\underset{\sim}{\nabla}_{r} \cdot \underset{\sim}{D}^{T} \cdot \left[\underset{\sim}{\nabla}_{r} + \frac{1}{kT} (\underset{\sim}{\nabla}_{r} V)\right] + \underset{\sim}{M} \cdot \underset{\sim}{D}^{R} \cdot \left[\underset{\sim}{M} + \frac{1}{kT} (\underset{\sim}{M} V)\right]\}\, P(\underset{\sim}{r}, \Omega, t) \tag{50}$$

$$\underset{\sim}{\nabla}_{r} \cdot \underset{\sim}{D}^{T} \cdot \underset{\sim}{\nabla}_{r} = D_{\perp}^{T} \nabla_{r}^{2} + (D_{\parallel}^{T} - D_{\perp}^{T}) \frac{\partial^{2}}{\partial z^{2}} \tag{51a}$$

$$\underset{\sim}{M}\cdot\underset{\sim}{D}^R\cdot\underset{\sim}{M} = -\left[D_\perp^R \; M^2 + (D_\parallel^R - D_\perp^R)M_z^2\right] \tag{51b}$$

In the above equations, $V(r,\Omega)$ is the anisotropic pseudo-potential implicitly defined from the relation:

$$P(r,\Omega) = Z^{-1}\int \{dX_1^N\} \exp\left[- \; V\{X^N\} / kT\right] \tag{52}$$

$\underset{\sim}{\nabla}_r$ is the gradient operator in Cartesian coordinates:

$$\underset{\sim}{\nabla}_r = \underset{\sim}{i}\frac{\partial}{\partial x} + \underset{\sim}{j}\frac{\partial}{\partial y} + \underset{\sim}{k}\frac{\partial}{\partial z} \tag{53}$$

and M^2, M_z are expressed in terms of the Euler angles as:

$$M^2 = -\left[\frac{\partial^2}{\partial\beta^2} + \cot\beta\,\frac{\partial}{\partial\beta} + \frac{1}{\sin^2\beta}\left(\frac{\partial^2}{\partial\alpha^2} + \frac{\partial^2}{\partial\gamma^2}\right) - 2\,\frac{\cos\beta}{\sin^2\beta}\,\frac{\partial^2}{\partial\alpha\partial\gamma}\right] \tag{54a}$$

$$M_z = -\,i\,\frac{\partial}{\partial\gamma}$$

In isotropic systems $V(r,\Omega) = 0$, and the solutions of the diffusion equations are well known (5).

The general solution for the conditional probability $P(\underset{\sim}{r}_0, \underset{\sim}{r}, t)$ expressed in terms of the principal components $D^T_{\alpha\alpha}$ of the translational diffusion tensor, is:

$$P(\underset{\sim}{r}_0,\underset{\sim}{r},t) = \prod_\alpha \left(4\pi\, D^T_{\alpha\alpha}\, t\right)^{-\frac{1}{2}} \exp\left(- \frac{(r_\alpha - r_\alpha^0)^2}{4D^T_{\alpha\alpha}\; t}\right) \tag{55}$$

which, for spherical molecules becomes

$$P(\underset{\sim}{r}_0,\underset{\sim}{r},t) = \left(4\pi\, D^T\, t\right)^{-\frac{3}{2}} \exp\left(- \frac{|\underset{\sim}{r} - \underset{\sim}{r}_0|^2}{4D^T\; t}\right) \tag{56}$$

The solution for the angular part in the case of axially symmetric molecules is given in terms of the Wigner functions (or generalized spherical harmonics) as

$$P(\Omega_0,\Omega,t) = \sum_{jpq} \left(\frac{2j+1}{8\pi^2}\right) D^j_{pq}(\Omega)\; D^{j*}_{pq}(\Omega_0)\; \exp\left(-\alpha^j_{pq}\; t\right) \tag{57}$$

where

$$\alpha^{j}_{pq} = D^{R}_{\perp}\, j(j+1) + (D^{R}_{\parallel} - D^{R}_{\perp})\, q^2 \tag{58}$$

The experimental determination of translational diffusion coefficients invariably implies the choice of an external reference frame (XYZ), whereas the $\underset{\sim}{D}^T$ tensor is diagonal in a molecular axis system (xyz). The transformation of the differential operator $\partial/\partial z$ (∇^2_r is an invariant) induces an angular dependence in the translational diffusion operator:

$$\partial/\partial z = \sum_i A_{zi}\, \partial/\partial X_i \tag{59}$$

$$\nabla_r \cdot \underset{\sim}{D}^T \cdot \nabla_r = D^T_{\perp} \nabla^2_r + (D^T_{\parallel} - D^T_{\perp}) \sum_{ij} A_{zi} A_{zj} \frac{\partial^2}{\partial X_i\, \partial X_j} \tag{60}$$

$$A_{zX} = \sin\beta \cos\alpha, \quad A_{zY} = \sin\beta \sin\alpha, \quad A_{zZ} = \cos\beta.$$

In liquid crystal phases, measurements can be performed along directions with different orientations relative to that of preferential alignment.

SOLUTION OF THE DIFFUSION EQUATIONS IN LIQUID CRYSTALLINE PHASES

We have seen in the previous paragraph that the roto-translational diffusion equations are generally coupled even if the roto-translational coupling tensors $\underset{\sim}{D}^{RT}$ and $\underset{\sim}{D}^{TR}$ are (or are assumed to be) zero. This effective coupling arises because experiments refer to laboratory (not to molecular) reference systems, or because the distribution function $P(r, \Omega)$ does not factorize in some liquid crystalline phases. Let us discuss briefly a few typical situations.

Nematic phases

In nematic phases $V = V(\beta,\gamma)$, (15) with β,γ the Euler angles specifying the orientation of the director in the molecular frame; the orientational potential reduces to $V(\beta)$ for cylindrical molecules. Therefore $P(r,\Omega)$ is independent of the position, and the diffusion equation is factorized. The Eq. (51b) is appropriate to describe the rotational diffusion; in the laboratory frame, however, the expression for R^T given in Eq. (60) implies a coupling with the rotational motion. A simplification occurs if we can take averages over fast rotational motions, and this appears to be the case in the NMR experiment where mean square translational displacements are actually measured over relatively long time intervals, during which the molecules are free to perform several rotations. As an example, molecules with D^R of the order of 10^9 sec^{-1} and D^T about 10^{-5} cm^2 sec^{-1} perform 10^4 rotations in the time required to travel about 10^{-5} cm. (The observation time is usually determined by the sequences of pulse experiments).

After taking the average over the rotational motions, the translational diffusion operator becomes:

$$\overline{R^T} = \overline{D^T_{\perp lab}}\,\nabla^2_r + (\overline{D^T_{\parallel lab}} - \overline{D^T_{\perp lab}})\,\frac{\partial^2}{\partial Z^2} \tag{61}$$

where the components in the laboratory system of the rotationally averaged translational diffusion tensor are:

$$\overline{D^T_{\perp lab}} = \tfrac{1}{3}\,\mathrm{Tr}\,\underset{\sim}{D}^T - \tfrac{1}{3}\,\mathrm{Tr}(\underset{\sim}{S}\cdot\underset{\sim}{D}^T) \tag{62a}$$

$$\overline{D^T_{\parallel lab}} = \tfrac{1}{3}\,\mathrm{Tr}\,\underset{\sim}{D}^T + \tfrac{2}{3}\,\mathrm{Tr}(\underset{\sim}{S}\cdot\underset{\sim}{D}^T) \tag{62b}$$

where $\underset{\sim}{S}$ is the ordering matrix (1a).

Note that $\underset{\sim}{D}^T{}_{lab}$ is axially symmetric, reflecting the uniaxiality of the nematic phase.

The full solution of the rotational problem has been discussed in detail elsewhere (1b). It is based on the representation of the operator R^R on the Wigner functions set. Recently, the non-orthogonal set of functions $P(\Omega)^{\frac{1}{2}}D^j_{pq}(\Omega)$ has been proposed to ensure more efficient convergence at relatively high orderings (16).

To conclude it should be noted that the roto-translational terms R^{TR} are ineffective in nematic phases as they are in isotropic liquids, as long as one is interested in the pure rotational motions. This situation applies to NMR ^{13}C or ^{2}H relaxation, where the relaxation mechanism is provided only by the rotational diffusion.

Smectic phases

Even for the most simple conceivable case of axially symmetric molecules in the uniaxial smectic-A phase, the anisotropic potential has the rather complicated form (17) given by the equation:

$$U(\zeta,\cos\beta) = - \quad [a\, P_2(\cos\beta) + b \cos\zeta\, P_2(\cos\beta) + c \cos\zeta] \quad (63)$$

ζ being (2π Z/d), with Z the displacement perpendicular to the smectic planes, and d the layer spacing.

Thus, P(Z,β) is not factorized, and in general the complete roto-translational diffusion equation should be handled. As simplifying assumptions one might neglect the R^{TR} coupling term, and average over the fast rotational motions. Still, the solution of the rotational problem cannot be disentangled from translational diffusion along the direction of the normal to the smectic planes, whereas a two-dimensional free diffusion occurs in the planes.

In principle, the roto-translational diffusion equation could be solved by using as basis functions the products of Wigner functions $D^J_{pq}(\alpha\beta\gamma)$ and periodic "lattice" functions exp(ikζ). However, a complete solution of this problem has not been given as yet. Some confusion also persists on the meaning of the translation coefficient for the diffusion parallel to the smectic director, measured in spin-echo experiments. The quantity obtained in this way is actually related to the mean square displacements in the Z-direction, and only in absence of translational order is it the same as the diffusion tensor component $\overline{D^T_\parallel}$ in the laboratory frame, as given by Eq. (62b). Because of the periodic structure of the phase, there is no simple relationship between the experimentally determined quantity $D^T_{\parallel exp}$ and $\overline{D^T_\parallel}$. An approximate relation can however be found by taking into account the characteristic time scale of the spin-echo experiments, where the observation time is much longer than the time required for the particle to travel across a smectic layer. One might then assign to the variable ζ a discrete set of values labelling the potential minima in the smectic layers, and assume a random walk model for the diffusion across the layers. In the coarse grained time scale of a spin-echo measurement, the motion could be described as a free diffusion, with an effective diffusion constant $D^T_{\parallel exp} = \frac{1}{2} d^2 w$, w being the rate of jumps between adjacent minima. The jump rate can be calculated in terms of the rotationally averaged quantity $\overline{D^T_\parallel}$ and the strength of the periodic potential (18). Asymptotic solutions, valid for relatively high potential values, have the simple form, for c > 1:

$$w = (4\pi \; \overline{D^T_\parallel} \;/\; d^2)\; c \exp(-2c) \quad (64)$$

$$D^T_{\parallel exp} \;/\overline{D^T_\parallel} = 2\,\pi\, c \exp(-2c) \quad (65)$$

This result may explain why it is found experimentally that

$(D^T_\parallel/D^T_\perp)$ is less than unity in smectics, while the opposite is generally true in nematics.

Cholesterics

In the cholesterics phase the rotational motions about the local director is assumed to be the same as in nematics. However, the translations across the nematic planes (parallel to the axis of helical structures) induce a new kind of coupling between the motions, and we shall examine this now in some detail.

Let us consider a single domain cholesteric phase, with the helix axis coincident with the laboratory Y-axis. Translations along this coordinate induce a rotation of the molecules, which tend to remain aligned to the local director, of an angle

$$\theta = \frac{2\pi}{p}\,(Y - Y_0) \tag{66}$$

p being the helical pitch. The differential operator $(\partial/\partial Y)$ in the expression for the translational diffusion operator in the laboratory system becomes

$$\partial/\partial Y = (\partial\theta/\partial Y)\ \partial/\partial\theta = (2\pi\ /p)\ \ \partial/\partial\theta \tag{67}$$

and it gives rise to the coupling term:

$$R^C = -(2\pi\ /p)^2\,\frac{\partial}{\partial\theta}\,D^T_{YY}\,P\,\frac{\partial}{\partial\theta}\,P^{-1} \tag{68}$$

Now the average over fast rotational motions is equivalent to choosing a projection operator of the type given in Eq. (31), in the form:

$$\mathcal{P}\,(\ldots) = P(\Omega)\int d\Omega\,(\ldots) \tag{69}$$

and to obtain a projected diffusion operator for the distribution $P(\theta,t)$ as $\mathcal{P}R^C\,\mathcal{P}$, neglecting the fast relaxing "memory" K term of Eq. (20)

Remembering that $P(\Omega)$ is an eigenfunction of R^R with eigenvalue zero, the only surviving term is:

$$\mathcal{P}\,R^C\,\mathcal{P} = -(2\pi\ /p)^2\,\frac{\partial}{\partial\theta}\,\overline{D^T_{YY}}\,\frac{\partial}{\partial\theta} \tag{70}$$

with $\overline{D^T_{YY}}$ given by Eq. (62b). This term gives rise to an important mechanism of spin relaxation in deuterium resonance (19).

SIDE-CHAIN DYNAMICS

The problem of chain motions is very complicated, and satisfactory solutions are obtained only in the case of short end groups.

In the case of methoxyl groups, the motion relative to the core of the mesogen molecule is a uniaxial rotation in a two-minima potential field. Note that the potential is determined essentially by the electronic distribution of the molecule itself, and it is negligibly perturbed by the intermolecular interactions of the liquid crystal phase. The local potential is expected to be appreciably larger than kT, and therefore asymptotic solutions of the problem can be derived. An interesting conclusion can thus be obtained: the motion described by the solution of the one-dimensional diffusion equation in the presence of a bistable potential V corresponds to a two-site random jump model, with a jump frequency characterized by an Arrhenius type temperature dependence, $\nu = \nu_0 \exp(-V/kT)$. This situation is similar to that already encountered when dealing with translational diffusion normal to smectic layers.

COLLECTIVE MOTIONS

At temperatures lower than the isotropic-liquid crystal transition, each deformation of the ordered structure will be opposed by a restoring force. An ordered phase has an elastic character with respect to curvature strains, and by symmetry considerations it is possible to obtain the minimum number of independent elastic constants for each phase. For a uniaxial liquid crystal there are three non vanishing elastic moduli, corresponding to the three distinct curvature strains of splay, twist and bend (20). The ratios of these elastic constants cannot be determined by symmetry arguments, but they have been related to the form of the intermolecular interactions, and can be measured experimentally. In nematics they are found to be of about the same order of magnitude, and for qualitative considerations they can be assumed to be of equal value.

Deformations of the ordered structure from the equilibrium configuration are generated by thermal fluctuations. The local ordering in a liquid crystal phase is usually specified by a normalized vector field $\underset{\sim}{n}(\underset{\sim}{r})$, the order director, which gives the direction of preferred alignment at the point $\underset{\sim}{r}$. Thermal fluctuations thus induce a time dependent disturbance:

$$\underset{\sim}{n}(\underset{\sim}{r},t) = \underset{\sim}{n}_0(\underset{\sim}{r}) + \delta\underset{\sim}{n}(\underset{\sim}{r},t) \qquad (71)$$

which can be analyzed in terms of normal modes, with amplitude:

$$\delta \underset{\sim}{n}(\underset{\sim}{q},t) = V^{-1} \int \delta \underset{\sim}{n}(\underset{\sim}{r},t)\, e^{-i\, \underset{\sim}{q}\cdot \underset{\sim}{r}}\, d^3r \tag{72}$$

In the limit of small amplitude deformations, $\delta \underset{\sim}{n}$ is orthogonal to $\underset{\sim}{n}_0$. The x and y components of $\delta \underset{\sim}{n}(\underset{\sim}{q})$ contribute with a quadratic term for each normal mode to the orientational free energy (20). Use of the equipartition theorem therefore gives:

$$\langle |\delta n_\alpha (\underset{\sim}{q})|^2\rangle = (kT/V)\, [K_\alpha q_\alpha^2 + K_3 q_3^2 + \chi_a H^2]^{-1} \tag{73}$$

where x, y, z axes are labelled 1, 2, 3 and K_1, K_2, K_3 are the splay, twist and bend elastic constants respectively. The term $\chi_a H^2$ takes into account the magnetic field contribution to the orientational free energy. The time evolution of these thermal distortions may be treated in the hydrodynamic limit, since they are localized. Each normal mode decays according to:

$$\delta \dot{n}_\alpha (\underset{\sim}{q}) = -\, \delta n_\alpha(\underset{\sim}{q}) / \tau_\alpha (\underset{\sim}{q}) \tag{74}$$

$$\tau_\alpha (\underset{\sim}{q}) = \eta_\alpha [K_\alpha q_\alpha^2 + K_3 q_3^2 + \chi_a H^2]^{-1} \tag{75}$$

where η_α is a Miesowicz viscosity coefficient (21).

The collective motions due to the order director fluctuations (DF) affect the spin relaxation rates, since they modulate the instantaneous director orientation with respect to the magnetic field. A complete description of molecular dynamics in an oriented phase requires therefore that both single particle variables and collective variables be considered (22).

In many cases these motions have different characteristic times, since a molecule reorients many times during a normal mode period. It is therefore possible to separate the motions according to their time scales, by averaging over the fast variables.

To a first approximation, the single particle motion can be supposed to be very fast with respect to the Larmor frequency. The spin hamiltonian can therefore be replaced by an averaged hamiltonian, which is axially symmetric around $\underset{\sim}{n}(\underset{\sim}{r},t)$ and the relaxation effects of the fast rotational motions can be neglected. Nuclear spins are then assumed to relax only because of the director fluctuations. If the molecule we are looking at contains a pair of equivalent $I = \frac{1}{2}$ spins, the partially averaged hamiltonian has the form:

$$H_{av} = H_0 + \sqrt{6}\omega_d \Sigma_m D^2_{m0}(\Omega_n) S_d^{(2,m)} \tag{76}$$

where: $S_d^{(2,m)}$ are the spherical components of the dipolar interaction; $\Omega_n(t)$ are the Euler angles of the time dependent transformation connecting the laboratory frame (Z along $\underset{\sim}{B}_0$) and the local director frame (z along $\underset{\sim}{n}(t)$); and ω_d is the magnitude of the averaged dipolar interaction:

$$\omega_d = \tfrac{1}{2}(3\cos^2\theta - 1)\,\overline{P_2}\gamma_1^2\,\hbar\, r_{12}^{-3} \tag{77}$$

Here $\bar{P}_2$ is the degree of orientational order of the molecule (assumed to be rigid and axially symmetric); θ is the angle between the internuclear vector r_{12} and the molecular axis. It is worth noting that the results obtained for this model system can be directly utilized for the case of a single nucleus of spin I = 1 with an axially symmetric quadrupolar tensor, by substituting $\tfrac{1}{2}e^2q\,Q\,\hbar^{-1}$ for the factor $\gamma_1^2\,\hbar\, r_{12}^{-3}$.

Since director fluctuations give rise to small deviations of $\underset{\sim}{n}$ from $\underset{\sim}{n}_0$, the correlation functions of the time dependent terms of the hamiltonian (76) can be computed by expanding the Wigner functions around $\Omega_n = \Omega_0$. Only linear terms are retained, and one obtains (χ being the angle between $\underset{\sim}{n}_0$ and $\underset{\sim}{B}_0$):

$$J_q(\omega) = 6\omega_d^2 \sum_p |d^2_{qp}(\chi)|^2 \; j_{p0}(\omega) \tag{78}$$

$$j_{p0}(\omega) = \delta_{p\pm1}\,\tfrac{3}{2}\int_0^\infty \overline{\delta\underset{\sim}{n}(t)^*\cdot\delta\underset{\sim}{n}(0)}\,\cos\omega t\,dt$$

$$= \delta_{p\pm1}\,\tfrac{3}{2}\int \sum_\alpha \overline{|\delta n_\alpha(\underset{\sim}{q})|^2}\;\tau_\alpha(\underset{\sim}{q})\,[1+\omega^2\tau_\alpha^2(\underset{\sim}{q})]^{-1}\,d^3q \tag{79}$$

The last integral is easily computed if $K_\alpha = K$ and $\eta_\alpha = \eta$, and if the term $\chi_a H^2$ can be neglected:

$$j_{p0}(\omega) = \delta_{p\pm1}\,\tfrac{3}{2}\,kT[\eta/8\pi^2 K^3\,\omega]^{\frac{1}{2}} \tag{80}$$

It should be noted that, even in the case of uniaxial liquid crystals, intramolecular interactions are modulated by translational diffusion, because director fluctuations break the instantaneous axial symmetry of the mesophase. The phenomenon has the same origin as the roto-translational coupling in cholesterics discussed before. The normal mode relaxation time is therefore shortened, and in the one constant approximation one has:

$$\tau^{-1}(\underset{\sim}{q}) = (K/\eta + D)\, q^2 \tag{81}$$

The spectral density becomes:

$$j_{po}(\omega) = \delta_{p\pm 1}\ (3kT/4\pi\ K)\ [(K/\eta + D)\ 2\omega]^{-\frac{1}{2}} \tag{82}$$

This spectral density function behaves incorrectly in the limits $\omega \to 0$ and $\omega \to \infty$ (a spectral density function should be summable in the interval 0, + ∞). While the first divergence is a direct consequence of the neglect of the terms $\chi_a H^2$, the high frequency anomaly is due to the limitations of the continuum model. In fact the wavelength of a director mode cannot be shorter than molecular dimensions, and so the introduction of a high frequency cutoff has been proposed (23).

Keeping in mind the limitations of the model, we obtain as a final result, after substituting Eq. (82) into Eq. (78):

$$J_q(\omega) = A(T)\ [P_2(\cos\theta)]^2\ f_q(\chi)\ \omega^{-\frac{1}{2}} \tag{83}$$

where

$$A(T) = \tfrac{9}{2}\ kT\ \bar{P}_2^2\ [\eta/8\pi^2K^3]^{\frac{1}{2}}\ \gamma_1^4\ \hbar^2/r_{12}^6 \tag{84}$$

$$f_0(\chi) = 3(1 - \cos^2 2\chi)/2 \tag{85a}$$

$$f_1(\chi) = 4\cos^4\chi - 3\cos^2\chi + 1 \tag{85b}$$

$$f_2(\chi) = 1 - \cos^4\chi \tag{85c}$$

Eqs. (83 - 85) show in an explicit manner the dependence of $J_q(\omega)$ upon the experimental parameters. In particular, the $\omega^{-\frac{1}{2}}$ dependence of the longitudinal relaxation rate is indeed a striking feature of T_1 measurements in nematics. It is expected that order fluctuation dominate the relaxation rate at low frequencies. This effect is well demonstrated by $T_{1\rho}$ measurements (24). The relaxation time in the rotating frame $T_{1\rho}$ is in fact given as:

$$1/T_{1\rho} = \tfrac{1}{8}\ \{3J_0(2\omega_1) + 5J_1(\omega_0) + 2J_2(2\omega_0)\} \tag{86}$$

where $\omega_1 = \gamma_1 B_1$ is the strength of the rf field and it is several orders of magnitude lower than ω_0. The value of $T_{1\rho}$ is therefore expected to be dominated by $J_0(2\omega_1)$, which is a symmetric function around χ = 45° (24,25).

Lastly, the relevance of the term $(P_2(\cos\theta))^2$ in Eq. (83) is to be noted. Since θ is the angle between the axial magnetic interaction tensor (dipole or quadrupole), and the long molecular axis, it can be easily realized that the DF contribution depends strongly upon the molecular geometry. If we consider for instance the phenyl rings in PAA or similar compounds, we see that for the interproton dipolar interactions ($\theta = 0$) the geometric factor $(P_2(\cos\theta))^2$ is nearly two orders of magnitude larger than for the deuteron quadrupolar interaction ($\theta \cong 60°$), whose principal axis lies along the CD bond.

REFERENCES

1. P.L. Nordio and U. Segre, *The Molecular Physics of Liquid Crystals*, G.R. Luckhurst and G.W. Gray eds. (Academic Press, London, 1978); a) *Magnetic Resonance Spectroscopy. Static Behaviour*; b) *Rotational Dynamics*; c) *Magnetic Resonance Spectroscopy. Dynamical Aspects*.

2. J.H. Freed, in *Electron Spin Relaxation in Liquids*, L.T. Muus and P.W. Atkins eds. (Plenum Press, New York, 1972).

3. C.P. Slichter, *Principles of Magnetic Resonance* (Springer-Verlag, Berlin Heidelberg New York, 1980).

4. H. Mori, *Progr. Theoret. Phys.* **33**, 423 (1965); R. Zwanzig, *Ann. Rev. Phys. Chem.* **16**, 67 (1965).

5. B.J. Berne, in *Physical Chemistry*, Vol. VIIIB, H. Eyring, D. Henderson and W. Jost eds. (Academic Press, New York, 1971).

6. J.T. Hynes and J.M. Deutch, in *Physical Chemistry*, Vol. XIB, H. Eyring, D. Henderson and W. Jost eds. (Academic Press, New York, 1975).

7. G.T. Evans, *Mol. Phys.* **36**, 65 (1978).

8. H. Goldstein, *Classical Mechanics* (Addison Wesley, Singapore, 1980).

9. M.E. Rose, *Elementary Theory of Angular Momentum* (Wiley, New York, 1957).

10. P. Barchewitz, *Spectroscopie Atomique et Moléculaire* (Masson, Paris, 1971).

11. G. Moro and P.L. Nordio, *Chem. Phys. Lett.* **93**, 429 (1982).

12. L.P. Hwang and J.H. Freed, *J. Chem. Phys.* **63**, 118 (1975).

13. P.G. Wolynes and J.M. Deutch, *J. Chem. Phys.* **67**, 733 (1977); J.A. Montgomery and B.J. Berne, *J. Chem. Phys.* **67**, 4589 (1977).

14. J. Happel and H. Brenner, *Low Reynolds Number Hydrodynamics* (Prentice Hall, Englewood Cliffs, 1965).

15. G.R. Luckhurst, *Molecular Field Theories of Nematics*, in *The Molecular Physics of Liquid Crystals*, G.R. Luckhurst and G.W. Gray eds. (Academic Press, London, 1978).

16. G. Moro and P.L. Nordio, *Chem. Phys. Lett.* **96**, 192 (1983).

17. W.L. McMillan, *Phys. Rev. A*, **4**, 1238 (1971); ibid., **6**, 936 (1972); R.L. Humphries and G.R. Luckhurst, *Mol. Phys.* **35**, 1201 (1978).

18. N.G. van Kampen, *Stochastic Processes in Physics and Chemistry*, (North Holland, Amsterdam, 1981).

19. Z. Luz, R. Poupko and E.T. Samulski, *J. Chem. Phys.* **74**, 5825 (1981).

20. P.G. De Gennes, *The Physics of Liquid Crystals*, (Clarendon Press, Oxford, 1974).

21. M.J. Stephen and J.P. Straley, *Rev. Mod. Phys.* **46**, 617 (1974).

22. J.H. Freed, *J. Chem. Phys.* **66**, 4183 (1977).

23. P. Ukleja, J. Pirs and J.W. Doane, *Phys. Rev.* **A14**, 414 (1976).

24. J.W. Doane, C.E. Tarr and M.A. Nickerson, *Phys. Rev. Lett.* **33**, 620 (1974).

25. P.L. Nordio and U. Segre, *Gazz. Chim. Ital.* **106**, 431 (1976).

DENSITY MATRIX FORMALISM FOR NMR STUDIES OF LIQUID CRYSTALS

Robert L. Vold

Department of Chemistry,
University of California,
San Diego.

INTRODUCTION

From an NMR spectroscopist's point of view liquid crystals are interesting because they exhibit the full range of spin interactions normally encountered in solids, while retaining many of the experimentally convenient features of liquids. In particular, the quenching of intermolecular interactions by virtue of rapid translational motion implies that the spin response to a complex pulse sequence involves only a small number of spins, and can therefore be calculated by essentially exact quantum mechanical procedures. Application of such pulse sequences to liquid crystals yields information which is, in many cases, difficult to obtain by traditional spectral techniques. By combining a quantum mechanical calculation of the pulse response with a statistical mechanical calculation of spin relaxation between pulses, one obtains a powerful new source of dynamical information about liquid crystals.

Density matrix formalism is needed to fully exploit pulsed NMR as a tool for studying liquid crystals. Unfortunately, the abstract appearance of density operator expressions combined with an abundance of superscripts and subscripts which sprout from most symbols in explicit matrix equations has led to a widespread belief that density matrix theory is "difficult". In fact, the notational complexity is not as hard to comprehend as that needed for a general description of liquid crystal ordering tensors, and the physical ideas underlying density matrix formalism are quite straightforward.

In this chapter we develop density matrix equations in a

J. W. Emsley (ed.), Nuclear Magnetic Resonance of Liquid Crystals, 231–252.

form which is especially useful for describing relaxation behaviour of liquid crystals. This includes a description of standard pulse techniques such as inversion-recovery (T_1) and spin echoes (T_2) as well as a collection of more specialized multiple quantum pulse sequences. We make no attempt at mathematical rigour, preferring instead to focus on a set of computational tools which can be used to assess the potential of new techniques for providing information about ordered fluids. An excellent treatment of basic density matrix theory is given by Slichter (1). Operator formulations of special relevance to liquid crystals have been developed and reviewed by Werbelow and Grant (2), density matrix calculations of relaxation have been reviewed by Vold and Vold (3), and density matrix descriptions of multiple quantum phenomena have been given by Ernst (4-6), Pines (7-9), and the Volds and their coworkers (10-12).

BASIC THEORY

Consider an assembly of spin systems such as the individual molecules (solvent or solute) in a liquid crystalline sample. Each system evolves in time in accord with its Schrödinger equation, which accounts for interactions of the spins with external fields and with each other. We are not interested in the exact quantum state of any particular molecule; all we can measure is properties of the whole ensemble. Such properties correspond to the ensemble average of the quantum mechanical expectation value:

$$\overline{\langle Q\rangle} = \sum_{j=1}^{N} \langle Q\rangle_j / N \; . \tag{1}$$

Here Q stands for the operator corresponding to a property associated with the spins of one system, the angular brackets denote the expectation value, and the bar reminds us that we are averaging over an ensemble of N systems (molecules).

At any time t, the state of a given system j is fully specified by its wavefunction $\psi^{(j)}(t)$. Thus the expectation value of Q for system j is

$$\langle Q\rangle_j = \langle \psi^{(j)}(t)|Q|\psi^{(j)}(t)\rangle \; . \tag{2}$$

It is instructive to expand each $\psi^{(j)}(t)$ in terms of an *arbitrary* complete set of time independent wavefunctions $\{\phi\}$:

$$\psi^{(j)}(t) = \sum_{k=1}^{M} C_k^{(j)}(t)\phi_k \; . \tag{3}$$

Note that for each system, all the time dependence is carried in the C-coefficients. Furthermore, the time independent basis functions ϕ_k are the same for all systems: system j differs from system j' only by virtue of a different set of C's.

Combining Eqs. (1)-(3), we obtain

$$\overline{\langle Q\rangle} = \sum_{j=1}^{N} \sum_{k=1}^{M} \sum_{\ell=1}^{M} C_k^{(j)*}(t)C_\ell^{(j)}(t)\langle\phi_k|Q|\phi_\ell\rangle/N \; . \tag{4}$$

Since neither Q nor the functions ϕ vary across the ensemble, it is appropriate to collect them into a single matrix element $Q_{k\ell}=\langle\phi_k|Q|\phi_\ell\rangle$, and to lump all the system dependent quantities in a second matrix with elements

$$\sigma_{\ell k} = \sum_{j=1}^{N} C_k^{(j)*}(t)C_\ell^{(j)}(t)/N = \overline{C_k^*C_\ell} \; . \tag{5}$$

Eq. (5) defines the elements of the spin density matrix σ relative to the basis $\{\phi\}$, and together with Eq. (4) yields the fundamental relation

$$\begin{aligned}\langle Q\rangle &= \sum_k \sum_\ell \sigma_{\ell k}(t)\langle\phi_k|Q|\phi_\ell\rangle \\ &= \sum_k \sum_\ell \sigma_{\ell k}(t)Q_{k\ell} \\ &= \mathrm{trace}(\sigma(t)Q).\end{aligned} \tag{6}$$

It is important to note that although Eq. (5) serves to define the density matrix, it is never used to compute density matrix elements. The whole point of statistical mechanics is that we may avoid explicit reference to states of individual systems (the $C_k^{(j)}(t)$) and deal only with states of the ensemble, i.e., with values of the average quantities $\sigma_{k\ell}$. Before examining ways of computing these quantities, it is useful to contemplate some illustrative cases.

Fig. 1 shows the appearance of the spin density matrix for three different states of an ensemble of molecules each containing a single deuteron.

$$\begin{bmatrix} 1+p & 0 & 0 \\ 0 & 1 & 0 \\ 0 & 0 & 1-p \end{bmatrix} \quad \begin{bmatrix} 1 & p/2^{\frac{1}{2}} & 0 \\ p/2^{\frac{1}{2}} & 1 & p/2^{\frac{1}{2}} \\ 0 & p/2^{\frac{1}{2}} & 1 \end{bmatrix} \quad \begin{bmatrix} 1 & 0 & -ip \\ 0 & 1 & 0 \\ ip & 0 & 1 \end{bmatrix}$$

(A) (B) (C)

Figure 1: *Density matrices for three different states of an ensemble of spins I=1. A) Thermal equilibrium. B) After a strong 90° pulse. C) After creation of pure double quantum coherence.*

Since there are three stationary states for each spin, the density matrix is of dimension 3x3. In Fig. 1A, the spins are at thermal equilibrium. This implies two fundamental conditions: the *phase* of the wavefunction for any system in the ensemble is uncorrelated with that of any other, and the ratio of populations of any pair of states is given by the corresponding Boltzmann factor. Referring to Eq. (5), it can be seen that the first condition is equivalent to requiring that all off-diagonal density matrix elements be zero. In order to obtain the eigenstate populations (diagonal density matrix elements) shown in Fig. 1A, it is assumed that the energy difference between any two adjacent states is a small fraction of the thermal energy kT. Then the Boltzmann factors which relate population ratios can be expanded to first order in powers of the quantity $\Delta E/kT = h\omega_0/kT$.

Using Eq. (6) with σ from Fig. 1A to calculate observable magnetization, we see that

$$M_x = \mathrm{trace}(\sigma I_x) = \sum_j \sigma_{jj} (I_x)_{jj} = 0 \qquad (7a)$$

$$M_y = \mathrm{trace}(\sigma I_y) = \sum_j \sigma_{jj} (I_y)_{jj} = 0 \qquad (7b)$$

$$M_z = \mathrm{trace}(\sigma I_z) = \sum_j \sigma_{jj} (I_z)_{jj} \qquad (7c)$$

$$= \sigma_{11} - \sigma_{33}$$

$$= (1 + p) - (1 - p) = 2p$$

$$\equiv M_0 \ .$$

All is in order: Fig. 1A is a simple visual indication of the well known fact that at equilibrium the magnetization lies along the static field. Note that the constant term in each diagonal density matrix element makes no contribution to any expectation value. It is usually dropped from further consideration, and one then deals with a "reduced" spin density matrix.

In Fig. 1B the ensemble has just experienced a strong 90° pulse, applied along the -y axis of the rotating frame. Again using Eq. (6) to compute expectation values, but this time with σ from Fig. 1B, we find

$$M_y = M_z = 0 \tag{8a}$$

$$\begin{aligned} M_x &= \sum_j \sum_k \sigma_{jk} (I_x)_{kj} \\ &= (\sigma_{12} + \sigma_{21})/\sqrt{2} + (\sigma_{23} + \sigma_{32})/\sqrt{2} \\ &= 2p \\ &= M_0 \ . \end{aligned} \tag{8b}$$

As expected, the density matrix in Fig. 1B corresponds to a transverse magnetization equal in magnitude to the equilibrium value M_0, but directed along the x axis. According to Eq. (5), this implies that phase coherence has been created in the ensemble. This takes place as individual molecules each absorb one photon simultaneously during the rf pulse.

A straightforward calculation of magnetization for ensemble 1C shows that *all* components of M are zero. This is surely not an equilibrium state, but where did the magnetization go? By analogy with Eq. (8) one can define a double quantum "magnetization" whose x-component, for example, is given by

$$M_x(13) = (\sigma_{13} + \sigma_{31})/\sqrt{2} \ . \tag{9}$$

It must be noted, however, that the analogy is not perfect. For example, it will be shown below that the double quantum magnetization defined in this fashion does *not* simply precess about an effective field in a frame rotating at the Larmor frequency or any multiple thereof. Most of the peculiar features of multiple quantum coherence can be traced to its description in terms of spherical tensors of rank greater than one.

The density matrices exhibited in Fig. 1 are all written using basis functions which are assumed to be eigenstates of the time independent Hamiltonian operator. Expectation values evaluated acording to Eq. (6) are of course independent of the basis in which the trace evaluation is carried out, but the choice of basis is crucial when we assign physical significance to particular density matrix elements. For example, the eigenstate basis for an A_2 spin system with two nuclei of spin 1/2 is

$$\begin{aligned} \psi_1 &= |\alpha\alpha\rangle \\ \psi_2 &= (|\alpha\beta\rangle + |\beta\alpha\rangle)/\sqrt{2} \\ \psi_3 &= (|\alpha\beta\rangle - |\beta\alpha\rangle)/\sqrt{2} \\ \psi_4 &= |\beta\beta\rangle . \end{aligned} \tag{10}$$

In this basis it is easy to show using Eq. (6) that the total complex transverse magnetization $M_x - iM_y$ is given by

$$M_x - iM_y = \mathrm{trace}(\sigma I_-) = (\sigma_{12} + \sigma_{23})/\sqrt{2} . \tag{11}$$

Naturally, we associate σ_{12} with one of the two transitions observable for A_2 in a liquid crystal solvent, and σ_{23} with the other. However, if the simple product basis rather than Eq. (10) were used to evaluate the trace, there would be four density matrix elements instead of two appearing in expressions for the transverse magnetization and the simple physical significance would be lost.

The best way to avoid such problems is to specify a set of operators whose expectation values, evaluated according to Eq. (6), can be associated with the individual transitions. The importance of such "fictitious spin-1/2" or "single transition" operators (13) has long been recognized, but general definitions and discussions of their properties have been lacking until quite recently. Vega and Pines (7) used single transition operators in their original description of double quantum coherence in I=1 spin systems, and Vega (14) extended the formalism to I=3/2. Wokaun and Ernst (5) have provided a complete set of definitions and commutation relations for single transition operators for N nuclei of spin-1/2, and Mehring et al. (15) have presented similar relations for the I=1 spin system.

It is easy to define single transition operators in terms of their matrix representations in the eigenstate basis: for the transition $|i\rangle \to |j\rangle$ one defines generalized Pauli spin matrices such that the only non-zero elements are ii, jj, ij or ji. For example, the complete set of single transition operators for one deuteron (15) is represented by the nine matrices

$$I_x(12) = \tfrac{1}{2}\begin{pmatrix} 0 & 1 & 0 \\ 1 & 0 & 0 \\ 0 & 0 & 0 \end{pmatrix} \quad I_x(23) = \tfrac{1}{2}\begin{pmatrix} 0 & 0 & 0 \\ 0 & 0 & 1 \\ 0 & 1 & 0 \end{pmatrix} \quad I_x(13) = \tfrac{1}{2}\begin{pmatrix} 0 & 0 & 1 \\ 0 & 0 & 0 \\ 1 & 0 & 0 \end{pmatrix}$$

$$I_y(12) = \tfrac{1}{2}\begin{pmatrix} 0 & -i & 0 \\ i & 0 & 0 \\ 0 & 0 & 0 \end{pmatrix} \quad I_y(23) = \tfrac{1}{2}\begin{pmatrix} 0 & 0 & 0 \\ 0 & 0 & -i \\ 0 & i & 0 \end{pmatrix} \quad I_y(13) = \tfrac{1}{2}\begin{pmatrix} 0 & 0 & -i \\ 0 & 0 & 0 \\ i & 0 & 0 \end{pmatrix}$$

$$I_z(12) = \tfrac{1}{2}\begin{pmatrix} 1 & 0 & 0 \\ 0 & -1 & 0 \\ 0 & 0 & 0 \end{pmatrix} \quad I_z(23) = \tfrac{1}{2}\begin{pmatrix} 0 & 0 & 0 \\ 0 & 1 & 0 \\ 0 & 0 & -1 \end{pmatrix} \quad I_z(13) = \tfrac{1}{2}\begin{pmatrix} 1 & 0 & 0 \\ 0 & 0 & 0 \\ 0 & 0 & -1 \end{pmatrix}$$

By applying Eq. (6) to each of these nine operators, it is possible to determine which particular density matrix elements are needed to describe the x, y, or z components of magnetization for each transition. Other, more powerful uses of single transition operators involve manipulations based on their commutation properties together with their relation to other operators such as spherical irreducible tensor components. General operator equations of this variety tend to be complicated, and we defer illustrative examples to the next section.

It is possible, and occasionally useful, to express the spin density operator as a linear combination of single transition operators:

$$\sigma(t) = \sum_{\alpha} \sum_{j<k} a_{\alpha}(jk) I_{\alpha}(jk) \ . \tag{12}$$

Here the $a_\alpha(jk)$ are (real valued) functions of time, corresponding to component α (x, y or z) of magnetization associated with transition $|j\rangle \rightarrow |k\rangle$. A major problem faced by users of density matrix formalism is that for most systems of chemical interest the number of spin states, and hence the dimensionality of the density matrix, is unmanageably large. Representation of σ by single transition operators in no way alleviates this problem since the number of operators needed in Eq. (12) is equal to the total number of elements in the density matrix. In such cases it is useful to seek alternate representations, with operators chosen to emphasize particular features of interest. For example, spherical irreducible tensor operators are a natural choice for discussions of pulse flip angle effects (16), and various combinations of spin angular momentum operators (chosen in part on group-theoretical grounds) have proven useful (17-20) describing spin-lattice relaxation phenomena.

In what follows, we shall introduce new operator representations as needed for visualizing free precession, response to pulses, and the creation and detection of multiple quantum coherence. The discussion will be based on operator solutions to the equation of motion

$$\frac{d\sigma}{dt} = i[\sigma, H] . \tag{13}$$

The Hamiltonian operator includes time *independent* interactions of all kinds among the spins and between spins and static fields, as well as time *dependent* interactions with (pulsed) rf fields. Eq. (13) can easily be derived by inserting the definition of σ, Eq. (5), into the time dependent Schrödinger equation. Relaxation effects are not included at this point. As indicated in the next chapter, extension of Eq. (13) to accommodate relaxation is straightforward though computationally messy.

It is easy to verify by direct substitution that when H is time independent, the solution to Eq. (13) is given by

$$\sigma(T) = \exp(-iHt)\sigma(0)\exp(iHt) . \tag{14}$$

The exponential operators appearing in this expression can always be defined by power series expansion, but this offers little in the way of explicit solutions. For simple spin systems, such that H can be diagonalized in closed form, the exponential matrices can be evaluated directly using standard procedures of linear algebra. However, neither method leads to satisfying physical pictures. The phenomena described in subsequent sections were selected in part to illustrate how, by decomposing H into a sum of appropriate operators and making use of commutator relations, it is possible to recover simple physical pictures of seemingly esoteric spin behaviour.

FREE PRECESSION

In absence of perturbing rf fields, each density matrix element σ_{jk} oscillates harmonically in time with a frequency corresponding to the energy difference between spin states $|j\rangle$ and $|k\rangle$. This fact, which is easy to verify by transcribing Eq. (13) in the eigenbasis of the static Hamiltonian, is equivalent to stating that the expectation values of single transition operators evolve in time according to the equations

$$\langle I_z(jk)\rangle_t = \text{constant} \tag{15a}$$

$$\langle I_x(jk)\rangle_t = \langle I_x(jk)\rangle_0 \cos \omega_{jk}t - \langle I_y(jk)\rangle_0 \sin \omega_{jk}t \tag{15b}$$

$$\langle I_y(jk)\rangle_t = \langle I_u(jk)\rangle_0 \cos \omega_{jk}t + \langle I_x(jk)\rangle_0 \sin \omega_{jk}t\ . \tag{15c}$$

The formal proof of Eq. (15) is surprisingly lengthy and will not be given here, but it is evident that one way to visualize free precession in a complex spin system involves vectors rotating about the field (z) axis; one such vector for each set of operators $I_x(jk)$, $I_y(jk)$, and $I_z(jk)$. Operator formalism is scarcely needed to establish this picture, but it is instructive to inquire more closely into precession following a single, strong 90° pulse applied at exact resonance. The I=1 spin system will be considered explicitly for illustrative purposes.

We assume that the spins experience a quadrupolar interaction which is small compared with the Zeeman interaction. Thus in a coordinate system rotating at the Larmor frequency ω_0 about the space fixed z axis,

$$H = (\omega_Q/3)(3I_z^2 - I\cdot I) \tag{16}$$

where $2\omega_Q$ is the doublet separation in the high resolution spectrum. As will be shown in the next section, a 90° pulse along the -x axis of the rotating frame produces a density matrix σ proportional to I_y so that Eq. (14) becomes

$$\sigma(t) = \exp\{-(i\omega_Q t/3)(3I_z^2 - I\cdot I)\}I_y \exp\{(i\omega_Q t/3)(3I_z^2 - I\cdot I)\}. \tag{17}$$

It is convenient at this point to introduce the "quadrupolar" operators

$$Q_x = I_xI_z + I_zI_x = \sqrt{2}(I_x(12) - I_x(23)) \tag{18a}$$

$$Q_y = I_yI_z + I_zI_y = \sqrt{2}(I_y(12) - I_y(23)) \tag{18b}$$

$$Q_z = I_z^2 - (I\cdot I)/3 = \frac{2}{3}(I_z(12) - I_z(23))\ , \tag{18c}$$

so that (17) becomes simply

$$\sigma(t) = \exp\{-i\omega_Q tQ_z\}I_y \exp\{i\omega_Q tQ_z\} \tag{19}$$

$$= I_y \cos \omega_Q t - Q_x \sin \omega_Q t\ .$$

The last equality in Eq. (19) follows from a very useful relation (21) among any three operators P,Q and R which obey the commutation rule $[P,Q] = i\kappa R$ (with cyclic permutations):

$$\exp(-i\theta P)Q\exp(i\theta P) = Q\cos\kappa\theta + R\sin\kappa\theta \ . \tag{20}$$

It can be shown by direct matrix multiplication that Q_z, I_y and Q_x satisfy the necessary commutation rules. A simple way to do this is to represent each operator by a 3x3 matrix. The physical picture of free precession implied by Eq. (19) emphasizes different features than that based on Eq. (15). Both pictures are shown in Fig. 2.

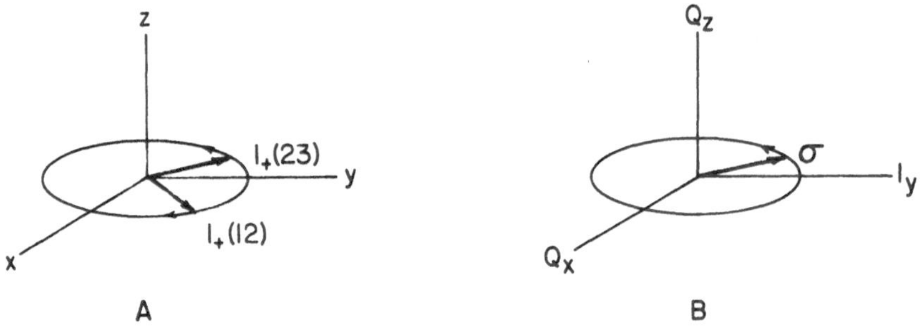

Figure 2: *Vector pictures of free precession for I=1. A) In ordinary spin space. B) In a fictitious spin space.*

Fig. 2A shows the two counter-rotating vectors in ordinary spin space which correspond to expectation values of the single transition operators. Fig. 2B illustrates the same motion, expressed as motion of a *single* vector with components $\langle Q_x \rangle$ and $\langle I_y \rangle$ rotating in a fictitious 3-dimensional spin space spanned by Q_x, I_y and Q_z. (The third axis in this space is irrelevant for present purposes since the vector moves in a plane perpendicular to it). Fig. 2A emphasizes the well known fact that free precession for I=1 consists of two rotating magnetization vectors, while Fig. 2B emphasizes the less well known fact that correlation between these vectors, as measured by $\langle Q_x \rangle$, grows with time at the expense of ordinary transverse magnetization. Pines and coworkers (22-24) have demonstrated in an elegant series of experiments that growth of such spin correlation is a necessary precursor to the creation of multiple quantum coherence.

RF PULSES: FLIP ANGLES AND PHASE SHIFTS

During an rf pulse, the oscillating magnetic field and local fields generated by spin interactions exert competing forces. For liquid crystal samples, the local fields arise primarily from dipolar and/or quadrupolar interactions. The general formulation of spin response to such complicated perturbations is quite complex, and we again focus attention on the simplest system which exhibits most of the essential features: the I=1 spin system with a small residual quadrupolar interaction.

The Hamiltonian operator which describes the spin response to a pulse is

$$H = -(\omega_0 - \omega)I_z - \omega_1 I_x + (\omega_Q/3)(3I_z^2 - I\cdot I) . \qquad (21)$$

Here $\omega_1 = \gamma B_1$ is the amplitude of the pulse and ω is its carrier frequency. Eq. (21) has been written in a coordinate system rotating with the pulse, so that the rf Hamiltonian $-\omega_1 I_x$ is time independent and the effective Zeeman interaction strength is $(\omega_0 - \omega)$.

By far the simplest situation occurs when the pulse is strong, so that both the Zeeman (off-resonance) and quadrupolar interactions can be ignored. Even in this simple situation, the existence of spin coupling produces unusual behaviour. For example, suppose that our I=1 spin system has been placed in a state such that $\sigma = I_x$ (eg, by applying a strong 90° pulse at exact resonance, along the rotating frame y-axis, and waiting for time $\tau = \pi/(2\omega_Q)$).

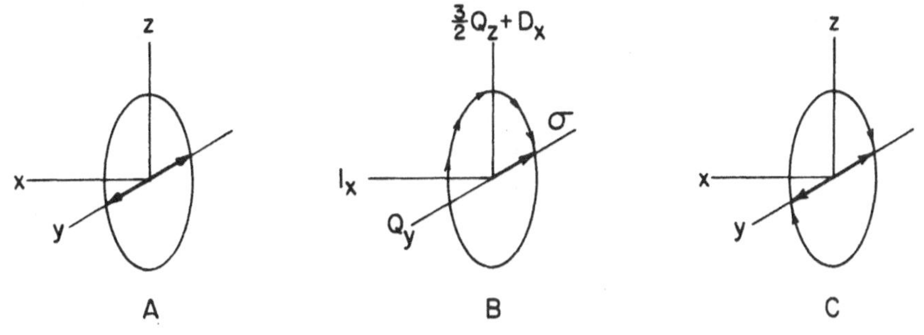

Figure 3: *Effect of a strong 90° pulse. A) Initial state, before the pulse. B) Rotation by 180° in an abstract space. C) Trajectories followed by the magnetization vectors.*

Intuitively, one would expect that a 90° pulse along x would realign the "double-headed arrow" (Fig. 2A) along z. The actual effect of such a pulse is quite different, and in order to visualize what happens we need first to establish commutation relations among a set of operators which includes Q_y (the initial density matrix, I_x (the rf Hamiltonian), and their commutator. By direct matrix multiplication it is easy to verify that the following commutation rule holds (with cyclic permutations):

$$[I_x, Q_y] = 2i(\tfrac{3}{2}Q_z + D_x) \tag{22}$$

where $D_x = I_x(13)$ is a double quantum operator whose significance will be discussed in the next section. It follows from Eq. (20) that

$$\exp(i\omega_1 I_x)\, Q_y \exp(-i\omega_1 I_x) = Q_y \cos(2\omega_1 t) - (\tfrac{3}{2}Q_z + D_x)\sin(2\omega_1 t)\ . \tag{23}$$

Thus after the 90° pulse Q_y has been "rotated" 180° rather than 90°, which is an inescapable manifestation of the quadrupolar character of the state described by $\sigma = Q_y$. This is shown in Fig. 3B as a simple rotation in the proper subspace, and in Fig. 3C as an equivalent rotation in ordinary spin space (where the double quantum coherence is invisible). Particularly clear theoretical calculations as well as direct experimental demonstrations of this behaviour have been presented by Emid et al. (25,26).

The response of coupled spins described by an arbitrary non-equilibrium but *diagonal* density matrix to a strong 90° pulse is closely related to the phenomenon described above. For example, suppose that the I=1 spin system has somehow been placed in a state such that one of the two equilibrium population differences has been inverted but not the other. This corresponds to an initial density matrix $\sigma(0)$ given by

$$\sigma(0) = \Delta\ \mathrm{diag}(-1,1,0) \tag{24}$$

$$= -(\Delta/2)(I_z + 3Q_z)$$

Here the *equilibrium* population differences (P_3-P_2) and (P_2-P_1) are both denoted by Δ, and (P_3-P_2) has been inverted. Intuitively, a strong 90° pulse applied to this system should yield a spectrum with one line "upside down" and the other line twice the equilibrium value. In practice, the result is dramatically different. As demonstrated in Eq. (24), the initial density matrix is composed of a part proportional to I_z and a part proportional to Q_z. A strong 90° pulse rotates the first part into I_y, but in view of Eq. (14), the fate of the second part is determined by evaluating the expression

$$\exp(i\omega_1 t_w I_x)\, Q_z \exp(-i\omega_1 t_w I_x) \ . \qquad (25)$$

Here the Zeeman and quadrupolar terms in Eq. (21) have been dropped, and the flip angle $\omega_1 t_w$ will shortly be set equal to 90°. *Cyclic* commutation relations involving I_x and Q_z are not readily obtainable, and it is simpler in this case to evaluate the rotation operator directly. After a rather lengthy series of manipulations, one obtains, for the special case of a 90° pulse,

$$\sigma(t_w) = (\Delta/2)(I_y + \tfrac{3}{2}Q_z + 3D_x) \qquad (26)$$

The physical result implied by this is that a 90° pulse "rotates" half of Q_z by 180° to $-Q_z$, while simultaneously transforming the other half into double quantum coherence, and the directly observable signal comes entirely from that part of $\sigma(0)$ proportional to I_z. Starting from a more general diagonal density matrix σ = diag (a,b,c) it can be shown that a 90° x-pulse *always* produces an observable signal proportional to $\langle I_y \rangle$. The implication, as originally recognized by Schaublin, et al. (27), is that a strong 90° pulse is incapable of revealing differential population inversion.

In their original discovery and description of such "pulse mixing" effects, Schaublin et al. (27) emphasized the importance of the non-linear nature of the exponential operator: there is really no reason to expect a *strong* pulse to produce a linear response such that each line intensity is proportional to a single population difference. They noted that for studies of spin lattice relaxation in which individual population differences are of interest, mixing effects could be avoided at the expense of signal/noise ratio by using observation pulses with small flip angles. More recently (16,28,29) the close connection between such mixing effects and the transformation properties of multipole operators has been recommended as a diagnostic technique for verifying the creation of unusual and potentially useful non-equilibrium spin states.

As a final and very useful example of mixing effects of strong pulses, consider the Jeener-Broekaert pulse sequence (30) 90_x-τ-45_y-t_1-45_y-ACQ. After the first pulse the density operator is just I_y, and if τ is properly selected then $\sigma = Q_x$ just prior to the second pulse. The second pulse establishes quadrupolar order in an I=1 spin system (the *diagonal* part of the density operator is Q_z. During the time t_1 quadrupolar order is gradually lost as Zeeman order (σ proportional to I_z) grows, and the exact state can be monitored by measuring sum and difference intensities in the spectrum obtained by Fourier transforming the signal acquired after the last pulse. As described in the following chapter, this technique is very well suited for determining individual motional spectral density parameters in liquid crystal media. Spiess (31,32) has used the same sequence to study very slow motions in solids.

Unfortunately, dipolar and quadrupolar couplings in liquid crystal samples are often so large that they cannot be entirely ignored during rf pulses. The commutation relations of the *full* rf Hamiltonian with simple spin operators are not yet well enough understood to yield simple physical pictures. In such cases, and also for strong pulses applied to complex, multispin systems, it is necessary if not desirable to resort to numerical solution of Eq. (14) with H given by Eq. (21). The results of such calculations are conveniently summarized by the "coherence transfer equation" (12):

$$\sigma_{pq} = \sum_{pq}\sum U^*_{rp}(\alpha,\phi)\, U_{sq}(\alpha,\phi)\, \sigma_{rs} \; . \tag{27}$$

Here $U(\alpha,\phi)$ stands for the propagator $\exp(iHt_w)$ where the pulse phase $\phi = 0$ for a pulse along the x-axis. U reduces to a simple rotation operator only when the pulse is strong. The coefficient in Eq. (26) which connects any density matrix element σ_{rs} with σ_{pq} is a measure of the efficiency with which the pulse converts the former coherence into the latter.

An important general result (5,33) can be established with the aid of the coherence transfer equation: if the transfer efficiency from σ_{rs} to σ_{pq} is denoted by $a_{pqrs}(0)$ for a pulse along the x-axis, then a pulse of the same (arbitrary) amplitude but with phase shift ϕ will have transfer efficiency

$$a_{pqrs}(\phi) = a_{pqrs}(0)\, \exp[i\phi(m_{pq} - m_{rs})] \; . \tag{28}$$

Here m_{pq} and m_{rs} refer to the magnetic quantum number differences between corresponding eigenstates of the static Hamiltonian operator, and they define the "order" of the coherence. Eq. (27) implies that if a pulse transfers coherence from a manifold of order n to a manifold of order m, then shifting the phase of the pulse by an amount ϕ will shift the phase of the transferred signal by $(m-n)\phi$. As described in the next section, this forms the basis of a powerful method for selective detection of coherences of any desired order.

MULTIPLE QUANTUM COHERENCE

Suppose that we have an assembly of coupled spins, existing in a state described by some arbitrary, non-equilibrium density matrix. Since the probe coil is sensitive only to dipole radiation, signal will be detected only if there is oscillating (precessing) transverse magnetization. According to the prescription given earlier, the (complex) signal is proportional to a collection of density matrix elements σ_{jk} such that $|j\rangle$ and $|k\rangle$ differ by one unit of angular momentum. These single quantum coherences thus correspond to the expectation values of single transition operators $I_x(jk) + I_y(jk)$, with $(m_j - m_k) = +1$. The other off diagonal elements of σ induce no voltage in the sample coil, but it is evident from Eq. (15b) and (15c) that if they could be detected, their precession frequencies might provide much useful spectral information. Since these multiple quantum coherences generate no directly observable signal, indirect detection methods are required. Fig. 4 illustrates the principle: create multiple quantum coherence by some appropriate pulse sequence, let it evolve freely (and invisibly) for some time t_1, and then transfer it back to transverse magnetization for observation during time t_2.

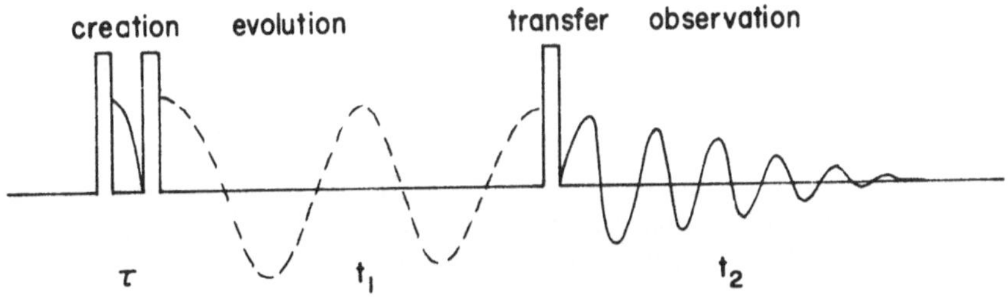

Figure 4: *General multiple quantum pulse sequence.*

The existence of multiple quantum coherence during t_1 is revealed by frequency and/or phase modulation of the signal detected during t_2.

The simplest example of multiple quantum coherence involves, of course, the I=1 spin system. Suppose that a pulse sequence 90_x-τ-90_{-x} is applied to this spin system. Referring to Eq. (19), we see that just prior to the second pulse σ is a linear combination of I_y and the quadrupolar operator ("double pointed arrow") Q_x. Using the well known commutation rules for

I_x, I_y, and I_z it follows from Eq. (20) that the part of σ proportional to I_y is simply converted back to I_z. The effect on Q_x of a strong pulse of duration t_w and amplitude ω_1 follows from the two *non-cyclic* commutation rules $[I_x,Q_x] = 2iD_y$ and $[I_x,D_y] = (i/2)\ Q_x$:

$$\exp(-i\omega_1 t_w I_x)\ Q_x\ \exp(i\omega_1 t_w I_x) = Q_x \cos(\omega_1 t_w) + 2D_y \sin(\omega_1 t_w). \quad (29)$$

Here $D_{x,y} = I_{x,y}$ (13) are double quantum operators whose matrix representations are given in Section 2. We see from Eq. (29) that the second pulse should have flip angle 90° to achieve maximum stimulation of double quantum coherence, and from Eq. (19) that the interval τ between the two excitation pulses should fulfill the condition $\omega_Q\tau = n\pi/2$ where n is an odd integer.

Postponing for the moment the question of how to observe multiple quantum coherence, the simple calculation of its creation outlined above raises several questions: 1) is this the only excitation method and if not, how does one obtain the strongest multiple quantum coherence? 2) can two pulses be used to excite coherence of order greater than two in suitably complex spin systems? What are the limits? 3) is it possible to design pulse sequences which will stimulate coherences of one, preselected order without creation of unwanted coherence of other orders? Pines and coworkers (8,9,34,35) have provided elegant answers to these questions.

One alternative to excitation by strong pulses is the application of a single, weak pulse at a point midway between two connected single quantum transitions. It can be shown (7,14) that such a pulse, with effective amplitude ω_1^2/ω_Q, rotates the equilibrium density operator smoothly from I_z to D_y. With most pulsed spectrometers now used to record liquid crystal spectra, the spectral coverage is not really uniform over the large spread of dipolar and especially quadrupolar splittings and it is likely that creation of double quantum coherence is the rule rather than the exception. *Optimal* stimulation of double quantum coherence by means of a single weak pulse applied to a complex system is not easy to investigate theoretically because the effective pulse flip angle is different for different splittings. Also, most spectrometers have limited capability for fine control of rf field strength. Stimulation techniques involving more than one pulse appear to be superior in practical applications.

It is evident that the first pulse of a two-pulse multiple quantum excitation sequence should be a 90° pulse, so that maximum single quantum coherence is created as a starting point.

It can also be shown very generally that maximum transfer is achieved when the flip angle of the second pulse is also 90°. Thus optimization of the two pulse sequence consists of choosing the pulse separation τ and the phases of the two pulses. In general, pairs of connected transitions will arrive at anti-parallel configurations after different, incommensurate time intervals. It follows that the choice of τ for multiple quantum excitation in a complex spin system must favour some coherences over others. We (12,36) and others (33,35) adopt the simple procedure of observing a single point of the free induction decay following a complete excitation/detection sequence, varying τ empirically for maximum signal. As noted in Section 3 the only constraint here is that multiple quantum *correlations* must be given time to evolve before the second pulse is applied. Thus to excite double quantum coherence in an assembly of partially ordered deuterons, τ can be on the order of the inverse of the quadrupolar splitting but to excite higher order coherence (necessarily involving two or more deuterons) τ must be much longer, on the order of the inverse of ^2H-^2H dipolar couplings (36).

Optimal selection of pulse phases for two-pulse multiple quantum stimulation is easy when only one double quantum coherence is involved. In this case, the two single quantum magnetizations achieve anti-parallel alignment along an axis oriented at angle $\Delta\tau$ in the xy plane of the rotating frame, where Δ is the offset of the carrier frequency from exact resonance. By applying the second pulse along precisely this axis, maximal double quantum excitation is achieved (7,12,37). Of course, it is not possible to satisfy this condition for a set of spins with internal chemical shifts, and the spectrometer may not be equipped with the requisite fine control of pulse phases. (As a rule of thumb, pulse phases for n-quantum spectroscopy should be settable in increments of (90/n)° or finer). If a 180° pulse is applied at the midpoint of the sampling interval, differential precession due to resonance offset is eliminated, and it is then possible to stimulate multiple quantum coherence using two pulses with identical phase.

Pines and coworkers (8,22,38,39) have drawn an important distinction between even-selective and odd-selective pulse sequences. For example, the sequence 90_x-$(\tau/2)$-180-$(\tau/2)$-90_{-x} will excite only coherence of even order, while changing the phase of the last pulse from -x to y leads to excitation of exclusively odd order coherence. The proof of this claim (too lengthy to reproduce here) relies on the fact that only even-quantum operators appear in density matrix formulae for the former sequence, and only odd-quantum operators are needed to describe the latter. In a nonselective two pulse sequence, the initial equilibrium density matrix is transformed into one in which off diagonal elements corresponding to coherences of all orders are non-zero. It is reasonable to expect, and borne out in practice (9), that use of selective excitation to suppress part of the potential coherence improves the stimulation efficiency of the remainder. In fact, by making clever use of time-reversed pulse sequences with sub-cycles chosen to be n-quantum selective, Warren et al. (8,9) were able to achieve remarkably efficient stimulation of very high order coherences.

High order multiple quantum spectra of liquid crystalline materials are of great potential interest because they have many fewer lines than 1-quantum spectra, and they are therefore easier to analyse. Pines and coworkers (9,34,38) have presented examples of readily interpretable multiple quantum spectra of oriented materials for which the single quantum spectra were hopelessly complicated and unresolved. Unfortunately, the multipulse sequences needed for efficient, selective excitation of high order multiple quantum spectra result in unavoidable sample heating. The well known sensitivity of dipolar and quadrupolar splittings to small changes of sample temperature is then a major source of error in such experiments. Until this problem is overcome, multipulse excitation sequences appear to be of limited general utility in liquid crystal spectroscopy.

Selective detection of multiple quantum coherence is much easier to achieve than selective excitation. The most widely used class of selective detection methods is based on the coherence transfer equation, Eq. (27). Suppose that a multiple quantum coherence σ_{rs} of order $\Delta m = (m_r - m_s)$ has been created by some particular sequence. If the phase of every pulse in the excitation sequence is advanced by ϕ, then the phase of the newly generated coherence will be advanced by $(m_r - m_s)\phi$. Thus to retain double quantum coherence while suppressing single quantum coherence, one co-adds two separate signals in which the excitation phases differ by 180°. Single quantum coherence then alternates in sign for the two experiments, and cancels, while the sign of the double quantum coherence remains unchanged and that signal accumulates.

In practice, application of this method requires evaluation of the phases of the desired coherence before and after all pulses in the sequence. For example, in the multiple quantum spin echo pulse sequence $90-\tau-90-(t_1/2)-180-(t_1/2)-90-ACQ$, suppose that a particular multiple quantum coherence σ_{rs} is transferred by the last pulse to a 1-quantum coherence σ_{rs}. The phase of this particular signal is given by

$$\phi(obs)=\phi_e(\Delta m) + 2\phi_r(\Delta m) + \phi_m(\Delta m-1) , \tag{30}$$

where subscripts e,r and m refer to the excitation, refocussing, and monitoring pulse phases respectively. The factor of 2 in front of ϕ_r arises because an ideal 180° pulse always transfers a coherence of order $+\Delta m$ to one of order $-\Delta m$. Clearly, the sign of double quantum coherence reverses upon incrementing the refocussing pulse phase by 45°. Thus if the monitoring pulse has constant phase, alternate addition and subtraction of signals will lead to accumulation of double quantum coherence alone. Further analysis in similar vein shows that incrementing the refocussing pulse phase by 90/n, combined with alternate addition and subtraction of n transients, cancels all coherences of order n-1 and lower. Suppression of other artifacts, caused for example by imperfect excitation pulses, requires phase cycling of the excitation pulses. It should be noted at this point that many commercial pulsed spectrometers impose additional, gratuitous phase shifts designed to eliminate artifacts associated with quadrature phase detection, and these must be included in the design of phase cycled sequences for multiple quantum spectroscopy. Finally, it must be remembered that the best of phase cycling procedures will *not* cancel imperfections arising from pulse mixing of coherences of the same order. Even with proper phase cycling, artifact suppression in multiple quantum spectroscopy requires high quality pulses with accurately set flip angles.

The precessional motion of multiple quantum coherence during the evolution time t_1 is interrupted significantly by the 180° pulse. Resonance offset, chemical shifts, and dipolar coupling to nonresonant spins are all refocussed by the 180° pulse, leaving a spectrum with narrowed lines and only quadrupolar and homonuclear dipole-dipole splittings. Especially when displayed as two-dimensional contour plots, so as to reveal the correlation between individual features of the single and multiple quantum spectral domain, such spectra would be useful adjuncts for liquid crystal structure determinations. Double quantum spectra, which are relatively easy to obtain with high sensitivity, are likely to be most useful for this purpose. Techniques analogous to those used for complex 2-dimensional single quantum spectra have not yet been highly developed, but the approach seems promising.

Since resonance offset is refocussed by the 180° pulse, it is reasonable to hope that linewidths in the multiple quantum domain can provide an accurate source of information about relaxation of the corresponding off-diagonal density matrix elements. This question is explored more fully in the next chapter.

REFERENCES

1. C.P. Slichter, *Principles of Magnetic Resonance*, Harper and Row, New York (1963).

2. L.G. Werbelow and D.M. Grant, *Adv. Magn. Reson.* **9**, 190 (1977).

3. R.L. Vold and R.R. Vold, *Progr. NMR Spectrosc.* **12**, 79 (1978).

4. W.P. Aue, E. Bartholdi and R.R. Ernst, *J. Chem. Phys.* **64**, 2229 (1976).

5. A. Wokaun and R.R. Ernst, *J. Chem. Phys.* **67**, 1752 (1977).

6. A. Wokaun and R.R. Ernst, *Mol. Phys.* **36**, 317 (1978).

7. S. Vega and A. Pines, *J. Chem. Phys.* **66**, 5624 (1977).

8. W.S. Warren and A. Pines, *J. Chem. Phys.* **74**, 2808 (1981).

9. W.S. Warren, D.P. Weitekamp and A. Pines, *J. Chem. Phys.* **73**, 2084 (1980).

10. R.L. Vold, R.R. Vold, R. Poupko and G. Bodenhausen, *J. Magn. Reson.* **38**, 141 (1980).

11. R. Poupko, R.L. Vold and R.R. Vold, *J. Phys. Chem.* **84**, 3444 (1980).

12. G. Bodenhausen, R.L. Vold and R.R. Vold, *J. Magn. Reson.* **37**, 93 (1980).

13. A. Abragam, *The Principles of Nuclear Magnetism*, University Press, Oxford, (1961).

14. S. Vega, *J. Chem. Phys.* **68**, 5518 (1978).

15. M. Mehring, E.K. Wolff and M.E. Stoll, *J. Magn. Reson.* **37**, 475 (1980).

16. L.G. Werbelow, *J. Magn. Reson.* **52**, 282 (1983).

17. L.G. Werbelow and D.M. Grant, *J. Chem. Phys.* **63**, 544 (1975).

18. L.G. Werbelow and D.M. Grant, *J. Chem. Phys.* **63**, 4772 (1975).

19. L.G. Werbelow and D.M. Grant, *J. Magn. Reson.* **20**, 554 (1975).

20. L.G. Werbelow, D.M. Grant, E.P. Black and J.M. Courtieu, *J. Chem. Phys.* **69**, 2407 (1978).

21. R.M. Wilcox, *J. Math. Phys.* **8**, 962 (1967).

22. Y.S. Yen, PhD Thesis, University of California, Berkeley, (1982).

23. A. Pines, paper presented at 23rd Exptl. NMR. Conference, Madison, Wisc. (1982).

24. Y.S. Yen, J.B. Murdoch, J. Tang, and A. Pines (to be published).

25. S. Emid, A. Bax, J. Konijnendijk, J. Smidt and A. Pines, *Physica* **96B**, 333 (1979).

26. S. Emid, J. Konijnendijk, J. Smidt and A. Pines, *Physica* **100B**, 215 (1980).

27. S. Schaublin, A. Hohener and R.R. Ernst, *J. Magn. Reson.* **13**, 196 (1974).

28. B.C. Sanctuary, T.K. Halstead and P. Osment, to be published.

29. A.D. Bain and J. Martin, *J. Magn. Reson.* **39**, 137 (1978).

30. J. Jeener and P. Broekaert, *Phys. Rev.* **157**, 232 (1967).

31. H.W. Spiess, *J. Chem. Phys.* **72**, 6755 (1980).

32. H.W. Spiess, *Colloid and Polymer Sci.* (1983) to be published.

33. A. Wokaun and R.R. Ernst, *Chem. Phys. Lett.* **52**, 407 (1977).

34. G. Drobny, A. Pines, S. Sinton, D. Weitekamp and D. Wemmer, *Faraday Division of the Chemical Society, Symposium* **13**, 49 (1979).

35. D.P. Weitekamp, J.R. Garbow and A. Pines, *J. Magn. Reson.* **46**, (1982).

36. D. Jaffe, R.L. Vold and R.R. Vold, *J. Magn. Reson.* **46**, 496 (1982).

37. G.L. Hoatson and K.J. Packer, *Mol. Phys.* **40**, 1153 (1980).

38. A. Pines, D. Wemmer, J. Tang, and S. Sinton, *Bull. Am. Phys. Soc.* **23**, 21 (1978).

39. D. Wemmer, PhD Thesis, University of California, Berkeley (1978).

NUCLEAR SPIN RELAXATION

Regitze R. Vold

Department of Chemistry,
University of California,
San Diego.

INTRODUCTION

The use of nuclear spin relaxation measurements in the study of liquid crystal dynamics goes back only to 1969, to work by Doane (1), Blinc (2), Cabane (3) and their coworkers. Practically all of the early work relied upon proton relaxation measurements (4,5), and several major mechanisms and relaxation pathways were identified for thermotropic nematic phases. However, since several different types of interactions are responsible for the overall relaxation of many chemically distinct protons it was quite difficult to assess the relative importance of the different motional processes, and selective deuteriation was soon recognized (6,7) as being necessary before substantive further progress could be made in the interpretation of relaxation rates. Deuterium (8-15) and ^{13}C (16,17) relaxation measurements are potentially the most useful for investigations of molecular motion at different sites in liquid crystal molecules. However, because of the need for complete proton decoupling which inevitably results in sample heating, ^{13}C relaxation measurements look less promising at this juncture. For this reason the techniques of deuteron relaxation are emphasized in this chapter with other nuclei treated only peripherally at the end.

For thermotropic nematic phases it appears fairly well established that one can expect to observe contributions to nuclear spin relaxation from restricted, but usually rapid, reorientation of individual molecules, from internal rotation within the molecules and, at low Larmor frequencies, also from long-range, quasi-coherent modes of motion such as director

J. W. Emsley (ed.), Nuclear Magnetic Resonance of Liquid Crystals, 253–288.

fluctuations. However, in the case of the more complex ordered phases, the cholesteric, the discoid and the various smectic phases (18,19) uncertainties remain about the motional processes and the hope is that the recently developed techniques for studying relaxation in coupled spin systems (20-26) will serve to increase our understanding of the dynamics of these systems.

In this chapter we shall review the derivation of the phenomenological expressions for the longitudinal and transverse relaxation behaviour of the spin density matrix for partially ordered, coupled spin systems subject to occasional perturbations in the form of strong or weak rf pulses. Relaxation rates for experimentally observable quantities (e.g. individual magnetizations or higher order coherences) will be expressed in terms of the spectral density functions which characterize the spectrum of molecular fluctuations. We will show that proper selection of relaxation experiments allows the determination of individual spectral densities of motion as a function of frequency and temperature, and that these spectral densities constitute the maximum amount of information one can hope to obtain from magnetic resonance experiments about the molecular motion. The three-level system of a partially ordered spin I = 1, the simplest system to show the effects of coupling, will serve as our case study. The experiments developed for measurements of different kinds of relaxation rates will be discussed, including such practical aspects as instrumental requirements and sensitivity to random and systematic errors. We will then discuss in a somewhat less detailed manner the type of information which can be obtained from relaxation experiments on multi-spin systems such as D_2 and CH_n.

THEORETICAL FRAMEWORK

For a partially ordered spin system we will write the Hamiltonian as

$$H = H_0 + H_1(t) + H'(t) \tag{1}$$

where the time-independent term, H_0, includes the Zeeman interaction with the static magnetic field, B_0, as well as residual, static averages of quadrupolar, dipolar and scalar couplings which characterize the steady state spectrum. $H_1(t)$ includes the effects of rf fields, and $H'(t)$ describes the relaxation interactions.

The time evolution of the reduced spin density operator σ in a typical pulsed NMR relaxation experiment can be treated in separate time intervals: during very short intervals when strong pulses are applied $H_1(t)$ dominates the Hamiltonian and the effects of relaxation during the pulse can be ignored. During the intervals between pulses $H_1 = 0$, and we can then write for the evolution of the density operator

$$d\sigma/dt = i[\sigma(t),H_0] - \int_0^t \overline{\{[\sigma_0,H'(t')],H'(t)\}}dt'. \quad (2)$$

The first term describes, as discussed in the preceding chapter, the free precession of the spin system. The second term, which may be derived using second order time dependent perturbation theory (27), is used to calculate the rates of relaxation once the perturbing Hamiltonian has been properly specified. For a partially ordered, uniaxial system the relaxation Hamiltonian may be written

$$H'(t) = \sum_{\mu}\sum_{M} A_{\mu}^{(L,M)}\, T_{\mu}^{(L,M)*} \quad (3)$$

where the first sum specifies the presence of several different relaxation mechanisms of rank L and the sum over M refers to projections of the spherical tensor operators A and T onto the space fixed z-axis along B_0. A_μ is the spin operator appropriate for the μ'th mechanism and the components of the interaction tensor T_μ in the laboratory frame are functions of time by virtue of the molecular motion. This fact may be conveniently expressed by a transformation of time-independent molecule fixed tensor components $T^{2,N}$ using Wigner rotation matrix elements

$$D^L_{M,N}(t) \equiv D^L_{M,N}(\alpha\beta\gamma):$$

$$H'(t)=\sum_{\mu MN}\sum\sum(-1)^N A_{\mu}^{(L,M)} T_{\mu}^{(L,-N)*}\{D^L_{M,N}(t) - \bar{D}^L_{M,N}\,\delta_{0,M}\} \quad (4)$$

The Wigner rotation elements are defined here using Rose's conventions (28) and the Euler angles $(\alpha\beta\gamma)$ are explicitly time-dependent. The form of Eq. (4) differs from $H'(t)$ for an isotropic system by having average values $\bar{D}^L_{M,N}$ of the Wigner rotation elements subtracted from the fluctuating functions. This reflects the fact that the molecular orientation fluctuates about a non-zero mean responsible for the quadrupolar and dipolar splittings characteristic of the NMR spectra of partially ordered systems, and it ensures that the associated correlation functions decay to zero such that their Fourier transforms can be defined.

For quadrupolar and dipolar relaxation, which are of most

interest in the present case, we deal with interactions of second rank (L = 2) for which $-2 \leqq M \leqq 2$. In the case of quadrupolar relaxation the spin operators in spherical irreducible tensor form are given by

$$A^{2,0} = [3I_z^2 - \mathbf{I}\cdot\mathbf{I}]/\sqrt{6} \tag{5a}$$

$$A^{2,\pm1} = [I_z I_\pm + I_\pm I_z]/2 \tag{5b}$$

$$A^{2,\pm2} = I_\pm^2/2 \tag{5c}$$

If for simplicity we now assume that the quadrupolar tensor is axially symmetric we may use $T^{2,-N} = T^{2,0}\delta_{0,N} = \sqrt{3/8}\ e^2qQ/\hbar$ and the expression for the relaxation Hamiltonian reduces to

$$H'(t) = \sum_M A_M\{D^2_{M,0} - \bar{D}^2_{M,0}\} \tag{6}$$

where

$$A_M \equiv T^{2,0}A^{2,M} = \sqrt{3/8}(e^2qQ/\hbar)A^{2,M} \tag{7}$$

For N = 0 the Wigner rotation matrix elements equal the spherical harmonics $C_M(t) \equiv C_{L,M}(\alpha\beta)$ as defined by Brink and Satchler (29).

$$C_0(t) = \tfrac{1}{2}[3\cos^2\beta(t) - 1] \tag{8a}$$

$$C_{\pm1}(t) = \pm\sqrt{3/2}\ \cos\beta(t)\ \sin\beta(t)\ \exp[\pm i\alpha(t)] \tag{8b}$$

$$C_{\pm2}(t) = \sqrt{3/8}\ \sin^2\beta(t)\ \exp[\pm 2i\alpha(t)] \tag{8c}$$

Inspection of the form of the spherical harmonics given in Eqs. (8) shows that $\bar{C}_1 = \bar{C}_2 = 0$ for a uniaxial phase aligned along B_0, which explains the use of the Kronecker-δ in Eq. (4); for biaxial phases this formulation is not appropriate. It should be noted that if the quadrupolar nucleus is not located on the principal axis of the molecular reorientation tensor the sum over N in Eq. (6) must be retained and a (time-independent) transformation to this frame from the principal axis system of the quadrupole coupling tensor is required in order to describe the molecular motion in simple terms.

In what follows we shall assume that the motion in the liquid crystal is fast enough for motional narrowing to apply. This implies that $H'(t)$ fluctuates rapidly compared with the precession which would result from a static interaction Hamiltonian H' in the absence of molecular motion; in other words, the motional frequencies must be high compared to the width of the rigid lattice spectrum. For the more rigid liquid crystalline phases this assumption may not be valid and the time evolution of the spin density operator must be evaluated using numerical integration methods, a subject which is outside the scope of this treatment. Freed (30,31) has provided a description of the slow motional regime for ESR spectra of highly ordered spin labels which is directly applicable to NMR lineshapes.

When motional narrowing applies we can follow Redfield (32) and express Eq. (2) in the form

$$d\sigma/dt = i[\sigma(t), H_0] - R\,[\sigma(t) - \sigma(\infty)]. \tag{9}$$

The form of the second term emphasizes that the density operator must return to equilibrium at $t = \infty$, and the relaxation super-operator R generates the integral of the double commutator in Eq. (2) with the upper limit t replaced by ∞. In any representation the evolution of individual density matrix elements may be written

$$d\sigma_{jk}/dt = i[\sigma(t), H_0]_{jk} + \sum_l \sum_m R_{jklm}[\sigma_{lm} - \sigma_{lm}(\infty)] \tag{10}$$

where the elements of the relaxation supermatrix are given by (32)

$$R_{jklm} = J_{jlkm}(\omega_{mk}) + J_{jlkm}(\omega_{jl}) - \delta_{km}\sum_n J_{nlnj}(\omega_{nj}) - \delta_{jl}\sum_n J_{nknm}(\omega_{kn}). \tag{11}$$

Eq. (11) introduces the spectral density $J_{jklm}(\omega)$ as a generalized transition probability for a process which involves the states j, k, l and m of a spin system exposed to fluctuations with frequency components ω. Expansion of the double commutator in Eq. (2) under motional narrowing conditions is described by Abragam (33) and Slichter (27) and the procedure leads to the following expression for $J_{jklm}(\omega)$,

$$J_{jklm}(\omega) = \int_0^\infty \overline{\langle j|H'(t)|k\rangle\langle l|H'(t+\tau)|m\rangle^*} \exp(-i\omega\tau)d\tau \quad (12)$$

$$= \sum_\mu \sum_M \langle j|A_{-M}^{(\mu)}|k\rangle\langle l|A_{-M}^{(\mu)}|m\rangle^* \, J_M^{(\mu)}(\omega).$$

The generalized transition probability is now given in terms of a one-sided Fourier transform of the correlation function for elements of the perturbing Hamiltonian. By substituting Eq. (3) into Eq. (12) we obtain the second equality which clearly shows the partition of the expression for $J_{jklm}(\omega)$ into (a) products of elements of angular momentum operators which contain the selection rules, and (b) spectral densities of motion for the fluctuating interactions. The former gives a quantitative measure of the interaction strength of the relaxation process, the latter a measure of the density of fluctuations at some frequency specified by the experiment. For uniaxial media it can be shown that cross products between Hamiltonian matrix elements of different M-value do not appear in Eq. (12), and we define the spectral density function for a given relaxation mechanism as the Fourier transform of the autocorrelation function

$$J_M(\omega) = \frac{1}{4\pi} \int_0^\infty \overline{[C_M(t) - \bar{C}_M][C_M^*(t+\tau) - \bar{C}_M^*]} \exp(-i\omega\tau)d\tau. \quad (13)$$

The particular separation of terms used in Eqs. (6) and (12) ensures that the spectral densities $J_M(\omega)$ refer to purely motional parameters, a procedure which facilitates the comparison of spectral densities determined in different types of experiments. Note also that the spherical harmonics have been normalized (29) such that $J_0(0) = \tau_c/5$ for fast motion in an isotropic medium.

From the form of Eq. (12) it can be seen that $J_{jklm}(\omega) = 0$ unless $M = m_j - m_k = m_l - m_m$. This requirement reduces the task of evaluating the elements of R by a substantial amount, as does the use of symmetry (34). Important general symmetry relations include

$$R_{jklm} = R_{lmjk} = R_{kjml} = R_{\Lambda j \Lambda k \Lambda l \Lambda m} \quad (14)$$

where Λ is the spin inversion operator (35). Note that the substitution of Eq. (3) into Eq. (2), or Eq. (12), leads to cross terms between different relaxation mechanisms. Their occurrence adds significantly to the complexity of relaxation calculations, which already grow rapidly with each additional nucleus in the spin system, but as discussed later, such cross terms may in fact be quite informative.

For a spin system with N energy levels Eq. (10) represents a system of N^2 coupled, linear differential equations. Fortunately, the vast majority of the coupling coefficients will equal zero or, in the case of the decay of off-diagonal density matrix elements (transverse relaxation), be insignificant compared to the precession terms. In general, one needs to identify from the N^2 relations in Eq. (10) some subset of equations which describes the time evolution of an observable $\langle Q \rangle$ of interest. The subset may be specified by the use of the trace relation

$$\langle Q \rangle = \mathrm{Tr}\{\sigma Q\} = \sum_p \sum_q \sigma_{pq} \langle q|Q|p \rangle \tag{15}$$

which for particular choices of $\langle Q \rangle$ leads to single, uncoupled differential equations. Such "normal modes" (23,36) thus have the advantage of showing exponential relaxation behaviour.

An alternative to the procedure described above consists of expanding the density operator (21,37)

$$\sigma = \sum_i a_i Q_i \tag{16}$$

in a set of operators $\{Q_i\}$ which ensures that the evaluation of the commutators in Eq. (2) leads to simple expressions for the time dependence of the coefficients, $a_i(t)$. However, the choice of operators is by no means straightforward; not only must a set of operators be chosen for each individual spin system, the proper choice, i.e. the set of operators which leads to uncoupled relaxation modes, also depends on the nature of the relaxation mechanism. In any case, the process of evaluating matrix elements of angular momentum operators gets replaced by evaluation of commutators of angular momentum operators, such that little, if any, saving in time is accomplished. For this reason, and because the procedure based on Redfield relaxation matrix elements is sufficiently general, we shall use Eqs. (4) - (15) in the remainder of this chapter.

As a final note it should be emphasized that the $J_M(\omega)$ defined in Eq. (13) are quantities which are available from experiment without reference to any molecular dynamics model; when measured as a function of temperature and frequency they are the best test of such models that NMR spectroscopy can provide. Calculations of $J_M(\omega)$ from Eq. (13) given a specific motional model are the subject of other chapters in this volume.

A CASE STUDY: ONE SPIN I = 1

Relaxation Calculations

As an example of how to determine individual spectral densities of motion, $J_M(\omega)$, from pulsed NMR experiments let us consider the simplest case of a coupled spin system, that of one partially ordered deuteron. A typical spectrum of a monodeuterated liquid crystal sample is shown in Figure 1.

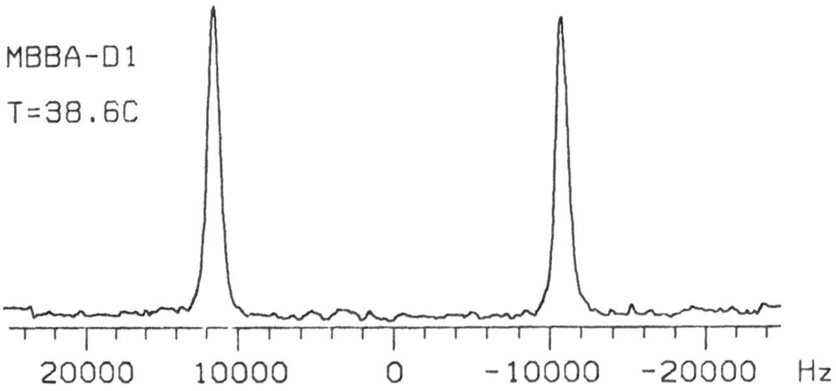

Figure 1. *38.4 MHz spectrum of p-methoxybenzylidene-n-butyl-aniline (MBBA) deuteriated in the linkage position. The lines are broadened by unresolved dipolar couplings with nearby protons, but relaxation of this deuteron may be treated as if it were an isolated spin I = 1.*

In order to analyze the relaxation behaviour and select informative experiments we follow a prescription suggested by the previous section.

a) Select a suitable basis for the spin system.

b) Use Eqs. (5) (or similar set for rank-1 interactions, if applicable) to evaluate matrix elements of the angular momentum operators A_M.

c) Calculate the products $\langle j|A_M|k\rangle\langle l|A_M|m\rangle$ and the generalized transition probabilities $J_{jklm}(\omega)$ from Eq.

(12).

d) Making use of the symmetry relations in Eq. (14), calculate the R_{jklm} from Eq. (11).

e) With the aid of Eq. (15) pick out the density matrix elements which correspond to a coherence or other observable of interest and construct an appropriate subset of Eq. (10) which describes the relevant experiment.

If the experiments are chosen wisely this procedure will generate simple, uncoupled differential equations for the density matrix elements with rate constants given in terms of simple linear combinations of the $J_M(\omega)$, although the analysis of non-exponential relaxation behaviour is quite straightforward (38).

Since the Zeeman interaction for deuterium is so much larger than the quadrupolar coupling, the appropriate basis for the spin density operator is the eigenstate basis of I_z, i.e. $|1\rangle = |+1\rangle$, $|2\rangle = |0\rangle$ and $|3\rangle = |-1\rangle$. Matrix representations in this basis were provided in the preceding chapter for numerous observables of interest in relaxation measurements.

Of the nine matrix elements which exist for each of the five second rank tensor operators in Eqs. (5) and (7) the only non-zero elements are: $\langle 1|A_0|1\rangle = \langle 3|A_0|3\rangle = 1$; $\langle 2|A_0|2\rangle = -2$; $\langle 1|A_{+1}|2\rangle = \langle 3|A_{-1}|2\rangle = -\sqrt{3}$; $\langle 2|A_{-1}|1\rangle = \langle 2|A_{+1}|3\rangle = \sqrt{3}$; $\langle 1|A_{+2}|3\rangle = \langle 3|A_{-2}|1\rangle = \sqrt{6}$. The values of the matrix elements are given in units of $\frac{1}{2}\pi(e^2qQ/h)$ and the identities are dictated by the inversion symmetry of the spin system.

By forming products of these elements for each value of M, recalling that cross products are forbidden by the orthogonality of the spherical harmonics, we obtain - out of 405 theoretically possible $J_{jklm}(\omega)$ - the 14 non-zero generalized transition probabilities listed in Table 1.

For a 3x3 spin density matrix the Redfield relaxation matrix is of dimension $3^2 \times 3^2$. However, the symmetry relations of Eq. (14) reduce the number of different R-elements for I = 1 to 17, and the rule $R_{jkkj} = 0$ further reduces that number to 15. Application of Eq. (11) shows that eight of these equal zero, leaving just seven different R_{jklm} to be evaluated; the 19 surviving non-zero elements of $\underset{\sim}{R}$ are listed in Table 2. It should be noted that in order to balance Eq. (10) we must require that

$$R_{jjjj} = \sum_{k \neq j} R_{jjkk} \quad (17)$$

a relation which serves as a useful check of the calculations.

Table 1. *Generalized Transition Probabilities* $J_{jklm} = J_{lmjk}$ *in units of* $\frac{\pi^2}{4}(e^2qQ/h)^2$ *for quadrupolar relaxation of a single spin* $I = 1$.

$$J_{1111} = J_{3333} = J_{1133} = J_0(0)$$
$$J_{2222} = 4J_0(0)$$
$$J_{1122} = J_{2233} = -2J_0(0)$$
$$J_{1212} = J_{2323} = J_{2121} = J_{3232} = 3J_1(\omega_0)$$
$$J_{1223} = J_{2132} = -3J_1(\omega_0)$$
$$J_{1313} = J_{3131} = 6J_2(2\omega_0)$$

For the spectrum in Figure 1 it is natural to consider measurements of the following quantities: the transverse decay rates of the single and double quantum coherences, i.e. the time evolution of the expectation values of I_+^{12}, I_+^{23} and I_+^{13} (the single transition operators are defined in the preceding chapter); the decay rate for the longitudinal difference magnetization (22), quadrupolar order or spin alignment (39,40), which is given by $\langle I_z^{12} - I_z^{23}\rangle$; the rate of recovery of individual z-magnetizations ($\langle I_z^{12}\rangle$ or $\langle I_z^{23}\rangle$) following selective inversion; and the "normal" spin-lattice relaxation rate of the sum magnetization ($\langle I_z^{13}\rangle = \langle I_z^{12} + I_z^{23}\rangle = \langle \tfrac{1}{2}I_z\rangle$) following non-selective inversion pulses. The time evolution of the desired quantity is obtained by application of the trace relation, Eq. (15), to Eq. (10), such that differential equations can be written for specific density matrix elements or combinations thereof.

Consider as the first example the non-selective spin-lattice experiment, typically performed by applying a strong inversion pulse to both lines and following the recovery with a strong 90° monitoring pulse. We seek the density matrix elements corresponding to the operator I_z and find that $\langle I_z\rangle$ =

$Tr\{\sigma I_z\} = \sigma_{11} - \sigma_{33}$. The first term in Eq. (10) is zero for all diagonal density matrix elements (populations do not precess) and we obtain

$$d(\sigma_{11} - \sigma_{33})/dt = (R_{1111} - R_{1133})[(\sigma_{11} - \sigma_{33}) - (\sigma_{11}(\infty) - \sigma_{33}(\infty))], \quad (18)$$

from which

$$R_1(\text{nsel}) = R_1(\text{sum}) = R_{1111} - R_{1133} \propto J_1(\omega_0) + 4J_2(2\omega_0). \quad (19)$$

Table 2. *Redfield Relaxation Elements for I = 1*

Diagonal Elements:

$$R_{1313} = R_{3131} = -\frac{3\pi^2}{2}(e^2qQ/h)^2[J_1(\omega_0)+2J_2(2\omega_0)]$$

$$R_{3333} = R_{1111} = -\frac{3\pi^2}{2}(e^2qQ/h)^2[J_1(\omega_0)+2J_2(2\omega_0)]$$

$$R_{2222} = -3\pi^2(e^2qQ/h)^2J_1(\omega_0)$$

$$R_{1212} = R_{2121} = R_{2323} = R_{3232} = -\frac{3\pi^2}{4}(e^2qQ/h)[3J_0(0) - 3J_1(\omega_0) - 2J_2(2\omega_0)]$$

Off-Diagonal Elements:

$$R_{1122} = R_{2211} = R_{3322} = R_{2233} = \frac{3\pi^2}{2}(e^2qQ/h)^2J_1(\omega_0)$$

$$R_{1133} = R_{3311} = 3\pi^2(e^2qQ/h)^2J_2(2\omega_0)$$

$$R_{1223} = R_{2132} = R_{2312} = R_{3221} = -\frac{3\pi^2}{2}(e^2qQ/h)^2J_1(\omega_0)$$

Similarly, for the decay of quadrupolar order created in a Jeener-Broekaert experiment (39-42) we seek the relaxation rate of $\langle Q_z\rangle = \frac{1}{3}\langle(3I_z^2 - I\cdot I\rangle = \frac{1}{3}(\sigma_{11} + \sigma_{33} - 2\sigma_{22})$ and obtain

$$R_1(Q_z) = R_1(\text{diff}) = R_{1111} + R_{1133} - 2R_{1122} \propto 3J_1(\omega_0) \qquad (20)$$

In the case of transverse relaxation experiments the precession term in Eq. (10) becomes important. Consider first a selective 90°-τ-180° spin echo experiment on one line of the spectrum (22), i.e. a measurement of the time evolution of $\langle I_+^{12}\rangle$. From Eq. (10) we obtain

$$d\sigma_{12}/dt = -i\omega_{12}\sigma_{12} + R_{1212}\sigma_{12} + R_{1223}\sigma_{23} \qquad (21)$$

an expression which includes coupling to σ_{23}. However, the other line was never pulsed and $\sigma_{23} = 0$ throughout the experiment such that the selective transverse relaxation rate is

$$R_2(\text{sel}) = -R_{1212} \propto 3J_0(0) + 3J_1(\omega_0) + 2J_2(2\omega_0). \qquad (22)$$

One should recognize that the 180° pulse used in the selective spin echo experiment reverses dephasing from both magnetic field inhomogeneities and a distribution of the order parameter (22) and is preferred over a 90° refocussing pulse because the 180° spin echo is twice as large (43).

A spin echo experiment on both lines requires the use of a 90° pulse for formation of a quadrupolar echo (44) and the time evolution of the total tranverse magnetization is then described by

$$d(\sigma_{12} + \sigma_{23})dt = -i(\omega_{12}\sigma_{12} + \omega_{23}\sigma_{23}) - (R_{1212} + R_{1223})(\sigma_{12} + \sigma_{23}). \qquad (23)$$

In the event that $\omega_{12} = \omega_{23}$, which happens in the isotropic phase where the quadrupolar splitting is zero, Eq. (23) is a simple Bloch equation for the transverse relaxation of the sum magnetization with a relaxation rate

$$R_2 = -(R_{1212} + R_{1223}) \propto 3J_0(0) + 5J_1(\omega_0) + 2J_2(2\omega_0). \qquad (24)$$

When differential precession occurs in the ordered phase, the two coupled differential equations for σ_{12} and σ_{23} may be treated using standard eigenvalue procedures. As described in detail for homonuclear dipolar interactions in isotropic proton spin systems (45,46), any cross relaxation terms between transverse magnetizations become non-secular (32) whenever the differential precession is large. For the partially ordered deuteron, R_{1223} effectively vanishes and the non-selective transverse decay rate is identical to the selective rate given in Eq. (22). However, if a spin lock is established and differential precession suppressed (24), the cross relaxation term is recovered and the decay rate is the same as that observed in the isotropic phase and given by Eq. (24). This phenomenon is related to the 3/2-effect for dipolar relaxation of "like" and "unlike" spins (33) which has been observed for several spin systems in the isotropic phase (45,47,48).

By similar analyses of other experiments several useful relations between observable relaxation rates and spectral densities of motion will materialize; they are collected in Table 3 together with a description of the pulse sequences. If quadrupolar relaxation, as defined above in Eqs. (5) - (8), is the only relaxation mechanism present the problem is overdetermined and only three experiments are required to determine $J_0(0)$, $J_1(\omega_0)$ and $J_2(2\omega_0)$. The three experiments must include a spin echo measurement of a single quantum transverse decay rate to allow determination of $J_0(0)$. Of course, if additional relaxation processes, such as intermolecular dipolar relaxation, paramagnetic interactions or chemical exchange are known to operate, additional spectral density terms will contribute to Eq. (11) and more than three experiments may allow the separation of the various contributions.

Relaxation Measurements

The question remains: which of the possible experiments suggested above lend themselves best to accurate determinations of the $J_M(M\omega)$? The choice among experiments is dictated in part by the characteristics of the particular spectrum and by the relative magnitudes of the spectral densities, which are not known a priori, and in part by the capabilities of the instrumentation at hand.

As already noted above, in order to determine $J_0(0)$ one of the experiments must be a single quantum spin echo experiment. If the quadrupolar splitting $2\nu_Q$ is large enough to permit the application of truly selective pulses, a selective spin echo experiment with a 180° refocussing pulse would be the experiment of choice for a single isolated deuteron. It can be shown (22,43) that reversible dephasing from static inhomogeneities in the magnetic field as well as temperature gradients will be refocussed in this experiment.

If the quadrupolar splitting is small, the non-selective spin echo experiment is preferable. A strong 90° refocussing pulse will generate a quadrupolar echo (44) which is unattenuated by either magnetic field or temperature inhomogeneities. The experiment does, however, suffer from sensitivity to pulse errors: an imperfect 90° pulse may be thought of as generating two kinds of echoes, one arising from

Table 3. *Pulse sequences and relaxation rates (in units of $\frac{3}{8}\pi^2(e^2qQ/h)^2$) for NMR experiments on a single ordered spin $I = 1$.*

EXPERIMENTAL	PULSE SEQUENCE[a]		RELAXATION RATE	REF.
	Populations (σ_{jj})			
Non-selective inversion recovery	$\pi - t_1 - \pi/2$-acq		$J_1 + 4J_2$	21,22,38,49
Selective inversion				
recovery of A	$\pi(A) - t_1 - \pi/2(A)$-acq		$\frac{5}{2}J_1 + J_2$[b]	21,22,38,49
recovery of B	$\pi(A) - t_1 - \pi/2(B)$-acq		$4J_1 - J_2$[b]	21,22,38,49
recovery of A&B	$\pi(A) - t_1 - \pi/4$-acq	sum	$J_1 + 4J_2$	14
		diff	$3J_1$	14
Quadrupolar order	$\pi/2_x - \tau - \pi/4_y - t_1 - \pi/4$-acq	sum	$J_1 + 4J_2$	14,15,39,40, 42,50,51
		diff	$3J_1$	14,15,39,40, 42,50,51
	Transverse Coherences (σ_{jk}, $j \neq k$)			
Selective 1Q spin echo	$\pi/2(A) - t_1/2 - \pi(A) - t_1/2$-acq		$\frac{1}{2}(3J_0+3J_1+2J_2)$	21,22,38
Non-selective 1Q spin echo	$\pi/2_x - t_1/2 - \pi/2_y - t_1/2$-acq		$\frac{1}{2}(3J_0+3J_1+2J_2)$	24,43
Non-selective spin lock	$(\pi/2)_x - (SL)_y$-acq		$\frac{1}{2}(3J_0+5J_1+2J_2)$	24
2Q Spin echo[c]	$\pi/2_x - \tau - \pi/2_x - t_1/2 - \pi - t_1/2 - \pi/2$-acq		$J_1 + 2J_2$	52-54

[a] A selective pulse on line A is indicated by (A) after the angle of the pulse. The phase of the pulse is indicated only when the relative phases are essential. Phase cycling, which is required in all experiments for suppression of artifacts, is not indicated, but is described in the original literature.

[b] The recovery of either line following selective inversion is weakly biexponential, and the table entries are initial recovery rates.

[c] An alternative excitation technique involves a suitably weak pulse (55) at the centre of the doublet.

transfer of σ_{jk} into $\sigma_{kj} = \sigma_{jk}^*$, the other from transfer of σ_{jk} into σ_{lm}, where m = Λj and l = Λk. Only the former transfer, which is maximized by a 90° pulse for quadrupolar coupled systems, accomplishes the desired refocussing effect, while the latter creates a faster decaying, modulated echo. The problem is not alleviated by phase cycling (56), a common cure for ailing spin echo experiments, since phase cycling cannot distinguish between coherences of the same order. It is possible instead (43) to separate pure quadrupolar echoes from those broadened by spurious dephasing via the use of 2D techniques, because the 2D experiment separates spin echo lines on the basis of precession frequency. 2D deuterium spin echo experiments, which have proven very useful in structural assignments (57,58), may consequently be the method of choice for the determination of $J_0(0)$. Unfortunately, in the case of mono-deuteriated liquid crystal molecules (see Figure 1) where the deuteron is scalar and dipolar coupled to a large number of heteronuclei (protons), either experiment will fail to give accurate R_2 values. Coupling to other spins in thermal equilibrium may in certain cases be refocussed by either technique, but if the heteronuclei are strongly coupled among themselves the echoes will be modulated (59,60). Worst of all, the presence of many different modulation frequencies will cause destructive interference and a spuriously short spin echo decay.

Sometimes it is desirable to remove certain transverse dephasing contributions, such as chemical exchange or translational diffusion, by spin locking the magnetization in the rotating frame, either through the use of continuous rf or closely spaced rf pulses. To the best of my knowledge the only application of a spin lock experiment on deuterium in a liquid crystal was performed by Ahmad et al. (24) on D_2O in a lyotropic system. A $\pi/2_x-(\tau-\pi/2_y-\tau)_n$ pulse sequence was used to spin lock the two deuteron lines in a quadrupolar equivalent of the Carr-Purcell-Meiboom-Gill (61,62) sequence. The experiment was successful because the quadrupolar splitting was small, and informative because the variation of the pulse spacing allowed water exchange rates to be measured. For thermotropic liquid crystals where the splittings are large, a deuteron spin locking experiment would be quite difficult to perform because of the power requirements. And whether rf is applied continuously or in the form of closely spaced pulses a large amount of heat needs to be dissipated; the concomitant rise in temperature would be intolerable for liquid crystalline samples. Similar difficulties are associated with proton decoupling in transverse relaxation experiments; for that reason very few deuteron T_2 measurements have been carried out on thermotropic liquid crystals, and even then only on solutes (38,52). It is unfortunate that reliable methods for determination of single quantum transverse decay rates in the presence of unresolved couplings have yet to be developed because the information gathered about low frequency molecular motion would be very valuable.

For a single ordered deuteron the transverse relaxation rate of the double quantum coherence may be measured easily using the pulse sequence listed in Table 3. Since only one 2Q coherence ($\langle I_+^{13} \rangle$) exists there is no need for a 2D spectral display. The experiment resembles the simplest T_2 measurement in that the double quantum precession is unaffected by the quadrupolar interaction (55) and is fully refocussed by a 180° pulse (54). A different linear combination of spectral densities, $J_1(\omega_0) + 2J_2(2\omega_0)$, is obtained in the 2Q spin echo experiment and may conveniently be combined with the result of a non-selective spin-lattice relaxation measurement for a determination of $J_1(\omega_0)$ and $J_2(2\omega_0)$ (63). However, since the experiment measures a transverse dephasing rate it is subject to all the field instability problems which plague ordinary 1Q transverse relaxation measurements. If the offending interactions are of a form which result in perturbations of $\langle I_z \rangle$, such as field drift, diffusion in a magnetic field gradient, or chemical exchange between chemically nonequivalent sites, then the 2Q spin echo, which is twice as sensitive to such effects (54), will be rapidly attenuated. It is, of course, possible to take advantage of this sensitivity of the multiple quantum spin echo in pulsed field gradient measurements of translational diffusion constants (64, 65), but if a determination of $J_1(\omega_0)$ and $J_2(2\omega_0)$ for molecular dynamics studies is the goal, then it is preferable to measure the longitudinal relaxation rates of the spin system; such experiments are much less susceptible to systematic errors.

If a bare minimum of precautions are taken the measurement of spin-lattice relaxation rates in ordinary liquids is one of the simplest experiments to execute successfully. The non-selective inversion-recovery experiment on partially ordered deuterons is almost as straightforward, provided that phase cycling is used to suppress interference from 1Q and 2Q coherence created by an imperfect inversion pulse (66). This reliability of the non-selective spin-lattice relaxation measurement makes it very attractive, but it provides only one linear combination of $J_1(\omega_0)$ and $J_2(2\omega_0)$ and a second experiment is necessary in order to separate the two spectral densities. In our own laboratory, whenever the quadrupolar splitting is sufficiently large, we use a selective 180° pulse on one of the lines in the spectrum followed by a selective 90° pulse on the same line to obtain the second equation for $J_1(\omega_0)$ and $J_2(2\omega_0)$. The dynamic range of the two inversion recovery experiments is good, the non-selective experiment yielding a change in magnetization of four units, the selective one yielding two. The fully selective inversion recovery is inherently biexponential

(22,38)

$$<I_z^{12}(t)> = <I_z^{12}(\infty)>[1 - \tfrac{1}{2}\exp(-R_1(\mathrm{sum})t) - \tfrac{3}{2}\exp(-R_1(\mathrm{diff})t)], \quad (25)$$

but in practice the expression listed in Table 3 for the initial rate of recovery is valid within experimental error over nearly one decade of recovery.

The main objection against using these two experiments as the source of $J_1(\omega_0)$ and $J_2(2\omega_0)$ is that two experiments must be performed. This objection may be overcome by the use of a non-selective monitor pulse which permits a simultaneous record to be obtained for the time evolution of both sum and different magnetizations. Unfortunately, the application of a 90° monitor pulse, which normally would be used to achieve maximal signal intensity, has the effect of mixing the intensities of the two lines. The formalism for such non-linear properties of strong pulses was developed for scalar coupled spin systems by Schäublin et al. (67), and for a partially ordered deuteron the degree of mixing by a pulse of flip angle α is (22)

$$<I_y^{12}(\tau_+)> = \sin\alpha[<I_z^{12}(\tau_-)>\cos^2(\alpha/2) + <I_z^{23}(\tau_-)>\sin^2(\alpha/2)] \quad (26a)$$

$$<I_y^{23}(\tau_+)> = \sin\alpha[<I_z^{12}(\tau_-)>\sin^2(\alpha/2) + <I_z^{23}(\tau_-)>\cos^2(\alpha/2)]. \quad (26b)$$

In these expressions τ_- and τ_+ refer to instances of time just before and after the monitor pulse. It is clear that the use of a small flip angle optimizes the relative difference between $<I_y^{12}>$ and $<I_y^{23}>$ at the cost of overall signal intensity. As in the case of the Jeener-Broekaert (41) experiment discussed below, α = 45° is the optimal choice for the monitor pulse (14), but the dynamic range of the recovery functions as well as the signal-to-noise ratio of the spectra are lower than those of the two separate experiments described above.

When the quadrupolar splitting is small it is difficult to create a large difference magnetization by selective pulsing, regardless of whether selectivity is obtained from weak on-resonance or audio-modulated pulses (38) or via tailored excitation (14,68). In this situation the difference magnetization may be generated using the Jeener-Broekaert pulse sequence (see Table 3) originally designed for measurements of dipolar order in solids (41). An example of such an experiment is presented in Figure 2, and the spin dynamics for one deuteron (14,39,40,42,69) is reviewed in the preceding chapter.

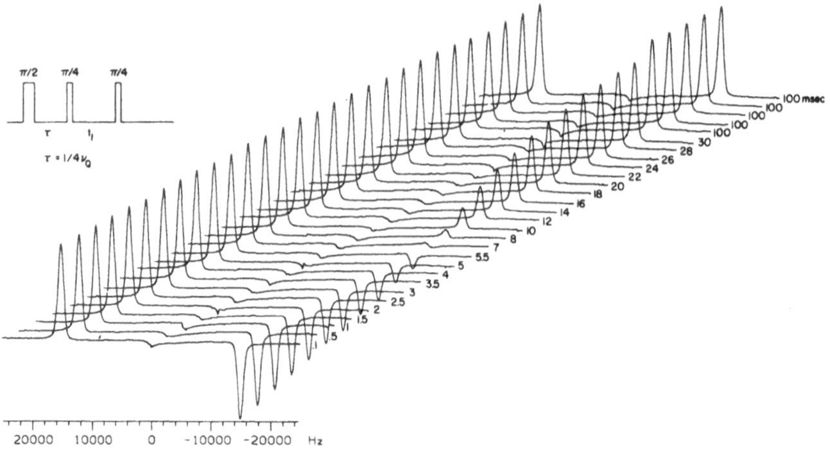

Figure 2. *Stacked plot of a Jeener-Broekaert experiment on linkage deuterated MBBA obtained at 38.4 MHz using phase cycled excitation and quadrature phase detection. The quadrupolar splitting at this temperature is 30,210 Hz, so that the theoretically optimum time between excitation pulses is 16.6 μs. A shorter value, 9 μs, was found to be better in practice because the 90° pulse was rather long (11 μs). The feature near zero is caused by imbalance in the dual phase sensitive detection circuitry.*

The experiment has been exceptionally useful in the study of very slow motion in deuterated solids (40), regular (70) and liquid crystalline (71) polymers, and is the method of choice for simultaneous determination of $J_1(\omega_0)$ and $J_2(2\omega_0)$ in liquid crystals (14,42,50,51).

The relative merits of the different techniques for determination of $J_1(\omega_0)$ and $J_2(2\omega_0)$ have been discussed (14,38,42) in some detail. For example, Beckmann et al. (14) have found that the dynamic range of the Jeener-Broekaert experiment exceeds that for selective inversion followed by non-selective detection. It is obviously more convenient to carry out one experiment rather than two, but the combination of a non-selective spin-lattice relaxation measurement with the fully selective inversion recovery experiment is the superior method for determining $J_1(\omega_0)$ and $J_2(2\omega_0)$ when the signal-to-noise in the spectra is low. For relaxation measurements on solutes in low concentration in liquid crystalline solvents, expecially at low magnetic field strengths, the loss of signal accompanying the 45° monitoring pulse is prohibitive, since a Jeener-Broekaert experiment may last 48 hrs or more. Under such circumstances, the additional adjustments required to set up two experiments is well worth the effort, since more accurate results are obtained in shorter time (38,50).

Instrumental Requirements

In the discussion presented above it was tacitly assumed that adequate instrumentation is available for these kinds of experiments. It remains to list what features of an NMR spectrometer must be present for satisfactory determinations of individual spectral densities of motion. Some of the features described below are incorporated in current commercial spectrometers, but more often than not the physical chemist is required to customize a spectrometer to permit particular experiments to be performed. The requirements for NMR relaxation work in liquid crystals are:

Adequate and adjustable pulse power for tailored excitation. A truly "strong" pulse for use in deuterium NMR of liquid crystal samples is beyond the capabilities of all present day spectrometers. A C-D bond with order parameter S = 0.7 (example: p-diethynylbenzene-d_2 in smectic 4-octyl-4'-cyanobiphenyl (72)) gives rise to a quadrupolar splitting in excess of 200 kHz. In order to invert both magnetizations of such a doublet to within 95%, we must apply a 0.77T (= 500 kHz for deuterium) rotating field for 1 μsec. For a 0.5 ml solenoid coil with a Q of 100 such a π-pulse requires more than 10 kW! Fortunately, perfect inversion is not required for good spin-lattice experiments; the use of phase cycling (56) will eliminate phase distortions from excitation of unwanted single and double quantum coherence (66), and for very large splittings modulated pulses may be used. Centre band suppressed audio modulation may be obtained very simply using a double balanced mixer and a gated audio oscillator which retains the proper phase from pulse to pulse and the resulting modulated pulses will cleanly invert the two line deuterium spectrum when $\nu_{mod} = \nu_Q$ (38). Modulated pulses obtained in this fashion may also be applied in selective pulse experiments, a technique which is particularly useful when combined with computer controlled switching of the transmitter frequency during a pulse sequence. However, the use of carefully tailored excitation sequences (68,73) for use in selective inversion-recovery and Jeener-Broekaert experiments (14) presents a much more elegant solution.

Flexible pulse programmer. The longest pulse sequence listed in Table 3 requires ten intervals per acquisition, but the entries in Table 3 make no mention of the extreme demands imposed by tailored excitation techniques (68), selective excitation of multiple quantum coherences (74), or any kind of double resonance experiment. In order to take advantage of the large amount of information inherent in liquid crystal spectra a pulse programmer with plenty of RAM and fast pre-loadable timers is an essential piece of equipment.

Programmable phase shifter. The ability to switch rf phase in less than 100 nsec is necessary in any experiment using composite pulses or a spin locking experiment for $T_{1\rho}$ measurements. Even if composite pulses are not used, fast phase switching is necessary during delays between pulses in order to accommodate the high rates of precession encountered in ordered spectra. In most cases quadrature phases are sufficient, as long as they can be changed according to some algorithm between acquisitions, but a digital, programmable phase shifter with 1.4° resolution (8 bits) is mandatory for selective detection (75) in multiple quantum spectroscopy.

Software for 2DFT Spectroscopy. Many kinds of multiple quantum spectroscopy (76) do not require two-dimensional display of the spectra. The simple measurement of the transverse relaxation of the 2Q coherence for a single ordered deuteron is an example, but if more than one multiple quantum coherence is present, 2D representation is required in order to assign individual lines and to measure their widths (52,77). In addition, as discussed above, measurements of 1Q spin echo decay rates benefit from the separation afforded by the 2-dimensional display.

Adequate temperature control. Perhaps the single most important spectrometer accessory for liquid crystal work is a very good temperature controller, because quadrupolar and dipolar splittings are strong functions of temperature. Temperature coefficients $d(\nu_Q)/dT \sim 200$ Hz/°C are common for deuterium spectra, and they may exceed 2000 Hz/°C in certain cases. The demands of 2D spectroscopy are most severe: temperature changes during the course of an experiment leads to modulations in the interferogram and lineshape distortions in the 2D spectrum. The temperature stability required for these experiments may exceed ± 0.005°C (77), but since temperature drift and inhomogeneity across the sample translates directly into line broadening conventional 1D experiments also demand tight control of temperature. Fortunately, it is not difficult to build a temperature controller using microprocessor technology (78) which will provide such control. Provided the NMR probe is designed with special attention to air flow rate and air flow pattern, it is not difficult to maintain a homogeneous sample temperature to within a few millidegrees in the presence of rapid pulsing, although continuous broadband decoupling and spin locking fields cannot be accommodated.

Results for systems with a single deuteron.

Up to this point no comments have been made about results derived from any of these experiments, or about any possible interpretation of the spectral densities in terms of molecular dynamics models. It is important to emphasize, that for this purpose it must be possible to carry out relaxation experiments at different field strengths in order to determine both $J_1(\omega_0)$ and $J_2(2\omega_0)$ as a function of frequency; the spectral densities of motion for both liquid crystal solvents and solutes are inherently frequency dependent due to the presence of several different kinds of slow motion. In Figures 3 and 4 are shown results for two nematic systems chosen to illustrate the different kinds of information which can be obtained from quite similar phases.

Figure 3 illustrates the use of a highly ordered solute (50) to probe the director fluctuations in Merck Licristal Phase 5. Since the order parameter for the acetylenic deuterons of the solute is large, the relaxation is strongly affected by contributions from director fluctuations. This is apparent from the observed $\omega_0^{-\frac{1}{2}}$ frequency dependence of $J_1(\omega_0)$ (79), which, however, drops off more rapidly at low temperatures because of the cutoff in the coherent modes (80). For this system $J_2(2\omega_0)$ is essentially frequency independent which shows not only that director fluctuations do not contribute to $J_2(2\omega_0)$ in agreement with the original theory (79), but also that the reorientation of this solute is comparatively fast.

The results presented for linkage deuterated MBBA in Figure 4 illustrate the relative insensitivity of deuteron relaxation behaviour to director fluctuations, when the C-D bond lies close to the magic angle. As observed previously by Rutar et al, (11), not only are the contributions from director fluctuations small on an absolute scale because of the S^2 dependence, but the molecular reorientation is relatively slow and therefore contributes strongly. We see that slow reorientation of the long axis of the molecule shows up in both $J_1(\omega_0)$ and $J_2(2\omega_0)$, and the results shown in Figure 4 can in fact be accounted for by assuming a symmetric top model for the reorientation of MBBA with minor contributions from director fluctuations included in the nematic phase (81). The isotropic phase results, obtained by a combination of deuteron spin-lattice relaxation rates and ^{13}C-^{2}H lineshape analysis (82), can be explained solely in terms of symmetric top reorientation. It is noteworthy that recent measurements of the transverse relaxation rates in isotropic MBBA (83) show a pronounced critical divergence of $J_2(0)$ above T_C. The data may be interpreted in terms of translational diffusion through critically growing, locally ordered structures, and it appears that the reorientation of the long axis of MBBA is strongly coupled to this translational motion.

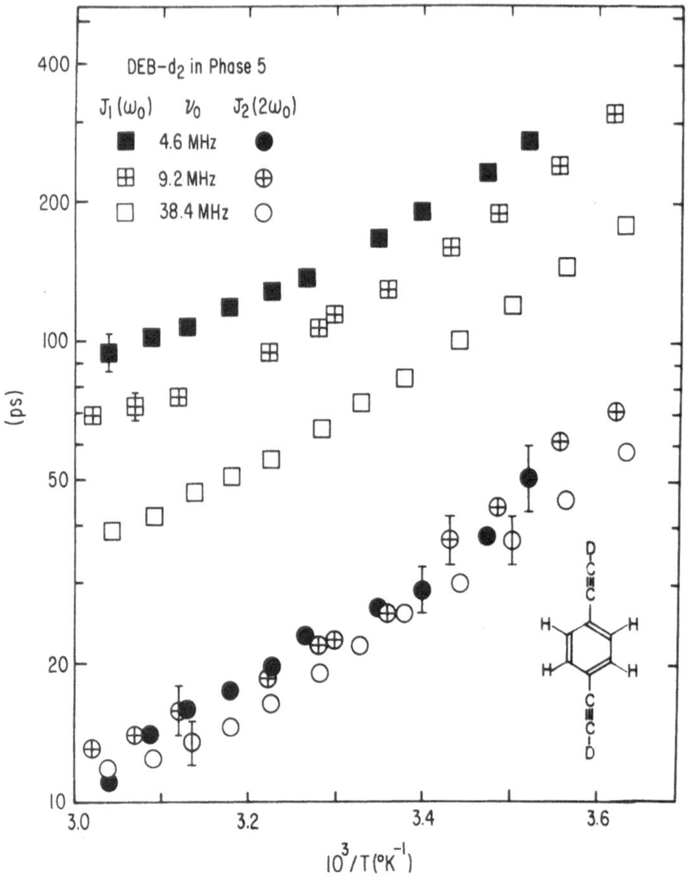

Figure 3. *Spectral densities $J_1(\omega_0)$ and $J_2(2\omega_0)$ determined at three fields (50) for a nematic Phase 5 solution of p-diethynyl-benzene deuteriated in the two acetylenic positions. The I→N transition occurs at 63.7°, and the order parameter S varies from ca 0.24 to 0.40 over the temperature range shown.*

Sample Rotation.

In addition to getting information about the frequency dependence of individual spectral densities of motion from field dependent relaxation measurements, one can exploit the sensitivity of the relaxation expression to angular factors in sample rotation experiments and extend the range of accessible values of the $J_M(\omega)$. Nuclear spin relaxation rates in

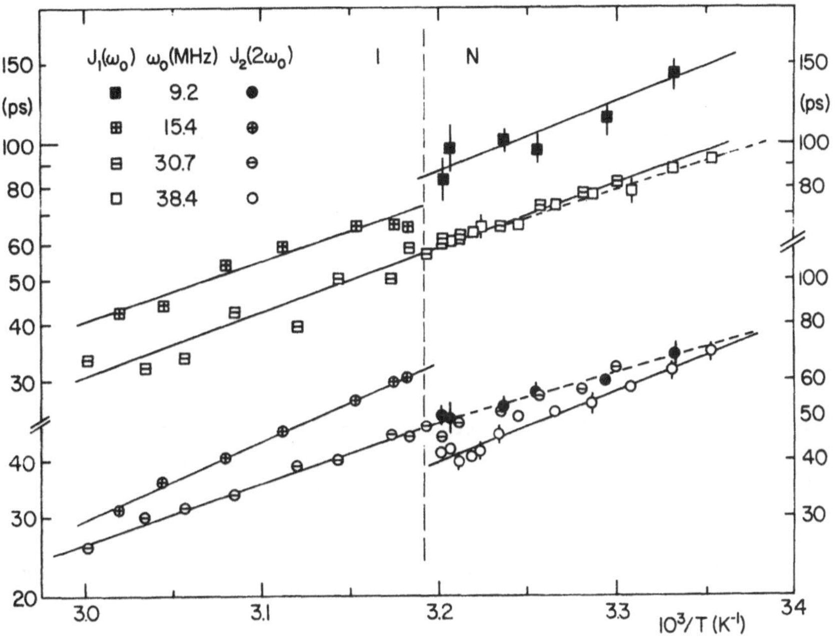

Figure 4. *Spectral densities $J_1(\omega_0)$ and $J_2(2\omega_0)$ determined as a function of temperature and frequency for MBBA deuteriated in the linkage position. The slightly impure material has a I→N phase transition range of 0.2° at 40.1°C. The order parameter for the C-D bond is close to 0.1 in the nematic phase.*

anisotropic samples are orientation dependent if the motional process responsible for the relaxation is strongly anisotropic. Proton relaxation arising from director fluctuations is such a process, as demonstrated by Doane, Tarr and their coworkers (80,84) in measurements on nematic liquid crystals oriented in both magnetic and electric fields. Without going into any detailed derivation, the dependence of the spectral densities upon the relative orientation of the director and the magnetic field lies buried in the transformation from the molecule fixed coordinate system (containing the quadrupolar coupling tensor) to the laboratory fixed frame. Useful formulae for the angular dependence of the nuclear spin relaxation rates have been derived by Doane and coworkers (80,85,86) and by Tarr et al. (87). Related expressions for ESR lineshapes were first derived by Luckhurst and Sanson (88). In the following expressions (15) for spectral densities, β refers to the angle between the director (in a uniaxial phase) and the external field, and terms of the form $J_M(M\omega)$ stand for values at $\beta = 0$.

$$J_0(\beta,0) = \tfrac{1}{4}(3\cos^2\beta - 1)^2\, J_0(0) + 3\cos^2\beta\,\sin^2\beta\, J_1(0) + \tfrac{3}{4}(\cos^2\beta - 1)^2 J_2(0) \quad (27a)$$

$$J_1(\beta,\omega_0) = \tfrac{3}{2}\cos^2\beta\,\sin^2\beta\, J_0(\omega_0) + \tfrac{1}{2}(4\cos^4\beta - 3\cos^2\beta + 1)J_1(\omega_0) + \tfrac{1}{2}(1 - \cos^4\beta)J_2(\omega_0) \quad (27b)$$

$$J_2(\beta,2\omega_0) = \tfrac{3}{8}(\cos^2\beta - 1)^2\, J_0(2\omega_0) + \tfrac{1}{2}(1 - \cos^4\beta)J_1(2\omega_0) + \tfrac{1}{8}(\cos^4\beta + 6\cos^2\beta + 1)J_2(2\omega_0). \quad (27c)$$

Eqs. (27) show that measurements of J_0, J_1 and J_2 as a function of the angle between the sample director and the magnetic field allow information to be obtained about these same functions at different frequencies.

In the past such experiments have been performed on nematic samples where the director orientation was altered using ac electric fields (80,85), but individual spectral densities were not determined. The simplest phases to explore using sample rotation are those fairly rigid smectics (19,88,89) which do not re-orient back to the magnetic field after the rotation. We are currently investigating whether relaxation experiments as a function of sample rotation angle in chain deuteriated 4-butoxybenzylidene-p-n-octylaniline (40.8) are helpful in unravelling the complex internal motions and their contributions to the deuteron relaxation rates.

Courtieu et al. (90) have developed an ingenious method for achieving uniform director alignment in nematic samples by spinning the sample about an axis which makes an angle β with the external field at a rate which is faster than the rate of director reorientation. Their technique may be superior to those based on electric field alignment or on slow sample spinning (91,92), because the director alignment remains homogeneous over a wide range of rotation angles; spectral resolution actually *improves* as the magic angle is approached. The method has not yet been exploited in relaxation studies, but it is clearly very promising.

MULTISPIN SYSTEMS

Deuterium.

Formulae and procedures developed in the preceding section for a single deuteron are of greater use than implied by their derivation. This important simplifying feature of multispin deuterium relaxation arises because the quadrupolar mechanism causes each spin to relax independently of its neighbours. Complications can arise when the nuclear spin states are mixed by deuteron-deuteron dipolar coupling, but to the extent that this relatively small interaction is negligible, spectra of polydeuteriated molecules may be regarded as superpositions of effectively isolated one-spin quadrupolar doublets. A typical spectrum of a deuteriated liquid crystal is shown in Figure 5.

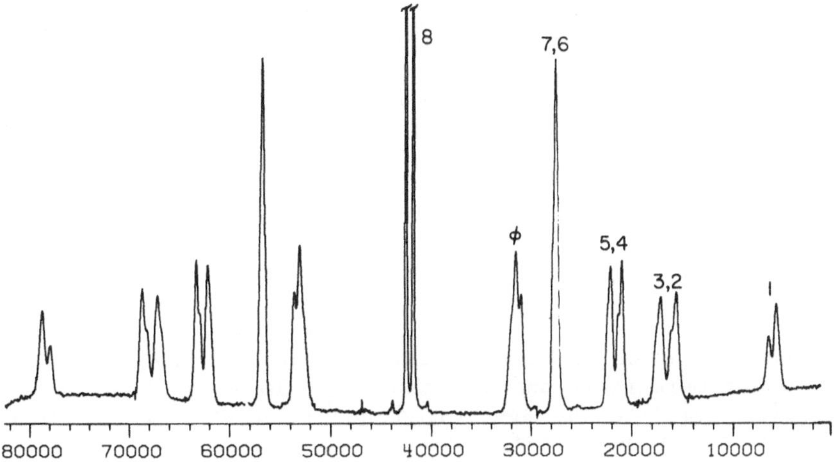

Figure 5. *38.4 MHz deuterium spectrum of 4-butoxybenzylidene-p-n-octylaniline-d_{21} (40.8) recorded at 59.4°C. The sample, kindly provided by M.E. Neubert and J.W. Doane, is deuteriated in the aniline ring and the octyl chain, and the peaks are labelled with increasing numbers going outwards from the aniline ring. The assignments for methylene groups 2,3 and 4,5 are tentative.*

The spectrum shows the familiar group of symmetrically arranged doublets with very little fine-structure from dipolar interactions. The aromatic (ϕ) and C_1 methylene deuterons, which reside in the most highly ordered part of the molecule, are the only ones with resolvable dipolar splittings. It is natural to treat the individual doublets as arising from isolated single deuterons, and many of the procedures described earlier may in fact be applied to extract spectral density data for each molecular site in liquid crystal molecules of this kind (14,15). The spectrum in Figure 5 is typical of a spin system with inversion symmetry, i.e. the right half is the mirror image of the left half. In this case we can define semi-selective pulses as those which achieve uniform excitation of a group of dipolar coupled lines on one side of the spectrum, and sum and difference magnetizations as those pertaining to equivalent resonances on either side of the centre. It has been shown (42) that whenever the quadrupolar splitting is large compared to the dipolar splittings the sum magnetization always recovers according to Eq. (19) and the difference magnetization generated by semi-selective pulses or the Jeener-Broekaert sequence always decays according to Eq. (20). However, if the quadrupolar splitting is similar to the dipolar splitting, which may happen for aromatic deuterons in liquid crystals, this simple picture breaks down (93); the spin-lattice relaxation rate for the sum magnetization still obeys Eq. (19), but a more detailed analysis is required for the decay of any difference magnetization.

It is natural to hope that deuteron spectra with resolvable fine-structure can provide more dynamical information, i.e. additional spectral density parameters, as a reward for carrying out the more complex density matrix analysis which is required to fully characterize relaxation pathways in such systems. Of particular interest are the cross correlation terms which appear in the relaxation expressions for spin systems where the states of two or more deuterons are mixed by (fully resolved) dipolar coupling (53,94). The spectral densities discussed earlier all arise from Fourier transformation of auto correlation functions, i.e. those which describe the behaviour of one interaction tensor. Inspection of Eq. (12) shows that if quadrupolar interactions for two or more deuterons are included in Eq. (3), as they must if the deuterons are dipolar coupled, then we obtain cross terms of the form

$$J^{c}_{jklm}(\omega) = \langle j\, A^{(1)}_{-M}|k\rangle\langle 1|A^{(2)}_{-M}|m\rangle^{*}\; J^{c}_{M}(\omega) \;; \tag{28}$$

$J^{c}_{M}(\omega)$ is the Fourier transform of the cross correlation function for the motion of two C-D bonds. Like the cross terms appearing in dipolar relaxation of scalar coupled spins (95-98) in isotropic fluids, the quadrupolar cross terms have proven very valuable in the analysis of solute reorientation (52,99) in nematic liquids. The detailed procedures by which this additional information may be extracted depend on the nature of the particular spin system. Techniques developed for D_2 (52) and D_3 spin systems (99) cannot be easily generalized to arbitrarily complicated sets of coupled deuterons and there appears to be no viable alternative to a complete density matrix analysis of each system. However, one general rule holds: any semi-selective perturbation applied to the spin system as described above yields no information about cross-correlation terms. It is usually not practical to apply fully selective pulses to generate difference magnetizations between individual lines on one side of a spectrum of dipolar coupled deuterons, but one might consider an asymmetric Jeener-Broekaert excitation in which a large difference magnetization is created between such lines. The analysis of such an experiment would probably be complicated because uneven excitation of several lines would have to be accounted for, and a possible alternative is to analyze single quantum and multiple quantum decay rates.

Whenever dipolar couplings are well resolved it is feasible to stimulate not only double, but also zero and higher order multiple quantum coherence for oriented deuteron systems (52,77,100). In favourable cases (52), the complete set of decay rates of the single and multiple quantum coherences provides an overdetermined set of linear equations from which all the spectral density parameters can be extracted. Unfortunately, as mentioned earlier, multiple quantum decay rates suffer from several sources of systematic error which may render them unsuitable for accurate analysis. Perhaps the most serious limitation is that while small, long range dipolar couplings may not be resolved, they certainly are a major source of line broadening in deuterium spectra of complex spin systems. These long range dipolar couplings may or may not be refocussed in multiple quantum spin echo spectra, but although proper density matrix analysis of such systems remains to be performed, experiences in our laboratory have not been encouraging.

In cases where high order multiple quantum coherence is either impossible to stimulate or unacceptably damped by instrumental artifacts, zero quantum decay rates may be useful. Zero quantum coherence can only occur between the spin states of two or more coupled deuterons (77) and to excite it, the dipolar coupling must be well resolved. Zero quantum coherence among any homonuclear set of coupled spins is insensitive to field inhomogeneity or to diffusion through external field gradients (101); it is therefore not necessary to use a refocussing pulse to measure the transverse decay rate for a ZQ coherence. Although a ZQ decay rate has so far been obtained only for one system, acetonitrile-d_3 in nematic Phase 5 (77), it appears to be a highly reliable way to obtain transverse relaxation information.

^{13}C-H Relaxation.

Relaxation of coupled ^{13}C-H spin systems offers complementary information to that available from deuterium studies, because even though some of the relevant tensors have coincident axes, the spectral densities which govern ^{13}C-H relaxation are evaluated at different (generally much higher) frequencies than those for deuterium. Moreover, for simple spin systems such as $^{13}CH_2$ and $^{13}CH_3$, the lines are narrow enough and the relaxation slow enough that highly selective perturbations of individual lines can be achieved with relative ease. Grant and coworkers (102,103) have made extensive use of these desirable attributes to develop elegant ways of following the evolution of normal modes, thereby obtaining the maximum information available from spin lattice relaxation measurements.

Proton-proton intermolecular dipolar interactions cannot be ignored in analyzing this type of relaxation data, even when observation is confined exclusively to ^{13}C. This occurs because proton relaxation by any mechanism causes population flow among levels inextricably mixed by dipolar (or scalar) coupling. (The analogous process for ^{2}H-^{1}H spin systems is usually too slow to compete with direct quadrupolar relaxation of deuterium (104)). A treatment of intermolecular proton-proton dipolar relaxation as a rank one random fluctuating field process shows that for a $^{13}CH_2$ group there are four spectral densities of motion which enter into the relaxation analysis: two auto correlations and two cross correlation terms. These terms so severely complicate the analysis of spin-lattice relaxation of ^{13}C-H systems that it has proven impossible to analyze the experiments without assuming a specific model of motion. The implication is that ^{13}C-H relaxation measurements constitute less definite tests of molecular motion models than are available from deuteron relaxation measurements. More discouraging, perhaps, is the fact that analysis of relaxation in coupled ^{13}C-H systems may be impossible for the liquid crystal molecules themselves.

Since it is not practical to determine individual spectral density parameters from measured ^{13}C relaxation rates, it is necessary to adopt specific motional models at an early stage of data analysis. The most commonly used model combines anisotropic rotational diffusion (105-108) with one or more slow processes such as director fluctuations (80,109) or "slowly relaxing local structures" (SRLS) (89,107). From the data available so far (99,103,110), including a limited comparison of results obtained by deuterium relaxation on similar systems, it appears that director fluctuations make a small but non-negligible contribution, primarily via the low frequency spectral density term $J_1(\omega_C)$. Contributions to ^{13}C relaxation from SRLS have yet to be unambiguously demonstrated (110), and additional data is required to settle this question.

Other Nuclei.

Dipolar coupling between ^{14}N and ^{13}C in nitrile groups is quite strong because of the short internuclear distance. It is never resolved in liquid crystals because of very rapid ^{14}N relaxation, but Bulthuis and coworkers (110-113) have found that the collapsed dipolar coupling produces a major contribution to ^{13}C transverse relaxation rates; the process is entirely analogous to scalar relaxation. Since direct detection of ^{14}N is very difficult (114) due to a combination of low sensitivity, wide lines and quadrupolar splittings in the MHz range, this approach to measurements of ^{14}N spin-lattice relaxation rates is quite useful.

In the beginning of this chapter it was pointed out that proton relaxation had played a major role in the early days of NMR relaxation measurements on liquid crystals and that the analysis of proton data was difficult because of the many different processes which were found to contribute to the relaxation rates. It is fitting to conclude by noting that proton relaxation measurements may prove to be useful once again in liquid crystal dynamics studies because of the very high fields which are now available. The spectral densities which govern relaxation pathways in coupled spin systems can then be studied in the range of 500 MHz (for J_1) to 1000 MHz (for J_2), for example, by using methods similar to those developed by Courtieu et al. (115,116) in their studies of proton relaxation behaviour in 2- and 3-proton spin systems for solutes in deuteriated liquid crystal solutions. For this purpose, deuteriated liquid crystal molecules which have been labelled with protons at a few well chosen sites would be especially informative.

CONCLUSIONS

No attempts have been made here to provide an exhaustive overview of what nuclear spin relaxation measurements have done and may do for molecular dynamics studies in thermotropic liquid crystals. The intention has been to supply some of the tools which allow modern relaxation experiments to be carried out and interpreted.

At this time deuteron relaxation measurements are clearly the best way to determine individual spectral densities of motion without any ambiguities and without resorting to molecular models in the analysis of the relaxation data.

Important questions remain to be answered about the feasibility of measuring accurate values of $J_0(0)$ for molecular reorientation of liquid crystal molecules, but the determination of $J_1(\omega_0)$ and $J_2(2\omega_0)$ is straightforward. Measurements of individual spectral densities as a function of temperature and frequency cannot by any stretch of the imagination be considered an easy undertaking, but commercial spectrometers are at least very close to providing all the capabilities needed to facilitate such measurements.

REFERENCES

1. J.W. Doane and J.J. Visintainer, *Phys. Rev. Lett.* **23**, 1421, (1969).
2. R. Blinc, D.L. Hogenbloom, D.E. O'Reilly, and E.M. Peterson, *Phys. Rev. Lett.* **23**, 969 (1969).
3. M. Weger and P. Cabane, *J. Phys. (Paris), Colloq.* **30**, C4-72 (1969).
4. C.G. Wade, *Ann. Rev. Phys. Chem.* **28**, 47 (1977).
5. J.W. Doane, in *Magnetic Resonance of Phase Transitions* (Eds. F.J. Owens, C.P. Poole, Jr. and H.A. Farach), Academic Press, New York, 1979, Ch. 4.
6. J.J. Visintainer, R.Y. Dong, E. Bock, E. Tomchuk, D.B. Dewey, A.-L. Kuo and C.G. Wade, *J. Chem. Phys.* **66**, 3343 (1977).
7. R.C. Matthews and C.G. Wade, *J. Magn. Reson.* **19**, 166 (1975).
8. B. Deloche and B. Cabane, *Mol. Cryst. Liq. Cryst.* **19**, 25 (1972).
9. R.D. Orwoll, C.G. Wade and B.M. Fung, *J. Chem. Phys.* **63**, 986 (1975).
10. J.W. Emsley, J.C. Lindon and G.R. Luckhurst, *Mol. Phys.* **32**, 1187 (1976).
11. V. Rutar, M. Vilfan, R. Blinc, and E. Bock, *Mol. Phys.* **35**, 721 (1978).
12. R.Y. Dong, J. Lewis, E. Tomchuk and E. Bock, *J. Chem. Phys.* **69**, 5314 (1978).

13. R.R. Vold and R.L. Vold, *Symposium on Liquid Crystals and Ordered Fluids*, 183rd Natl. A.C.S. Meeting, Las Vegas, Nevada (1982), in press.

14. P.A. Beckmann, J.W. Emsley, G.R. Luckhurst and D.L. Turner, *Mol. Phys.* **50**, 699 (1983).

15. T.M. Barbara, R.R. Vold and R.L. Vold, *J. Chem. Phys.* **79**, 6338 (1983).

16. M. Schwartz, P.E. Fagerness, C.H. Wang and D.M. Grant, *J. Chem. Phys.* **60**, 5066 (1974).

17. H. Hutton, E. Bock, E. Tomchuk, and R.Y. Dong, *J. Chem. Phys.* **68**, 940 (1978).

18. R. Blinc, M. Vilfan, M. Luzar, J. Seliger and V. Zagar, *J. Chem. Phys.* **68**, 303 (1978).

19. R.Y. Dong, *J. Chem. Phys.* **75**, 2621 (1981).

20. W.M.M.J. Bovee, *Mol. Phys.* **29**, 1673 (1975).

21. J.P. Jacobsen, H.K. Bildsoe and K. Schaumburg, *J. Magn. Reson.* **23**, 153 (1976).

22. R.R. Vold and R.L. Vold, *J. Chem. Phys.* **66**, 4018 (1977).

23. L.G. Werbelow and D.M. Grant, *Adv. Magn. Reson.* **9**, 190 (1977).

24. S.B. Ahmad, K.J. Packer and J.M. Ramsden, *Mol. Phys.* **33**, 857 (1977).

25. R.L. Vold and R.R. Vold, *Progr. NMR Spectrosc.* **12**, 79 (1978).

26. R.L. Vold and R.R. Vold, *Israel J. Chem.* in press.

27. C.P. Slichter, *Principles of Magnetic Resonance*, Harper and Row, New York (1963).

28. M.E. Rose, *Elementary Theory of Angular Momentum*, Wiley, New York (1957).

29. D.M. Brink and G.R. Satchler, *Angular Momentum*, Oxford Univ. Press (Clarendon), London (1975).

30. C.F. Polnaszek, G.V. Bruno and J.H. Freed, *J. Chem. Phys.* **58**, 3185 (1973).

31. G. Moro and J.H. Freed, *J. Chem. Phys.* **74**, 3757 (1981).

32. A.G. Redfield, *Adv. Mag. Res.* **1**, 1 (1965).

33. A. Abragam, *The Principles of Nuclear Magnetism*, University Press, Oxford, (1961).

34. R.A. Hoffman, *Adv. Mag. Res.* **4**, 88 (1970).

35. P.L. Corio, *The Structure of High-Resolution NMR Spectra*, Academic Press, New York, (1967).

36. M.D. Zeidler, *Ber. Bunsenges. Phys. Chem.* **72**, 481 (1968).

37. H. Bildsoe, J.P. Jacobsen and K. Schaumburg, *J. Magn. Reson.* **23**, 137 (1976).

38. R.R. Vold, R.L. Vold and N.M. Szeverenyi, *J. Phys. Chem.* **85**, 1934 (1981).

39. S.B. Ahmad and K.J. Packer, *Mol. Phys.* **37**, 47 (1979).

40. H.W. Spiess, *J. Chem. Phys.* **72**, 6755 (1980).

41. J. Jeener and P. Broekaert, *Phys. Rev.* **157**, 232 (1967).

42. R.L. Vold, W.H. Dickerson and R.R. Vold, *J. Magn. Reson.* **43**, 213 (1981).

43. R.L. Vold and R.R. Vold, *J. Magn. Reson.* **42**, 173 (1981).

44. J.H. Davis, K.R. Jeffrey, M. Bloom, M.I. Valic and T.P. Higgs, *Chem. Phys. Lett.* **42**, 390 (1976).

45. R.L. Vold and S.O. Chan, *J. Chem. Phys.* **56**, 28 (1972).

46. R.L. Vold, *J. Chem. Phys.* **56**, 3210 (1972).

47. R.L. Vold and R.R. Vold, *J. Chem. Phys.* **61**, 2525 (1974).

48. H.E. Simon and R.L. Vold, *J. Magn. Reson.* **24**, 399 (1976).

49. J.P. Jacobsen and K. Schaumburg, *J. Magn. Reson.* **24**, 173 (1976).

50. W.H. Dickerson, R.R. Vold and R.L. Vold, *J. Phys. Chem.* **87**, 166 (1983).

51. T.C. Wong and K.R. Jeffries, *Mol. Phys.* **46**, 1 (1982).

52. R.L. Vold, R.R. Vold, R. Poupko and G. Bodenhausen, *J. Magn. Reson.* **38**, 141 (1980).

53. D. Jaffe, R.L. Vold and R.R. Vold, *J. Magn. Reson.* **46**, 496 (1982).

54. G. Bodenhausen, R.L. Vold and R.R. Vold, *J. Magn. Reson.* **37**, 93 (1980).

55. S. Vega and A. Pines, *J. Chem. Phys.* **66**, 5624 (1977).

56. G. Bodenhausen, R. Freeman and D.L. Turner, *J. Magn. Reson.* **27**, 511 (1977).

57. J.W. Emsley and D.L. Turner, *J. Chem. Soc. Faraday 2.* **77**, 1493 (1981).

58. J.W. Emsley and D.L. Turner, *Chem. Phys. Lett.* **82**, 447 (1981).

59. A. Kumar and R.R. Ernst, *Chem. Phys. Lett.* **37**, 839 (1976).

60. R.R. Vold and R.L. Vold, *J. Chem. Phys.* **64**, 320 (1976).

61. H.Y. Carr and E.M. Purcell, *Phys. Rev.* **94**, 630 (1954).

62. S. Meiboom and D. Gill, *Rev. Sci. Instr.* **29**, 688 (1958).

63. G. Bodenhausen, N.M. Szeverenyi, R.L. Vold and R.R. Vold, *J. Am. Chem. Soc.* **100**, 6265 (1978).

64. J.F. Martin, L.S. Selwyn, R.R. Vold and R.L. Vold, *J. Chem. Phys.* **76**, 2632 (1982).

65. D. Zax and A. Pines, *J. Chem. Phys.* **78**, 6333 (1983).

66. R.R. Vold and G. Bodenhausen, *J. Magn. Reson.* **39**, 363 (1980).

67. S. Schäublin, A. Höhener and R.R. Ernst, *J. Magn. Reson.* **13**, 196 (1974).

68. B.L. Tomlinson and H.D.W. Hill, *J. Chem. Phys.* **59**, 1775 (1973).

69. S.B. Ahmad and K.J. Packer, *Mol. Phys.* **37**, 59 (1979).

70. H.W. Spiess, *Colloid and Polymer Sci.* **261**, 193 (1983).

71. H.W. Spiess, *Israel J. Chem.* (in press).

72. L.S. Selwyn, R.R. Vold and R.L. Vold, *J.Chem.Phys.*, in press.

73. G. Bodenhausen, R. Freeman and G.A. Morris, *J. Magn. Reson.* **23**, 171 (1976).

74. W.S. Warren, D.P. Weitekamp and A. Pines, *J. Chem. Phys.* **73**, 2084 (1980).

75. A. Wokaun and R.R. Ernst, *Chem. Phys. Lett.* **52**, 407 (1977).

76. G. Drobny, A. Pines, S. Sinton, D. Weitekamp and D. Wemmer, *Faraday Division of the Chemical Society*, Symposium **13**, 49 (1979).

77. D. Jaffe, R.R. Vold and R.L. Vold, *J. Magn. Reson.* **46**, 475 (1982).

78. R.L. Vold and R.R. Vold *J. Magn. Reson.* **55**, 78 (1983).

79. P. Pincus, *Solid State Commun.* **7**, 415 (1969).

80. J.W. Doane, C.E. Tarr and M.A. Nickerson *Phys. Rev. Lett.* **33**, 620 (1974).

81. J.F. Martin, R.L. Vold and R.R. Vold, to be published.

82. J.F. Martin, R.L. Vold and R.R. Vold, *J. Magn. Reson.* **51**, 164 (1983).

83. J.F. Martin, R.R. Vold and R.L.Vold, *J.Chem.Phys.*, in press.

84. C.E. Tarr, M.A. Nickerson and C.W. Smith, *Appl. Phys. Lett.* **17**, 318 (1970).

85. J.W. Doane and D.S. Moroi, *Chem. Phys. Lett.* **11**, 339 (1971).

86. P. Ukleja, J. Pirs and J.W. Doane, *Phys. Rev.* **A14**, 414 (1976).

87. C.E. Tarr, F. Vosman and L.R. Whalley, *J. Chem. Phys.* **67**, 868 (1977).

88. G.R. Luckhurst and A. Sanson, *Mol. Phys.* **24**, 1297 (1972).

89. W.-J. Lin and J.H. Freed, *J. Phys. Chem.* **83**, 379 (1979).

90. J. Courtieu, D.W. Alderman, D.M. Grant and J.P. Bayles, *J. Chem. Phys.* **77**, 723 (1982).

91. F.M. Leslie, G.R. Luckhurst and H.J. Smith *Chem. Phys. Lett.* **13**, 368 (1972).

92. J.W. Emsley, J.C. Lindon and G.R. Luckhurst, *Chem. Phys. Lett.* **19**, 345 (1973).

93. P.R. Luyten, R.R. Vold and R.L.Vold, *J.Magn.Reson.*, in press.

94. R. Poupko, R.L. Vold and R.R. Vold, *J. Magn. Reson.* **34**, 67 (1979).

95. J.S. Blicharski and H. Schneider, *Ann. Phys.* **22**, 306 (1969).

96. L.G. Werbelow and D.M. Grant, *J. Chem. Phys.* **63**, 544 (1975).

97. C.L. Mayne, D.M. Grant and D.W. Alderman, *J. Chem. Phys.* **65**, 1684 (1976).

98. R.L. Vold and R.R. Vold and D. Canet, *J. Chem. Phys.* **66**, 1202 (1977).

99. D. Jaffe, R.L. Vold and R.R. Vold, *J. Chem. Phys.* **78**, 4852, (1983).

100. S. Hsi, H. Zimmermann and Z. Luz, *J. Chem. Phys.* **69**, 4126 (1979).

101. W.P. Aue, E. Bartholdi and R.R. Ernst, *J. Chem. Phys.* **64**, 2229 (1976).

102. J.M. Courtieu, C.L. Mayne and D.M. Grant, *J. Chem. Phys.* **66**, 2669 (1977).

103. E.P. Black, J.M. Bernassau, C.L. Mayne and D.M. Grant, *J. Chem. Phys.* **76**, 265 (1982).

104. J. Voigt and J.P. Jacobsen, *J. Chem. Phys.* **78**, 1693 (1983).

105. P.L. Nordio and P. Busolin, *J. Chem. Phys.* **55**, 5485 (1971).

106. P.L. Nordio, G. Rigatti and U. Segre, *J. Chem. Phys.* **56**, 2117 (1972).

107. C.F. Polnaszek and J.H. Freed, *J. Phys. Chem.* **79**, 2283 (1975).

108. J.M. Bernassau, E.P. Black and D.M. Grant, *J. Chem. Phys.* **76**, 253 (1982).

109. J.H. Freed, *J. Chem. Phys.* **66**, 4183 (1977).

110. P.R. Luyten, J. Bulthuis, W.M.M.J. Bovee and L. Plomp, *J. Chem. Phys.* **78**, 1712 (1983).

111. H.A. Lopes Cardozo, J. Bulthuis and C. Maclean, *J. Magn. Reson.* **33**, 27 (1979).

112. H.A. Lopes Cardozo, J. Bulthuis, J.H. Freed and W.M.M.J. Bovee, *Chem. Phys. Lett.* **60**, 335 (1979).

113. P.R. Luyten, H.A. Lopes Cardozo and J. Bulthuis, *Chem. Phys. Lett.* **87**, 496 (1982).

114. B. Cabane and W.G. Clark, *Phys. Rev. Lett.* **25**, 91 (1970).

115. J. Courtieu, J. Julien, N. Thoi Lai, A. Guillois, P. Gonord, S.K. Kan and C.L. Mayne, *J. Chem. Phys.* **72**, 953 (1980).

116. J. Courtieu, N.T. Lai, C.L. Mayne, J.M. Bernassau and D.M. Grant, *J. Chem. Phys.* **76**, 257 (1982).

MULTIPLE QUANTUM NMR IN LIQUID CRYSTALLINE PHASES

Gary Drobny

Department of Chemistry,
University of Washington,
Seattle.

INTRODUCTION

Nuclear magnetic resonance has been used extensively as a probe of molecular ordering and structure in liquid crystalline phases. In rigid molecules dissolved in mesogenic solvents, magnetic interactions are functions of geometric factors and order parameters. The geometric factors relate elements of a magnetic interaction tensor in some principal axis system (pas) to a molecule-fixed frame, and so provide information on molecular structure, while order parameters describe anisotropic molecular motions (1). For molecules in uniaxial phases the order tensor is traceless and symmetric and so is composed of five independent elements (2). A further reduction of the number of independent nonzero ordering tensor elements is possible through the judicious selection of a molecular-fixed coordinate system.

Nonrigid molecules have the added complication that different geometric configurations are characterized by distinct ordering tensors. For example, if a molecule dissolved in a uniaxial solvent interconverts between N configurations, there will be 5N unknowns. Furthermore, each unknown will consist of the product of a configurational probability and an element of the order tensor of that configuration. It is a basic problem in the NMR of oriented systems to measure a large enough number of interactions with sufficient accuracy to allow a complete determination of all 5N unknowns. In this chapter we describe a new approach to this problem, namely the measurement of proton dipolar couplings by multiple quantum NMR. In the following section we will quantify the problem of the averaging of

J. W. Emsley (ed.), Nuclear Magnetic Resonance of Liquid Crystals, 289–314.

magnetic interaction tensors and discuss the various NMR methods used to probe internal motion and ordering in mesogenic phases. We will then consider, in some detail, the response of a system of dipolar-coupled spin ½ nuclei to a four pulse multiple quantum experiment and we will emphasize in particular the increase in sensitivity achieved by obtaining the full two dimensional spectrum. Finally, the spectral simplification achieved by multiple quantum NMR will be considered. As an example, the number of transitions in the multiple quantum spectra of an oriented four-methylene chain will be calculated, and experimental proton multiple quantum spectra of oriented n-hexane-d_6 will be presented. The sensitivity of these multiple quantum spectra to several simple models of chain motion will be considered.

THE AVERAGING OF MAGNETIC INTERACTION TENSORS

Spin interaction Hamiltonians may be expressed as scalar products of second rank tensors (3,4)

$$H = \sum_{L,m} (-1)^m A^{(L,-m)} T^{(L,m)} \tag{1}$$

where A is a tensor that involves spatial degrees of freedom and T involves spin angular momentum operators. The usual truncation of H at high field results in a Hamiltonian of the form

$$H = \sum_{L} A^{(L,0)} T^{(L,0)} \quad . \tag{2}$$

Now the tensor A as written in Eq. (2) is usually defined in a laboratory-fixed coordinate system in which the z axis is parallel to the direction of the magnetic field. In a well-ordered nematic liquid crystalline phase this direction is usually also parallel to the director. We are interested tensors for which L = 2 so the relationship between $\tilde{A}^{(2,0)}$ in the director frame and the principal values of A in a molecular frame is,

$$\tilde{A}^{(2,0)} = \sum_{p,q} (-1)^{-q} A^{(2,p)} D^2_{p,q} (\Omega_0)\, \bar{D}^2_{-q,0} (\Omega_1) \tag{3}$$

In Eq. (3), Ω_0 is the solid angle relating the principal axis frame to a molecule-fixed frame and Ω_1 relates the molecular-fixed frame to the director frame. The terms $\bar{D}^2_{-q,0}$ (Ω_1) are motionally averaged Wigner rotation matrix elements and are the second-rank order parameters defined by Zannoni in chapter 1.

If the molecule is also interconverting between N distinct geometric configurations, Eq. (3) becomes

$$\tilde{A}^{(2,0)} = \sum_N P_N \sum_{p,q} A_N^{(2,p)} D^2_{p,q}(\Omega_0^N)\, \bar{D}^2_{-q,0}(\Omega_1^N)\,(-1)^{-q}\ , \qquad (4)$$

which justifies the qualitative claim of the preceding section, namely that there will be 5N unknowns of the form $P_N\bar{D}^2_{-q,0}(\Omega_1)$, where P_N is a configurational probability. Therefore, in order to determine this set of unknowns, at least 5N measurements $\tilde{A}^{(2,0)}$ must be obtained from the NMR spectrum.

If the number of configurations is small the problem may, in principle, be tractable, but for many molecules, and for oriented alkyl chains in particular, N can be a very large number. For example, an alkyl chain with 3 internal bonds has 27 rotational isomers, and even though many of the configurations are related by symmetry (there are only 9 independent types), the number of unknowns is still very large.

The deuterium nuclear quadrupolar coupling is by far the most extensively used probe of internal motion and molecular ordering in liquid crystalline phases. However, the number of quadrupolar couplings is determined by the number of distinct nuclear types, and there is rarely a sufficiently large number of such couplings to completely determine the set of N ordering tensors. For example, the deuterium NMR spectrum of perdeuteriated n-hexane consists of three doublets, which is an insufficient amount of information to determine the ordering tensors of all configurations of n-hexane.

A simplification of Eq. (4) may be achieved by assuming that molecular reorientation is independent of molecular configuration and utilizing an average ordering tensor. This average tensor may in turn be reduced to a single parameter (6). While such assumptions do make the problem more tractable, several authors have remarked that the use of average ordering tensors is not valid in general (7-11) and is critically dependent upon the choice of molecular-fixed axes. The set of unknowns in Eq. (4) may also be reduced by restricting the number of configurations with nonzero probability (12-13), which may be justified by the tendency of the anisotropic solvent to elongate and align the chain (14).

Any quantitative analysis of chain ordering in uniaxial liquids will therefore involve the measurement of a distinct ordering tensor for each molecular configuration, and even if the number of configurations is restricted to some subset of extended forms, the quadrupolar couplings will usually provide insufficient information for the analysis of a multiple bond chain.

The proton-proton direct dipolar interaction is another useful probe of mesogenic ordering. The advantage of this method can be appreciated by considering that while n distinct deuterium nuclei will have n quadrupolar couplings, n spin $\frac{1}{2}$ nuclei will have $n!/(n-2)!2!$ dipolar couplings. Therefore, the number of dipolar couplings will clearly exceed the number of quadrupolar couplings for $n>3$. However, the proton NMR spectra of molecules dissolved in mesogenic phases are usually quite complex except for the most trivial of systems and so dipolar couplings must be extracted by an iterative fitting of the NMR spectra (15). Most methods involve the fitting of spectral frequencies and intensities by a linear least squares algorithm using the dipolar couplings as adjustable parameters. In recent years, iterative algorithms of considerable sophistication have been developed (16-18) that have enabled the analysis of relatively complex spectral systems (19-22). However, in the limit of large n and low molecular symmetry, the number of spectral transitions may be so large as to seriously degrade resolution to the point that spectral fitting becomes difficult or even impossible.

When spectral complexity hinders analysis, an obvious solution is isotopic labelling. For example, one might measure heteronuclear dipolar couplings from the local field spectra of ^{15}N or ^{13}C labels (23). Alternatively, the number of protons may be reduced by selective deuteriation.

Spectral simplification may also be obtained without the necessity of isotopic labelling by the method of multiple quantum spectroscopy (24,25). The utility of multiple quantum NMR lies in the fact that there are far fewer transitions in the very high multiple quantum orders than there are in the single quantum spectrum. In fact, the number of m quantum transitions in a system of n spin $\frac{1}{2}$ nuclei is, neglecting symmetry, (25)

$$\frac{(2n!)}{(n+m)!\,(n-m)!} \qquad (5)$$

for $\Delta m \neq 0$. A repeated application of Eq. (5) shows that the number of $\Delta m = n-2$ quantum transitions is $n(2n-1)$, the number of $\Delta m = n-1$ quantum transition is $2n$, and there can be only a single n quantum transition. It is clear that for large n, the n-1 and n-2 quantum orders will be far less complex than the single quantum spectrum.

Having explained the utility of multiple quantum spectroscopy, we will next consider the manner in which multiple quantum transitions may be observed.

MULTIPLE QUANTUM NMR

Time Domain Cross Sections

We will concern ourselves primarily with the four-pulse multiple quantum experiment in which a 180 degree pulse has been applied at the centre of the multiple quantum evolution period to remove dephasing due to a magnetic field inhomogeneity (see Fig. (1).

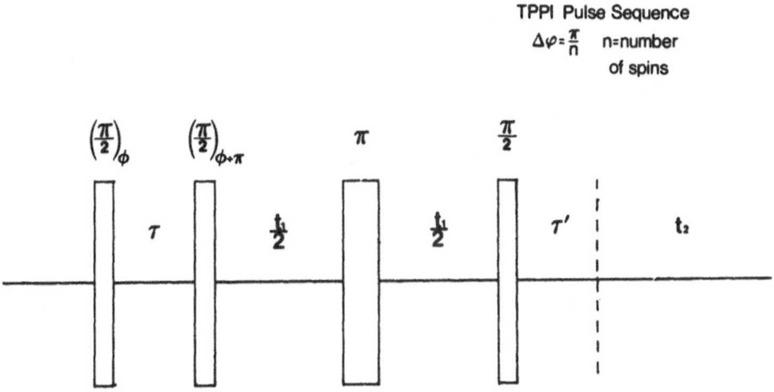

Figure 1. *The pulse sequence used to effect a* Δm *dependent separation of the multiple quantum transitions. The phase of the first two pulses is incremented by* $\Delta\phi = 2\pi/2n$ *as* t_1 *is incremented by* Δt_1. *The offset for the nth multiple quantum order is* $\Delta\omega(n) = n\Delta\phi/\Delta t_1$.

The response of a system of spin ½ nuclei to such a pulse sequence is, in the density operator formalism:

$$\rho(\tau + t_1 + \tau' + t_2) =$$

$$e^{-iHt_2}\, e^{-iH\tau'}\, R_x(\pi/2)\, e^{-iHt_1/2}\, R_x(\pi)\, e^{-iHt_1/2}$$

$$R_{\phi+\pi}(\pi/2)\, e^{-iH\tau}\, R_\phi(\pi/2)\, \rho(0)\, R_\phi^{-1}(\pi/2)\, e^{iH\tau}$$

$$\cdot \; R^{-1}_{\phi+\pi}(\pi/2)\, e^{iHt_1/2}\, R_x^{-1}(\pi)\, e^{iHt_1/2}\, R_x^{-1}(\pi/2)$$

$$\cdot \; e^{iH\tau'}\, e^{iHt_2}. \tag{6}$$

where the preparation period is of length τ, the evolution period is of length t_1, the mixing period is of length τ', and the detection period is of length t_2. We have also used the definitions

$$R_x(\theta) = e^{i\theta I_x} \tag{7}$$

$$R_\phi(\theta) = e^{i\theta I_\phi} \tag{8}$$

$$I_\phi = e^{-i\phi I_z}\, I_x\, e^{i\phi I_z} \tag{9}$$

and

$$\rho(0) = \frac{1}{Tr(1)}\, e^{-\beta\gamma B_0 I_z} \tag{10}$$

The expression for H, the rotating frame Hamiltonian is

$$H = H_z + H_D$$

$$= -\sum_i \Delta\omega_i I_{zi} + \sum_{i<j} D_{ij}(I_{zi}I_{zj} - \tfrac{1}{4}(I_{+i}I_{-j} + I_{-i}I_{+j})). \tag{11}$$

Now the multiple quantum transition frequencies may be obtained from the projection of the two dimensional Fourier transform onto the ω_1 axis. But from the projection theorem of two dimensional spectroscopy (26), it is known that a given time domain cross section and the projection of the phase sensitive two dimensional transform onto a parallel frequency axis form a Fourier pair. This means that the multiple quantum transition frequencies may be obtained from the Fourier transform of the cross section $t_2 = 0$, but with a considerable loss in sensitivity. We will return to this point later.

It is assumed for the present that only the signal amplitude at $t_2 = 0$ is recorded. Furthermore, it can be explicitly assumed in Eq. (11) that the chemical shift difference between nuclear types is negligible compared to the dipolar couplings. Then the expression for the density operator given in Eq. (6) simplifies to

$$\rho(\tau + t_1 + \tau) =$$

$$e^{-i\mathcal{H}\tau'} R_x(\pi/2)\, e^{-i\mathcal{H}_D t_1} e^{-i\phi I_z} e^{-i\mathcal{H}_{xx}\tau}$$

$$\cdot\ \rho(0)\, e^{i\mathcal{H}_{xx}\tau} e^{i\phi I_z} e^{i\mathcal{H}_D t_1} R_x^{-1}(\pi/2)\, e^{i\mathcal{H}\tau'} \tag{12}$$

where

$$\mathcal{H}_{xx} = e^{-i\pi/2 I_x} \mathcal{H}\, e^{i\pi/2 I_x} . \tag{13}$$

The NMR signal at $t_2 = 0$ is given by the expression

$$S(t_1) = S_x(t_1) + iS_y(t_1)$$

$$\propto \mathrm{Tr}(\rho(\tau+t_1+\tau')I_x) + i\ \mathrm{Tr}(\rho(\tau+t_1+\tau')I_y). \tag{14}$$

Concentrating our attention on $S_y(t_1)$, the calculation is greatly simplified by adding a virtual pulse operator $R_x(\pi/2)$ at $t_2 = 0$ and calculating the trace of density operator with I_z. The expression for $S_y(t_1)$ is given by

$$S_y(t_1) \propto \mathrm{Tr}\left(e^{-i\mathcal{H}_D t_1} e^{-i\phi I_z} e^{-i\mathcal{H}_{xx}\tau} I_z \right.$$

$$\left. \cdot\ e^{i\mathcal{H}_{xx}\tau} e^{i\phi I_z} e^{i\mathcal{H}_D t_1} e^{i\mathcal{H}_{xx}\tau'} I_z\, e^{-i\mathcal{H}_{xx}\tau'} \right) , \tag{15}$$

where $\rho(0)$ has been expanded and only the term linear in I_z retained (high temperature approximation). Use has also been made of the invariance of the trace to cyclic permutations. Now it can be shown (25,27) that the preparation density operator assumes the form

$$U = e^{-i\mathcal{H}_{xx}\tau} I_z e^{i\mathcal{H}_{xx}\tau} =$$

$$\sum_{k,p.c} \left(a^{k,c}_{2p} D^{k,c}_{2p} \cos\Delta\omega\tau + a^{k,c}_{2p+1} T^{k,c}_{2p+1} \sin \Delta\omega\tau \right) \tag{16}$$

where T^k_q is an operator corresponding to a coherent state for which $q = \Delta m$, k is the rank of the tensor, and c completes the definition of the operator basis. The action of the operator $e^{-i\phi I_z}$ is to introduce a Δm dependent phase term into the preparation operator, that is

$$e^{-i\phi I_z} U e^{i\phi I_z} =$$

$$\sum_{k,p,c} \left(a^{k,c}_{2p} T^{k,c}_{2p} e^{i2p\phi} \cos \Delta\omega\tau + a^{k,c}_{2p+1} T^{k,c}_{2p+1} e^{-i(2p+1)\phi} \sin \Delta\omega\tau \right) . \quad (17)$$

Now for the pulse sequence given in Figure (1), $\tau = \tau'$ and V, the mixing density operator is identical to U. Given these facts together with Eq. (15), an expansion of Eq. (13) in an eigenbasis of *H* gives

$$S_y(t_1) = \sum_{\alpha\beta} e^{-i\omega_{\alpha\beta} t_1} \cdot$$

$$\left[\sum_{k,p,c} (a^{k,c}_{2p})^2 \cos^2 \Delta\omega\tau \, (T^{k,c}_{2p})^2_{\alpha\beta} \, e^{-i2p\phi} + (a^{k,c}_{2p+1})^2 \sin^2 \Delta\omega\tau \, (T^{k,c}_{2p})^2_{\alpha\beta} \, e^{-i(2p+1)\phi} \right] \quad (18)$$

Some discussion of Eq. (18) is necessary at this point. The coefficients $a^{k,c}_q$ are functions of the dipolar couplings. Now in Eq. (18), the square of complex terms introduce phase differences among multiple quantum coherent states. These phase differences may be removed by time reversal in the mixing period (28). If the additional phase term ϕ is incremented as a function of t_1 by the amount $2\pi/2n$, a separation of multiple quantum orders in ω_1 results. This phase cycling of the preparation pulse phase is necessary whenever an echo pulse is applied in t_1 since such a pulse will refocus dephasing due to resonance offset, collapsing the multiple quantum orders.

From Eq. (18) it is also clear that since the coefficients $a^{k,c}_q$ are functions of the dipolar couplings multiplied by τ, the preparation time, not all multiple quantum coherent states may be strongly prepared for a given τ. Therefore, it is necessary to repeat the multiple quantum experiment for a range of τ values in order to assure preparation of all possible transitions. A systematic method for selecting appropriate τ values has been described elsewhere (28).

Projections of Two Dimensional Spectra

We have already stated that a time domain cross section and a projection of the phase sensitive two-dimensional transform onto an axis parallel to the cross section axis constitute a Fourier pair. The implication of this relationship to the sensitivity of the nonselective multiple quantum experiment will now be considered.

If the entire t_2 signal is recorded for each point taken in t_1, the result is a two dimensional time domain matrix, and the expression for the y component of the NMR signal is obtained by substituting Eq. (6) into Eq. (14).

$$S_y t_1,t_2) = Tr(\rho(t_1,t_2)I_y)$$

$$= \sum_{ij} \sum_{k\ell} e^{-i\omega_{ij}t_2} e^{-i\omega_{k\ell}t_1} Z_{ij,k\ell}$$

where

$$Z_{ij,k\ell} = (R_xV)_{ik}(R_xV)^{-1}_{\ell j}(UI_zU\dagger)_{k\ell}(I_y)_{ji}. \tag{19}$$

Now for the four-pulse multiple quantum experiment under consideration, the coefficients $Z_{ij,k\ell}$ will be complex numbers. This means that the phases of the single quantum coherent states ij connected to a given multiple quantum state $k\ell$ will, in general, differ and will therefore add destructively upon projection onto ω_1.

There are several ways in which destructive interference can be avoided. One method is to produce a mixing operator V that is the adjoint of U (28). This will reduce $Z_{ij,k\ell}$ to squared modulus form and the projection spectrum will be pure absorptive (26,28). A second method is to project the two dimensional magnitude spectrum (26). In the absolute magnitude spectrum all intensities are positive and therefore destructive interference is avoided upon projection onto ω_1, but the lineshape will no longer be pure absorptive.

We will now compare the results of multiple quantum experiments carried out by the time domain cross section method and the absolute magnitude projection method. In both cases a nonselective multiple quantum pulse sequence was applied to a sample of oriented benzene. In Fig. (2) is shown the absolute magnitude proton multiple quantum NMR spectrum of the $t_2 = 0$ cross section. The cross section consisted of 4096 points. Before transformation, the t_1 signal was apodized (sine-squared) and zero-filled to 16384.

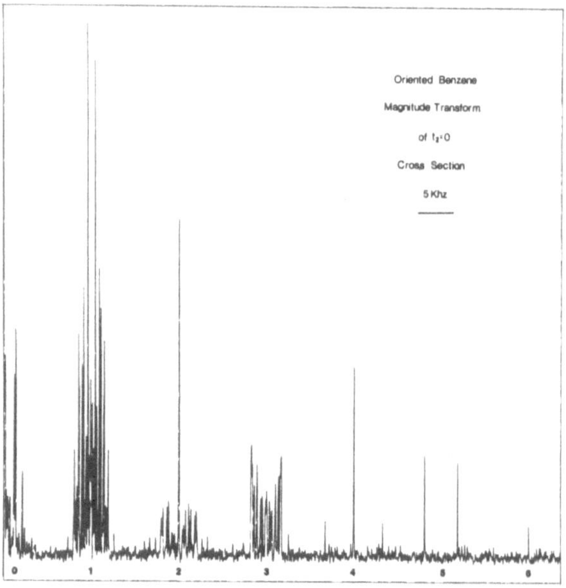

Figure 2. *The absolute magnitude transform of the time domain cross section $t_2 = 0$, for oriented benzene. The cross section consisted of 4096 points in t_1, and four transients were accumulated for each point in t_1 to improve signal-to-noise. Before Fourier transformation, the cross section was zero-filled to 16384.*

In Fig. (3) is shown a two dimensional absolute magnitude spectrum of oriented benzene. The time domain matrix consisted of 1024 points in t_1 and 256 complex points in t_2. Sine-squared apodizations were applied in t_1 and t_2, followed by zero filling to 4096 and 1024, respectively. In practice, resolution in ω_1 is emphasised and therefore multiple quantum time domain matrices are highly rectangular, and the transforms of such matrices are not easily viewed in their entirety. Suffice it to say that the data to be compared to Fig. (2) were obtained from a matrix

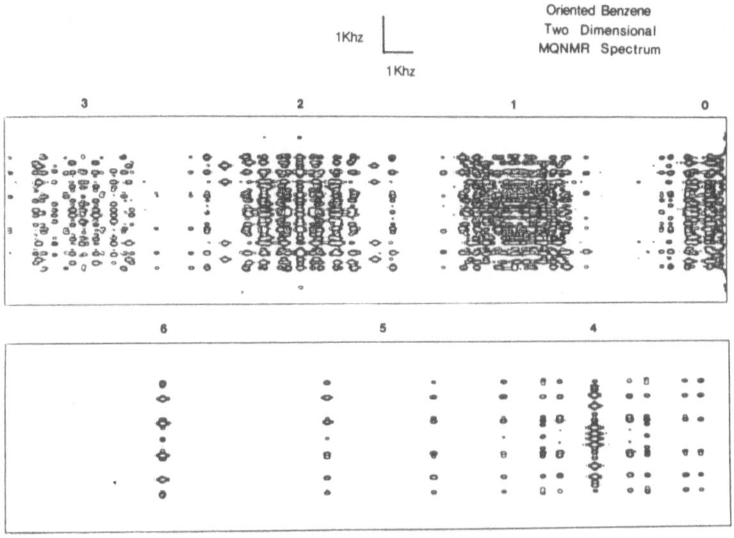

Figure 3. *A two dimensional absolute magnitude of oriented benzene. The time domain matrix consisted of 1024 points in t_1 and 512 points in t_2. Before t_1 and t_2 transformations, sine-squared apodization was performed followed by zero-filling to 4096 in t_1 and 2048 in t_2.*

composed of 4096 points in t_1 and 128 complex points in t_2. The matrix was sine-squared apodized in t_1 and t_2, followed by zero filling to 16384 and 512, respectively. After Fourier transformation in t_1 and t_2, the absolute magnitude spectrum was calculated. The projection may be appreciated from Fig. (3) as a summation of all ω_2 cross sections and was performed nonselectively, that is, all cross sections were weighted equally, regardless of how much signal was present in each. Therefore, the resultant projection, shown in Fig. (4), represents a "worst case" projection. Figs. (2) and (4) are scaled according to the most intense single quantum transitions. As a reference, the signal-to-noise ratio of the six quantum line in the projection case is about seven times that of the cross section case.

We conclude therefore that whenever there exists resolution in ω_2, the signal to noise of the projection of the absolute magnitude two dimensional spectrum will be superior to that of the magnitude transform of the time domain cross section t_2 = 0.

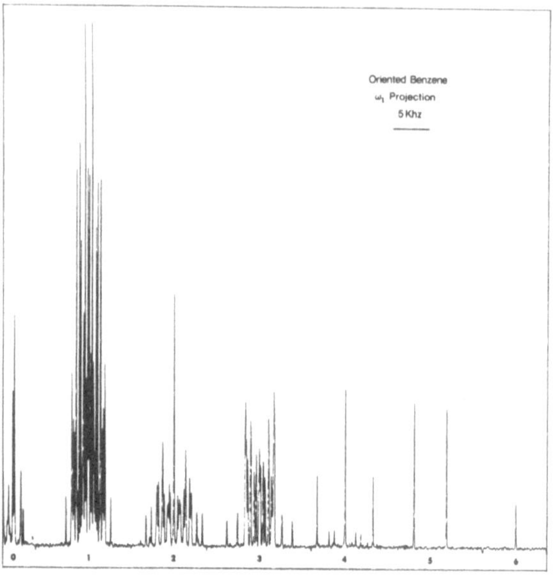

Figure 4. *The projection onto ω_1 of the two dimensional absolute magnitude spectrum of oriented benzene. As in the case of the cross section experiment, four transients were accumulated in order to improve signal-to-noise. 4096 points were accumulated in t_1 and 128 complex points in t_2. After appropriate sine-squared apodization in t_1 and t_2, each t_2 row was zero-filled to 512 and each t_1 column was zero-filled to 16384, and transformed in t_1 and t_2. After calculating the absolute magnitude spectrum, the projection onto ω_1 was carried out without scaling.*

THE COMPLEXITY OF THE MULTIPLE QUANTUM SPECTRA OF AN ORIENTED CHAIN

In this section we will consider the complexity of the various multiple quantum orders of an oriented four methylene chain. The number of transitions in each multiple quantum order may be calculated by group theoretical methods. First we assume a rotational isomeric model (30) of internal chain motion. We also assume that the probability of a trans to gauche (+)

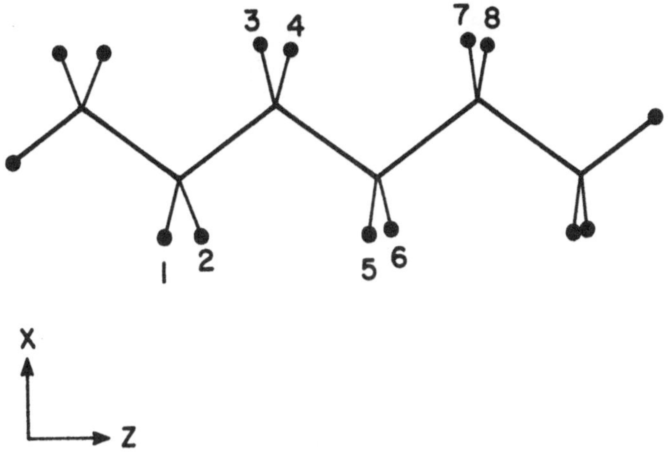

Figure 5. *The nuclear labelling convention used with n-hexane-d_6.*

interconversion equals a trans to gauche (-) interconversion, and the ends of the chain are assumed to be indistinguishable. The symmetry group is composed of those permutations and permutation inversions that leave the NMR Hamiltonian invariant(31), and using the nuclear labelling convention shown in Fig. (5), the symmetry operations are

$$O_1 = (12)(34)(56)(78) \equiv \sigma_h \tag{20a}$$

$$O_2 = (18)(27)(36)(45) \equiv i \tag{20b}$$

$$O_3 = (17)(28)(35)(46) \equiv C_2 \tag{20c}$$

$$O_4 = E \tag{20d}$$

The group is of order 4 and is isomorphous to C_{2h}. To obtain the energy level diagram, it is necessary to calculate the characters of each symmetry operation in the direct product basis of each Zeeman manifold. The result is given in table 1.

The energy level diagram may be obtained by decomposing the reducible representations into irreducible components using the formula

$$a_j = h^{-1} \sum_k N_k \ \chi^{(j)} (C_k)^* \ \chi(C_k) \tag{21}$$

Table 1. *Reducible character table of the NMR group of the eight protons of a four-methylene chain.*

	E	C_2	σ_h	i
$\Gamma(m = \pm 4)$	1	1	1	1
$\Gamma(m = \pm 3)$	8	0	0	0
$\Gamma(m = \pm 2)$	28	4	4	4
$\Gamma(m = \pm 1)$	56	0	0	0
$\Gamma(m = \pm 0)$	70	0	0	0

where a_j is the number of times that the jth irreducible representation occurs within Γ, h is the order of the group (h = 4), $\chi^{(j)}(C_k)$ is the character of an operation of the class C_k in the jth irreducible representation, and $\chi(C_k)$ is the character of the same operation in a reducible representation. Using the fact that $N_k = 1$ for all k, and applying Eq. (21) we obtain the following decompositions:

$$\Gamma(m = \pm 4) = A_g \quad (22a)$$

$$\Gamma(m = \pm 3) = 2A_g \oplus 2A_u \oplus 2B_g \oplus 2B_u \quad (22b)$$

$$\Gamma(m = \pm 2) = 10A_g \oplus 6A_u \oplus 6B_g \oplus 6B_u \quad (22c)$$

$$\Gamma(m = \pm 1) = 14A_g \oplus 14A_u \oplus 14B_g \oplus 14B_u \quad (22d)$$

$$\Gamma(m = \pm 0) = 22A_g \oplus 16A_u \oplus 16B_g \oplus 16B_u \quad (22e)$$

which yield the energy level diagram directly of table 2.

Assuming that the chemical shift difference between protons on different methylene groups is small, lines will occur in pairs within each order. The number of pairs in each multiple quantum order is given in table 3.

In table 3, we have neglected to count those transitions in the even orders below eight that occur at multiples of $\Delta\omega$ since those frequencies will contain no coupling information. It is also clear from the data in table 3 that the five, six, and seven quantum orders are far simpler than the single quantum order. In a system of eight protons, there are twenty-eight

Table 2. *n-Hexane-d_6 energy level diagram.*

m	A_g	A_u	B_g	B_u
4	1			
3	2	2	2	2
2	10	6	6	6
1	14	14	14	14
0	22	16	16	16
-1	14	14	14	4
-2	10	6	6	6
-3	2	2	2	2
-4	1			

Table 3. *The number of transitions for each multiple quantum order of the eight spin chain. Transitions occur listed as doublets since the chemical shift difference is negligible between methylene groups. Transitions at multiples of $\Delta m < 8$ occur in all even orders. For example, the $\Delta m = 6$ order contains an eight-fold degenerate transition at $6\Delta\omega$.*

Δm	
8	1 line at $8\Delta\omega$
7	2 pairs
6	14 pairs
5	70 pairs
4	224 pairs
3	546 pairs
2	944 pairs
1	1430 pairs
0	1049 pairs

Table 4. *Classes of dipolar couplings that are identical by symmetry for an eight spin chain.*

$$D_{12} = D_{78}$$
$$D_{17} = D_{28}$$
$$D_{18} = D_{27}$$
$$D_{34} = D_{56}$$
$$D_{35} = D_{46}$$
$$D_{36} = D_{45}$$
$$D_{13} = D_{24} = D_{57} = D_{68}$$
$$D_{14} = D_{23} = D_{58} = D_{67}$$
$$D_{15} = D_{26} = D_{37} = D_{48}$$
$$D_{16} = D_{25} = D_{38} = D_{47}$$

dipolar couplings, but from the symmetry of the problem the number of independent couplings is reduced to ten. Couplings that are identical by symmetry are listed in table 4.

MULTIPLE QUANTUM NMR EXPERIMENTS ON ORIENTED CHAIN

Having demonstrated by a group theoretical calculations that the multiple quantum spectra of high order for an oriented four-methylene chain are far simpler than the single quantum spectrum, we will now turn to experimental results for a particular system of interest, n-hexane. In particular we will consider the proton multiple quantum spectra of the eight methylene protons, so the system of actual interest is methyl-deuteriated n-hexane (n-hexane-d_6). All multiple quantum experiments were performed with deuterium decoupling during the preparation, evolution, and mixing periods. Oriented samples were prepared by dissolving n-hexane-d_6 in the nematogen Eastman 15320 to produce about a 30% mole ratio solution.

All experiments on n-hexane were performed by the time domain cross section method, using the four-pulse sequence described earlier, with 28 transients accumulated for each t_1 value to improve signal to noise. 4096 t_1 values were accumulated with a timing increment of 10 microseconds. Furthermore, as implied by Eq. (16), the preparation and mixing operators are dependent upon the dipolar couplings between protons, so one cannot expect to observe all multiple quantum transitions with a single preparation/mixing time. In practice, therefore, it is necessary to perform several multiple quantum experiments with varying preparation/mixing times. For n-hexane-d_6, fifteen experiments were performed for preparation times varying from two to fifteen milliseconds. The resulting multiple quantum free induction decays were Fourier transformed and the absolute magnitude spectra were added together. The resulting multiple quantum spectrum is shown in Fig. (6), and

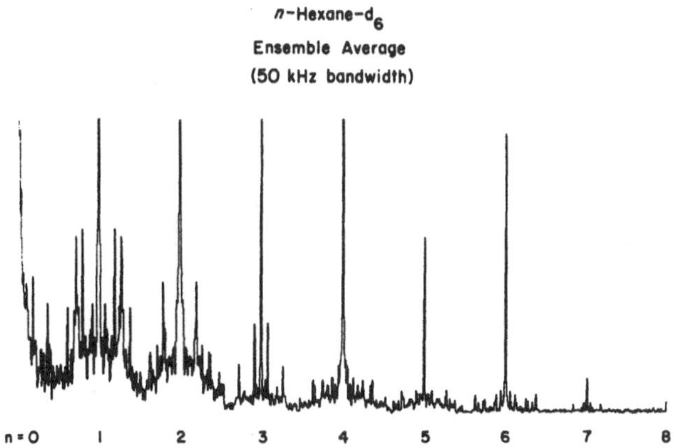

Figure 6. *The proton multiple quantum spectra of n-hexane-d_6 obtained by the pulse sequence shown in Fig. (1). 4096 points were accumulated in t_1 for which Δt_1 = 10 usec and $\Delta\phi$ = 22.5°. The spectrum shown is the sum of fifteen absolute magnitude spectra obtained for preparation/mixing times varying from one to fourteen milliseconds. Each multiple quantum experiment involved the accumulation of twenty transients per point in t_1.*

the six and seven quantum regions are shown in expanded form in Fig. (7). In the six quantum region, eleven of the expected fourteen transitions are observed together with an intense central transition, resulting from m = 3 to m = -3 transitions. The seven quantum region has two pairs, and the central peak is an artifact resulting from the phase shifting of the first two pulses.

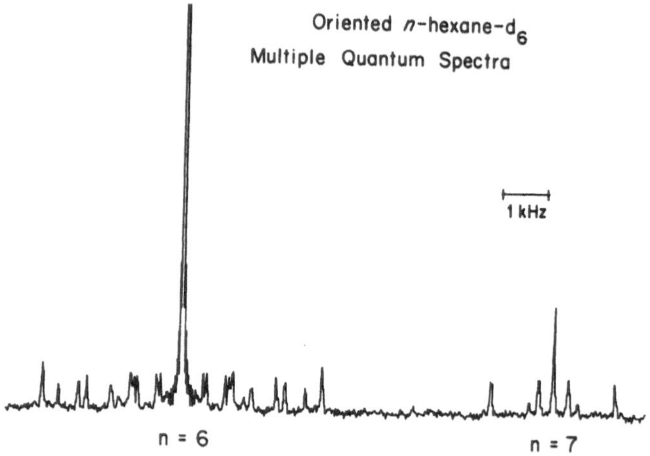

Figure 7. *An expansion of the six and seven quantum regions. In the six quantum region at least eleven of the fourteen predicted pairs are observed together with an eight-fold degenerate central transition. The central line in the seven quantum region is an artifact produced by the phase shifting of the preparation pulses.*

SPECTRAL SIMULATIONS

As stated earlier, it is a basic problem in the NMR of liquid crystalline systems to extract enough information from the spectrum with sufficient accuracy to allow a determination of all 5N unknowns. Since there are at most 10 independent dipolar couplings (see table 4) it will be impossible to perform a rigorous inversion of Eq. (4) with the proton dipolar couplings alone. A solution to the problem is to augment the data set with heteronuclear couplings, by carbon-13 labelling for instance. Another approach is to reduce the number of unknowns by making several simplifying assumptions. For example, the number of configurations might be restricted to a subset of "most probable" configurations, or the form of the order tensor might be simplified. Pursuing this idea, let us test the sensitivity of the six and seven quantum spectra to motional modelling by:

(i) simplifying the order tensor to a single parameter S_{zz} that is independent of configuration; and

(ii) restricting the number of configurations.

Assumption (i) amounts to reducing the ordering to a scaling of the spectrum. Our approach, therefore, will be to attempt to fit the six and seven quantum spectral transitions by a judicious selection of configurational probabilities.

We will consider three models of chain motion, keeping in mind that we are not attempting a rigorous analysis but simply attempting to demonstrate the sensitivity of multiple quantum transition frequencies to chain configuration.

(i) Model 1: only the all-trans and the single gauche states are populated. S_{zz} is scaled for an optimal fit;

(ii) Model 2: all configurations not involving adjacent gauche isomerizations of opposite sense (g±g∓) are included, with probabilities weighted according to the number of gauche isomerizations. In addition, $P(g^{\pm})/P(t) = \frac{1}{2}$ and S_{zz} is scaled for an optimal set;

(iii) Model 3: all configurations not involving adjacent gauche isomerizations are included with probabilities weighted as in model 2. S_{zz} is adjusted for optimal fit.

Table 5. *Calculated configurational probabilities for the motional models of n-hexane-d_6.*

Configuration	Probability for the three models		
	1	2	3
ttt	.25	.16	.20
tg±t	.125	.08	.10
ttg± (= g±tt)	.125	.08	.10
g±tg∓	0	0.04	0.05
tg±g± (= g±g±t)	0	0.04	0
g±tg±	0	0.04	0.05
g±g±g±	0	0.02	0

Table 6. *Calculated averaged dipolar coupling constants for the motional models of n-hexane-d_6 in units of hertz.*

$\tilde{D}_{(ij)}$	Model 1	Model 2	Model 3
12	1642.20	847.10	1434.47
13	113.77	37.20	4.43
14	69.81	40.30	-23.20
15	-1018.56	-764.90	-915.11
16	561.49	-621.60	-639.74
17	-301.54	0363.70	-420.00
18	-160.26	-384.00	-152.75
23	69.81	40.30	-23.20
24	113.77	37.20	4.43
25	-561.49	-621.60	-639.74
26	-1018.56	-764.90	-915.11
27	-160.26	-384.00	-152.75
28	-301.54	-363.70	-420.00
34	2080.48	2252.00	2161.37
35	67.70	5.60	89.97
36	14.75	-30.60	29.79
37	-1018.56	-764.90	-915.11
38	-561.49	-621.60	-639.74
45	14.75	-30.60	29.79
46	67.70	5.60	89.97
47	-561.49	-621.60	-639.74
48	-1018.56	-764.90	-915.11
56	2080.48	2252.00	2161.37
57	113.77	37.20	4.43
58	69.81	40.30	-23.20
67	69.81	40.30	-23.20
68	113.77	37.20	4.43
78	1642.20	847.10	1434.47

In the following simulations, a carbon-carbon bond length of 1.54 Angstroms and a carbon-hydrogen bond length of 1.10 Angstroms are assumed. Rotational isomers are further assumed to be related by 120° rotations (32). Table 5 shows the set of configurational probabilities for each model. Using these probabilities and a value of S_{zz} = .12, proton dipolar couplings were calculated for each model. The results are given in table 6. Finally, using these couplings, six and seven quantum spectra were calculated and the transition frequencies for each model are listed in table 7. Calculated spectra are shown in Figs. (8), (9), and (10).

Table 7. *Calculated six and seven quantum transition frequencies of the motional models of n-hexane-d_6**

	Model 1	Model 2	Model 3
$\Delta m = 6$			
	±2257	±2223	±2219
	±2006	±2136	±2050
	±1922	±1724	±1770
	±1369	±1435	±1504
	±1253	±1435	±1323
	±1015	±1378	±1178
	±997	±1268	±1051
	±655	±720	±958
	±477	±712	±881
	±417	±609	±622
	±314	±589	±471
	±257	±227	±414
	±137	±191	±218
	-	-	55
$\Delta m = 7$			
	±1063	±1080	±1053
	±40	±644	±270

* In units of hertz, measured from the centre of the multiple quantum order.

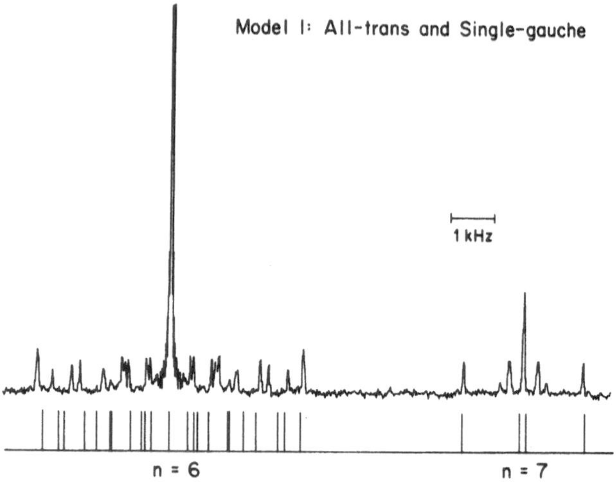

Figure 8. *Calculated six and seven quantum spectra compared to experimental data for a model assuming population of the all-trans and single gauche states only. A value of* S_{zz} *= .12 was assumed.*

DISCUSSION

Comparing the data in table 7 and the experimental six and seven quantum transition frequencies in table 8, it is clear that model 3 best fits the data, with an average deviation of 48 hertz. Given the digital resolution of the experiment, this amounts to about two points in an 8192 point spectrum. Of course a thorough analysis of the data must involve an accurate determination of dipolar couplings by an iterative fitting of multiple quantum transition frequencies and a fitting of intensities. An iterative fitting of intensities is difficult from a practical standpoint due to the complex form of the preparation and mixing operators. However, a satisfactory approach might be to iteratively fit the transition frequencies and check the final solution by direct calculation of transition intensities. This approach has been successfully employed in determining the dihedral angle in a cyanobiphenyl liquid crystal from the 5, 6, and 7 multiple quantum spectra of the aromatic ring protons (33).

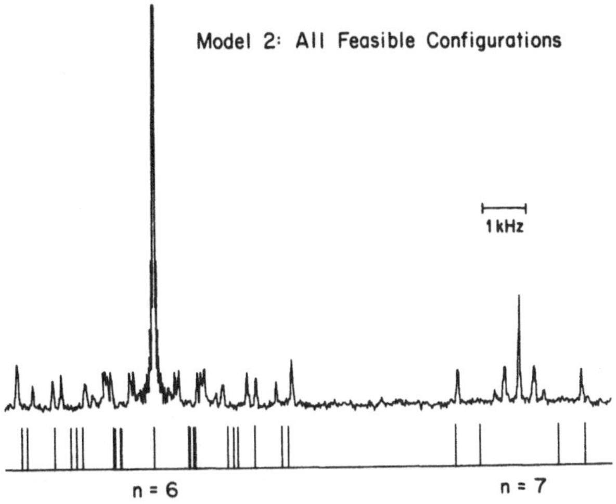

Figure 9. *Calculated six and seven quantum spectra for a model assuming population of all configurations except those involving gauche bond isomerizations of opposite sense (e.g., tg±g+, g±g+g±). We call this the "feasible" configuration model.*

CONCLUSION

We have demonstrated the feasibility of observing multiple quantum transitions of very high order in systems of oriented polymethylene chains. This method enables the use of proton dipolar couplings as a probe of chain configurations, but avoids the necessity of analyzing and simulating highly complicated spectra. This is a distinct advantage over single quantum methods. Furthermore, we have shown that multiple quantum transition frequencies are very sensitive to configuration in polymethylene chains.

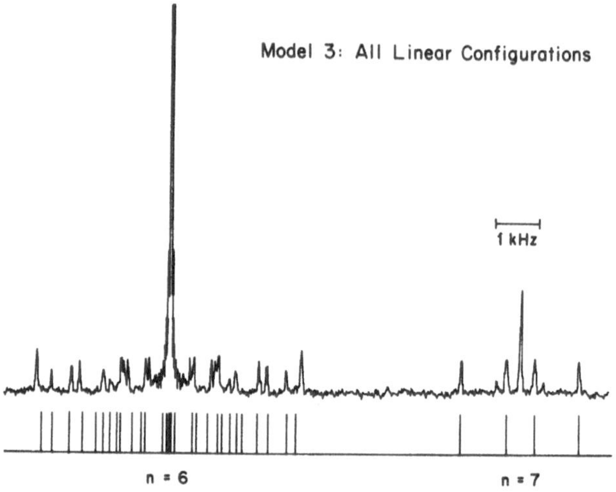

Figure 10 *Calculated six and seven quantum spectra for a model assuming population of all configurations except those involving adjacent gauche isomerizations. This model is called the linear or extended configuration model.*

Table 8. *Experimental six and seven quantum transition frequencies**

$\Delta m = 6$	$\Delta m = 7$
±2298 hz	±1044 hz
±2046	±252
±1698	
±1554	
±1254	
±1146	
±1026	
±846	
±798	
±714	
±414	
±354	

* Frequencies are measured relative to the centre of the multiple quantum orders ($6\Delta\omega$ and $7\Delta\omega$).

REFERENCES

1. A. Saupe, *Z. Naturforsch* **19A**, 161 (1964).

2. L.C. Snyder, *J. Chem. Phys.* **43**, 4041 (1965).

3. W. Haeberlin, *High Resolution in Solids, Selective Averaging* (Academic Press, New York, 1976).

4. M. Mehring, *High Resolution NMR in Solids* (Springer-Verlag, New York, 1976).

5. J.W. Emsley and J.C. Linden, *NMR Spectroscopy Using Liquid Crystal Solvents* (Pergamon Press, New York, 1975), Ch. 6.

6. S. Hsi, H. Zimmerman, and Z. Luz, *J. Chem. Phys.* **69**, 4126 (1978).

7. E.E. Burnell and C.A. de Lange, *J. Magn. Reson.* **39**, 46 (1980).

8. J.W. Emsley and G.R. Luckhurst, *Mol. Phys.* **41**, 19 (1980).

9. E.E. Burnell and C.A. de Lange, *Chem. Phys. Lett.* **76**, 268 (1980).

10. E.E. Burnell, C.A. de Lange, and O.G. Mouritsen, *J. Magn. Reson.* **50**, 188 (1982).

11. J.W. Emsley, G.R. Luckhurst, and C.P. Stockley, *Proc. R. Soc. Lond.* A **381**, 117 (1982).

12. J.W. Emsley, G.R. Luckhurst, and C.P. Stockley, *Mol. Phys.* **38(5)**, 1687 (1979).

13. E.T. Samulski, *Ferroelectrics* **30**, 83 (1980).

14. A. Ben Shaul, Y. Rabin, and W.M. Gelbart, *J. Chem. Phys.* **78**, 4303 (1983).

15. P. Diehl, E. Fluck, and R. Kosfeld (ed.), *NMR: Basic Principles and Progress* (Springer-Verlag, New York, 1972), Vol. 6.

16. P. Diehl, S. Sykora, and J. Vogt, *J. Magn. Reson.* **19**, 67 (1975).

17. V. Lueg and G. Hägele, *J. Magn. Reson.* **26**, 505 (1977).

18. D.S. Stephenson and G. Binsch, *J. Magn. Reson.* **37**, 395 (1980).

19. E.E. Burnell and P. Diehl, *Mol. Phys.* **24**, 489 (1972).

20. J.W. Emsley, et al., *J. Chem. Soc. Perkin II*, 1542 (1975).

21. D.S. Stephenson and G. Binsch, *Mol. Phys.* **43**, 697 (1981).

22. G. Chidichimo, A. Liguori, and M. Longeri, *J. Magn. Reson.* **51**, 438 (1983).

23. A. Höhener, L. Müller, and R.R. Ernst, *Mol. Phys.* **38**, 909 (1979).

24. A. Wokaun and R.R. Ernst, *Mol. Phys.* **36**, 317 (1978).

25. A. Pines, D. Wemmer, J. Tang and S. Sinton, *Bull. Am. Phys. Soc.* **21**, 23 (1978).

26. K. Nagayama, P. Bachmann, K. Wütrich and R.R. Ernst, *J. Magn. Reson.* **31**, 133 (1978).

27. G. Drobny, A. Pines, S. Sinton, D. Weitekamp, and D. Wemmer, *Faraday Symposium* **13**, 49 (1979).

28. Y.S. Yen and A. Pines, *J. Chem. Phys.* **78**, 3579 (1983).

29. D.P. Weitekamp, J.R. Garbow, and A. Pines, *J. Magn. Reson.* **46**, 529 (1982).

30. P.J. Flory, *Statistical Mechanics of Chain Molecules* (John Wiley and Sons, New York, 1969).

31. H.C. Longuet-Higgins, *Mol. Phys.* **6**, 445 (1963).

32. G. Drobny, D.P. Weitekamp, and A. Pines, submitted for publication.

33. S. Sinton and A. Pines, *Chem. Phys. Lett.* **76**, 263 (1980).

DYNAMICS OF MOLECULAR PROCESSES BY NMR IN LIQUID CRYSTALLINE SOLVENTS

Z. Luz

Isotope Department,
The Weizmann Institute of Science,
Rehovot 76100 Israel.

INTRODUCTION

The effect of chemical exchange processes on NMR lineshapes in solution was discovered (1) in 1953 very shortly after the discovery of the chemical shift and spin-spin splitting phenomena. Since then this effect has been extensively used to study dynamic processes such as isomerization, ring inversion, hindered rotation, proton transfer, and ligand exchange in isotropic liquids (2). The basic idea of the method is that dynamic processes that modulate the transition frequencies at rates comparable to characteristic splitting in the spectrum will affect its lineshape. The interactions that are usually considered in liquids are the chemical shift and the indirect spin-spin coupling. For protons the spin-spin splittings are of the order of several hertz while the chemical shifts in conventional magnetic fields are of the order 10 to 10^3 Hz. This limits the range over which dynamic processes can be studied to $1-10^6$ sec^{-1}. Other nuclei such as ^{13}C or ^{19}F have larger coupling and consequently allow wider dynamic ranges to be studied.

The purpose of the present article is to review dynamic studies performed in liquid crystalline solvents rather than in isotropic liquids. The main difference between the two media arises from the presence in liquid crystal solvents of anisotropic splittings due to dipole-dipole or quadrupolar interactions, which are absent in isotropic liquids. It is shown that the use of liquid crystals as solvents may have certain advantages for dynamic studies over the use of isotropic liquids, but there are also severe limitations. The method will

J. W. Emsley (ed.), Nuclear Magnetic Resonance of Liquid Crystals, 315–342.

be demonstrated using the ring inversion process of s-trioxane (3), the bond shift reaction in cyclooctatetraene (4) and the bond rearrangement in bullvalene (5) by proton NMR spectroscopy, and using the ring inversion reactions in cyclohexane (6) and p-dioxane (7) by deuterium NMR spectroscopy. The theory of dynamic lineshape as applied to anisotropic liquids is then discussed including the use of group theory to factorize the lineshape problem (8).

DYNAMIC ^{1}H NMR IN LIQUID CRYSTALS

Ring Inversion in s-trioxane (3)

In an isotropic solution at sufficiently low temperatures (below -70°C) the proton NMR of s-trioxane consists of an AB quartet for the axial and equatorial hydrogens. On increasing the temperature the quartet coalesces because of the ring inversion reaction

Typical spectra for a 0.3 wt% solution of s-trioxane in Freon are shown in Fig. (1).

In Fig. (2) are shown proton NMR spectra of solutions of s-trioxane in liquid crystalline solvents at temperatures ranging between -20°C and +80°C. The sharp features in these spectra are due to the s-trioxane solute while the broad asymmetric background signals (in particular the large hump slightly off-centre in the -20°C to +30°C spectra) are due to the liquid crystal solvent. The high temperature spectrum (i.e. above 80°C) of s-trioxane in liquid crystalline solvents was studied earlier by Cocivera (9). Its structure was interpreted in terms of the dipole-dipole interactions between the various nuclei in the molecule, averaged over the two chair conformations. As may be seen in Fig. (2), on cooling to below 80°C changes in the lineshape occur which as for the isotropic solutions are attributed to the inversion reaction between the two chair forms. However, since the dipolar interactions are much larger than the chemical shift (δ) and the spin-spin splitting (J), the range at which dynamic effects on the lineshapes are observed, occurs at considerably higher temperatures in the liquid crystalline solvents than in the isotropic liquids. To derive kinetic parameters from the liquid

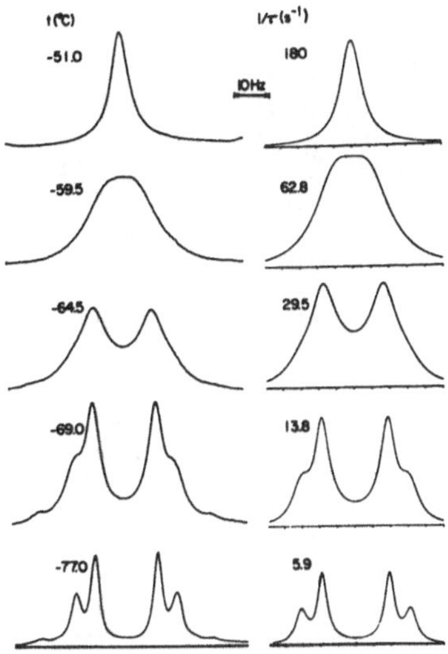

Figure 1. *Left: Experimental ^1H NMR spectra of a 0.3 wt% solution of s-trioxane in Freon at the temperatures indicated in the figure. Right: Corresponding simulated dynamic spectra using the parameters $J=6.3Hz$, $\delta=25.1Hz$, $1/T_2=5.0$ sec^{-1} and $1/\tau$ values as indicated in the figure.*

crystalline spectra they are compared with theoretical lineshapes computed with appropriate magnetic parameters and various values for the specific rate, $1/\tau$, of the interconversion process. The computation method is a straightforward extension of the Kaplan-Alexander density matrix theory (10-12) in which anisotropic interactions such as dipole-dipole and quadrupole couplings are added to the spin hamiltonian. It should be noted however, that although it is often safe to assume that isotropic interactions are (very nearly) independent of temperature the anisotropic interactions in liquid crystalline solutions are proportional to the molecular ordering parameters which are usually very sensitive to the temperature. In comparing experimental and calculated spectra appropriate scaling due to this effect must be employed.

A sequence of theoretical dynamic "anisotropic" spectra for s-trioxane are shown in Fig. (3). These were calculated from the dipolar interactions derived by Cocivera (9) using an orientational order parameter of -0.2, and other parameters as

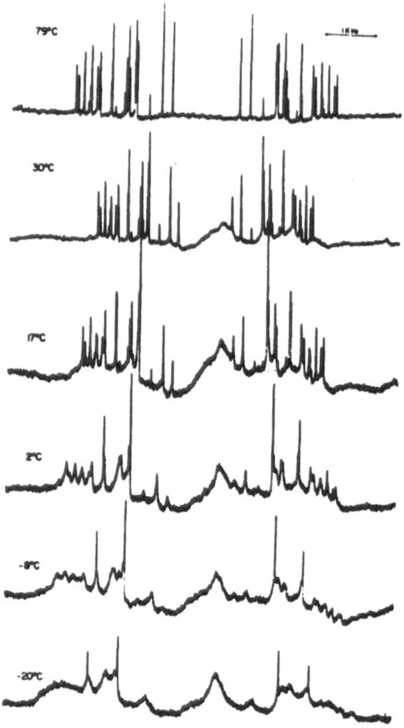

Figure 2. *Proton NMR spectra of s-trioxane in liquid crystalline solutions at different temperatures. The -20 to +30°C spectra are of a 2.8 wt% solution in Merck "phase V licristal". The 79°C spectrum corresponds to a 2.0 wt% solution in dihexyloxyazoxybenzene.*

indicated in the figure caption. It may be seen that the main features of the experimental spectra are reproduced by the computed lineshapes quite well. A more detailed comparison between experimental and theoretical spectra is shown in Fig. (4).

As indicated above, to derive the specific rate from such comparisons the order parameter appropriate to the experimental results must be determined. This was made by comparing splittings of characteristic features in the simulated and experimental spectra. Particularly useful for this purpose are the invariant transitions which remain sharp over the whole

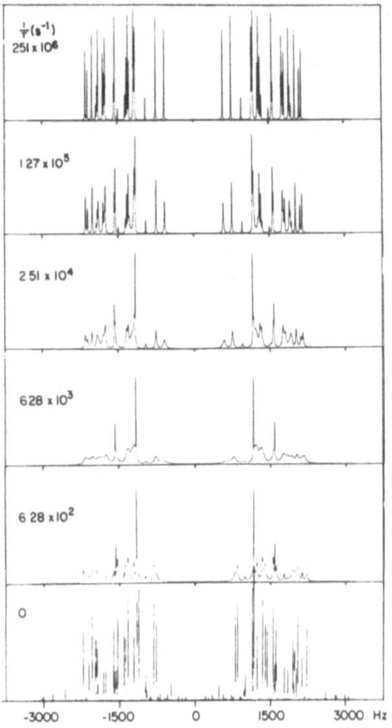

Figure 3. *Simulated dynamic spectra of s-trioxane in anisotropic solvents. The calculations were done using the parameters* $J=-6.3Hz$, $\delta=25.0Hz$, $D_{12}=-394.1Hz$, $D_{14}=965.1Hz$, $D_{45}=-77.5Hz$, $D_{15}=-49.9Hz$, $1/T_2=15.7\ sec^{-1}$ *and* $1/\tau$ *values as indicated in the figure.*

dynamic range. The specific rates so obtained are plotted in Fig. (5) versus the reciprocal absolute temperature, from which the following activation parameters were derived: $\Delta H^{\#} = 11.1$ kcal/mol, $\Delta S^{\#} = 2.5$ e.u. and $1/\tau(300°K) = 2x10^5 sec^{-1}$. Also given in Fig. (5) are the results from the Freon solution. It may be seen that both sets of data lie on essentially the same Arrhenius curve, suggesting that intramolecular rearrangements of the type exhibited by s-trioxane are apparently not significantly affected by the nature of the solvent.

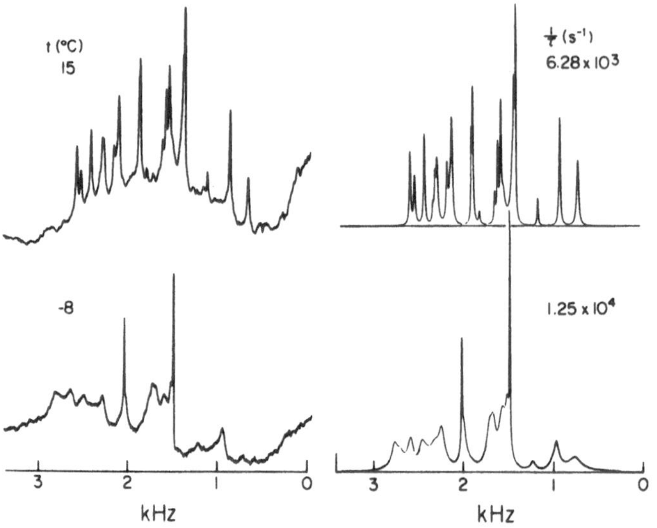

Figure 4. *Comparison of experimental and simulated dynamic* ^1H *NMR spectra of s-trioxane in anisotropic liquids. The spectra are from Figs. (2) and (3) and are drawn on an expanded scale. Only the low frequency halves are shown.*

Bond Shift in Cyclooctatetraene (4)

A similar study was performed on solution of cyclooctatetraene (COT) undergoing the bond shift reaction.

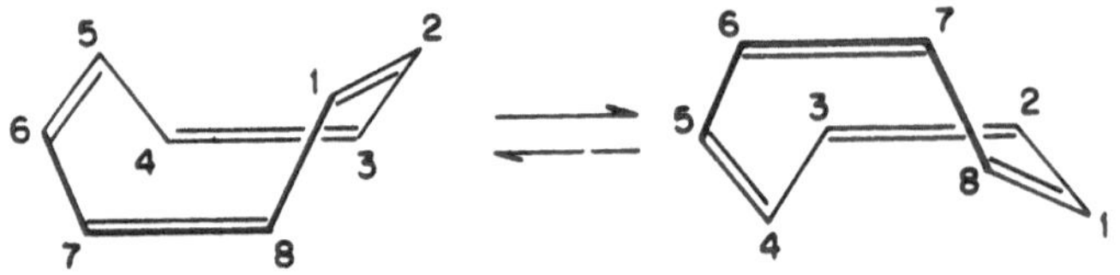

This case is particularly interesting because the equivalence of all protons in the COT molecule means that its spectrum in an isotropic solvent is not affected by the bond shift reaction. In anisotropic solvents on the other hand, the dipolar interaction between the protons results in a complicated spectrum whose

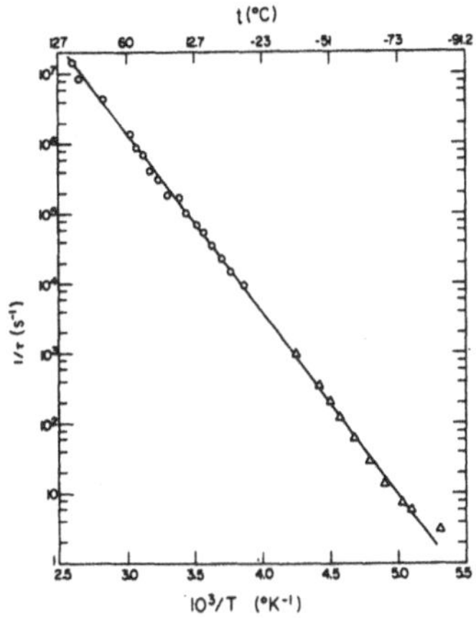

Figure 5. *An Arrhenius plot of the specific rate 1/τ for the ring inversion of s-trioxane. The triangular symbols at the low temperature part of the figure correspond to results obtained from an isotropic solution (Freon). The circles correspond to the two liquid crystalline solutions described in the caption to Fig. (2).*

lineshape is strongly temperature dependent (13) (cf. left column in Fig. (6)). This temperature dependence is attributed to the bond shift process indicated above. Over the temperature range -40°C to +170°C the spectrum may be seen to change from that corresponding to a static COT molecule up to that of a rapidly interconverting molecule. From the intermediate range, kinetic parameters may be derived in essentially the same way as explained for the case of s-trioxane. However, the computations in this case are considerably more complicated and were done using a group theoretical method as described later. Examples of spectra that approximately match the experimental results are reproduced in the right column of Fig. (6). By comparing these and similar traces with the experimental results the specific

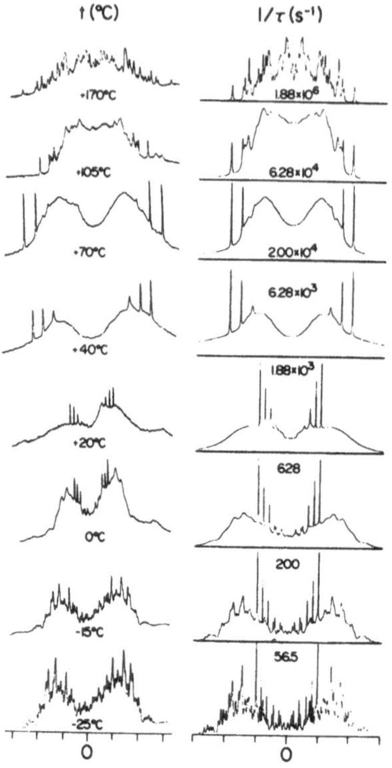

Figure 6. *Experimental (left) and simulated (right) ^{1}H NMR spectra of cyclooctatetraene (COT) in anisotropic solvents. The parameters used for the simulation are given in ref. 4. The spacing between frequency markers are as follows. For the experimental spectra they are 500 Hz for the bottom four tracings, 250 Hz for the next three, and 125 Hz for the topmost trace. For the simulated spectra they correspond to 1000 Hz for the bottom four traces, and 500 Hz for the four upper ones.*

rates for the bond shift at various temperatures could be determined. An Arrhenius plot of these results is shown in Fig. (7). The figure also contains earlier results of $1/\tau$ from proton NMR of deuteriated COT ($C_8H_2D_6$) recorded under conditions of deuteron decoupling. The two sets of data are in close agreement and yield the following activation parameters $\Delta H^{\#}$ = 10kcal/mol, $\Delta S^{\#}$ = -9.7 e.u. and $1/\tau(300°C) = 1.2x10^3 sec^{-1}$.

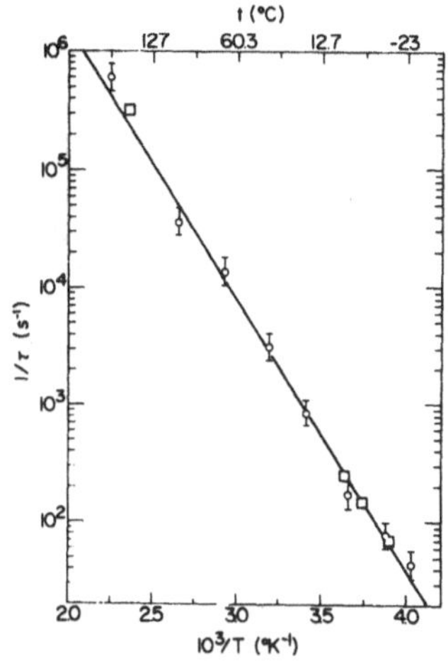

Figure 7. *Arrhenius plot for the bond shift process in cyclooctatetraene. The circles are for normal COT, while the squares are for deuteriated* $C_8H_2D_6$ *species.*

Bond Rearrangement in Bullvalene

Another example that exhibits dynamic NMR spectra in liquid crystalline solvents is that of bullvalene (14). This molecule undergoes a degenerate rearrangement of the type

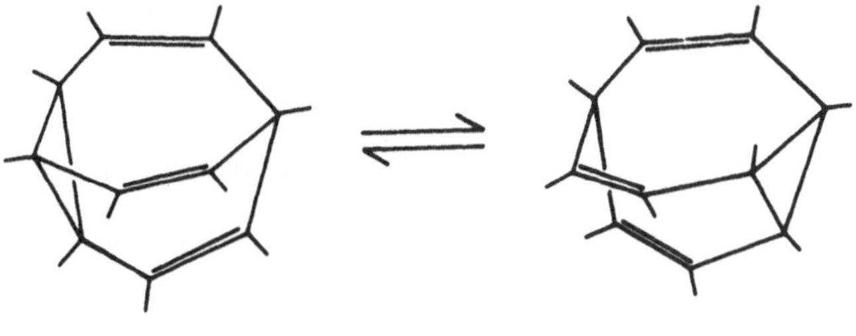

It was studied previously by dynamic ^1H and ^{13}C NMR in isotropic solvents (15). Its dynamic spectrum in a liquid crystal solvent (5) is shown in Fig. (8). This case is considerably more complicated and its quantitative interpretation has not yet been completed. There are ten protons per molecule and symmetry factorization still leaves large matrices to be diagonalized.

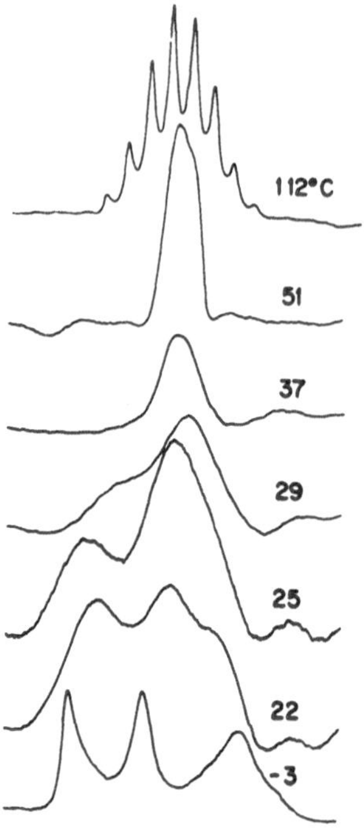

Figure 8. *Proton NMR spectra of a solution of bullvalene in phase V at different temperatures as indicated in the figure.*

DYNAMIC DEUTERIUM NMR IN LIQUID CRYSTALS

In the above section we discussed dynamic ^1H NMR spectra of molecules dissolved in liquid crystalline solvents. In this section we extend the discussion to deuterium NMR in solutions of deuteriated molecules. The dominant anisotropic splitting in this case is the deuterium quadrupolar interaction which is often several tens of kilohertz wide (16). This may allow measurements of specific rates as high as $10^9 sec^{-1}$ corresponding to a lifetime of a nsec.

Ring Inversion of Cyclohexane (6)

The ring inversion of cyclohexane

was the subject of many previous studies using proton NMR in normal liquids of specifically deuteriated molecules (17,18). However for performing the experiments it was necessary to cool the system down to below -60°C in order to arrive at the dynamic range of the proton spectrum. In deuterium NMR the dynamic range corresponds to considerably higher temperatures.

In Fig. (9) are shown deuterium NMR spectra of cyclohexane-d_{12} in a liquid crystal solvent at various temperatures. The interpretation of the spectra is quite straightforward: at low temperatures (below -20°C) they consist of four relatively sharp lines, representing two symmetric doublets. The outer doublet corresponds to the axial deuterons while the inner doublet to the equatorial ones. At this temperature range the inversion rate is slower or of the order of the spectral linewidth and thus its effect on the spectrum is small. The resolution is apparently not sufficient however to observe structure due to dipolar interaction between the deuterons. It will therefore be neglected in the following discussion. The spacing between the components of the quadrupole doublets in the deuterium spectra is given by

$$\tilde{\nu}_Q = \frac{3}{2} \frac{e^2qQ}{h} \frac{1}{2} (3 \cos^2\alpha - 1)S \qquad (1)$$

where S is the order parameter of the molecular symmetry axis and α the angle between this axis and the C-D bond. For cyclohexane $\alpha_{ax} = 2.6°$ and $\alpha_{eq} = 109.7°$, thus

$$\tilde{\nu}_Q^{ax}/\tilde{\nu}_Q^{eq} = -3.03 \qquad (2)$$

in very close agreement with the experiments.

As the temperature is raised above -30°C line broadening caused by the ring inversion reaction sets in. The effect of this process is to switch between the axial and equatorial resonances. Since however $\tilde{\nu}_Q^{ax}$ and ν_Q^{eq} have opposite signs the switching pairs are the low frequency axial component with the high frequency equatorial one and vice versa. Consequently a broad structureless signal is obtained in the intermediate

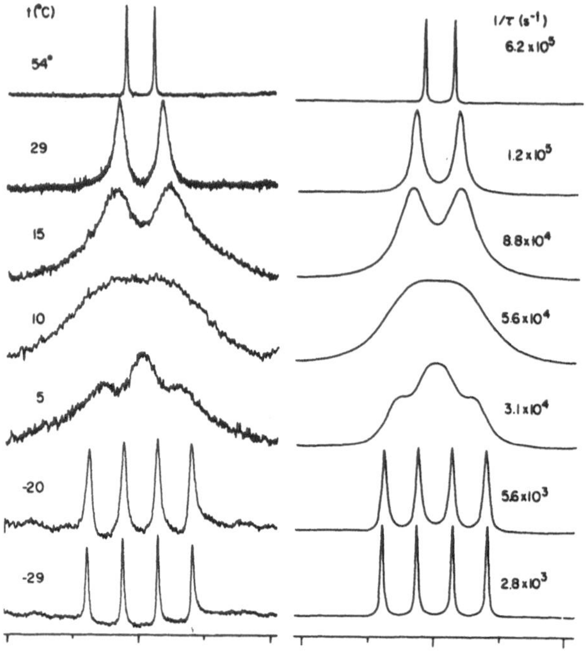

Figure 9. *Dynamic deuterium NMR spectra of C_6D_{12} in anisotropic solvents. The experimental spectra (left) correspond to a 2.6 wt% solution in phase V. The simulated spectra (right) for the particular temperatures indicated on the left were calculated using the order parameter of Fig. (10) and the specific rates as indicated in the figure. The frequency spacing between ticks is 16.7 kHz for the lowest three traces in each column, and 10 kHz for the upper four.*

range. At high temperatures a single doublet emerges whose splitting corresponds to the average splitting of the two deuterons.

$$\tilde{\nu}_Q^{av} = \tfrac{1}{2}(\tilde{\nu}_Q^{ax} + \tilde{\nu}_Q^{eq}) = 0.335\ \tilde{\nu}_Q^{ax} \tag{3}$$

where in the second equality Eq. (2) was used.

In order to derive kinetic parameters from the experimental

spectra they were compared with simulated lineshapes such as are shown e.g. in the right column of Fig. (9). For the simulation it was necessary to determine the order parameter S. In the low temperature region this was obtained directly from the experimental spectra. Likewise at high temperatures it could be

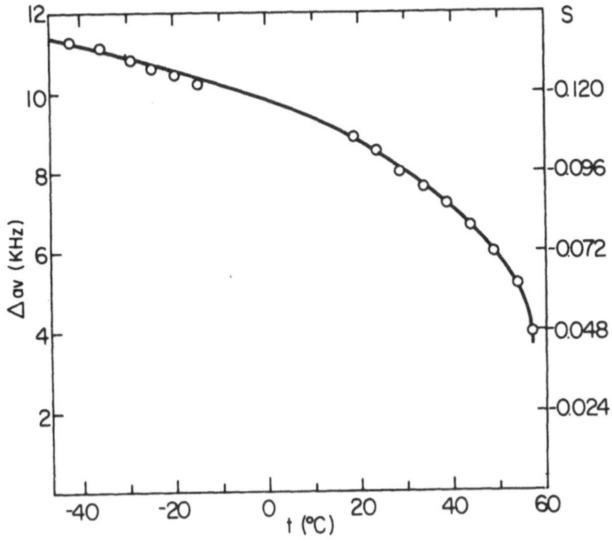

Figure 10. *The temperature dependence of the average quadrupolar splitting of the cyclohexane deuterons in the solution described in the caption to Fig. (9). The low temperature points were obtained by averaging axial and equatorial splittings measured from the experimental spectra, taking into account their opposite signs. The high temperature data were obtained directly from the experimental results. The right hand coordinate corresponds to the order parameter of the molecular symmetry axis.*

determined from the average splitting using Eq. (3), while in the intermediate range interpolation between the two extreme ranges was employed. For the solution of cyclohexane-d_{12} whose spectra are depicted in Fig. (9) the average quadrupole splitting, $\tilde{\nu}_Q^{av}$, so obtained are plotted in Fig. (10).

In Fig. (11) are plotted the specific rates for the ring inversion of cyclohexane versus reciprocal absolute temperature obtained from various NMR experiments. The results span an

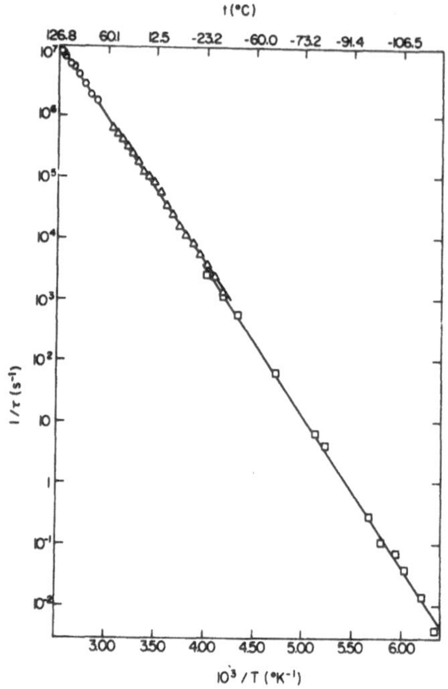

Figure 11. *Arrhenius plot of the specific rate for inversion of cyclohexane. The low temperature squares are from proton NMR in CS_2 solutions. The triangles and circles correspond to the liquid crystalline solvents, phase V and dihexyloxyazoxybenzene respectively.*

overall range of nine decades in $1/\tau$, all lying on essentially the same Arrhenius line. As for s-trioxane there does not seem to be a significant difference between the results in isotropic or anisotropic solvents. The kinetic parameters derived from the latter results are: $\Delta H^{\#}$ = 10.3 kcal/mole, $\Delta S^{\#}$ = 0.5 e.u., $1/\tau$ (300°k) = 1.5×10^5 sec^{-1}.

Ring Inversion in p-Dioxane (7)

The ring inversion of p-dioxane

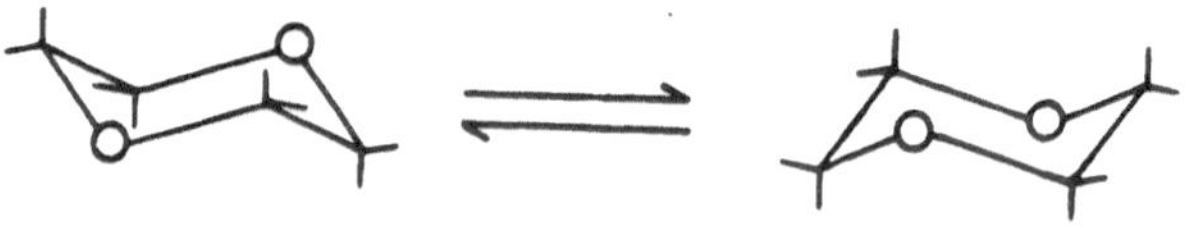

was also studied by ^{2}D NMR in liquid crystalline solvents. For this particular case this method is especially useful because the small chemical shift difference between the axial and equatorial protons means that not very reliable kinetic results could be obtained from proton NMR in isotropic liquids (19). Deuterium NMR spectra of p-dioxane-d_8 in two different liquid crystals are shown in Figs. (12) and (13). The spectra exhibit

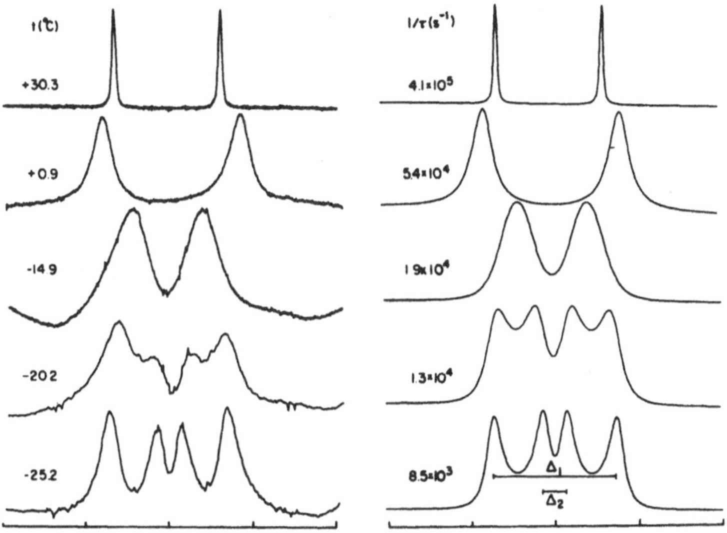

Figure 12. *Deuterium NMR spectra of p-dioxane-d_8 in liquid crystals. The experimental spectra (left) correspond to a 3.0 wt% solution in phase V. The simulated spectra (right) were calculated using the order parameters shown in Fig. (14). The frequency spacings between ticks are 12.5 KHz for the bottom three spectra in each column and 4 KHz for the upper two.*

typical dynamic effects similar to those observed in the case of cyclohexane-d_{12}. However, a conspicuous difference exists between the two sets of spectra. While in solution I (Fig. (12)) the inversion process results in the separate coalescence of the low field pair and high field pair of lines, in solution II (Fig. (13)) the coalescence involves the low field component of the low field pair with the low field component of the high field pair and similarly for the high field components of the two pairs. These results show that the axial and equatorial quadrupolar interactions have the same sign in the spectra of solution I and the opposite signs in solution II. In the case of the cyclohexane molecule which has a C_3 symmetry axis the relative sign of $\tilde{\nu}_Q^{ax}$ and $\tilde{\nu}_Q^{eq}$ is uniquely determined by the molecular geometry to be negative (Eq. 2). However, for p-dioxane whose symmetry is C_{2h}, three independent order parameters determined the values of $\tilde{\nu}_Q$, and no apriori prediction can be made on the magnitude or the sign of $\tilde{\nu}^{ax}/\tilde{\nu}^{eq}$.

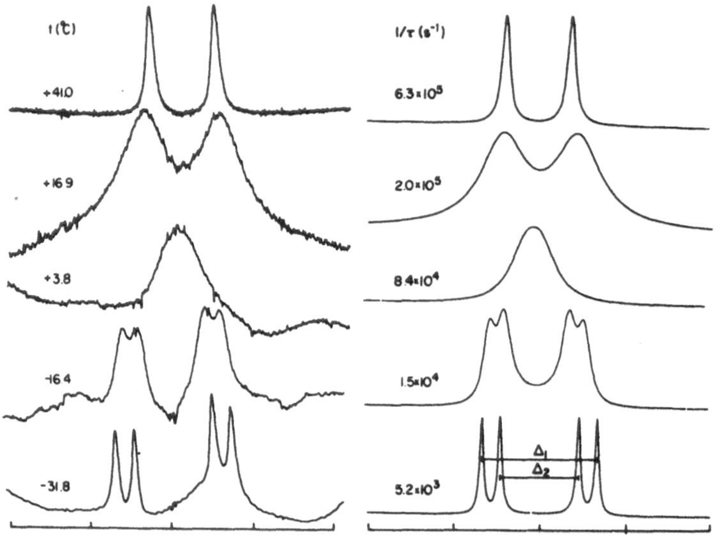

Figure 13. *As in Fig. (12) for a 3.0 wt% solution of p-dioxane-d_8 in Merck "1565 TNC" liquid crystal mixture. The frequency spacings correspond to 25 KHz for the bottom three traces and 4 KHz for the upper two.*

To derive kinetic parameters from the experimental spectra it was necessary to know both $\tilde{\nu}_Q^{ax}$ and $\tilde{\nu}_Q^{eq}$ over the whole dynamic range. They could however be determined only at the low temperature region where separate signals are observed. Their values at higher temperatures were obtained by a linear extrapolation of the low temperature splittings (cf. Figs. (14) and (15)). Using these parameters, theoretical lineshapes were computed (see right columns in Figs. (12) and (13)) and by comparison with the experimental spectra the specific rate at various temperatures was determined. An Arrhenius plot of the kinetic parameters so obtained is shown in Fig. (16) from which the following kinetic parameters were obtained: $\Delta H^{\#}$ = 9.64

kcal/mol; $\Delta S^{\#} = 0.1$ e.u. $1/\tau(300°K) = 3.16 \times 10^5$ sec^{-1}.

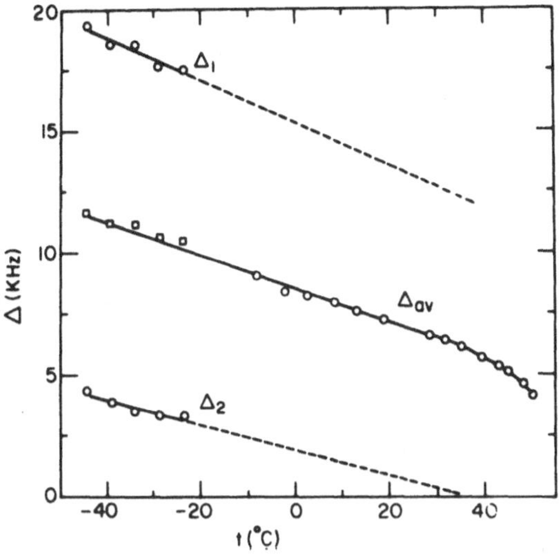

Figure 14. *Plots of the quadrupolar splittings for the same solution as described in the caption to Fig. (12).* Δ_1 *and* Δ_2 *correspond respectively to the outer and inner doublets and* Δ_{av} *to the average splitting of the two. The dashed lines are extrapolation of the low temperature results of* Δ_1 *and* Δ_2.

Bond Rearrangement in Bullvalene

In the section discussing dynamic ^{1}H NMR in liquid crystals preliminary results on solutions of bullvalene were reported. Considerably simpler spectra are obtained for deuterium NMR of deuteriated bullvalene. From the molecular geometry it may be noted that there are four different types of deuterons and indeed at low temperatures (below ~−15°C) four distinct quadrupole doublets are observed with relative intensities 1:3:3:3 (see Fig. (17)) (20). As the temperature is raised the whole spectrum broadens and eventually coalesces into a single doublet. This is indeed the expected behaviour since the Cope rearrangement reshuffles and equalizes all the different positions in the bullvalene molecule. Consequently in the fast exchange limit a single doublet with a splitting corresponding to the weighted average quadrupole interaction is expected. The traces on the right hand side of Fig. (17) are computed spectra

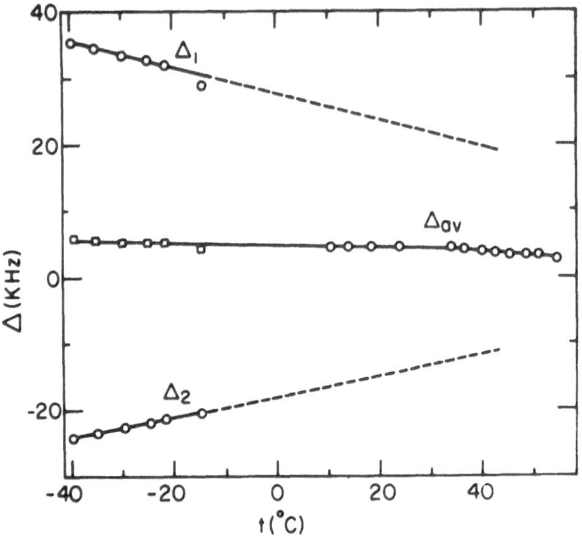

Figure 15. *As in Fig. (14) but for the solution described in the caption to Fig. (13). The computation of* Δ_{av} *was made using opposite signs for* Δ_1 *and* Δ_2.

with specific rates as indicated in the figure. It may be seen that they reproduce very closely the experimental results. The kinetic analysis of this system has been completed recently and is described in reference (20).

Ring Inversion of Bridged Perinaphthalenes

The ring inversion of bridged perinaphthalenes was studied in a number of homologues by proton NMR in isotropic solvents (21). The dynamic process in this case averages the inequivalent protons of the methylene groups. This effect is also observed in the deuterium NMR of homologues deuteriated in the aliphatic methylene groups (22). An example is shown in Fig. (18). It corresponds to a 3wt% solution of $-CD_2SCD_2$-bridged perinaphthalene in Phase V. At around 0°C and above a single doublet is observed since all the deuterons are equivalent. However on cooling the doublet components broaden and eventually (at ~ - 60°C) turn into a broad structureless signal. It was checked that this broadening is indeed due to incomplete averaging of the methylene (deuteron) interaction by observing the signal of a probe molecule (C_6D_6) which was added to the

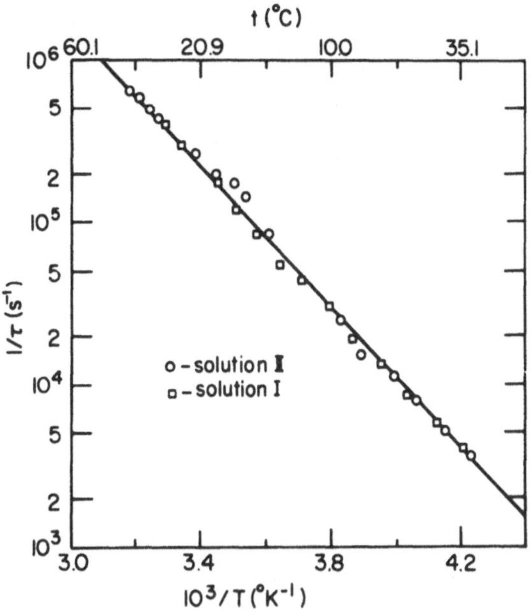

Figure 16. *Arrhenius plot of the specific rate for the ring inversion of p-dioxane determined from the ²D NMR results in the solutions described in the captions to Figs. (12) and (13).*

same solution. The linewidths of the benzene-d_6 probe were considerably narrower and broadened only very slightly in the temperature range over which the measurements were made. In this case however the activation energy of the reaction is very low (< 6.3 kcal/mole (21) so that its rate remains fast even at -60°C and consequently the spectrum of the "frozen molecule" could not be obtained. For that reason no quantitative analysis of the spectra can be made.

DYNAMIC LINESHAPES IN LIQUID CRYSTALLINE SOLUTIONS

The theory of dynamic NMR lineshape for liquid crystalline solution is equivalent to that for isotropic liquids (2,12). However, the presence of the direct dipolar interactions, which are in general larger than the chemical shifts, makes it necessary to employ the more general density matrix formalism rather than the simpler Bloch-McConnell type theory. Also, since the interactions involve all magnetic nuclei in the molecule, the dimension of the problem (i.e. the number of coupled equations that need be solved) increases very steeply with the number of magnetic nuclei in the molecule. The density matrix formalism was first developed by Kaplan (10) and Alexander (11) and later generalized and put into a form which is more suitable for computer programming by Binsch (23). It has extensively been used for isotropic liquids but only very recently has it been applied to anisotropic liquids (3,6,8).

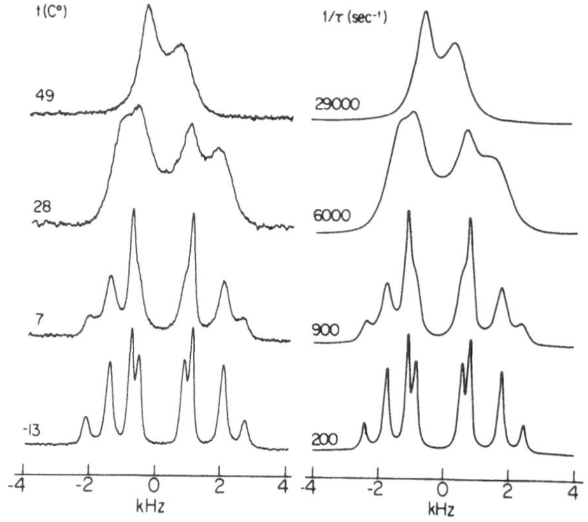

Figure 17. ^{2}D *NMR spectra of deuteriated bullvalene. The spectra on the left are experimental and correspond to a 3 wt% solution in Phase V. The spectra on the right are computed with the specific rates indicated in the figure.*

In the present section we discuss topics related to the lineshape simulation in liquid crystals. We first consider molecules containing only nuclei with spin $I = \frac{1}{2}$. The hamiltonian then contains dipolar interaction terms in addition to the chemical shift and scalar coupling. In this connection we also comment on the application of group theory to simplify the simulation of dynamic lineshapes. Finally, we consider molecules containing nuclei with spin $I > \frac{1}{2}$. The hamiltonian then also contains quadrupolar interaction terms. Specifically we shall consider dynamic NMR spectra of deuterons in deuteriated molecules.

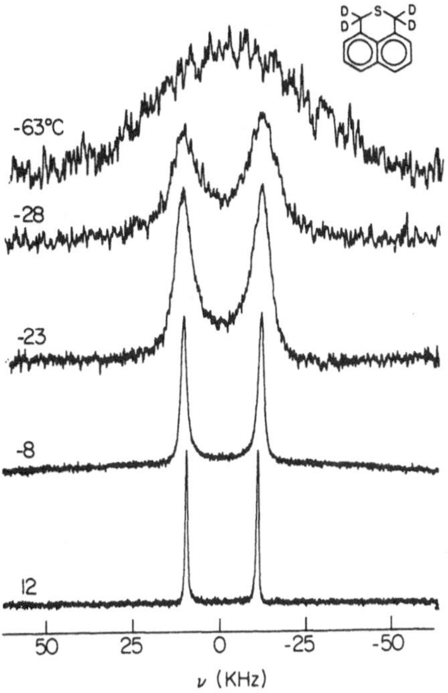

Figure 18. ^2D *NMR spectra of methylene deuteriated (3 wt% in Phase V) at different temperatures.*

Dynamic Lineshape for $I = \frac{1}{2}$ Systems (3)

The spin hamiltonian for this system is:

$$H = H^0 + H^S + H^{rf} \tag{4}$$

where, using standard symbols

$$H^0 = -\sum_i \omega_0^i I_z^i \tag{5}$$

$$H^S = \sum_i \delta_i I_z^i + \sum_{i<j} J_{ij} \left[I_z^i I_z^j + \tfrac{1}{2}(I_+^i I_-^j + I_-^i I_+^j)\right] + \tag{6}$$

$$\sum_{i<j} 2D_{ij} \left[I_z^i I_z^j - \tfrac{1}{4}(I_+^i I_-^j + I_-^i I_+^j)\right]$$

$$H^{rf} = \omega_1 \sum_i (I_x^i \cos \omega t + I_y^i \sin \omega t). \tag{7}$$

The equation of motion for the elements of the density matrix ρ in the presence of an intramolecular exchange process characterized by the exchange operator P and a specific rate $1/\tau$ is

$$\frac{d\rho}{dt} = -i[H,\rho] + \frac{1}{\tau}[P^{-1}\rho P-\rho] - \frac{\rho_{od}}{T_z} - \frac{\rho_{diag}}{T_1} \tag{8}$$

where od and diag refer to off-diagonal and diagonal elements respectively. The NMR spectrum $I(\omega)$ is proportional to

$$\mathrm{Im}[\mathrm{Tr}(\rho^{\omega} \sum_i I_-^i)] = \mathrm{Im}[\sum_{\ell k} \rho_{k\ell}^{\omega} (\sum_i I_-^i)_{\ell k}] \tag{9}$$

where the superscript ω refers to the rotating coordinate system. Under steady state and no saturation condition Eq. (8) gives for $\rho_{k\ell} \equiv \rho_{\ell k}$:

$$[i(\omega_0-\omega)- \frac{1}{T_2}]\rho_{k\ell} + i(\sum_r \rho_{kr} H_{r\ell}^s - \sum_t H_{kt}^s \rho_{t\ell}) \tag{10}$$

$$+ \frac{1}{\tau}(\sum_{nm} P_{kn}^{-1} \rho_{nm} P_{m\ell} - \rho_{k\ell}) = iC(\sum_i I_+^i)_{k\ell}$$

where C is a constant proportional to the rf field intensity. We note that because of Eq. (9) only $\rho_{k\ell}$ are required for which $M_k - M_\ell = 1$, and since Eqs. (10) couple only $\rho_{k\ell}$ belonging to the same M_k and M_ℓ manifolds the various subspectra corresponding to the different $M_\ell \rightarrow M_{\ell-1}$ transitions can be calculated separately.

It is convenient to express the set of Eqs. (10) in a matrix form

$$[i(\omega_0-\omega)\bar{\bar{1}} + (i\bar{\bar{L}} + \bar{\bar{P}})]\bar{\rho} = iC\bar{I}_+ \tag{11}$$

where $\bar{\bar{1}}$ is the unit matrix, $\bar{\bar{L}}$ and $\bar{\bar{P}}$ are matrices with elements

$$L_{k\ell,nm} = H^{s}_{m\ell}\,\delta_{kn} - H^{s}_{kn}\,\delta_{\ell m} \tag{12}$$

$$P_{k\ell,nm} = (1/\tau)\,(P^{-1}_{kn}\,P_{m\ell} - \delta_{kn}\,\delta_{m\ell}) - (1/T_2)\,\delta_{kn}\delta_{m\ell}$$

and $\bar{\rho}$ and $\bar{I}_{\pm}$ are vectors with elements $\rho_{k\ell}$ and $(\sum_i I^{i}_{\pm})$ respectively. From Eqs. (9) and (11) we finally obtain a compact form for $I(\omega)$ suitable for computer programming.

$$I(\omega) \propto \mathrm{Re}\,\{(\bar{I}_{-}\bar{\bar{S}})\,[I(\omega_0-\omega)\,\bar{\bar{1}} + \bar{\bar{\lambda}}]^{-1}(\bar{\bar{S}}^{-1}\bar{I}_{+})\} \tag{13}$$

where $\bar{\bar{S}}$ is the matrix that diagonalizes $(i\bar{\bar{L}}+\bar{\bar{P}})$, and $\bar{\bar{\lambda}}$ is a diagonal matrix of the resulting eigenvalues.

Application of Group Theory (8)

The procedure described in the previous section was used to compute e.g. the dynamic proton NMR spectrum of s-trioxane, a molecule with six protons. In this case the spectrum consists of six subspectra corresponding to the $M \rightarrow M'$ transitions, $\pm 3 \rightarrow \pm 2$; $\pm 2 \rightarrow \pm 1$; $\pm 1 \rightarrow 0$. The dimensions of the various subspectra are respectively 6, 90, and 300. When a larger number of protons are involved the dimension of some of the subspectra may become prohibitively large.

In certain cases the block of equations corresponding to particular subspectra can be further factorized using symmetry properties of the molecules and of the dynamic process under consideration. This factorization results from the fact that not all symmetry species intermix by the dynamic process. To determine which symmetry species do intermix the correlation diagram between the symmetry group, H, of the hamiltonian of the static molecule and the group G of the average hamiltonian (corresponding to the fast exchange spectrum) are considered. These diagrams show the relation between the symmetry species in the two groups and in particular they indicate which representations of H are intermixed by dynamic process.

To achieve the symmetry factorization of Eqs. (10) they are written in the basis of symmetry adopted functions ψ^{γ}_{k} of the group H. This procedure will split-up the $M \rightarrow M-1$ blocks into group sub-blocks consisting only of basis functions belonging to representations which intermix by the dynamic process. Eqs. (10) then become:

$$[i(\omega_0-\omega) - \frac{1}{T_2}]\rho^{\gamma\gamma}_{k\ell} + i(\sum_r \rho^{\gamma\gamma}_{kr}H^{\gamma\gamma}_{r\ell} - \sum_t H^{\gamma\gamma}_{kt}\,\rho^{\gamma\gamma}_{t\ell}) \tag{14}$$

$$+ \frac{1}{\tau}(\sum_\delta{}' \sum P^{-1\gamma\delta}_{kn}\,\rho^{\delta\delta}_{nm}\,P^{\delta\gamma}_{m\ell} - \rho^{\gamma\gamma}_{k\ell}) = iC(\sum_i I^i_+)^{\gamma\gamma}_{k\ell}$$

where matrix elements of operators are indicated by

$$X^{\gamma\delta}_{k\ell} = \langle\psi^\gamma_k \mid X \mid \psi^\delta_\ell\rangle \tag{15}$$

and the primed summation is only over those representations δ which intermix with γ. In multidimensional representation a further factorization is possible which allows to consider only a single set of partners for each representation.

To appreciate the power of the method consider the case of cyclooctatetraene (COT) discussed earlier. When symmetry considerations are not applied the $0\rightarrow \pm 1$ blocks are of dimension 3920x3920, while with symmetry, each of these blocks factorizes into group sub-blocks of dimensions 91, 49, 126 and 224. In fact, the simulation of the dynamic spectra of COT (Fig. (6)) was made possible only by applying the symmetry consideration just discussed.

Dynamic Lineshapes for I=1 nuclei (6)

When the nuclear spin is larger than ½ the quadrupole interaction has to be considered. The hamiltonian H^s (Eq. 6) then becomes:

$$H^s = \sum_i \delta_i I^i_z + \sum_i \tilde{\nu}^i_Q\,[(I^i_z)^2 - \tfrac{1}{3}\,I\,(I+1)]$$

$$+ \sum_{i<j} J_{ij}[I^i_z I^j_z + \tfrac{1}{2}\,(I^i_+ I^j_- + I^i_- K^j_+)] \tag{16}$$

$$+ \sum_{i<j} 2D_{ij}\,[I^i_z I^j_z - \tfrac{1}{4}\,(I^i_+ I^j_- + I^i_- I^j_+)]$$

For deuterons the quadrupolar interaction usually dominates the spectrum while dipolar interactions are rarely observed except for spacially very close pairs of nuclei. To demonstrate the evolution of the dynamic lineshape of quadrupole nuclei in the presence of dipolar interaction we reproduce in Fig. (19) calculated spectra for two interchanging I=1 nuclei which have a dipolar coupling. The spectra were calculated using Eq. (13) for two different sets of parameters as given in the caption to the figure.

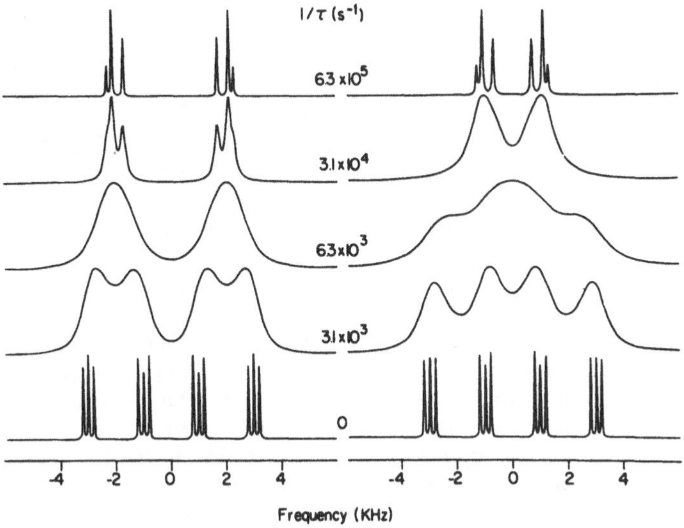

Figure 19. *Simulated dynamic deuterium NMR spectra for two inequivalent deuterons undergoing exchange. The parameters used for the computation (in Hz) are J=0, δ=0, D=-100,* $1/T_2=15$, $\tilde{\nu}_Q^2=3000$, $|\tilde{\nu}_Q^1|=1000$. *For the spectra on the left the latter parameter was taken to be positive, while for the right column it was assumed negative.*

SUMMARY AND CONCLUSIONS

It has been demonstrated that dynamic NMR in liquid crystals is feasible and in certain cases may be superior to using isotropic solvents. The main advantages of the method are: (i) The large dipolar and quadrupolar splittings means that considerably larger dynamic ranges may be achieved in liquid crystalline solvents. This point was clearly demonstrated using both ^1H and ^2D NMR. (ii) In certain cases a dynamic process is not manifested in the spectrum because it does not modulate any interaction or because it modulates spin spin interaction between equivalent nuclei. In liquid crystalline solution there is always dipolar splitting and any process that modulates the distances between the magnetic nuclei will affect the lineshape when the specific rate of the reaction matches the corresponding splitting. (iii) In certain cases the evolution of the dynamic spectrum does not reveal unequivocally the details of the reaction mechanism. This is particularly true for first order spectra. However, for "strong interaction" spectra which is essentially always the case in liquid crystalline solutions the lineshapes are more sensitive to the mechanistic route of the reaction.

Despite these advantages the use of liquid crystalline solvents has certain limitations. (i) Liquid crystals are limited to specific types of compounds which are not always suitable as solvents. (ii) The temperature range is limited to the stability range of the liquid crystalline mesophase. As yet no liquid crystals are known which are suitable for work below -40°C. (iii) The complexity of the NMR spectrum in liquid crystals increases very steeply with the number of magnetic nuclei in the molecule. In fact for molecules with more than eight protons an interpretation of the spectrum becomes essentially impractical because of the large number of peaks and excessive overlap between them. Even the simulation of such spectra becomes extremely difficult and time consuming. The latter difficulty may partly be resolved using group theory or by using approximate methods to compute the lineshape. It should however be emphasized that in the case of quadrupole nuclei the quadrupolar splittings often dominate the spectrum while dipolar interactions may be within the linewidth, and an interpretation of the spectrum is then quite straightforward.

ACKNOWLEDGEMENT

This chapter is based on an article by the author in the Journal of the Israel Chemical Society **23** (3), 305-315 (1983), and is reproduced here with minor alteration and additions with the permission of the Editorial Board.

REFERENCES

1. H.S. Gutowsky, D.W. McCall and C.P. Slichter, *J. Chem. Phys.* **21**, 279 (1953).

2. For a general review see "*Dynamic Nuclear Magnetic Resonance Spectroscopy*" edited by L.M. Jackman and F.A. Cotton (Academic, New York, 1979).

3. Z. Luz, R. Naor and E. Meirovitch, *J. Chem. Phys.* **74**, 6627 (1981).

4. R. Naor and Z. Luz, *J. Chem. Phys.* **76**, 5662 (1982).

5. R. Naor, *Thesis*, The Feinberg Graduate School of the Weizmann Institute of Science, Rehovot 1981.

6. R. Poupko and Z. Luz, *J. Chem. Phys.* **75**, 1675 (1981).

7. M.E. Mosely, R. Poupko and Z. Luz, *J. Magn. Reson.* **48**, 354 (1982).

8. Z. Luz and R. Naor, *Mol. Phys.* **46**, 891 (1982).

9. M. Cocivera, *J. Chem. Phys.* **47**, 3061 (1967).

10. J. Kaplan, *J. Chem. Phys.* **28**, 278 (1958); **29**, 462 (1958).

11. S. Alexander, *J. Chem. Phys.* **37**, 967 (1962); **37**, 974 (1962).

12. J.I. Kaplan and G. Fraenkel, "*NMR of Chemically Exchanging Systems*" Academic Press, New York, 1980.

13. Z. Luz and S. Meiboom, *J. Chem. Phys.* **59**, 1077 (1973).

14. C.S. Yannoni, *J. Am. Chem. Soc.* **92**, 5237 (1970).

15. M. Saunders, *Tetrahedron Lett.* 1699 (1963); J.F.M. Oth., K. Mullen, J.M. Gilles and G. Schroder, *Helv. Chim. Acta* **57**, 1415 (1974); H. Nakanishi and O. Yamamoto, *Tetrahedron Lett.* 1803 (1974); H. Gunther and J. Ulmin, *Tetrahedron Lett.* **30**, 3781 (1974); Y. Huang, S. Macura and R.R. Ernst, *J. Am. Chem. Soc.* **103**, 5327 (1981).

16. J.W. Emsley and J.C. Lindon, "*NMR Spectroscopy Using Liquid Crystal Solvents*" (Pergamon, New York, 1975), Chap. 6.

17. For a review see F.A.L. Anet and R. Anet, Ref. 2, Chap. 14.

18. D. Hofner, S.A. Lesko and G. Binsch, *Org. Mag. Reson.* **11**, 179 (1978).

19. F.R. Jensen and R.A. Neese, *J. Am. Chem. Soc.* **93**, 6329 (1971); F.A.L. Anet and J. Sandstorm, *Chem. Commun.* 1558 (1971).

20. R.Poupko, Z.Luz and H.Zimmermann, *J.Am.Chem.Soc.* (in press).

21. J.E. Anderson, F.S. Jørgensen and T. Thomsen, *J. Chem. Soc. Chem. Commun.*, 333 (1982).

22. I. Belsky, R. Poupko and Z. Luz, private communication.

23. G. Binsch, *J. Am. Chem. Soc.*, **91**, 1304 (1969).

DISCOTIC LIQUID CRYSTALS AND THEIR CHARACTERIZATION BY DEUTERIUM NMR

Z. Luz, D. Goldfarb

Isotope Department,
The Weizmann Institute of Science,
76100 Rehovot, Israel

and

H. Zimmermann

Max-Planck-Institut für Medizinische Forschung,
D-6900 Heidelberg, Germany.

INTRODUCTION

In the present chapter we summarize experimental NMR studies performed on various discotic compounds. The approach used, as in many other studies of liquid crystalline systems, has been deuterium NMR of specifically deuteriated mesogens, as well as of deuteriated probe molecules dissolved in the various mesophases. We first review some chemical aspects related to discotic compounds, followed by the classification of the discotic mesophases as known to date. The basic equations used in the interpretation of the spectra will then be given with particular emphasis on the effect of the mesophase symmetry. Finally the experimental results are discussed under the following subheadings: (a) uniaxial phases, (b) biaxial phases, (c) NMR of probe molecules, and (d) diffusion measurements. The presentation follows very closely and sometimes overlaps the discussion in the review article recently published in the Israel Journal of Chemistry (1).

CHEMICAL STRUCTURE AND CLASSIFICATION OF DISCOTIC MESOPHASES

Discotic liquid crystals are formed by compounds consisting of

J. W. Emsley (ed.), Nuclear Magnetic Resonance of Liquid Crystals, 343–377.

planar disc-like molecules (hence the name discotic mesophases). The existence of discotic mesophases was anticipated from theoretical considerations and from analogy with similar phases observed in pyrolitically treated carbonatous materials (2). However it was not until 1977 (3) and early 1978 (4) when the first neat discotic compounds were prepared and their mesomorphic character demonstrated (3-6). Several review articles have recently appeared describing the structure and properties of these mesophases (7-9).

Table 1. *Homologous series and abbreviations used.*

Rigid core	Side chain	Abbreviation
benzene	$-O-C(=O)-C_nH_{2n+1}$	BHAn
	$-SO_2C_nH_{2n+1}$	-
triphenylene	$-O-C_nH_{2n+1}$	THEn
	$-O-C(=O)-C_nH_{2n+1}$	THAn
	$-O-C(=O)-C_6H_4-C_nH_{2n+1}$	THnB
	$-O-C(=O)-C_6H_4-O-C_nH_{2n+1}$	THOnB
truxene	$-O-C(=O)-C_nH_{2n+1}$	TxHAn
	$-O-C(=O)-C_6H_4-O-C_nH_{2n+1}$	TxHOnB
cyclohexane	$-O-C(=O)-C_nH_{2n+1}$	-

The molecules of discotic compounds usually consist of a central aromatic core to which a number (usually six) of aliphatic side chains are covalently linked via certain groups. The overall molecular symmetry is usually trigonal (or higher) but molecules with lower symmetries which exhibit discotic mesophases are also known. Typical cores of discotic compounds are benzene, triphenylene, and truxene (see Fig. (1)), while the side chains are alkyl radical bonded to the centre core via ether, carboxyl, or benzoate bridges (see Table 1 for structural formulae of side chains and abbreviated names of the liquid crystal homologues). Very recently two new types of discotic

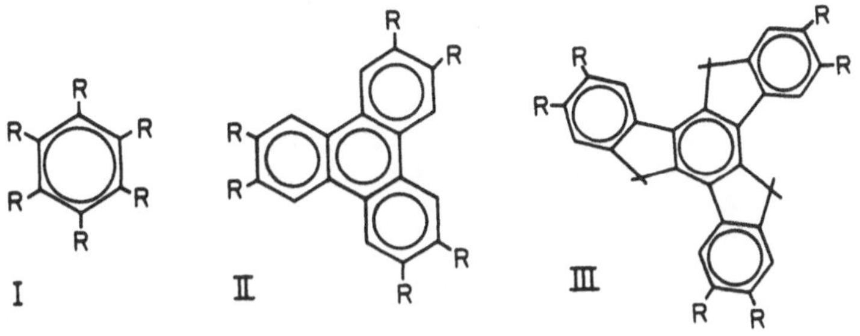

Figure 1. *Molecular structural formulas of the benzene triphenylene and truxene cores.*

compounds were prepared (10). One type consists of the hexaalkylsulfonobenzene series which exhibit discotic mesophases when the alkyl chains contain 7 to 15 carbon atoms per chain. The second class consists of hexasubstituted cyclohexane, thus representing the first example of a discotic mesogen with a saturated core.

As yet at least eight discotic mesophases have been identified (11-13). Two more phases are known to exist but their structures have not yet been determined (14-15). In Fig. (2) are given schematic diagrams of some of the known discotic phases, as determined by X-ray diffractometry and optical microscopy (12,13,16-20). Most discotic mesophases have columnar structures: the mesogen molecules are stacked into columns which in turn are arranged in characteristic two-dimensional arrays. These phases (usually labelled by capital D) are classified according to the symmetry of the 2-dimensional unit cell of the mesophase, and according to the way the molecules are stacked within the columns. Thus the indices h, r and ob refer respectively to hexagonal, rectangular and oblique arrangements

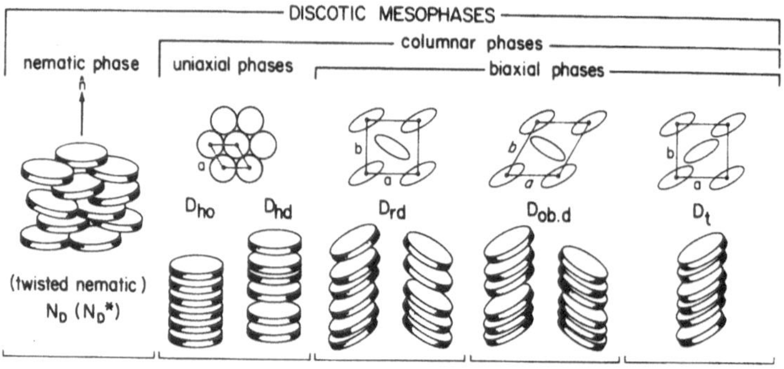

Figure 2. *Schematic diagrams, and structures of the two dimensional unit cells, of some discotic mesophases.*

of columns, while the letters o and d refer to phases in which the molecules are respectively ordered or disordered within the columns. Phases in which the molecular symmetry axis is on the average tilted with respect to the columnar axis are referred to as tilted discotics, D_t (10). In addition to the columnar phases a discotic nematic, N_D, phase is known as well as a discotic cholesteric phase N_D^*. In the latter case the side chains contain asymmetric carbon atoms. Clearly the N_D and D_h phases are uniaxial while D_r and D_t must be biaxial. The two phases labelled D_o and D_x (which occur in certain triphenylene and truxene esters) have not yet been identified (14,15) and it is not clear whether they represent new phases or are a manifestation of existing ones. Finally it should also be noted that another classification for the discotic mesophases has been suggested which is based on the notation used for the conventional smectics (7,8).

Amongst the discotic liquid crystals, some exhibit only a single type of mesophase but many are polymorphic and exhibit a sequence of different phases. Inverted and reentrant mesophase were also observed (13,22,24). Examples of phase diagrams for several homologous series are shown in Fig. 3. In addition to the X-ray,(13,16-19) optical microscopy (6,7,12,14-16,21-27), and miscibility (28-29) studies referred to above, thermodynamic, IR, (30-31) magnetic susceptibility (32) and disclination defects (33-36) were also investigated. Very little theoretical work specific to discotic liquid crystals has appeared so far. Feldkamp et al. (37) and Chandrasekhar (38) used McMillan's approach to study the phase transformation

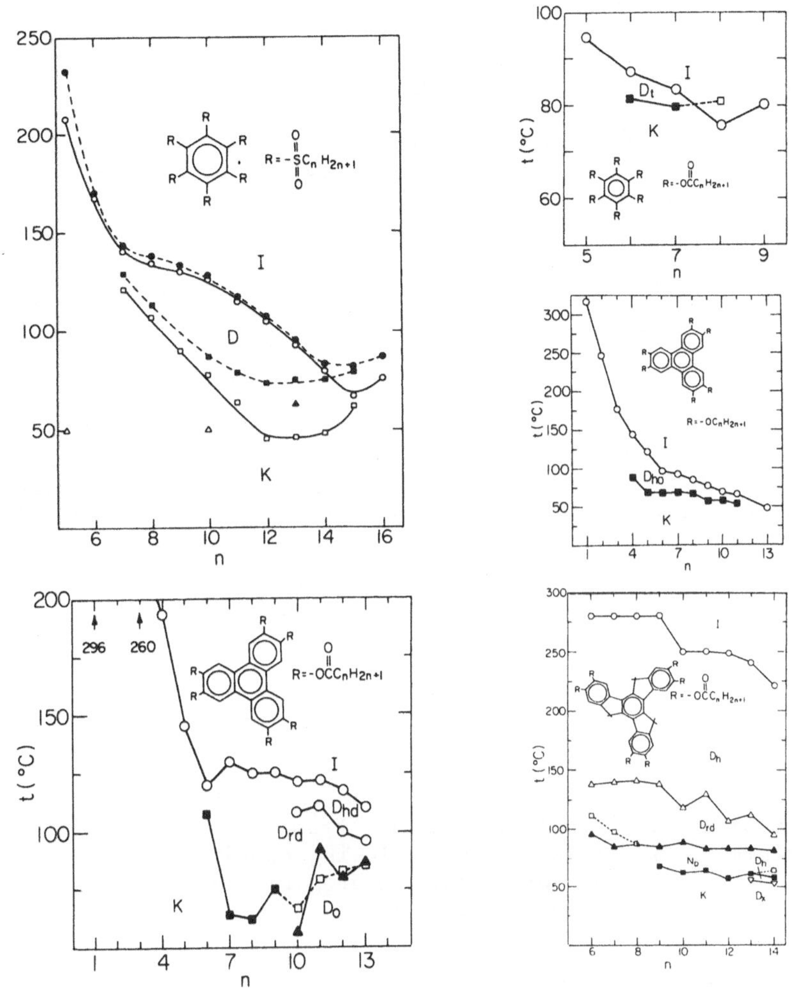

Figure 3. *Phase diagrams for the homologous series (a) benzene hexaalkanoates, (b) benzenehexaalkylsulfones, (c) hexaalkyloxytriphenylenes, (d) triphenylenehexaalkan-oates, (e) truxenehexaalkanoates.*

between the discotic columnar and nematic phases and the normal liquid. The possibility of biaxial nematic in mixtures of discotic and rod-like mesogens was considered by Gelbart and coworkers (39) but experimental evidence for such a phase does not exist as yet.

DEUTERIUM NMR IN DISCOTIC MESOPHASES

As indicated above most of the NMR work performed so far on discotic liquid crystals employed deuterium (40-45), although a few papers on ^{13}C and ^{1}H were also published (46-47). All these studies indicated that molecular motion (both rotational and translational) of the mesogen molecules is fast on the NMR timescale. This statement applies a fortiori to small probe molecules dissolved in the mesophase. Consequently a particular molecular species in a liquid crystalline domain will yield a characteristic spectrum reflecting the average interaction over the distribution in this domain. Therefore the general theories derived for high resolution NMR in the conventional liquid crystals (48-53) apply also to discotic mesophases. For deuterium NMR studies (I=1) in liquid crystals it is usually sufficient to consider, besides the Zeeman energy only the nuclear quadrupolar term

$$\mathcal{H}^{sec} = -\nu_0 I_Z + F^{(2,0)} \sqrt{\tfrac{3}{2}} \; (I_Z^2 - \tfrac{2}{3}) \qquad (1)$$

where the superscript, sec., indicates that only the secular part of the hamiltonian is considered, ν_0 is the Larmor frequency, and $F^{(2,0)}$ is the average quadrupolar tensor component in the direction of the magnetic field, Z. The quantity $F^{(2,0)}$ can be obtained from the deuterium quadrupolar interaction in the molecular frame by a series of transformations involving the following coordinate systems (see Fig. (4)): (i) The principal quadrupolar axes a,b,c at the deuterium nucleus. In this system

$$F_P^{(2,0)} = (\tfrac{3}{2})^{\frac{1}{2}} e^2qQ/2h; \quad F_P^{(2,\pm1)} = 0;$$

$$F_P^{(2,\pm2)} = F_P^{(2,0)} \, \eta_P/\sqrt{6} \qquad (2)$$

where η_P is the asymmetry parameter; (ii) a molecular fixed frame x', y', z' chosen according to the molecular symmetry; (iii) the axes x,y,z fixed in the liquid crystal mesophase according to the symmetry of the mesophase; (iv) the laboratory coordinate system with Z along the magnetic field direction. The result for $F^{(2,0)}$ is,

$$F^{(2,0)} = \tfrac{1}{2}\,(3\cos^2\theta_0-1)\;F_{LC}^{(2,0)}+(-\sqrt{\tfrac{3}{8}}\sin\,2\theta_0\exp(-i\phi_0)F_{LC}^{(2,1)}+c.c.)$$

$$+\;(\sqrt{\tfrac{3}{8}}\sin^2\theta_0\exp(-i2\phi_0)F_{LC}^{(2,2)}+c.c.) \tag{3}$$

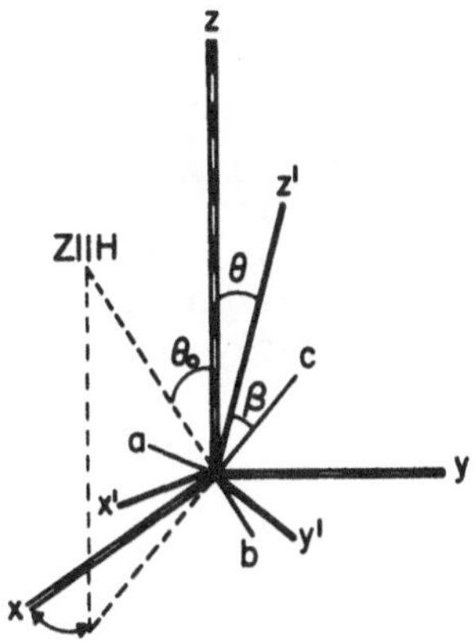

Figure 4. *Interrelation between coordinate systems: The principal quadrupolar tensor frame (a,b,c); The molecular fixed frame (x',y',z'); The liquid crystalline frame (x,y,z). Z is the direction of the magnetic field. The polar angles between the various coordinate systems are also indicated.*

where θ_0 and ϕ_0 are the polar angles of the magnetic field in the liquid crystal (x,y,z) coordinate system, and

$$F_{LC}^{(2,\ell)} = F_P^{(2,0)} \Sigma f_m \bar{D}^2_{m\ell} \quad (4)$$

$$f_m = D^2_{0m}(\alpha\beta\gamma) + (\eta_P/\sqrt{6})[D^2_{2m}(\alpha\beta\gamma) + D^2_{-2m}(\alpha\beta\gamma)] \quad (5)$$

In these equations the $D^2_{km}(\alpha\beta\gamma)$ are Wigner matrices with the corresponding Euler angles α, β, γ, that transform the principal coordinate system a, b, c to the molecular x', y', z' axes, and the order parameters $\bar{D}^2_{m\ell}$ are defined by (see chapter 1)

$$\bar{D}^2_{m\ell} = \int P\,(\phi\theta\psi)D_{m\ell}(\phi\theta\psi)\sin\theta d\theta d\phi d\psi \quad (6)$$

where $P(\phi\theta\psi)$ is the singlet orientational probability distribution of the molecules in the liquid crystalline phase. The hamiltonian of Eq. (1) shows that the NMR spectrum of a single deuteron in a single liquid crystalline domain consists of a doublet centered around ν_0 and separated by:

$$\tilde{\nu}_Q(\theta_0,\phi_0) = \sqrt{6}\, F^{(2,0)} \quad (7)$$

where $F^{(2,0)}$ is given by Eq. 3.

Symmetry considerations involving both the nature of the molecules and of the liquid crystalline phases can considerably simplify the computation of $F^{(2,0)}$ by restricting the number of independent values of $\bar{D}^2_{m\ell}$. The molecular symmetry enters, in our notation, through the index m. Thus for molecules with a C_n (n>3) symmetry axis, taking this axis parallel to z', only $\bar{D}^2_{m\ell}$ with m=0 will differ from zero. For "orthorhombic" molecules, i.e. having C_{2V} or D_2 symmetry and taking x', y', z', to coincide with the molecular symmetry axes, only $\bar{D}^2_{0\ell}$ and $\bar{D}_{2\ell}=\bar{D}_{-2\ell}$ will differ from zero. Similarly the symmetry of the mesophase will impose restriction on the index ℓ of the $\bar{D}^2_{m\ell}$. For example for uniaxial or hexagonal phases taking the unique axis as z only $\bar{D}^2_{m0}$ (and thus only $F_{LC}^{(2,0)}$) can appear in the equations. For orthorhombic phases such as rectangular discotics, taking x, y, z to coincide with the principal phase axes, both $\bar{D}^2_{m0}$ and $\bar{D}^2_{m2}=\bar{D}^2_{m-2}$) will appear, while for monoclinic phases (e.g. D_t) taking z to coincide with the unique axis of the phase, three sets of parameters will appear i.e. $\bar{D}^2_{m0}$, $\bar{D}^2_{m2}$, $\bar{D}^2_{m-2}$. In Table 2 are summarized the non vanishing $\bar{D}^2_{m\ell}$ for a number of symmetries.

Table 2. *Non vanishing order parameters $\bar{D}^2_{m\ell}$ for molecules and mesophases of various symmetries. In the present notation the indices m and ℓ refer respectively to the molecular and mesophase axes.*

phase symmetry / molecular symmetry	Cubic (S_D, BP, Iα)	Uniaxial, Hexagonal (N, S_A, N_D, D_{ho})	Orthorhombic (S_E, D_{rd})	Monoclinic (S_c, D_t)
Cubic	–	–	–	–
C_n(n>3)	–	$\bar{D}^2_{00}$	$\bar{D}^2_{00}$, $\bar{D}^2_{00}=\bar{D}^2_{0-2}$	$\bar{D}^2_{00}$, $\bar{D}^2_{02}$, $\bar{D}^2_{0-2}$,
C_{2v}, D_2, D_{2h}	–	$\bar{D}^2_{00}$, $\bar{D}^2_{00}$, $\bar{D}^2_{20}=\bar{D}^2_{-22}$	$\bar{D}^2_{02}=\bar{D}^2_{0,-2}$ $\bar{D}^2_{20}=\bar{D}^2_{-20}$ $\bar{D}^2_{22}=\bar{D}^2_{-2-2}$	$\bar{D}^2_{00}$, $\bar{D}^2_{02}$, $\bar{D}^2_{0-2}$, $\bar{D}^2_{20}=\bar{D}^2_{20}$, $\bar{D}^2_{-2-2}=\bar{D}^2_{2-2}$ $\bar{D}^2_{22}=\bar{D}^2_{-22}$
C_s, C_2, C_{2h}	–	$\bar{D}^2_{00}$, $\bar{D}^2_{20}$, $\bar{D}^2_{-20}$,	$\bar{D}^2_{00}$, $\bar{D}_{02}=\bar{D}^2_{0-2}$ $\bar{D}^2_{20}$, $\bar{D}^2_{22}=\bar{D}^2_{2-2}$ $\bar{D}^2_{-20}$, $\bar{D}^2_{-22}=\bar{D}^2_{-2-2}$	$\bar{D}^2_{00}$, $\bar{D}^2_{02}$, $\bar{D}^2_{0-2}$, $\bar{D}^2_{20}$, $\bar{D}^2_{22}$, $\bar{D}^2_{2-2}$, $\bar{D}^2_{-20}$, $\bar{D}^2_{-22}$, $\bar{D}^2_{-2-2}$,

In the principal coordinate axes of the average quadrupolar tensor Eq. (7) becomes

$$\tilde{\nu}_Q(\theta_0,\phi_0) = 2\tilde{\nu}_Q^{LC}[(\tfrac{1}{2})(3\cos^2\theta_0-1) + (\tfrac{1}{2})\tilde{\eta}^{LC}\sin^2\theta_0\cos^2\phi_0] \qquad (8)$$

where

$$\tilde{\nu}_Q^{LC} = (\tfrac{3}{4})\ (e^2qQ/h)\sum_m f_m \bar{D}^2_{m0}$$

$$\tilde{\eta}^{LC} = \sqrt{6}\ \sum f_m \bar{D}^2_{m2}/\sum f_m \bar{D}^2_{m0}$$

Clearly in uniaxial and hexagonal phases $\tilde{\eta}^{LC}$ must vanish.

If the sample is not a single liquid crystalline domain, but rather consists of a distribution of domains, the deuterium NMR spectrum will correspond to a superposition of doublets weighted by the distribution $P(\theta_0\phi_0)$. Assuming a Lorentzian form for each of the doublet components with a characteristic linewidth of $1/T_2$, the overall lineshape $I(\nu)$ will be given by:

$$I(\nu) = \int L[\nu,\tilde{\nu}_Q(\theta_0\phi_0),\ 1/T_2]P(\theta_0\phi_0)d\theta_0 d\phi_0 \qquad (9)$$

where

$$L[\nu,\tilde{\nu}_Q(\theta_0\phi_0)1/T_2] = \sum_{g=\pm 1} \frac{1/T_2}{[\nu-(\tfrac{1}{2})g\tilde{\nu}_Q(\theta_0\phi_0)]^2+(1/T_2)^2}\ . \qquad (10)$$

In our study of discotic liquid crystals three major types of distributions were encountered: (i) An isotropic distribution of domains i.e. $P(\theta_0\phi_0)=\sin\theta_0 d\theta_0 d\phi_0$. We refer to such a sample as a "powder sample":

$$I\ (\nu)^{powder} = \int L[\nu,\tilde{\nu}_Q(\theta_0,\phi_0)1/T_2]\sin\theta_0 d\theta_0 d\phi_0\ . \qquad (11)$$

(ii) Another commonly encountered distribution consists of a distribution in which the major principal axes of the liquid crystalline domains, z, lie in a plane perpendicular to the original field direction, Z, i.e. it is formed with $\theta_0=\pi/2$ for all domains, but ϕ_0 is equally distributed in the azimuthal plane:

$$I(\nu)^{planar} = \int L(\nu,\tilde{\nu}_Q(\pi/2,\phi_0),\ 1/T_2]d\phi_0\ . \qquad (12)$$

(iii) A modification of case (ii) occurs when the liquid crystalline z axis is distributed about $\theta_0=\pi/2$. Assuming this distribution to be Gaussian with a characteristic width, σ, one obtains for $I(\nu)$:

$$I(\nu) = \int L[\nu,\tilde{\nu}_Q(\theta_0\phi_0),1/T_2]\exp[-(\theta_0-\pi/2)^2/2\sigma^2]\sin\theta_0 d\theta_0 d\phi_0 \ . \quad (13)$$

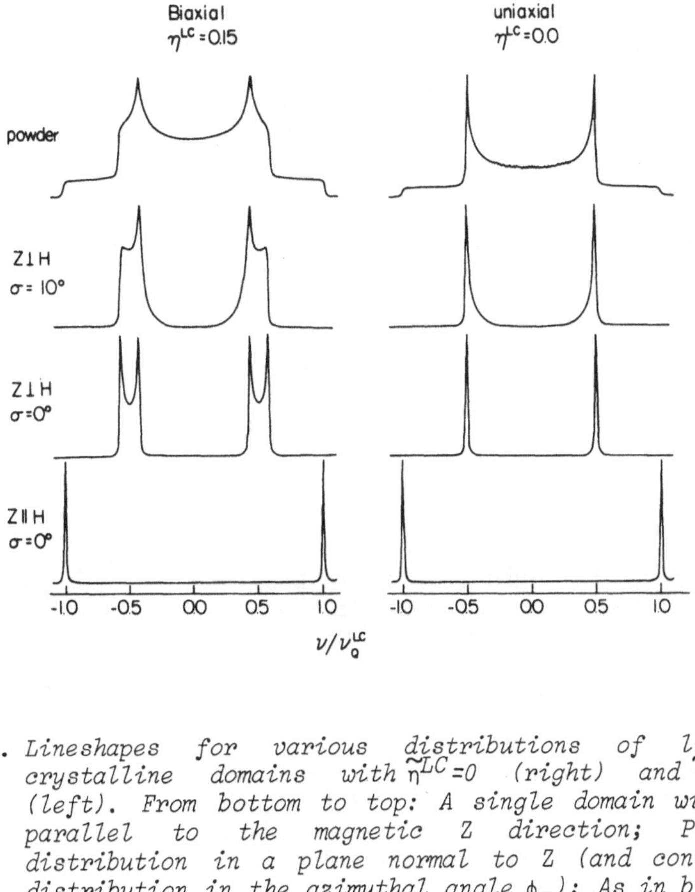

Figure 5. *Lineshapes for various distributions of liquid crystalline domains with* $\tilde{\eta}^{LC}=0$ *(right) and* $\tilde{\eta}^{LC}\neq 0$ *(left). From bottom to top: A single domain with z parallel to the magnetic Z direction; Planar distribution in a plane normal to Z (and constant distribution in the azimuthal angle* ϕ_0*); As in b with a Gaussian distribution of the domain z axes about* $\theta_0=\pi/2$ *with* $\sigma=10°$*; An isotropic powder of domains.*

Examples of deuterium NMR spectra calculated with the above

distribution for two types of liquid crystalline phases namely an axially symmetric phase and a biaxial phase are given in Fig. (5). Information about the average magnetic parameters, $\tilde{\nu}_Q^{LC}$ and $\tilde{\eta}^{LC}$ and about the distribution $P(\theta_0,\phi_0)$ may be obtained by comparing experimental and computer simulated spectra.

The nature of the liquid crystalline phases can be studied either by direct observation of the NMR signals from the mesogen molecules, or by recording the spectra of probe molecules dissolved in the mesophase. The latter approach has the advantage that various probes with particular molecular symmetries can be chosen. Since the probe resonances are usually sharp, and (due to their smaller ordering parameters) are spread over narrow frequency ranges, the signal to noise ratio of their spectra is often considerably better than that obtained from labelled mesogen molecules. On the other hand probe molecules give only qualitative information about the nature of the mesogen structure, and no quantitative information on the ordering parameters of the mesogen. Furthermore, the presence of the probe might modify the nature of the phase diagram by forming lyotropic phases. In the following we shall discuss NMR experiments made with labelled mesogen molecules as well as with probes.

UNIAXIAL DISCOTIC MESOPHASES

Alignment in a magnetic field

Amongst the discotic mesophases described in the introduction, the D_{ho}, D_{hd} and N_D are uniaxial. Of these the D_{ho} phase which appears in the THEn series has been studied most extensively (40-43, 46-47). As indicated in Fig. (3), the phase diagram of the THEn homologues (n=5-11) is:

$$\text{Solid} \longrightarrow D_{ho} \longrightarrow I$$

On slowly cooling a THEn compound inside a magnetic field from its isotropic phase, down to the D_{ho} phase a liquid crystalline sample is obtained with a two dimensional distribution of domains in which the directors lie in a plane perpendicular to the external magnetic field. This is a consequence of the fact that its anisotropic magnetic susceptibility, $\Delta\chi = \chi_\parallel - \chi_\perp$, is negative. Once the transformation to D_{ho} is completed, the distribution of domains is frozen-in and will not change. This observation was confirmed by rotating this sample about an axis perpendicular to the external magnetic field and recording the spectrum at various angles of rotations. The results are shown in Fig. (6). At 0° a single doublet is observed because the major axes (director) of all domains are similarly inclined relative to the external field (i.e. π/2). However at other angles, and in particular at 90° a powder-like spectrum is obtained, typical of a planar distribution of domains. The traces on the right hand side of Fig. (6) were calculated using this model and introducing a small Gaussian distribution as discussed for case (iii) above (40).

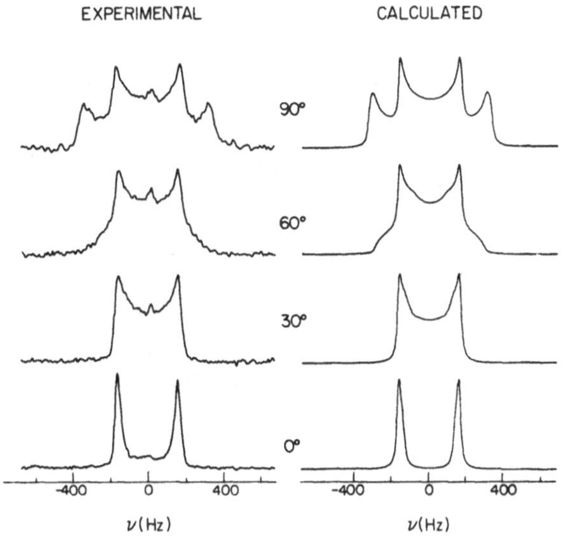

Figure 6. *Deuterium NMR spectra of a 6.6 wt.% solution of C_6D_6 in the D_{ho} phase of THE6 (T=56.5°C). The spectra on the left are experimental and were obtained from a sample consisting of a planar distribution of domains with the magnetic field normal to the plane (0°). The other spectra were obtained after rotating the sample about an axis perpendicular to the magnetic field by respectively 30°, 60° and 90°. The spectra on the right are theoretical and were computed using the parameters $\tilde{\nu}_Q$=313Hz $1/T_2$=10Hz and σ=9°.*

When a sample of THEn is allowed to cool from the isotropic phase inside a magnetic field, while spinning the sample about a direction perpendicular to the field, a single domain of the D_{ho} phase is obtained whose director lies along the sample's spinning axis. This phenomenon can readily be understood on the basis of the above observation i.e. that the director aligns perpendicular to the field direction. During the spinning those domains which nucleate with their director along the spinning axis will not feel any magnetic torque, while those which are formed in a general direction will experience alternating torques which will tend to align them parallel to the spinning axis.

Molecular Order Parameter

In discussing order parameters of discotic liquid crystals it is convenient to distinguish between the ordering of the aromatic part of the molecules and that of the methylene groups in the side chains. The former may be thought of as a measure of the molecular ordering within the columns while the latter is essentially a measure of the chains disorder due to their flexibility. The molecular order parameter of THEn molecules in the mesophase region can be determined very accurately from the deuterium NMR of molecules labelled in the aromatic site. Since the phase is uniaxial and the molecules have trigonal symmetry the quadrupolar splitting of an aromatic deuteron is given by:

$$\tilde{\nu}_Q^{ar} = -\tfrac{3}{4}(e^2qQ/h)^{ar}\, \bar{D}^2_{00}\, (3\cos^2\theta_0 - 1)/2 \qquad (14)$$

where it is assumed that the principal deuterium quadrupole tensor is axially symmetric with its major axis pointing along the C-D bond direction. Taking $\theta_0 = \pi/2$ (vide supra) and $(e^2qQ/h)^{ar} = 183$ kHz the parameter $\bar{D}^2_{00}$ can readily be calculated from the observed splitting.

The temperature dependence of $\tilde{\nu}_Q^{ar}$ measured in THE6 and the derived order parameter ($S = \bar{D}^2_{00}$) are shown in Fig. (7). It may be seen that the ordering is quite high (around 0.9) and changes very little with temperature (40). Similar values for $\bar{D}^2_{00}$ are obtained for other members of the THEn series (43). The behaviour is typical of smectic like mesophases in the conventional thermotropic liquid crystals but the degree of ordering is exceptionally high.

The $\bar{D}^2_{00}$ order parameter of THE5 was also determined from ^{13}C NMR of the aromatic carbons anisotropic chemical shift. The results obtained from these experiments (0.85 ± 0.1) agree well with those obtained from deuterium NMR (46).

Chain Conformation in the Mesogen

Once the molecular order parameter of the rigid aromatic core was determined the ordering of the side chains could also be studied via the deuterium quadrupolar splittings of the methylene and methyl groups. The question posed was whether the side chains are fully stretched in the aromatic plane ("diablo"-like), bent out of this plane ("octopus"-like) or completely disordered (25). X-ray studies could not provide information on this matter besides stating that the chains are probably "liquid-like" and inclined 20° out of the aromatic plane (19).

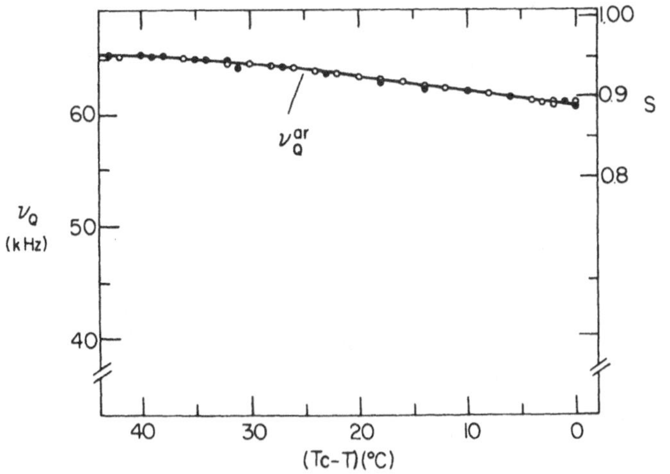

Figure 7. *Plot of the quadrupole splitting* $\tilde{\nu}_Q$ *of the aromatic deuterons of THE6-*d_6 *in a sample consisting of a planar distribution of domains with the magnetic field normal to the plain* ($\theta_0 = \pi/2$). *On the right coordinate are shown the corresponding* $\bar{D}^2_{00}$=*S values.*

In Fig. (8) the absolute values of $\tilde{\nu}_Q$ for the aliphatic deuterons in chain perdeuterated THEn (n=5-8) are plotted versus the carbon number in the chains. To interpret these results a model was employed according to which there is fast dynamic equilibrium between the various trirotameric states of the methylene segments in the side chains. The observed splittings are then the weighted averages of the methylene deuteron splittings in the various conformers. The problem is thus reduced to finding a distribution of conformations that yields the experimentally observed splittings. The data obtained from the deuterium NMR alone are not sufficient to determine such a distribution uniquely, but certain general conclusions may nevertheless be drawn. In particular the presence of certain conformations can be ruled out from the distribution while others appear to be necessarily included in order to explain the results. Perhaps the most important conclusions that emerge from

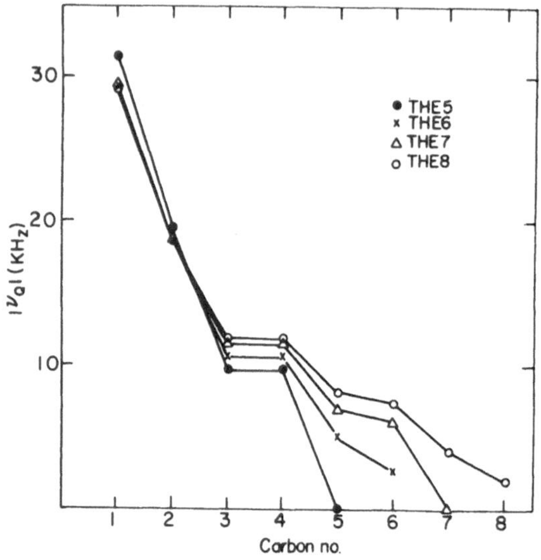

Figure 8. *Plot of the quadrupolar splittings of the various methylene deuterons in the mesophase region of THEn versus the carbon position along the chain.*

this analysis is that the chains deviate considerably from the all-trans conformation and are quite disordered. In addition it seems that an important fraction of the distribution consists of chains in which the O-C_1 bonds are oriented out of the aromatic plane.

Measurements on one member of this series (THE5) were also performed using ^{13}C NMR (46). While the conclusions regarding the ordering of the aromatic core were consistent with the ^2D study, those concerning the side chains were not. In particular, the ^{13}C NMR results seem to indicate that a major fraction of the side chains are coplanar with the aromatic ring contrary to the conclusions drawn from the deuterium NMR study.

In this connection it is interesting to mention the relaxation studies (T_1 and $T_{1\rho}$) performed on the proton resonances of two members of the THEn series (46,47). The results for THE5 were found to be temperature as well as frequency dependent. In general the relaxation times increase discontinuously on going from the D_{ho} mesophase to the isotropic liquid, but the effect becomes smaller at higher frequencies.

The results could be interpreted in terms of a theoretical model (53) which considered molecular translational and rotational diffusion, chain isomerization, as well as director fluctuation. Similar measurements were performed on the THE8 protons. As in THE5, T_1 is frequency dependent but no discontinuity was found on going from the D_{ho} to the isotropic phase. The difference in the behaviour is probably due to the higher ratio of aliphatic to aromatic protons in THE8 compared to THE5, resulting in a larger overall mobility of the protons in THE8.

In the above section we have described experiments performed on the D_{ho} phase of the THEn series. As indicated there are at least two more axially symmetric discotic mesophases i.e. the discotic nematic N_D and the disorder hexagonal phase D_{hd}. So far no studies on the neat mesogen in these phases have been performed, although measurements on probes dissolved in them were done. We shall refer to these results in a later section.

BIAXIAL DISCOTIC MESOPHASES

In this section we describe NMR results obtained for two biaxial discotic phases: the orthorhombic D_{rd} phase found in certain triphenylene and truxene hexaalkanoates and the monoclinic D_t phase of the benzene hexa-alkanoates series.

The D_{rd} Mesophase

The D_{rd} phase occurs in homologues of the THAn series. For example for n=6-9 the phase diagram is

$$\text{Solid} \longrightarrow D_{rd} \longrightarrow \text{I}$$

The TxAHn compounds and the higher homologues of THAn are polymorphic and exhibit in addition to D_{rd} other phases (see Fig. (3)). Several deuteriated probes dissolved in the D_{rd} phase of THAn were studied, including acetonitrile (CD_3CN) benzene (C_6D_6), mesitylene [s-$C_6D_3(CD_3)_3$] and p-xylene [p-$C_6D_4(CD_3)_2$] (44). Examples of a C_6D_6 spectrum are shown in Fig. (9). The spectrum corresponds to a powder sample and the typical features due to the asymmetry parameter ($\tilde{\eta}^{LC}$) are clearly seen. From the definitions following Eq. (8) and remembering that D_{rd} is orthorhombic, and that the benzene has a D_{6h} symmetry, the expressions for the appropriate magnetic parameters are:

$$\tilde{\nu}_Q^{LC} = (\tfrac{3}{4})\ (e^2qQ/h)^{ar}\ f_0\bar{D}^2_{00} = -(\tfrac{3}{8})(e^2qQ/h)^{ar}\ \bar{D}^2_{00} \qquad (15)$$

$$\tilde{\eta}^{LC} = \sqrt{6}\ \bar{D}^2_{02}/\bar{D}^2_{00}\ , \tag{16}$$

where

$$\bar{D}^2_{00} = <(\tfrac{1}{2})(3\cos^2\theta-1)> = S^{z\ z}_{z'z'} \tag{17}$$

measures the tendency of the molecular z axis to orient along the liquid crystal z' axis and

$$\bar{D}^2_{02} = \sqrt{(\tfrac{3}{8})}\ <\sin^2\theta(\cos^2\psi-\sin^2\psi)> = \sqrt{6}\ (S^{x\ x}_{z'z'} - S^{y\ y}_{z'z'}) \tag{18}$$

measures the preference of the molecular z axis to align in the mesophase x'z' plane relative to that in the y'z' plane.

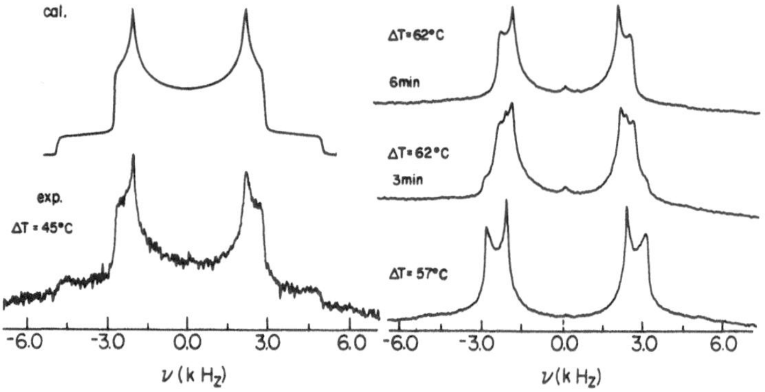

Figure 9. *Deuterium NMR spectra of C_6D_6 solution (4.6 wt.%) in the D_{rd} mesophase of THA9. On the left are shown experimental and computed isotropic powder spectra. The parameters for the computation were $\tilde{\nu}_Q^{LC} = 5.0kHz$, $\tilde{\eta}^{LC}=0.14$, $1/T_2=25Hz$. The spectra on the right are of a planar distribution of domains of the same solution obtained by slowly cooling the sample in a magnetic field. The trace at ΔT=57°C (i.e. 57°C below the clearing point) corresponds to the D_{rd} phase, the middle trace was recorded 3 minutes after setting the temperature at ΔT=62°C and the upper spectrum after 3 more minutes. These two spectra show the evolution of an unknown monotropic phase from the D_{rd} phase.*

From the analysis of the spectra in Fig. (9) $|\tilde{\nu}_Q^{LC}|$ and $|\tilde{\eta}^{LC}|$ could be derived yielding (using Eqs. (15) and (16) $|\bar{D}^2_{00}|$ = 0.036, $|\bar{D}^2_{02}|$ = 0.002. It should be noted that these results apply to the solute benzene and they do not bear simple relations to the ordering of the mesogen molecules. Nevertheless such spectra can be used to characterize mesophases, in particular whether they are uniaxial or biaxial. Spectra of probe molecules are also sensitive to phase transition in polymorphic liquid crystals, as may be seen for example in the series of spectra on the right hand side of Fig. (9). They correspond to a planar distribution of domains in a THA9 sample containing C_6D_6 as a solute. The $\Delta T = 57°C$ spectrum is of the D_{rd} phase while the upper $\Delta T = 62°C$ spectrum is of an unidentified new monotropic phase.

When the symmetry of the probe molecule is lower than uniaxial, interactions from several inequivalent nuclei are required in order to derive all the ordering parameters. As an example consider the powder spectra obtained from perdeuteriated mesithylene and p-xylene dissolved in the D_{rd} phase of THA9 (Fig. (10)). The results show that while in mesithylene $\tilde{\eta}^{LC}$ for

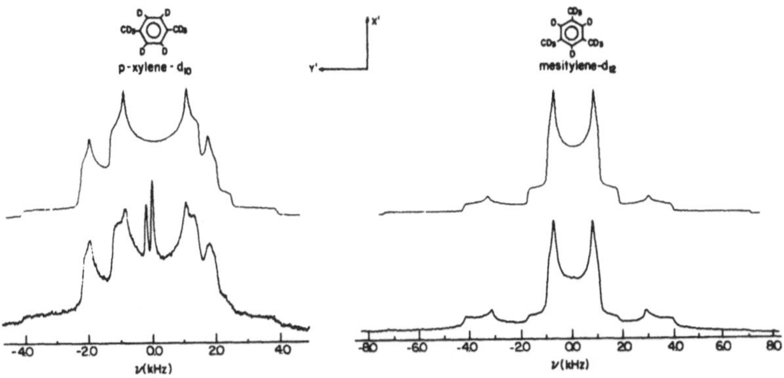

Figure 10. *Deuterium NMR spectra of perdeuteriated p-xylene (4.7 wt.%) and mesithylene (4.7 wt.%) dissolved in the D_{rd} phase of THA9. The lower traces are experimental while the upper traces are theoretical and were computed using the parameters $\tilde{\eta}^{LC}(CD_3)=0.17$, $\tilde{\eta}^{LC}(ar)=0.07$ for the p-xylene and $\tilde{\eta}^{LC}(CD_3)=\tilde{\eta}^{LC}(ar) = 0.13$ for mesithylene. The sharp singlet at the centre of the p-xylene spectrum is due to a small amount of an isotropic solution.*

the methyl and the aromatic deuterons are identical within experimental accuracy ($\tilde{\eta}^{LC} = 0.13$), they differ considerably in p-xylene (0.17 and 0.07 for the methyl and aromatic deuterons respectively). These results are consistent with the above discussion: for a uniaxial probe (i.e. mesithylene) $\tilde{\eta}^{LC} = \sqrt{6}\, \bar{D}^2_{02}/\bar{D}^2_{00}$ and is thus the same for all nuclei in the molecule. The p-xylene molecule has D_{2d} symmetry, and consequently in an orthorhombic liquid crystal assuming $\eta_p = 0$:

$$\tilde{\nu}_Q^{LC} = (\tfrac{3}{4})(e^2qQ/h)[f_0\bar{D}^2_{00} + 2(\mathrm{Re}f_2)\bar{D}^2_{20}] \tag{19}$$

$$\tilde{\eta}^{LC} = \frac{\sqrt{6}[f_0\bar{D}^2_{02} + 2(\mathrm{Re}f_2)\bar{D}^2_{22}]}{[f_0\bar{D}^2_{00} + 2(\mathrm{Re}f_2)\bar{D}^2_{20}]} \tag{20}$$

Since the geometrical parameters f_0 and f_2 differ for different nuclei $\tilde{\eta}^{LC}$ will in general vary from one nucleus to another, as indeed found for p-xylene.

In certain cases the NMR data may also provide information on the structure of the mesophase, as well as changes in the structure occurring in the vicinity of, or during a phase transition. As an example we consider the phase sequence $N_D \rightarrow D_{rd} \rightarrow D_{ho}$ which occurs in the phase diagram of the higher homologues of the TxHAn series (see diagram in Fig. (3)) (45). An interesting question related to this sequence is the correlation between the principal axes in the D_{rd} and D_{ho} phase. In Fig. (11) are shown spectra of C_6D_6 (1.7wt%) in the various phases of TxHA13. The bottom spectrum, recorded at 57°C, corresponds to the nematic phase and since $\Delta\chi$ for this phase is negative it corresponds to a planar sample with the magnetic field perpendicular to the plane, i.e. $\theta_0 = \pi/2$ for all domains. As the temperature is raised to above 65°C, the D_{rd} phase is formed and as expected spectra corresponding to a typical biaxial planar sample (see trace corresponding to 80°C) is obtained. At around 103°C this sample transforms to the D_{ho} phase, however the lineshape of the resulting spectra depend on the condition at which the transition was allowed to occur.

The two traces corresponding to 105°C in Fig. (11) were taken under two different experimental conditions. In one experiment the sample was taken out of the magnetic field at about 100°C and heated to above the $D_{rd} \rightarrow D_{ho}$ transition before placing it back into the spectrometer probe-head which was heated during that time to 105°C. The resulting spectrum is shown on the right hand side of the figure. When the sample was allowed to undergo the $D_{rd} \rightarrow D_{ho}$ transition within the magnetic field the spectrum corresponding to the 105°C trace on the left hand side of the figure was obtained. The results are quite distinct and can be explained by assuming that when the phase

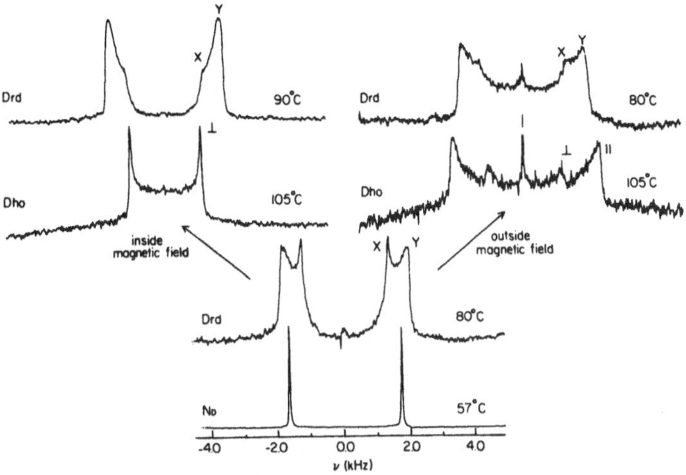

Figure 11. *Deuterium NMR spectra of a 1.7 wt% C_6D_6 solution in TxHA 13. The bottom spectrum is of the nematic phase at 57°C. The next spectrum was obtained after heating the sample to 80°C and corresponds to a planar distribution of biaxial domains of the D_{rd} phase. The 105°C trace on the right was obtained by heating the sample outside the magnetic field to above the $D_{rd} \to D_{ho}$ transition and then placing it back in the NMR probe at the designated temperature. The left 105°C trace was obtained after heating the sample to this temperature inside the magnetic field. The upper 80°C traces were obtained by cooling the corresponding samples back to the D_{rd} phase inside the magnetic field.*

transition occurs inside the magnetic field, magnetic torques reorient the liquid crystalline domains, while when the phase transition occurs outside the magnetic field the various domains retain their orientation in both the D_{rd} and D_{ho} phases.

Referring to the spectra on the right hand side of Fig. (11) we may conclude that the unique axis of the D_{ho} phase must correspond to one of the perpendicular (x or y) directions of the D_{rd} phase. Additional experiments indicated that this is most likely the x axis. A possible interpretation (45) of these results is to assume that in the orthorhombic D_{rd} phase the molecules in alternating columns are tilted in opposite directions. The fast diffusion of molecules means that the columnar direction remains a principal symmetry axis of the phase. However the molecular tilt is apparently sufficiently large that the most negative χ is no longer in the columnar direction as it is for the D_{ho} phase but rather perpendicular to it and to the molecular tilt axis. Thus in the D_{rd} phase the columnar direction corresponds to the x axis, and the axis about which the molecules are tilted corresponds to the y direction.

As the transition D_{rd} to D_{ho} is approached from the low temperature D_{rd} phase, the molecular tilt angle decreases monotonically until it reaches the D_{ho} phase. Thus if a D_{rd} sample with a planar xy distribution is allowed to worm up into the D_{ho} phase outside the magnetic field a two dimensional distribution of domains with columns both parallel and perpendicular to the external field will be obtained. This is indeed seen in the right 105°C trace of Fig. (11). If however the transition takes place inside the magnetic field, when the molecular tilt angle becomes sufficiently small, magnetic torques will reorient the domains to the perpendicular direction resulting in a spectrum as shown in the left 105°C spectrum of the figure.

This example demonstrates that information about the molecular stacking in the discotic phases can be obtained even if probe molecules are observed. Moreover the results for the TxHA13 suggest that the $D_{ho} \rightarrow D_{rd}$ transition in TxHAn is most probably second order and occurs by a continuous tilt of the molecular plane relative to the columnar axis with opposite tilt in alternating columns.

The D_t Mesophase

This biaxial mesophase is characterized by the directors in the columnar structures being uniformly tilted with respect to the columns' axes (See Fig. (2)). The D_t phase appears in members of the BHAn series or their mixtures (see Table 2 and Fig. (3)). The anisotropic magnetic susceptibility of these compounds is quite small and it is therefore difficult to align them even on slow cooling from the isotropic liquid at relatively high fields (44). Examples of spectra obtained (at 6.3T) from powder (bottom) and partially aligned (top) samples of the D_t phase of a 1:1 mixture of BHA6 and BHA7 containing a low concentration of C_6D_6 are shown in Fig. (12). Note that in the partially aligned sample the canonical features corresponding to the largest quadrupolar splitting are enhanced. This is because the magnetic anisotropy of this mesophase is positive, although as indicated above, quite small. This is probably a consequence of the larger ratio of aliphatic to aromatic carbons in the molecule compared to the triphenylene and truxene derivatives.

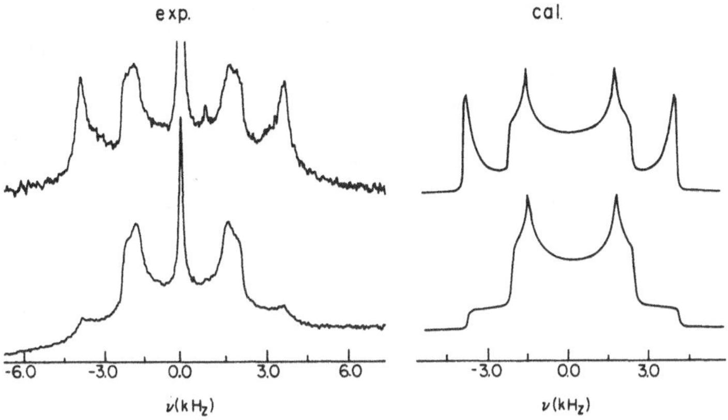

Figure 12. *Deuterium NMR spectra of C_6D_6 (3.0 wt.%) dissolved in the D_t phase of a 1:1 mixture of BHA6 and BHA7. The spectra on the left are experimental and correspond to a powder (bottom) and a partially aligned (upper) sample. On the right are the corresponding simulated spectra. They were calculated using $\tilde{\nu}_Q^{LC} = 3.9$kHz, $\tilde{\eta}^{LC} = 0.16$, $1/T_2 = 25$Hz. The partially aligned sample was calculated as a superposition of 85% powder and 15% single domain mixture. The single domain was assumed to be aligned along the magnetic field direction with a Gaussian distribution in θ_0 of $\sigma=20°$.*

Note that in the D_t phase which is monoclinic, three order parameters are required, $\bar{D}^2_{00}$, $\bar{D}^2_{02}$ and $\bar{D}^2_{0-2}$ even for axially symmetric molecules, consequently (Eq. (4)) there are three values of $F^{(2,2)}_{LC}$ different from zero: $F^{(2,0)}_{LC}$, $F^{(2,2)}_{LC}$ and $F^{(2,-2)}_{LC}$. In Cartesian coordinates these parameters are:

$$F^{(2,0)}_{LC} = \sqrt{\tfrac{3}{2}}\,(eQ/2h)\ V_{zz};\quad F^{(2,\pm 2)}_{LC} = (eQ/4h)\ (V_{xx}-V_{yy}\pm i2V_{xy}) \tag{21}$$

where the V_{ii} are the electric field gradients in the principal coordinate system of the average quadrupole interaction tensor. Thus in a monoclinic phase the z symmetry axis is a common axis for all tensorial properties but the x and y coordinates are in general not. The rotation angle ε from the x,y,z system to the canonical coordinates of the particular tensor is:

$$\varepsilon = \tfrac{1}{2}\tan^{-1}\left\{\frac{\mathrm{Im}[f_0\,\bar{D}^1_{02} + f_1\,(\bar{D}^2_{12} - \bar{D}^2_{1,-2}) + f_2\,(\bar{D}^2_{22} - \bar{D}^2_{2,-2})]}{\mathrm{Re}[f_0\,\bar{D}^2_{02} + f_1\,(\bar{D}^2_{12} + \bar{D}^2_{1,-2}) + f_2\,(\bar{D}^2_{22} + \bar{D}^2_{2,-2})]}\right\} \quad (22)$$

and the asymmetry parameter is

$$\tilde{\eta}^{LC} = (\sqrt{6}\ \mathrm{Re}\ \Sigma\ f_m\ \bar{D}^2_{m2})/(\Sigma f_m\ \bar{D}^2_{m0}\ \cos 2\varepsilon) \quad (23)$$

In the special case of uniaxial molecules these expressions become:

$$\varepsilon = \tfrac{1}{2}\tan^{-1}[\mathrm{Im}\ \bar{D}^2_{02}/\mathrm{Re}\ \bar{D}^2_{02}] \quad (24)$$

$$\tilde{\eta}^{LC} = \sqrt{6}\ (\mathrm{Re}\ \bar{D}^2_{02}/\bar{D}^2_{00})/\cos 2\varepsilon \quad (25)$$

Consequently for such molecules the bulk average principal axes for all tensors are colinear and their asymmetry parameter is the same.

DISCOTIC MESOPHASES AS SOLVENTS

In the previous sections we described the characteristic properties of some of the discotic mesophases using deuteriated mesogen molecules or deuteriated probes, dissolved in the mesophase. In the present section we discuss the properties of the discotic mesophases as solvents by studying the dependence of the orientational order of the solutes on such parameters as temperature, the chemical structure of the mesogen molecules, and the "crystallographic" structure of the discotic mesophases.

In Fig. (13) are shown deuterium NMR spectra of a C_6D_6 solution in the D_{ho} phase of THE6 at different temperatures within the mesophase region. The splittings, $|\tilde{\nu}_Q|$, are plotted in Fig. (14) as function of the temperature difference from the clearing point T_c and several features emerge. First the splittings are quite small, despite the fact that the order parameter of the host mesogen molecules is very high. Secondly the splittings are strongly temperature dependent; upon cooling from the clearing point they first decrease, reach zero and then increase again. This is unexpected since as shown above the order parameter of the host molecules was found to be almost independent of temperature. The point at which the quadrupole splitting vanishes does not correspond to any particular discontinuity in the phase structure. It must therefore mean that the quadrupole splitting changes continuously and that the "zero splitting point" corresponds to the temperature at which $\tilde{\nu}_Q$ (i.e. $\bar{D}^2_{00}$) changes sign.

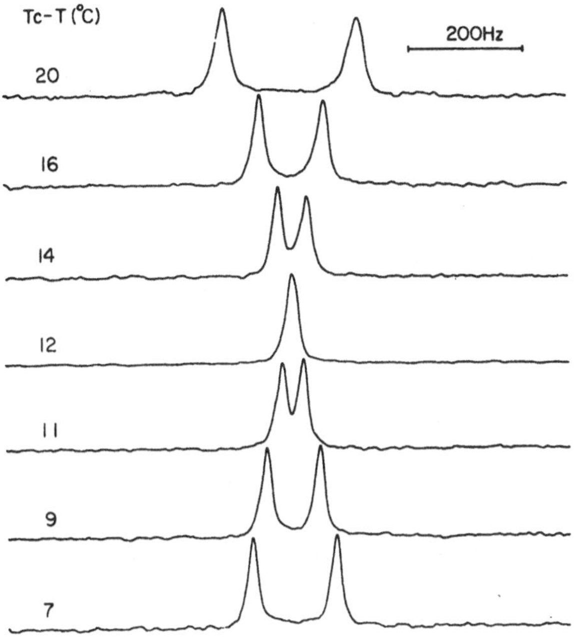

Figure 13. *Deuterium NMR spectra of a C_6D_6 (5.0 wt.%) solution in the D_{ho} mesophase of THE6 at different temperatures below the clearing point (T_c).*

To explain this behaviour a model was proposed according to which the benzene molecules occupy two solvation sites, which are in fast dynamic equilibrium. The observed splitting is then the weighted average of the splittings in the two sites:

$$\tilde{\nu}_Q = P^I\tilde{\nu}_Q^I + P^{II}\tilde{\nu}_Q^{II} \tag{26}$$

where P^i and $\tilde{\nu}_Q^i$ are respectively the corresponding fractional populations and splittings in the two sites. If it is further assumed that the two $\tilde{\nu}_Q$ have opposite signs, the results of Fig. (13) can readily be explained in terms of a temperature dependent equilibrium of P^I/P^{II}. Clearly at the temperature for which $\tilde{\nu}_Q = 0$

$$P^{I}/P^{II} = |\tilde{\nu}_Q^{II}/\tilde{\nu}_Q^{I}| \quad (27)$$

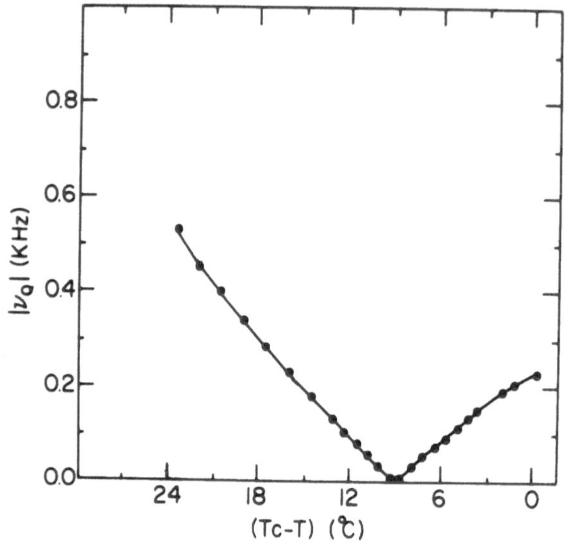

Figure 14. *A plot of the* C_6D_6 *deuterium quadrupolar splitting as function of temperature for a solution containing 6.6 wt.% benzene-*d_6.

It is natural to assume that one of the solvation sites (say I) corresponds to benzene molecules intercalated within the columns, parallel to the aromatic triphenylene core (and thus $\bar{D}^{2\,I}_{0\,0}>0$) while the second site (II) corresponds to the aliphatic region between the columns with apparently perpendicular orientation (i.e. $\bar{D}^{2\,II}_{0\,0}<0$). (Here $\bar{D}^{2\,I}_{0\,0}$ and $\bar{D}^{2\,II}_{0\,0}$ are the benzene ordering parameters in the two sites). It may also be assumed that site I is more stable and is therefore more populated at lower temperatures, while at higher temperature, entropy effect shifts the equilibrium into the more disordered region II of the side chains. Consequently the results of Fig. (13) and (14) would mean that the sign of $\tilde{\nu}_Q$ at temperatures below the zero point is positive, and negative above it (see corresponding plot in Fig. (15)). A crude analysis of the results gave for the enthalpy $\Delta H=H^{II}-H^{I}$ and entropy $\Delta S=S^{II}-S^{I}$ differences between the sites ~50KJmol^{-1} and ~3 e.u. respectively (41).

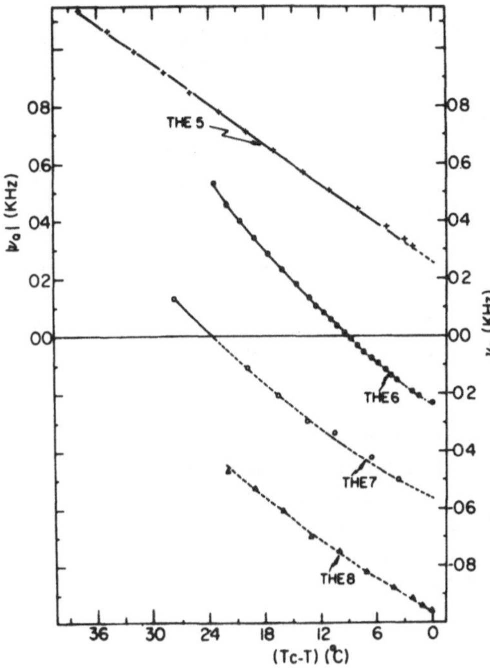

Figure 15. *Deuterium quadrupolar splitting of C_6D_6 dissolved in the D_{ho} mesophases of THEn (n=5-8) as a function of temperature. The dashed curve corresponds to the algebraic values (right scale) of the splitting under the assumption that $\tilde{\nu}_Q$ becomes more negative with increasing temperatures.*

Further support for the two solvation site model was obtained by comparing the quadrupolar splitting of C_6D_6 dissolved in a series of THEn with n=5,6,7 and 8. The results are plotted in Fig. (15) assuming that with increasing temperature the average splittings become more negative. The lines for the various solvents are nearly parallel to each other and indeed become more negative the larger the chain length of the solvent. These results are consistent with the above model in which site II is associated with the aliphatic region of the columnar phase.

A similar two site behaviour was found for other probes dissolved in the D_{ho} mesophase of the THEn series (41). An unexpected behaviour was observed in perdeuteriated triphenylene. Its NMR spectrum in THE6 consists of two doublets due to deuterons in position α and β respectively. The splittings of both deuterons is very similar over the whole mesophase region and as for benzene they are temperature dependent and pass through zero. However the temperature at which the zero crossing occurs differs for both doublets by about 1°C. This is inconsistent with Eq. (27) which predicts that for uniaxial molecules having several inequivalent deuterons the crossing of the zero line of all quadrupolar splittings should occur at the same temperature. This prediction is based on the assumption that the molecular geometry and hence the principal deuterium quadrupolar splittings in both solvation sites is the same. The discrepancy must therefore suggest that even for such rigid molecules as triphenylene the e^2qQ/h parameter may be site dependent (42).

The model was also supported by an ESR study of a series of paramagnetic spin probes dissolved in the D_{ho} phase of THE6. Here too, most spectra can be interpreted in terms of the two site model and at least in one case [3-spiro(N-4',4'-dimethyloxazolidine)-5-cholestone] distinct spectra for the two solvation sites are observed (54).

The existence of two solvation sites is characteristic for columnar phases having well defined aromatic and aliphatic regions. The partitioning of solute molecules between the two sites may however differ for ordered and disordered columnar phases. It is expected that in the latter phases where the mesogen molecules are not regularly stacked within the columns, the fractional population of solutes in the intercalation site should be higher compared with the ordered phases since the intercalation does not significantly perturb the columnar structure. This was in fact observed in experiments involving C_6D_6 solutions in D_{rd} and D_{hd} mesophases. The results for the D_{rd} phase of THA8 are shown in Fig. (16) along with those for the D_{ho} phase of THE8. It may be seen that the splittings are considerably higher in the disordered phase and very little temperature dependent (43).

Finally we refer to the nematic discotic phase. In this case there are no distinct solvation regions (although there may still be specific molecular interactions), and a normal Maier-Saupe type behaviour for solute molecules is expected. This is indeed observed experimentally as may be seen from results obtained in the N_D phase of C_6D_6 in a 2:1 mixture of THA8:TH8B (Fig. (16)) (43).

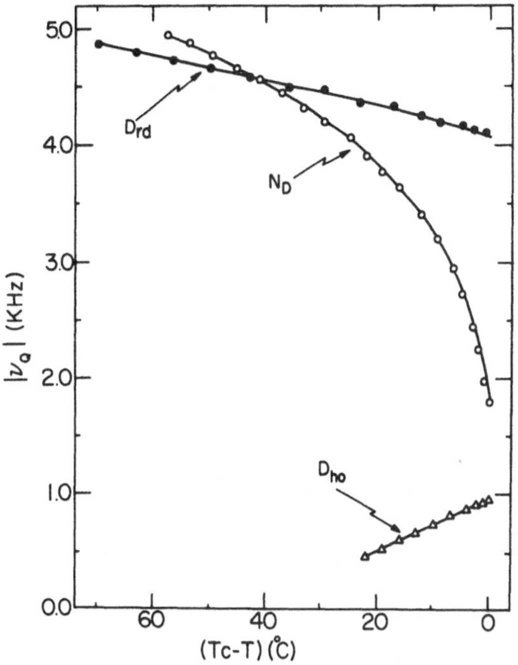

Figure 16. *The measured magnitude of* $\tilde{\nu}_Q$ *for* C_6D_6 *solutions in the* D_{rd} *phase of THA8, (6.2 wt.%),* D_{ho} *phase of THE8 (4.6 wt.%) and* N_D *phase of a 1:1 mixture of THA8 and TH8B (5.2 wt%).*

MOLECULAR TRANSLATIONAL DIFFUSION IN DISCOTICS

We end this chapter by describing some recent measurements on the molecular translational diffusion in discotic mesophases (55). Because of the unique structure of the columnar phases and the shape of the molecules we expect the diffusion of the mesogen molecules to be highly anisotropic i.e. to depend on the direction with respect to the liquid crystal axes. Reliable results were obtained in the mesophases of THE8 and THA10. The measurements were made on the residual methyl proton signals in chain perdeuteriated compounds using the pulsed field gradient method. The results are summarized in Fig. (17). They refer to samples with a planar distribution of domains as shown in the insert. Consequently for uniaxial phases the results for diffusion along z (the magnetic field direction) refer to $D_\perp$,

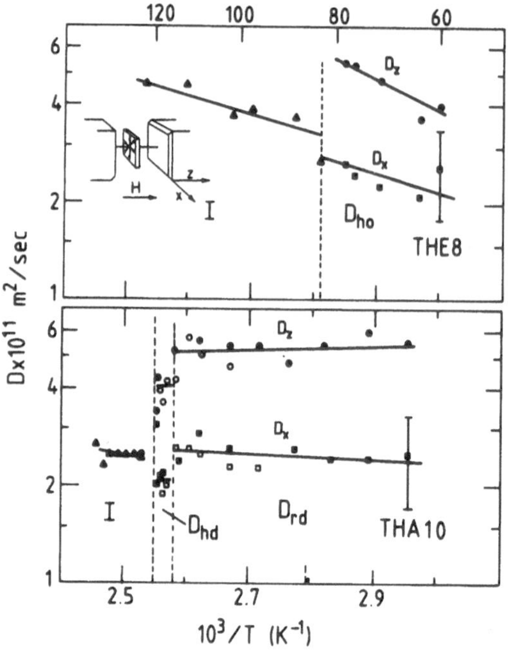

Figure 17. *Self diffusion measurement in the various mesophases of THE8 and THA10. The insert shows the structure and alignment of the sample. The circles and squares correspond respectively to diffusion along the magnetic field direction (z) and perpendicular to it, while triangles refer to the isotropic liquid. Open and filled symbols refer to different experimental runs. The vertical bars represent the uncertainty in the result for the single measurements.*

the diffusion coefficient perpendicular to the director, while the result along x corresponds to an average of $D_{\parallel}$ and $D_{\perp}$. Thus, $D_{\perp}=D_z$, and $D_{\parallel}\cong 2D_x-D_z$. It may be seen that in all cases, $D_x \cong D_z/2$ indicating that $D_{\parallel} << D_{\perp}$. On the other hand the order of magnitude of the diffusion coefficients in the mesophase is very similar to that found in the isotropic liquid and it is of the order of magnitude found in other liquids of similar molecular weight. The results clearly show that there is relatively free lateral diffusion from one column to another, but that the longitudinal diffusion is slow and probably limited by the longitudinal diffusion of the whole columns.

The situation with regard to diffusion of solute molecules is however quite different. Measurements were made on solutions of benzene in the D_{ho} phase of THE6. Since the solute order parameter is small the C_6H_6 signal is unsplit and relatively sharp, allowing reliable diffusion experiments to be made. The results for three samples with different domain structures, as indicated in the inserts, are shown in Fig. (18). In general the diffusion of benzene is about an order of magnitude faster than

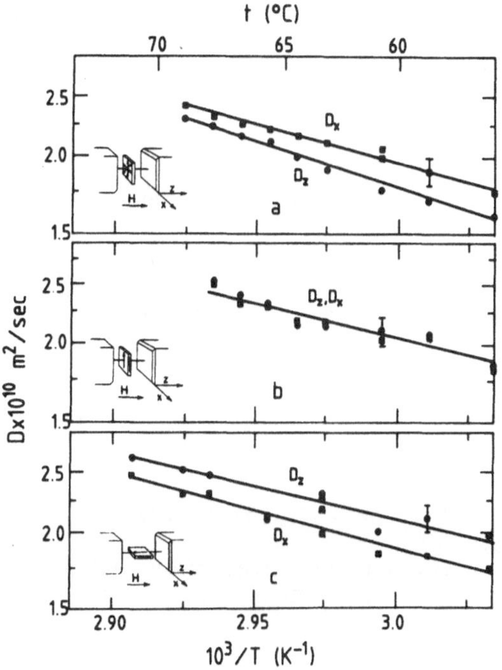

Figure 18. *Experimental results for the three sets of experiments on the diffusion of benzene dissolved in THE6. The texture and alignment of the sample used in each set is depicted in the corresponding inserts.*

that of the self diffusion of the mesogen molecules and the degree of anisotropy is much smaller. However note that in contrast to the self diffusion results, $D_{\parallel}$ is larger (by about 10%) than $D_{\perp}$. These results are consistent with the two-site model discussed in the previous section. There it was shown that benzene molecules in the D_{ho} phase occupy predominantly the intercolumnar site II of the aliphatic chains. In this site the probe molecules are very weakly ordered and considerably more mobile than in site I. The results also show that in this site there is apparently a slight preference for the molecules to diffuse parallel to the columns rather than in the perpendicular direction. This is in contrast to the mesogen molecules which in order to undergo translational diffusion need to detach from the columnar structure.

ACKNOWLEDGMENT

This work was supported by the National Council for Research and Development, and by a DFG Project in partnership with Professor K. Praefcke of the Technische Universitat, Berlin.

REFERENCES

1. D. Goldfarb, Z. Luz and H. Zimmermann, *Israel J. Chem.* **23**, 341 (1983).

2. P.G. de Gennes, in "*The physics of liquid crystals*", Clarendon Press, Oxford, 1974, p. 38; E.I. Kats, *Zh. Eksp. Teor. Fiz.* **75**, 1819 (1978); *Sov. Phys. J.E.T.P.* **48**, 916 (1978).

3. S. Chandrasekhar, B.K. Sadashiva and K.A. Suresh, *Pramana* **9**, 471 (1977).

4. J.C. Dubois, *Ann. Phys.* **3**, 131 (1978).

5. J. Billard, J.C. Dubois, N.H. Tinh and A. Zann, *Nouv. J. Chim.* **2**, 535 (1978); H. Gasparoux, C. Destrade and G. Fug, *Mol. Cryst. Liq. Cryst.* **59**, 109 (1980).

6. N.H. Tinh, C. Destrade and H. Gasparoux, *Phys. Lett.* **72A**, 251 (1979).

7. J. Billard, in "*Liquid Crystals of One and Two Dimensional Order*", Ed. W. Helfrich and G. Heppke, (1980), p. 383.

8. J.C. Dubois and J. Billard, *Proc. of the Las Vegas A.C.S. Meeting Liq. Cryst. and Ordered Fluid.* Vol. 4, Eds. A.C. Griffin and J.F. Johnson.

9. C. Destrade, H. Gasparoux, P. Foucher, N.H. Tinh, J. Malthete and J. Jacques, *J. Chim. Phys.* **80**, 137 (1983); A.M. Levelut, *J. Chim. Phys.* **80**, 149 (1983).

10. W. Poules and K. Praefcke, *Chemiker Ztg.* **107**, 310, 374 (1983); K. Praefcke, W. Poules, B. Scheuble, R. Poupko and Z. Luz, Z. *Naturforsch.* (in press); B. Kohne and K. Praefcke, *Ang. Chem. Int. Ed. (Engl.)*, **23**, 82 (1984).

11. C. Destrade, N.H. Tinh and H. Gasparoux, *Mol. Cryst. Liq. Cryst.* **71**, 111 (1981).

12. L. Mamlok, J. Malthete, N.H. Tinh, C. Destrade and A.M. Levelut, *J. Phys. Lett.* **43**, L-641 (1982).

13. J. Billard, J.C. Dubois, C. Voucher and A.M. Levelut, *Mol. Cryst. Liq. Cryst.* **66**, 115 (1981).

14. C. Destrade, M.C. Mondon-Bernaud and N.H. Tinh, *Mol. Cryst. Liq. Cryst. Lett.* **49**, 169 (1979).

15. N.H. Tinh, J. Malthete and C. Destrade, *Mol. Cryst. Liq. Cryst. Lett.* **64**, 291 (1981).

16. F.C. Frank and S. Chandrasekhar, *J. de Phys.* **41**, 1285 (1980).

17. A.M. Levelut, *Proceeding Int. Liq. Cryst. Conf. Bangalore*, Heyden and Son, London (1979), p. 21.

18. A.M. Levelut, F. Hardouin, H. Gasparoux, C. Destrade and N.H. Tinh, *J. Phys.* **42**, 147 (1981).

19. A.M. Levelut, *J. de Phys. Lett.* **40**, L-781 (1979).

20. S. Chandrasekhar, B.K. Sadashiva, K.A. Suresh, N.V. Madhusudana, S. Kumar, R. Shashidhar and G. Venkatesh, *J. de Phys. Colloque* C3, **40**, C3-120 (1979).

21. J. Malthete, C. Destrade, N.H. Tinh and J. Jacques, *Mol. Cryst. Liq. Cryst. Lett.* **64**, 223 (1981).

22. C. Destrade, H. Gasparoux, H. Babeau and N.H. Tinh, *Mol. Cryst. Liq. Cryst.* **67**, 37 (1981).

23. N.H. Tinh, J. Malthete and C. Destrade, *J. de Phys. Lett.* **42**, L-417 (1981).

24. C. Destrade, P. Foucher, J. Malthete and N.H. Tinh, *Phys. Lett.* **88A**, 187 (1982).

25. C. Destrade, M.C. Mondon and J. Malthete, *J. de Phys. Colloque* C3, **40**, C3-17 (1979).

26. A. Queguiner, A. Zann, J.C. Dubois and J. Billard, *Proc. Int. Liq. Cryst. Conf. Bangalore*, Heyden, London (1980).

27. C. Destrade, J. Malthete, N.H. Tinh and H. Gasparoux, *Phys. Lett.* **78A**, 82 (1980).

28. J. Billard and B.K. Sadashiva, *Pramana* **13**, 309 (1979).

29. R.E. Goozner and M.M. Labes, *Mol. Cryst. Liq. Cryst. Lett.* **56**, 75 (1979).

30. M. Sorai, K. Tsuji, H. Suga and S. Seki, *Mol. Cryst. Liq. Cryst.* **59**, 33 (1980).

31. M. Sorai and H. Suga, *Mol. Cryst. Liq. Cryst.* **73**, 47 (1981).

32. G. Sigaud, M.F. Achard, C. Destrade and N.H. Tinh, in "*Liquid Crystals of One and Two Dimensional Order*", Ed. W. Helfrich and G. Heppke, (1980), p. 383.

33. P. Oswald and M. Kleman, *J. de Phys.* **42**, 1461 (1981).

34. M. Kleman and P. Oswald, *J. de Phys.* **43**, 655 (1982).

35. Y. Bouligand, *J. de Phys.* **41**, 1307 (1980).

36. P. Oswald, *J. de Phys. Lett.* **42**, L-171 (1981).

37. G.E. Feldkamp, M.A. Handschy and N.A. Clark, *Phys. Lett.* **85A**, 359 (1981).

38. S. Chandrasekhar, *Phil. Trans. R. Soc. London,* **A309**, 93 (1983).

39. Y. Rabin, W.E. McMullen and W.M. Gelbart, *Mol. Cryst. Liq. Cryst.* **89**, 67 (1982); W.M. Gelbart, *J. Phys. Chem.* **86**, 4298 (1982).

40. D. Goldfarb, Z. Luz and H. Zimmermann, *J. de Phys.* **42**, 1303 (1981).

41. D. Goldfarb, Z. Luz and H. Zimmermann, *J. de Phys.* **43**, 421 (1982).

42. D. Goldfarb, Z. Luz and H. Zimmermann, *J. de Phys.* **43**, 1255 (1982).

43. D. Goldfarb, Z. Luz and H. Zimmermann, *J. Chem. Phys.* **78**, 7065 (1983).

44. D. Goldfarb, R. Poupko, Z. Luz and H. Zimmermann, *J. Chem. Phys.*, **79**, 4035 (1983).

45. D. Goldfarb, I. Belsky, Z. Luz and H. Zimmermann, *J. Chem. Phys.*, **79**, 6203 (1983).

46. V. Rutar, R. Blinc, M. Vilfan, A. Zann and Y.C. Dubois, *J. Phys.* **43**, 761 (1982); M. Vilfan, G. Lahajnar, V. Rutar, R. Blinc, B. Topic, A. Zann and J.C. Dubois, *J. Chem. Phys.* **75**, 5250 (1981).

47. A.F. Martins and A.C. Ribeiro, *Portugal Phys.* **11**, 169 (1980).

48. L.C. Snyder, *J. Chem. Phys.* **43**, 4041 (1965).

49. H.R. Falle and G.R. Luckhurst, *J. Mag. Reson.* **3**, 161 (1970).

50. Z. Luz and S. Meiboom, *J. Chem. Phys.* **59**, 275 (1973).

51. J.W. Doane, in "*Magnetic Resonance of Phase Transitions*", Eds. F.J. Owens, C.P. Poole Jr. and H.A. Farach, Academic Press, New York (1979), p. 171.

52. C. Zannoni, "*The Molecular Physics of Liquid Crystals*", Ed. G.R. Luckhurst and G.W. Gray, Academic Press (1979), p. 51.

53. S. Zumer and M. Vilfan, *Mol. Cryst. Liq. Cryst.* **70**, 39 (1981).

54. E. Meirovitch, Z. Luz and H. Zimmermann, *J. Phys. Chem.* (in press).

55. R.Y. Dong, D. Goldfarb, M.E. Moseley, Z. Luz and H. Zimmermann, *J. Phys. Chem.* (in press).

MEASUREMENT OF ORIENTATIONAL ORDERING BY NMR

J.W. Emsley

Department of Chemistry,
University of Southampton, UK.

RIGID SOLUTE MOLECULES

In earlier chapters it has been established that the NMR experiment measures partial averages of the component along the field of a number of second-rank properties. The relationship between $\tilde{A}_{\parallel}$, the partial average of the component along the director of a particular second-rank interaction, is related to $A_{\alpha\beta}$, the components in a molecular frame and $S_{\alpha\beta}$, elements of the Saupe ordering matrix by

$$\tilde{A}_{\parallel} = A_0 + \tfrac{2}{3} \sum_{\alpha\beta} S_{\alpha\beta} A_{\alpha\beta} \; . \qquad (1)$$

A_0, the isotropic average, and $A_{\alpha\beta}$ should be known in order that measurement of $\tilde{A}_{\parallel}$ can lead to information on $S_{\alpha\beta}$. Clearly, to completely characterize S there must be at least as many observable, independent values of $\tilde{A}_{\parallel}^{i}$, the superscript referring to sites in the molecule, as there are unknown elements of S. This is easily achieved for many rigid molecules and it is for this reason that our discussion of orientational ordering commences with solute molecules. These can be chosen so that the principal axes of S can be located with certainty and such that NMR can give precise measurements of the two independent principal elements, S_{zz} and $S_{xx} - S_{yy}$. The disadvantage of such studies is that their interpretation requires modelling solute-solvent interactions rather than the interactions between pure mesogens.

J. W. Emsley (ed.), Nuclear Magnetic Resonance of Liquid Crystals, 379–412.

Choice of Interaction

In theory the interactions which can be measured are the quadrupolar, $\tilde{q}_i$, the dipolar $\tilde{D}_{ij}$, the shielding tensor, $\tilde{\sigma}_i$, or the indirect spin coupling $\tilde{J}_{ij}$. In practice the values of $\tilde{J}_{ij}$ cannot be derived easily from spectra and have not been used to probe orientational ordering. Also, shielding tensors are known for only a very small number of nuclei in particular molecular sites; to obtain these data requires careful and time consuming studies on single crystals. The two remaining interactions, however, are eminently suitable for studying orientational order, each having particular advantages as well as disadvantages.

Dipolar Coupling

The dipolar coupling between a pair of nuclei with gyromagnetic ratios γ_i and γ_j is given by

$$\tilde{D}_{ij} = -\gamma_i\gamma_j h S_{ij}/4\pi^2\langle r_{ij}^3\rangle \quad . \tag{2}$$

S_{ij} is the order parameter for the direction of the vector r_{ij} between the two nuclei. To evaluate S_{ij} from $\tilde{D}_{ij}$ it is necessary to know $\langle r_{ij}^3\rangle$, the angular brackets denote averaging over small amplitude vibrational motion. For many molecules it is quite practical to carry out this average from a knowledge of the normal coordinates, as discussed earlier by Diehl. The difference between r_{ij}^3 and its average $\langle r_{ij}^3\rangle$ is, however, often small enough ($\sim$ 1%) to be safely ignored. This will not be true for molecules with T_d or O_h symmetry whose orientational ordering is manifested only as a consequence of the lower symmetry of particular vibrational modes (1,2,3). Apart from deciding whether to use r_{ij}^3 or $\langle r_{ij}^3\rangle$ it is clearly necessary to know the magnitude and direction of r_{ij}, and it is often possible to determine ratios of these distances by the techniques discussed in Chapter 7. There have been, however, very few systematic studies of ordering in this way, because until recently it has been a tedious task to record and analyse the complex spectra given by dipolar coupled nuclei. When the recording of such spectra became possible on solutes at concentrations of about 1% there was still no rush to acquire data, probably because it is not obvious what is gained by such a study. In this chapter I aim to explain why the measurement of ordering of solutes can be a sensitive probe into the factors determining how molecules interact to give long range orientational ordering.

As an example of the use of dipolar couplings to study S let us consider the specific example of paradinitrobenzene

(PDNB). The proton spectrum, shown in Fig. (1) of a sample in a

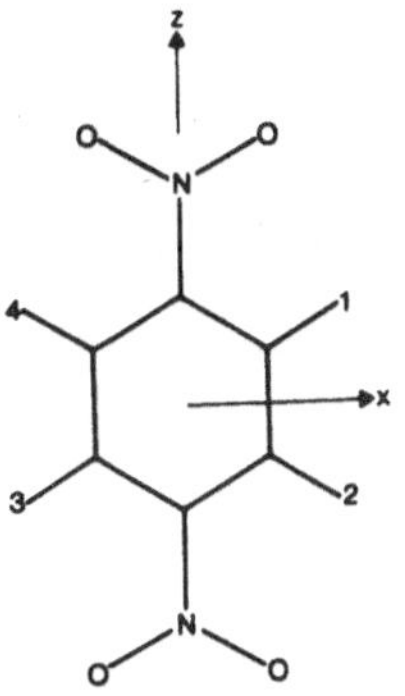

nematic phase such as in Phase 5 is simple to analyse and yields three dipolar couplings $\tilde{D}_{12}$, $\tilde{D}_{14}$ and $\tilde{D}_{13}$. Assuming a structure for the molecule as planar with D_{2h} symmetry means that x y z as

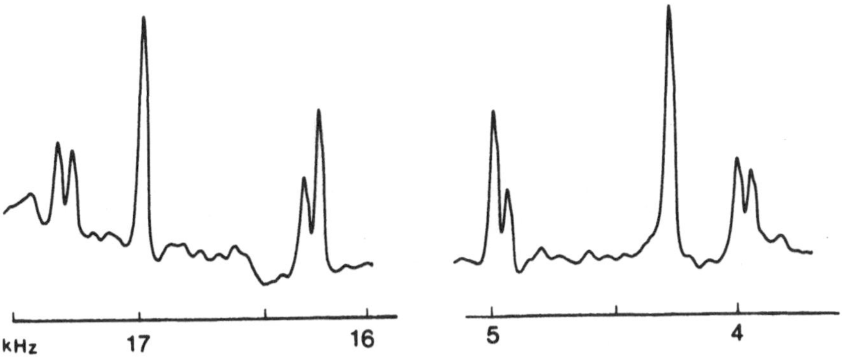

Figure 1. *Proton spectrum of a sample of paradinitrobenzene dissolved in the nematic phase of Phase 5.*

shown are principal axes for S. The principal components are obtained from:

$$S_{zz} = -\tilde{D}_{12} 4\pi^2 \langle r_{12}^3 \rangle / \gamma_H^2 h$$

$$S_{xx} = -\tilde{D}_{14} 4\pi^2 \langle r_{14}^3 \rangle / \gamma_H^2 h$$

and

$$S_{yy} = -S_{xx} - S_{zz}.$$

This yields the data shown in Fig. (2) for the variation of S_{zz} with $S_{xx} - S_{yy}$ for PDNB in three liquid crystal solvents: Phase 5, which is a mixture of alkyloxyazoxybenzenes, E5, a mixture of alkylcyanobiphenyls and 4,4'-di-n-heptylazoxybenzene (HAB). We will return later to discuss why the data is presented in this way.

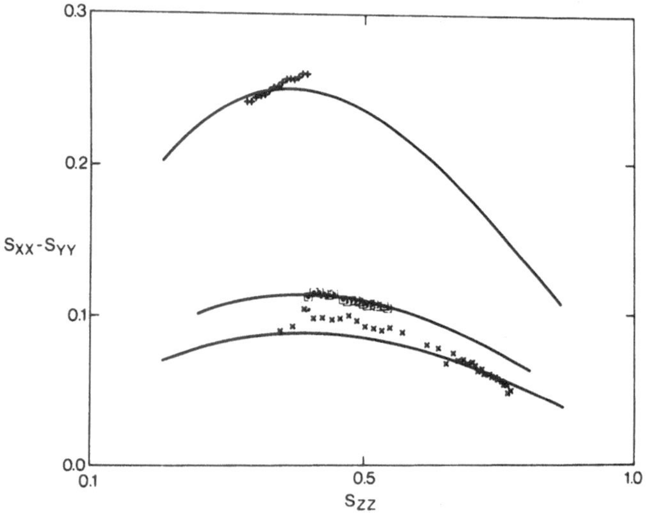

Figure 2. *Variation of $S_{xx} - S_{yy}$ with S_{zz} for paradinitrobenzene in Phase 5 (x), E5 (+) and HAB (◙). The continuous curves are the variations calculated with λ = 0.2, 0.27 and 0.58.*

Quadrupolar Splittings

For more complex spin systems it is an advantage to study the deuteron spectra of deuteriated solutes. The spectra are considerably simplified compared with the proton spectra of the normal molecules because of the large splittings, $\Delta\tilde{\nu}$, produced by the quadrupolar interaction which is of the order of 10^2 greater than dipolar interactions between deuterons. The relationship between $\Delta\tilde{\nu}_i$ at a particular site in a molecule and the principal components of S is,

$$\Delta\tilde{\nu}_i = \tfrac{3}{2}\, q^i_{bb}\, [S_{zz}(\ell^2_{zbi} + \tfrac{1}{3}\eta_i(\ell^2_{zai} - \ell^2_{zci}))$$

$$+S_{xx}(\ell^2_{xbi} + \tfrac{1}{3}\eta_i(\ell^2_{xai} - \ell^2_{xci}))$$

$$+S_{yy}(\ell^2_{ybi} + \tfrac{1}{3}\eta_i(\ell^2_{yai} - \ell^2_{yci}))] \,, \qquad (3)$$

where a b c are the principal axes of the quadrupolar tensor for the deuteron at site i in the molecule, and ℓ_{xbi} etc. are the direction cosines of x with b. The asymmetry parameter $\eta_i = (q^i_{aa} - q^i_{cc})/q^i_{bb}$ and q^i_{bb} is the quadrupolar coupling constant. The evidence available from single crystal studies allows us to locate b to lie along the X-D bond direction. Furthermore, for C-D bonds the values of q^i_{bb} and η_i are found to lie within a few percent of one another for particular kinds of bonds. Thus, for aromatic compounds taking q^i_{bb} to be 185 kHz seems to be correct to within ± 5 kHz for compounds without strongly perturbing substituents. Also the asymmetry parameter is usually close to zero, and for strongly ordered systems, that is with $S_{zz} \gg (S_{xx} - S_{yy})$ can be safely neglected. Note, however, the importance of η in magnitude and sign when these conditions are not fulfilled (4). For aliphatic, hydrocarbon chains it is usual to take q^i_{bb} to be 168 kHz and $\eta = 0$. There is in fact a pressing need for studies to be made of the deuteron spectra of single crystals of aliphatic hydrocarbons in order to determine q^i_{bb} and η_i more accurately, a daunting but rewarding task.

The application of Eq. (3) to the determination of S_{zz} and $S_{xx} - S_{yy}$ is illustrated by a study of anthracene-d_{14}.

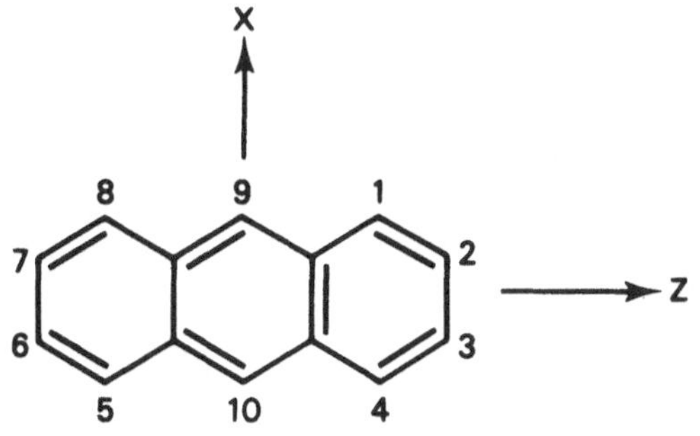

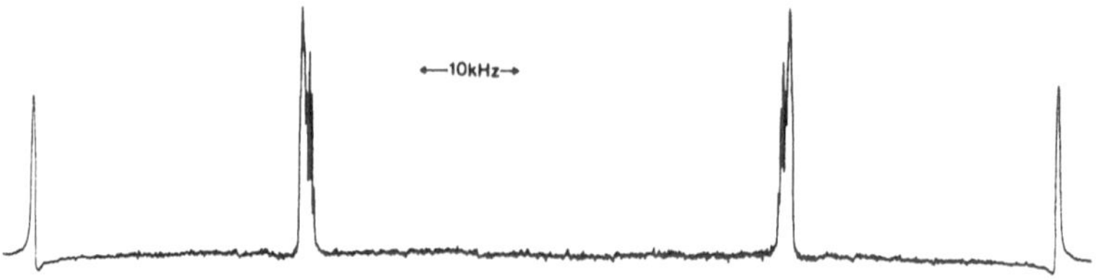

Figure 3. *Deuteron spectrum at 30.7 MHz of 2% anthracene-d_{10} in Phase 5.*

The deuteron spectrum of a 1% w/w solution in Phase 5 is shown in Fig. (3). The deuterons at positions 1, 4, 8 and 5 have a quadrupolar splitting differing from that of deuterons 9 and 10 by only about 2% and hence the spectrum from these six deuterons is more complex than the single broad lines in each component of the doublet from the set 2, 3, 6, 7. The spectrum from 1, 4, 5, 8, 9 and 10 is shown on an expanded scale in Fig. (4) and for this particular temperature the analysis yielded:

$$\begin{aligned}
\Delta\tilde{\nu}\ (2, 3, 6, 7) &= 104000\ \text{Hz}\\
\Delta\tilde{\nu}\ (1, 4, 5, 8) &= -49230\ \text{Hz}\\
\Delta\tilde{\nu}\ (9, 10) &= -48070\ \text{Hz}\\
\tilde{D}_{12} &= -64\ \text{Hz}\\
\tilde{D}_{19} &= -105\ \text{Hz}\\
\tilde{D}_{17} &= -6\ \text{Hz}\\
\tilde{D}_{18} &= -13\ \text{Hz}\ ;
\end{aligned}$$

all other dipolar couplings are negligibly small. The ability to extract dipolar couplings and quadrupolar splittings is important because it enables the relative signs of the $\Delta\nu_i$ to be established. Without this information there is always an uncertainty in the sign of the individual order parameters and in both the sign and magnitude of $S_{xx} - S_{yy}$.

Anthracene was chosen for study (5) because the location of the principal axes is known and a single crystal study has established values of q and η (6). With this data and assumptions about geometry it is possible to obtain S_{zz} and $S_{xx} - S_{yy}$ and this is shown in Fig. (5).

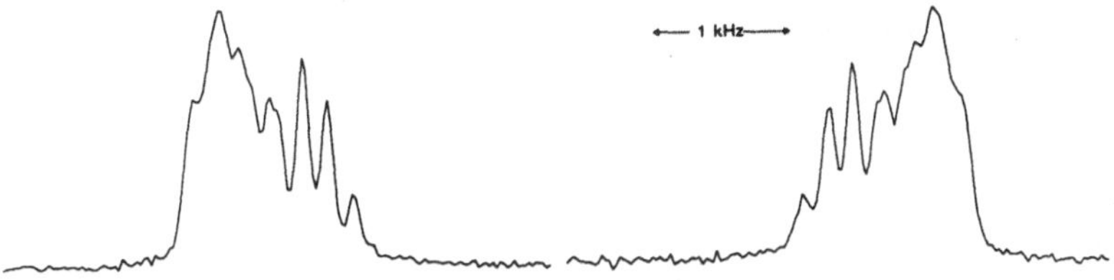

Figure 4. *Spectrum from deuterons 1, 4, 5, 8, 9 and 10 in anthracene-d_{10} dissolved in Phase 5.*

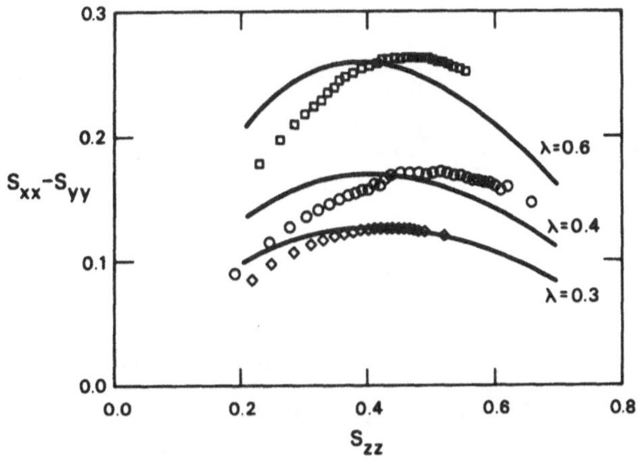

Figure 5. *Variation of S_{zz} with $S_{xx} - S_{yy}$ for anthracene-d_{10} dissolved in ZLI 1167 (▣), E9 (⊙) and Phase 5 (◈).*

Comparison with Theoretical Models

The elements of S are determined by the singlet orientational distribution function, which may be expressed in terms of U(ω) the potential of mean torque,

$$S_{ab} = Z^{-1} \int d\omega \ \tfrac{1}{2}(3\ell_a(\omega)\ \ell_b(\omega) - \delta_{ab}) \exp[-U(\omega)/kT] \ , \qquad (4)$$

where $\ell_a(\omega)$ is the direction cosine of axis a and the mesophase director and depends upon ω, the orientation of the director in the molecular frame. The orientational partition function is

$$Z = \int d\omega \exp\left[-U(\omega)kT\right] . \tag{5}$$

The derivation of $U(\omega)$ using the mean field approach for a single component system of non-axially symmetric molecules in a uniaxial mesophase has been described by Luckhurst, Zannoni, Nordio and Segre (7). An extension of the theory to give $U(\omega_2)$, the potential of mean torque for a biaxial solute molecule at infinite dilution in a uniaxial solvent has been provided by Emsley, Hashim, Luckhurst, Rumbles and Viloria (8); ω_2 is the orientation of the director in a frame fixed in the solute molecule. The general form of $U(\omega_2)$ is,

$$U(\omega_2) = \sum_{L\ \mathrm{even}} \bar{u}_{Lkk'}\, \bar{C}^{*}_{Lk}\, C^{*}_{Lk'}(\omega_2) \; ; \tag{6}$$

$C_{Lk'}(\omega_2)$ is a reduced spherical harmonic and $\bar{C}^{*}_{Lk}$ are ensemble averages of these functions for axes fixed in the solvent molecule, that is they are the complete set of solvent order parameters. The $\bar{u}_{Lkk'}$ are expansion coefficients averaged over the distance dependence of the solute-solvent pair potential. The solvent molecules are assumed to be rigid.

The potential of mean torque given by Eq. (6), however, contains too many terms to be useful and it is usual to make the drastic simplification of restricting the summation to the first term, i.e. $L = 2$. In this case we can write $U(\omega_2)$ as,

$$\begin{aligned} U(\omega_2) &= (\bar{u}_{200}\,\bar{C}_{20} + 2\bar{u}_{220}\,\bar{C}_{22})\, d_{00}(\beta_2) \\ &\quad + (2\bar{u}_{202}\,\bar{C}_{20} + 4\bar{u}_{222}\,\bar{C}_{22})\, d_{02}(\beta_2) \cos 2\gamma_2 \, . \end{aligned} \tag{7}$$

Here we use principal axes for solute and solvent and $d_{mn}(\beta_2)$ is an element of a reduced Wigner rotation matrix of rank 2. To simplify the notation Eq. (7) is written as

$$U(\omega_2) = -kT\,(a d_{00}(\beta_2) + b d_{02}(\beta_2) \cos 2\gamma_2) \tag{8}$$

where

$$a = -(\bar{u}_{200}\,\bar{C}_{20} + 2\bar{u}_{220}\,\bar{C}_{22})/kT \tag{9}$$

$$b = -(2\bar{u}_{202}\,\bar{C}_{20} + 4\bar{u}_{222}\,\bar{C}_{22})/kT \, . \tag{10}$$

The solute order parameters are averages of $d_{00}(\beta_2)$ and

$d_{00}(\beta_2)\cos 2\gamma_2$ and are,

$$S_{zz} = \overline{d_{00}(\beta_2)}$$

$$= 2\pi Z^{-1} \int_0^\pi d\beta_2 \sin\beta_2\, d_{00}(\beta_2)\, I_0[b\, d_{02}(\beta_2)]\exp[a\, d_{00}(\beta_2)] \;, \qquad (11)$$

and

$$S_{xx} - S_{yy} = 6^{\frac{1}{2}}\, \overline{d_{02}\;(\beta_2)\;\cos\,2\gamma_2}$$

$$= \sqrt{24}\,\pi Z^{-1} \int_0^\pi d\beta_2 \sin\beta_2\, d_{02}(\beta_2)\, I_1[b\, d_{02}(\beta_2)]\exp[a\, d_{00}(\beta_2)] \qquad (12)$$

The $I_n\,[b\,d_{02}(\beta_2)]$ are nth order modified Bessel functions. These equations are identical in form with those derived for a single component system (7), the difference lies in the definitions of the a and b coefficients.

Now, the solvent, liquid crystalline molecules can be reasonably assumed to have $\bar{C}_{20} >> \bar{C}_{22}$ and we may also assume $\bar{u}_{200} > \bar{u}_{202} \cong \bar{u}_{220} > \bar{u}_{222}$, so that the coefficients a and b can be written to good approximation as,

$$a = -\bar{u}_{200}\;\bar{C}_{20}\;/kT \qquad (13)$$

$$b = -2u_{202}\;\bar{C}_{20}\;/kT \qquad (14)$$

The ratio

$$\lambda = b/2a = \bar{u}_{202}\;/\;\bar{u}_{200} \qquad (15)$$

and should be independent of temperature. For a particular solute-solvent mixture, at low solute concentration, the variation of $S_{xx} - S_{yy}$ with S_{zz} is entirely dependent on λ. This means that at any particular temperature a measurement of $(S_{xx} - S_{yy})/S_{zz}$ is sufficient to determine λ.

The data shown in Figs. (2) and (5) has been used to test the prediction of the theory that the variation of $S_{xx} - S_{yy}$ with S_{zz} is determined by a single value of λ. Curves of constant λ are shown in both figures and clearly the prediction is followed with reasonable precision for PDNB in Phase 5, but there are noticeable deviations for this solute from the curves for E5 and HAB. The data for anthracene shown in Fig. (5) show

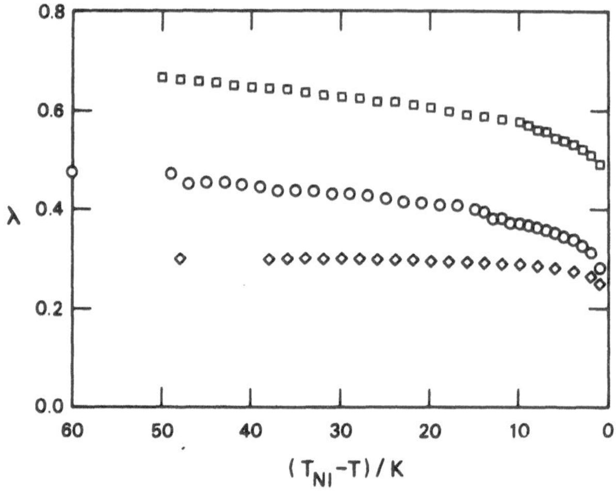

Figure 6. *Temperature dependence of λ for anthracene in E9 (⊙), Phase 5 (◈) and ZLI 1167 (⊡).*

much larger deviations, which is seen more clearly in Fig. (6) where we show the temperature dependence of λ for anthracene in each mixture. The temperature dependence of λ is small and could arise from a variety of sources, such as the neglect of solvent biaxiality and flexibility, the truncation of the expansion of $U(\omega_2)$, or perhaps more simply that the data are collected at constant pressure rather than constant volume.

The most striking feature shown by the data in Figs. (2) and (5) is the strong dependence of λ on solvent. This result conflicts with dispersive forces making the dominant contribution to $U(\omega_2)$. In this case the coefficients $\bar{u}_{2kk'}$ are related to the anisotropic electric polarisability tensors of solute $\alpha_2^{(2,k')}$ and solvent $\alpha_1^{(2,k)}$, thus

$$\bar{u}_{2kk'} = B_{12}\,\alpha_1^{(2,k)}\alpha_2^{(2,k')} \tag{16}$$

where B_{12} depends on the radial average of r_{12}^{-6}. The value of λ is now given exactly by,

$$\lambda = \alpha_2^{(2,2)} \Big/ \alpha_2^{(2,0)}$$

and depends only on the anisotropic polarisability of the solute. Clearly, therefore, the solute-solvent interactions for PDNB and anthracene in the liquid crystals studied cannot all be attributed to the dispersion interaction. This behaviour has been found for other solute-solvent systems by Catalano, Forte, Veracini and Zannoni (9). Having established that not all solutes interact with their solvents via dispersive forces does not of course rule out the possibility that such an interaction is dominant for some systems. To shed further light on this problem it will be necessary to study a wider range of solute-solvent systems. Those systems studied so far have a different λ for each solute-solvent pair!

NON-RIGID MOLECULES

We saw from the last section that the main impetus for studying the orientational ordering of solutes was that they could be chosen to be rigid. As soon as this restriction is no longer imposed then there is no reason to confine our attention to solutes and hence we will begin our discussion by considering the simpler system of a non-rigid, single component liquid crystal phase. There are two methods of approaching the problem of calculating the orientational ordering, or of any property which depends on ordering. The properties measured are usually an average over all the molecules when the system has reached thermal equilibrium. The averages can be evaluated by taking an ensemble average, in which case the problem is one of identifying the potential of mean torque. This approach of using equilibrium statistical mechanics has been described in detail by Emsley and Luckhurst (1) and is the one to be described here; an alternative approach to this problem has been described by Burnell and de Lange (10,11).

In the general case of a continuous variation of both internal, as described by a set of variables X, and external, as described by ω, coordinates the average value of a second rank tensor component along the director is,

$$\tilde{A}_{\parallel} = Z^{-1} \int dX\, d\omega A_{\parallel}(X,\omega) \exp[-U(X,\omega)/kT] \tag{18}$$

where

$$Z = \int dX\, d\omega \exp[-U(X,\omega)/kT]. \tag{19}$$

The potential of mean torque, $U(X,\omega)$, now depends upon both internal and external coordinates. Transforming from a director to molecular frame gives

$$A_{\|}(X,\omega) = \sum_{\alpha\beta}^{abc} A_{\alpha\beta}(X)\, \ell_\alpha \ell_\beta \;; \tag{20}$$

A(X) in the abc frame depends on the conformation of the molecule and ℓ_α, ℓ_β are the direction cosines of the director in this frame. Substitution of Eq. (20) into (18) gives,

$$\tilde{A}_{\|} = Z^{-1} \sum_{\alpha\beta}^{abc} \int dX\, d\omega\, A_{\alpha\beta}(X)\, \ell_\alpha \ell_\beta \exp[-U(X,\omega)/kT] \tag{21}$$

The evaluation of partial averages of second rank interactions using Eq. (21) is straightforward if $U(X,\omega)$ is known. In practice, however, it has proved useful in constructing models for $U(X,\omega)$ to make a division of this potential into $U_{int}(X)$, a part dependent only on the internal coordinates, and $U_{ext}(X,\omega)$ which depends on both X and ω,

$$U(X,\omega) = U_{int}(X) + U_{ext}(X,\omega). \tag{22}$$

This division of $U(X,\omega)$ simply states that the potential energies for internal rotation are to be considered as independent of the molecular orientation. It does not mean that P(X), the probability that the molecule is in a conformation with internal coordinates X irrespective of its orientation is independent of orientational ordering. This important point will be discussed later. Using Eq. (22) in (21) gives,

$$\tilde{A}_{\|} = Z^{-1} \sum_{\alpha\beta}^{abc} \int dX\, d\omega\, A_{\alpha\beta}(X)\, \ell_\alpha \ell_\beta \exp[-U_{int}(X)/kT] \cdot$$
$$\exp[-U_{ext}(X,\omega)/kT], \tag{23}$$

and defining an order parameter $\overline{\ell_\alpha \ell_\beta}\,(X)$ as,

$$\overline{\ell_\alpha \ell_\beta}(X) = Q(X)^{-1} \int d\omega\, \ell_\alpha \ell_\beta \exp[-U_{ext}(X,\omega)/kT], \tag{24}$$

where

$$Q(X) = \int d\omega \exp[-U_{ext}(X,\omega)]/kT \tag{25}$$

and

$$Z = \int dX\, Q(X) \exp[-U_{int}(X)/kT] \tag{26}$$

We now obtain $\tilde{A}_{\parallel}$ as,

$$\tilde{A}_{\parallel} = Z^{-1} \sum_{\alpha\beta}^{abc} \int dX\, A_{\alpha\beta}(X)\, \overline{\ell_\alpha \ell_\beta}\,(X) Q(X)\, \exp[-U_{int}(X)/kT] \quad (27)$$

This relationship can be compressed by noting that P(X) is

$$P(X) = Z^{-1}\, \exp[-U_{int}(X)/kT] \int d\omega\, \exp[-U_{ext}(X,\omega)/kT)] \quad (28)$$

$$= Z^{-1}\, Q(X)\, \exp[-U_{int}(X)/kT], \quad (29)$$

so that,

$$\tilde{A}_{\parallel} = \sum_{\alpha\beta}^{abc} \int dX\, P(X)\, A_{\alpha\beta}(X)\, \overline{\ell_\alpha \ell_\beta}(X) \;. \quad (30)$$

Eq. (30) is a general relationship for $\tilde{A}_{\parallel}$ which does not depend on making a particular division between internal and external coordinates and is entirely equivalent to Eq. (21). The reason for preferring to express $\tilde{A}_{\parallel}$ through P(X) and $\overline{\ell_\alpha \ell_\beta}(X)$ becomes apparent when particular molecular systems are considered. Thus, for a molecule which exists in n discrete conformations the integral over X can be replaced by a summation over n, and we obtain,

$$\tilde{A}_{\parallel} = A_0 + \tfrac{2}{3} \sum_n P_n \sum_{\alpha\beta}^{abc} S_{\alpha\beta}^{n}\, A_{\alpha\beta}^{n} \;. \quad (31)$$

The $S_{\alpha\beta}^{n}$ are elements of an ordering matrix for a molecule in the nth conformation with a normalised probability P_n.

Location of Axes

There is a special case of the use of Eq. (31) which concerns the rotation of molecular groups about a symmetry axis. We illustrate this with the example of anisole (see over):

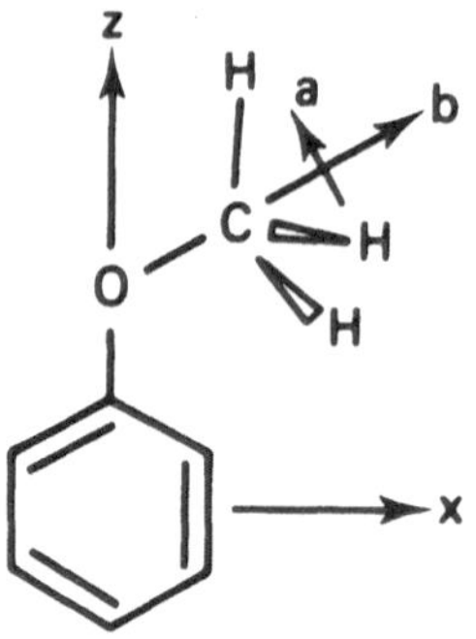

Consider first the $O-CH_3$ group and axes (abc). Rotation of the methyl protons about b is between symmetry related positions, and clearly this operation leaves the shape of the molecule unchanged. Similarly, rotation of the phenyl group about z is between identical forms of the molecule. Thus, although there are six rotational forms they are all identical and share a single ordering matrix and have equal statistical weights. In this case we can regard anisole as a rigid body with a single ordering matrix with three non-zero elements:

$$S_{aa},\ S_{bb}-S_{cc} \text{ and } S_{ab}.$$

Now, we may also locate axes in the phenyl ring, xyz and obtain S_{zz}, $S_{xx} - S_{yy}$ etc. In this case the rotations about b are still between forms with identical ordering matrices, but this is no longer true for the rotation about z. The rotation is relative to a fixed ring and hence has the effect of moving the OCH_3 group between two positions: A and B, (see over):

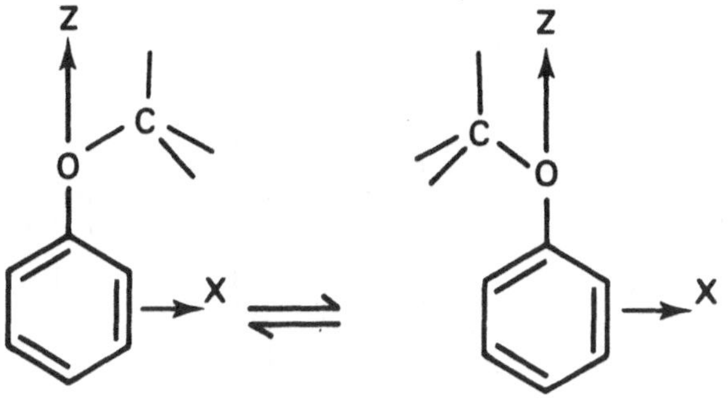

A B

These have different shapes so that $S^A \neq S^B$. However, $p^A = p^B$ and the two forms are related by an yz reflection plane. This means that

$$S^A_{zz} = S^B_{zz}$$

$$S^A_{xx} = S^B_{xx}$$

$$S^A_{yy} = S^B_{yy}$$

$$S^A_{xz} = -S^B_{xz}$$

Thus the two matrices are described by the magnitude of three numbers and their principal axes are located at angles of α and -α relative to a rotation about y. It is important, therefore, to choose axes with care when describing the ordering matrices of molecules undergoing internal rotations.

Local Ordering Matrices

In molecules which have several degrees of internal rotational freedom it is useful to introduce the local ordering matrix S^L to describe individual rigid groups. The location of the axes which diagonalize S^L depends on the symmetry of the rigid group and of the potential for internal rotation. Thus,

$$S^L_{\alpha\beta} = \sum_n P_n S^n_{\alpha\beta} \tag{32}$$

where S_n, as before, is the ordering matrix for the nth conformation and is referred to the same axes as S^L.

$$S_{xz}^n \qquad S_{xz}^n \qquad -S_{xz}^n \qquad -S_{xz}^n$$

$$S_{xy}^n \qquad -S_{xy}^n \qquad S_{xy}^n \qquad -S_{xy}^n$$

$$S_{yz}^n \qquad -S_{yz}^n \qquad -S_{yz}^n \qquad S_{yz}^n$$

$$(1) \qquad\qquad (2) \qquad\qquad (3) \qquad\qquad (4)$$

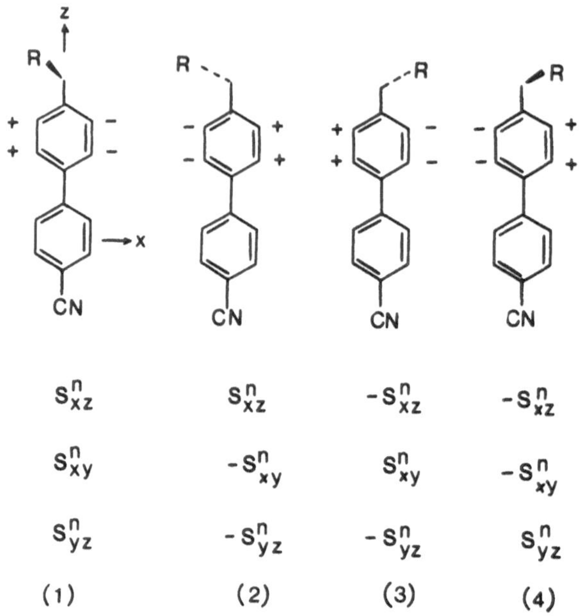

Figure 7. *The four possible conformations of a 4-n-alkyl-4'-cyanobiphenyl which have the alkyl chain, represented by -R in a fixed structural arrangement and the orientation of the second phenyl ring changes between four equivalent positions, generated by hindered rotation about z. The values of $S_{\alpha\beta}^n$ for these conformations are related to one another as shown.*

Let us now consider the local ordering matrices in a liquid crystal such as an alkylcyanobiphenyl. In Fig. (7) is shown the four possible conformations produced by 180° ring flips, and a 180° rotation of the alkyl chain, in both cases about the z-axis. The local order matrix for the ring A is diagonal in the xyz frame even though not one of the S^n is diagonal in this same frame. The same is true for ring B if xyz were located in this ring in the same way. Now let us consider a $-CH_2-$ or $-CD_2-$ group in R. As shown in Fig. (8), the only element of symmetry which can exist in a CD_2 group in an alkyl chain attached at one end to a phenyl, or some other group which destroys the chain symmetry, is a reflection plane. In this case the local ordering matrix has three non-zero elements S_{zz}, $S_{xx} - S_{yy}$ and S_{xz}. No one has yet succeeded in measuring three independent magnetic

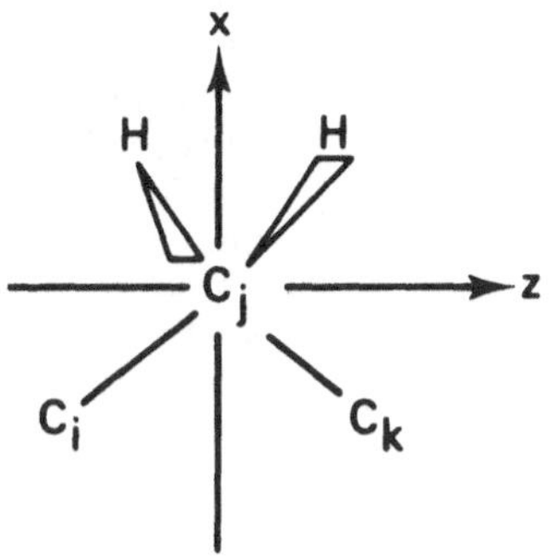

Figure 8. *A CH_2 group in an alkyl chain RX. Carbons i, j and k define a plane (xz) which is perpendicular to the CH_2 plane (xy).*

interactions for a CH_2 or CD_2 group in a chain. For the phenyl rings there have been several studies of local ordering matrices (12-16). These parallel the studies on solutes discussed earlier in that either deuteron quadrupolar or inter nuclear dipolar interactions are measured for nuclei in the rigid fragment. These kinds of experiment are summarised as follows.

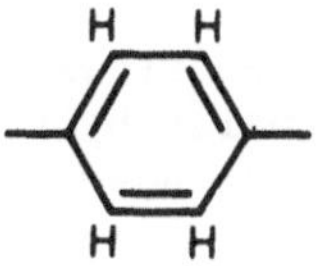

The remainder of the molecule is deuteriated and proton-deuteron dipolar couplings removed by either ^1H-^2H decoupling (12) or spin echo refocussing (17). The dipolar couplings provide the most precise measure yet of the local ordering matrix.

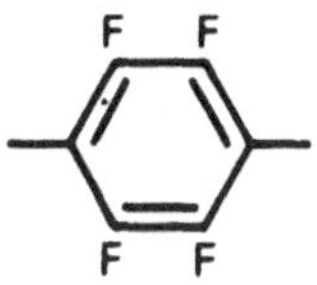

In this case the ^1H-^{19}F dipolar couplings are removed by spin-echo refocussing (15) to enable the inter fluorine dipolar couplings to be measured.

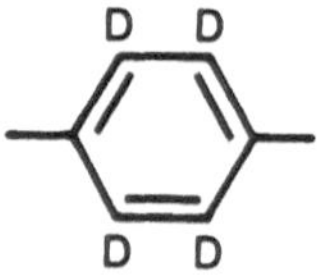

In favourable cases both ^2H-^2H dipolar and ^2H quadrupolar interaction can be obtained. Only the ortho deuterons couple sufficiently strongly to give a resolvable splitting in the deuteron spectrum. (16,18). The C-D vectors are, unfortunately, at an unfavourable angle (~60°) to the most ordered axis. This means that although S_{zz} is determined with reasonable precision the uncertainty in the geometry leads to very large percentage errors in $S_{xx} - S_{yy}$. Thus, it is difficult if not impossible to differentiate $S_{xx} - S_{yy}$ from zero.

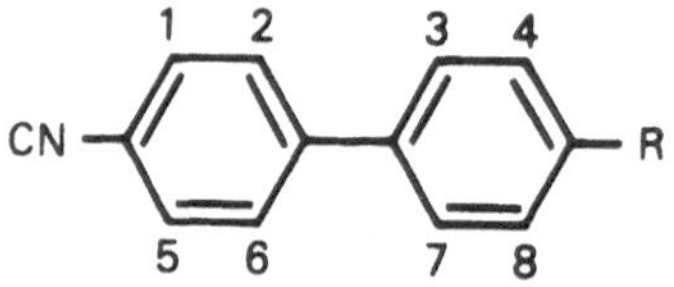

The spectrum from the eight protons in such compounds, with R deuteriated, can yield both structure and ordering information. The proton spectrum from 4-n-octyl-d_{17}-4'-cyanobiphenyl is shown in Fig. 9. Analysis of this spectrum has so far proved impossible, although it is a relatively simple task to extract approximate values of $(D_{12} + D_{34})/2$ and D_{23}, which is averaged by internal motion such that $D_{23} = D_{27} = D_{67} = D_{36}$. Making the assumption that the molecule is like a symmetrical 4,4' disubstituted biphenyl and is cylindrically symmetric gives a value of the inter-ring angle of approximately 30°.

Much more detailed information can be obtained from multiple quantum proton spectra (13). In this elegant experiment the spectral simplification is achieved principally by studying spectra with $\Delta m > 1$, examples of which are given in the chapter by Drobny.

The Ordering of Alkyl Chains

As discussed previously the local ordering matrices for $-CD_2-$ groups have not been determined by any technique of measurement. Deuteron resonance has, however, provided a wealth of data on the variation of orientational ordering in alkyl chains which can be used to provide critical tests of theories of orientational ordering in flexible molecules. Before discussing the theories we discuss first some general features of the spectra. Fig. (10) shows the deuteron spectrum of 4-n-pentyl-d_{11}-4'-cyanobiphenyl-d_4 (5CB-d_{15}). The first point to note is that assignment of the four CD_2 groups cannot be done by examination of this spectrum. The best method of assignment is the synthesis of specifically labelled molecules, but this is a difficult task and has been achieved for only a limited number

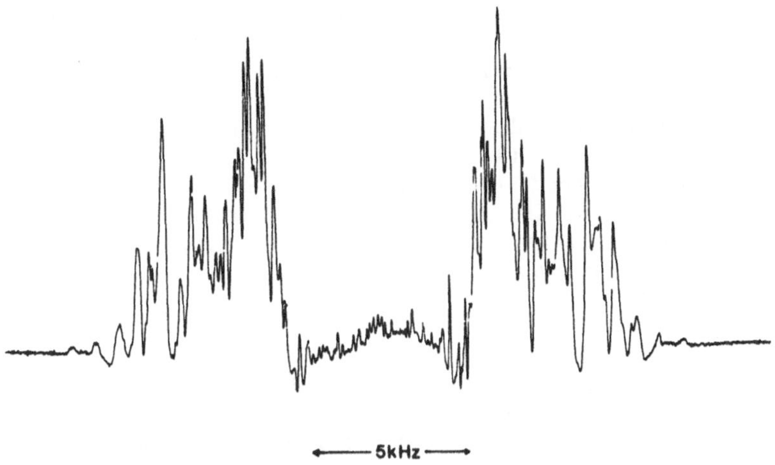

Figure 9. *Proton spectrum at 200 MHz of 4-n-pentyl-d_{11}-4'-cyano-biphenyl. Convolution difference techniques have been used to improve the resolution.*

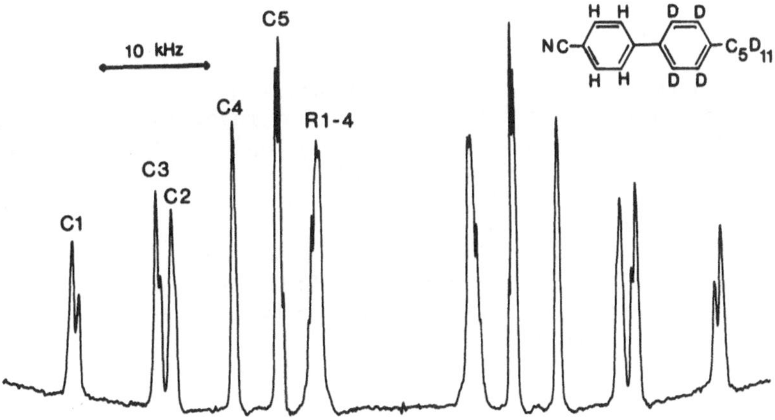

Figure 10. *30.7 MHz deuteron spectrum of 5CB-d_{15}. The peaks are labelled according to their assignment to sites in the molecule.*

of mesogens (19-21), and not for 5CB. The assignment for this molecule was achieved by a spin-echo decoupling experiment on 4-n-pentyl-d_{11}-4'-cyanobiphenyl (5CB-d_{11}). The CD_2 groups are broadened by unresolved dipolar coupling to the protons in the phenyl ring and this decreases as the distance of the methylene group from the ring increases (22). The assignment may also be achieved by measuring T_1 values of the deuterons. These relaxation times probably, but not certainly, increase along the chain reaching a maximum at the CD_3 group. A third, spectroscopic method of assignment uses the differences in chemical shift for CD_2 groups in the chain. At field strengths of 11.6T these shifts can be large enough to give an assignment of the quadrupole doublets, either directly by noting the centre point of each doublet, or more elegantly by a 2D experiment (23).

The number of assignments of CD_2 groups in mesogens is small and general conclusions are hard to draw. The assumption often made in the absence of an assignment that the quadrupolar splittings should decrease monotonically along an alkyl chain is certainly not universally true. This kind of behaviour has been definitely established in only one mesogen, 4-n-pentyloxy-benzylidene-4'-n-heptylaniline (21), and an alternation in magnitude of the splittings at some sites in the chain appears to be more common (19,20).

The assignment problem becomes even more acute when two deuteriated chains are present, which are difficult if not impossible to deuteriate separately. It is, however, possible in some cases to assign the resonances to the separate chains using a 2D auto correlation experiment. The experiment is a generalisation of selective population inversion and gives the pattern of coupled energy levels in a spin system. Its use for liquid crystal samples has been discussed by Emsley and Turner (24) who demonstrated its application with the deuteron spectrum from perdeuteriated para-azoxyanisole (PAA-d_{14}), whose structure and spectrum are shown in Fig. (11). The peaks A, A', B, B' and C, C' certainly arise from the aromatic deuterons whilst D, D' and E, E' are from the two nonequivalent OCD_3 groups. The assignment of the peaks to specific deuterons was solved by examination of the 2D spectrum obtained by recording free induction decays in the period t_2 at different values of the pulse spacing t_1, (see over):

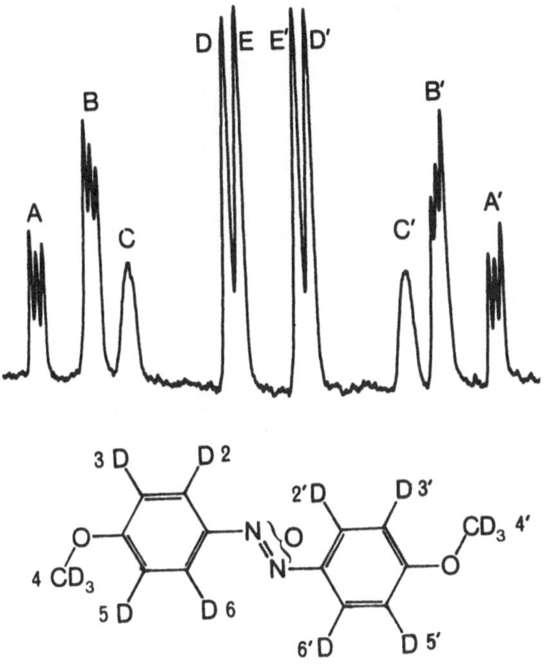

Figure 11. *The deuteron spectrum of PAA-d_{14} at 390K and 30.7 MHz (24).*

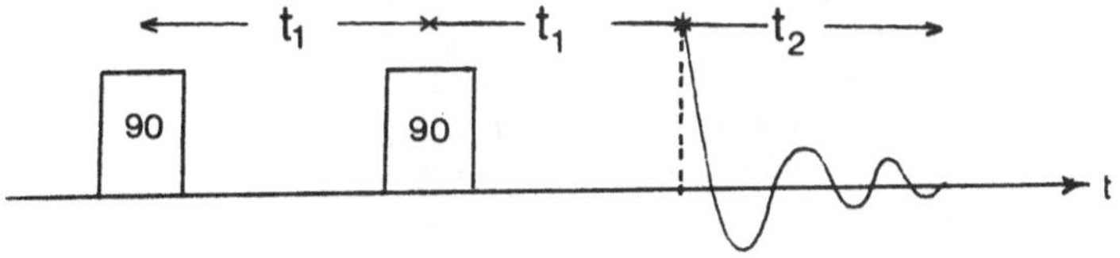

The signal $S(t_1, t_2)$ was Fourier transformed first with respect to t_2 and then t_1 to give the 2D spectrum shown in Fig. (12). The peaks along the section at $F_1 = 0$ and along the 45 degree section from top left to bottom right simply show the connections between the two halves of the quadrupolar doublets and hence contain the same information as the spectrum in Fig. (11). The remaining peaks, however, reveal the deuterons which

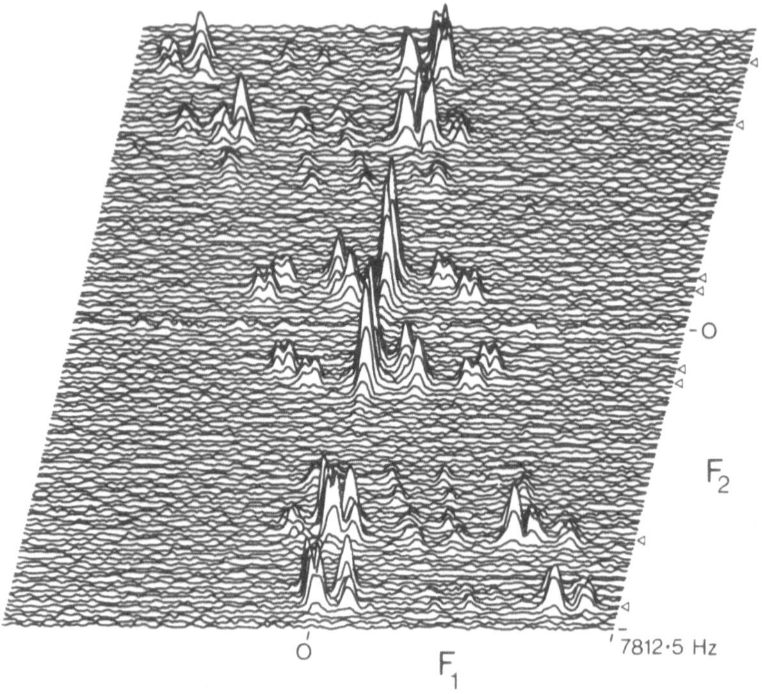

Figure 12. *The auto-correlation deuteron spectrum from PAA-d_{14} (24).*

are significantly dipolar coupled. This is more evident from the sections taken parallel to the F_1 axis at the points marked on Fig. (12), and shown in Fig. (13). Peaks in these cross-sections arise when the deuterons are dipolar coupled and their intensity reflects the magnitude of the coupling. The 2D spectrum unambiguously assigns the peaks in Fig. (11) to be:

A, A', B, B' (one half), E, E' from one anisole fragment, and
B, B' (one half), C, C', D, D' from the other.

The method was also applied to the much more complex spectrum given by 4,4'-di-n-octyloxyazoxybenzene-d_{42} whose deuteron spectrum is shown in Fig. (14). The 2D auto-correlation spectrum is now very complex and a contour plot is shown in Fig. (15) with the contour level chosen to reveal the off-diagonal peaks

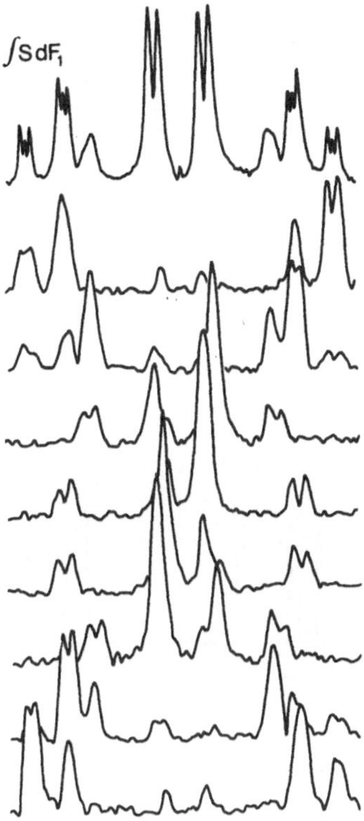

Figure 13. *Cross-sections from Fig. (12) taken parallel to the F_1 axis at the positions on F_2 marked by* ∇.

corresponding to the dipolar coupling between deuterons at chain position 6 and 8. It was found, in fact, that the dipolar coupling along each of the alkyl chains was larger between deuterons separated by four rather than three bonds. The 2D spectrum enabled the peaks to be assigned as shown in Fig. (14) and to confirm the assignment within the chains found by Boden et al (25) by synthesis of specifically labelled compounds and also to assign the peaks to a specific end chain (26).

Comparison With Theory

The number of conformations adopted by alkyl chains can become very large, thus in the five membered chain of 5CB there are 27 conformations and 729 for the chain in 8CB. Clearly to make any progress in understanding the orientational ordering in such

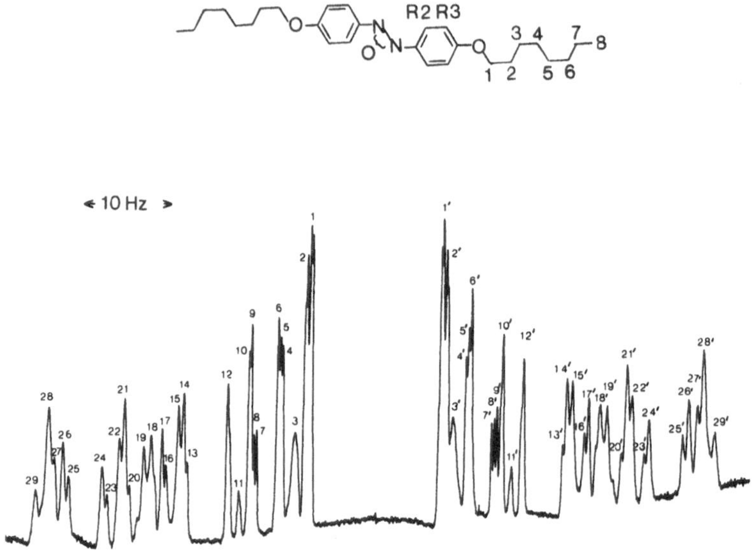

Figure 14. *Structure and deuteron spectrum of 4,4'-di-n-octyloxyazoxybenzene (26).*

mesogens via Eq. (31) requires a simple method of constructing $U_{int}(n)$ and $U_{ext}(n,\omega)$. The first attempt to do this was by Marcelja (27) whose basic ideas were developed later by our own group in Southampton (28,29), and are discussed by Luckhurst in chapter 3, and here we describe in more detail the results of calculations on the n-alkyl- and n-alkyloxycyanobiphenyls. Here the aim was to use the qudrupolar splittings, and hence S_{CD}, the order parameters of the C-D bonds in the alkyl chains, to determine the interaction parameters X_a and X_{cc} which define the interaction tensors $\underset{\sim}{\varepsilon}^n$ and hence the potential of mean torque. It was in fact found necessary to introduce an additional parameter X_{cd} for each C-D bond in the chain before very close agreement could be achieved for the compounds 5CB, 8CB, 6OCB and 8OCB. The interaction tensor elements $\varepsilon^n_{2,m}$ for the nth conformation are now:

$$\varepsilon^{n}_{2,m} = X_a \delta_{\theta m} + X_{cc} \sum_{j=1}^{N} C_{2,m}(\omega^{j}_{CC}) + X_{cd} \sum_{j=1}^{N-1} (C_{2,m}(\omega^{j}_{CD1})$$

$$+ C_{2,m}(\omega^{j}_{CD2})) + 3X_{cd}\, C_{2,m}(\omega^{N}_{CC})\, P_2 (\cos \phi) \;, \tag{33}$$

where ϕ is the CCD angle in the methyl group. For 5CB and 6OCB the calculated and observed quadrupolar splittings are in virtually exact agreement, whilst for 8CB and 8OCB there are small deviations between the two sets, as illustrated for 8CB in Fig. (16).

The information yielded by an analysis of this kind can be summarised in the following way.

a) The very close agreement achieved between experiment and theory is strong support for the assumptions made in constructing both $U_{int}(n)$ and $U_{ext}(n,\omega)$. The calculations are in fact more constrained than might appear to be the case in that the geometry of the chain is sensibly restricted by choosing only reasonable values of bond lengths and angles. The calculations on 5CB for example show good agreement only for the correct assignment of the similar splittings at carbon positions 2 and 3.

b) The temperature variation of X_a, X_{cc} and X_{cd} may be obtained and used to test mean field theories of these quantities (27, 29). In principal this could provide a link between the quantities measured by NMR, which can be used to test theories in which the temperature dependence of the interaction tensors need not be considered, and properties such as T_{NI} the nematic-isotropic transition temperature, the entropy of transition and the value of the ordering parameters at T_{NI}, in which temperature plays a central role.

c) The calculations yield values of S^n for all the molecular conformations, and it is noteworthy that both the magnitude of the biaxiality in S^n and the location of their principal axes vary considerably with conformational state.

d) The conformational distributions are predicted. At first sight it might seem that by fixing E_{tg} the conformation distribution is fixed (neglecting the small effects of $E_{g\pm g\mp}$), but in fact P_n is dependent on the orientational ordering. This important phenomenon arises because P_n depends upon both $U_{int}(n)$ and $U_{ext}(n,\omega)$, which can be seen by extending Eq. (28) to the case of discrete conformations,

$$P_n = Z^{-1} [\exp\{-U_{int}(n)/kT\}] \int d\omega \exp \{-U_{ext} (n,\omega)/kT\}. \tag{34}$$

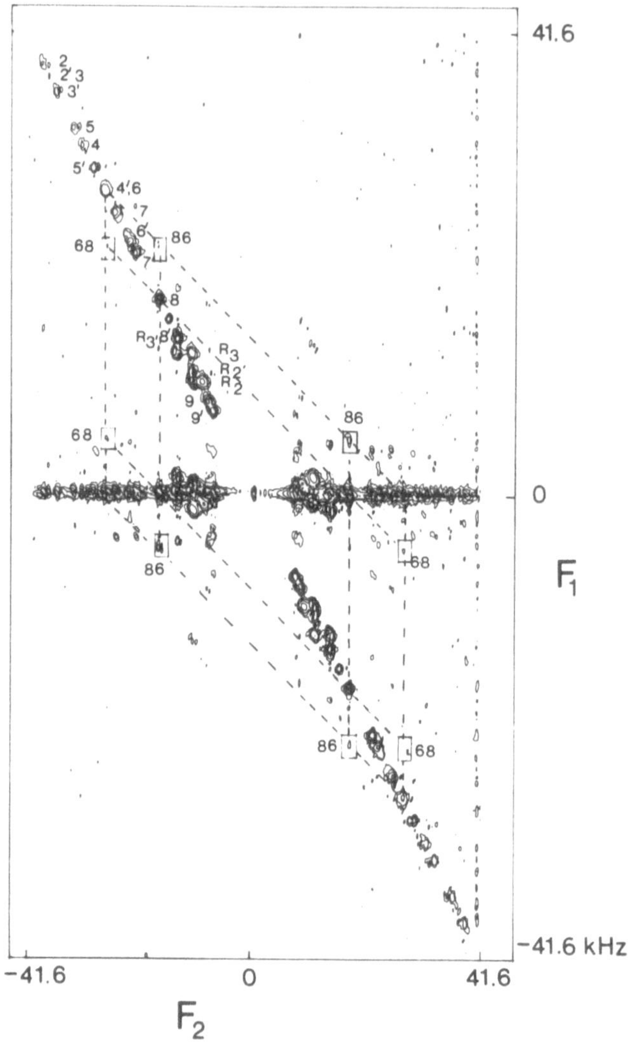

Figure 15. *Auto-correlation deuteron spectrum from 4,4'-di-n-octyloxyazoxybenzene showing the peaks connecting deuteron resonances from position 6 and 8 in one alkyl chain (26).*

The predicted changes in P_n for the conformations of the alkyl chains are appreciable (see chapter 3).

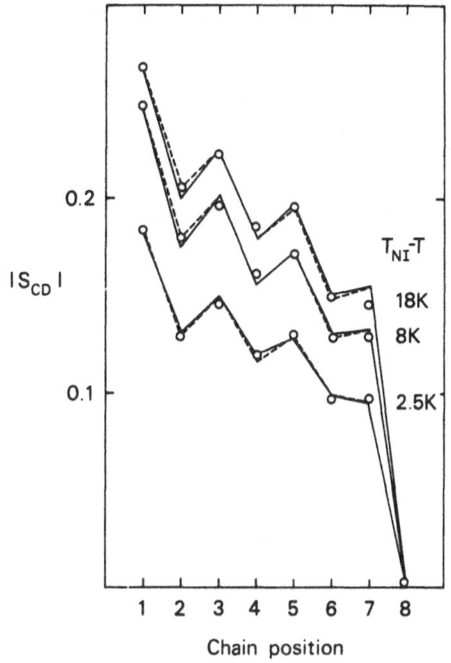

Figure 16. *Calculated S_{CDk} values with the potential of mean torque given by Eq. (33) and E_{tg} varied (---) or fixed at 4.3kJ mol^{-1} (——). The values of S_{CD} (open circles) derived from the observed quadrupolar splittings are shown for each of three temperatures (29).*

Non-rigid Solutes

The potential of mean torque for a non-rigid solute in a uniaxial liquid crystal solvent has the same form as that for a non-rigid pure phase, except that the interaction tensors are now describing solute-solvent interactions. Thus, Eq. (33) is still the form of $\varepsilon^{n}_{2,m}$, but X_a, X_{cc} and X_{cd} are replaced by X^s_a, X^s_{cc}, X^s_{cd}, the superscript s denoting that we are concerned with a potential of mean torque for a solute at low dilution (30).

We have tested this theory for alkanes dissolved in uniaxial solvents, again using deuteron quadrupolar splittings to derive values of S_{CD} which are then used to derive the interaction parameters. For alkanes X^{s}_{cc} and X^{s}_{cd} are the only parameters specifying the potential of mean torque, and in Fig. (17) we show the agreement obtained between theory and experiment for n-octane-d_{18} dissolved in the liquid crystal phase of Phase 5. The agreement is good but not exact, indicating that we do not yet fully understand how to predict either the conformational distribution in an alkane or the potential of mean torque. It will be interesting to extend calculations of this type to longer alkyl chains, and to simulate their behaviour in amphiphillic systems, such as model membranes, by restricting their movement at one end. This is

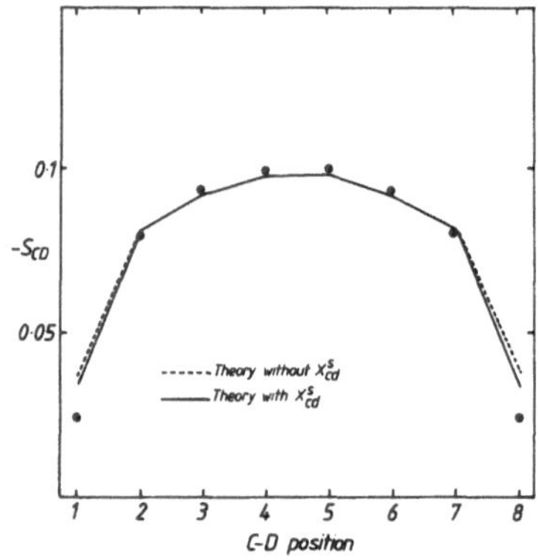

Figure 17. *Calculated and observed (●) values of S_{CD} for n-octane-d_{18} dissolved in Phase 5.*

done simply by having an X^{s}_{a} value for the first segment. Preliminary calculations of this kind, which set X^{s}_{cd} to zero and treat a nine-membered chain have been carried out (31) and give the order parameter profiles shown in Fig. (18). It is seen that the profile is very sensitive to the ratio $\lambda_{cc} = X^{s}_{cc}/X^{s}_{a}$, and the familiar "plateaux" profile can be reproduced without imposing any extra constraints on the conformations that can be adopted by the alkyl chain.

CARBON-13 STUDIES OF ORIENTATIONAL ORDER

The proton decoupled C-13 spectra of liquid crystals provide a chemical shift for each resolvable resonance. Writing δ_i as the difference in values of $\sigma_{\parallel}^{i}$, the component along the director of the shielding tensor for the carbon at the ith site, and the

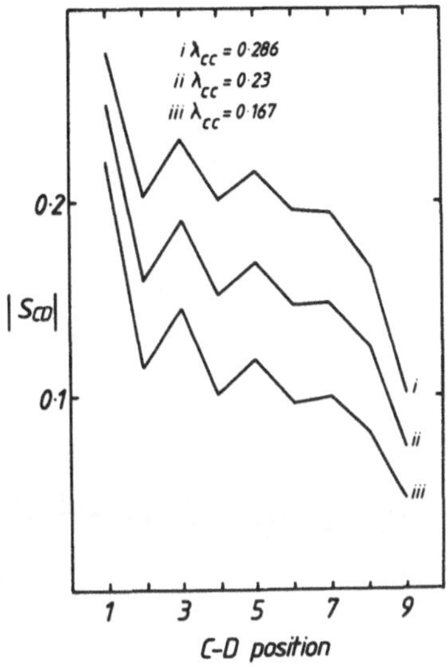

Figure 18. *Predicted variation of* S_{CD} *along a nine-membered alkyl chain attached to an axially symmetric head group whose symmetry axis is parallel to the first C-C bond.* $\lambda_{cc} = X^s_{cc}/X^s_a$ *(31).*

isotropic average shift, then

$$\sigma_i = \tfrac{2}{3} \sum_{\alpha\beta} S^i_{\alpha\beta}\, \sigma^i_{\alpha\beta} \;. \qquad (35)$$

$S^i_{\alpha\beta}$ is a local ordering matrix with respect to axes fixed in the rigid sub-unit containing the ith nucleus, and $\sigma^i_{\alpha\beta}$ are components of σ^i in this frame. The number of values of σ^i that can be measured for a typical mesogen is large, but unfortunately the $\sigma^i_{\alpha\beta}$ are unknown for most nuclei in most molecules. However, the values of δ_i can be used to monitor general phase behaviour and there are considerable advantages in studying a nucleus whose resonance can be observed at natural abundance. The first example of a study of this kind was by Pines and Chang (32) who monitored the ordering at T_{NI} of a number of di-n-alkyloxyazoxybenzenes.

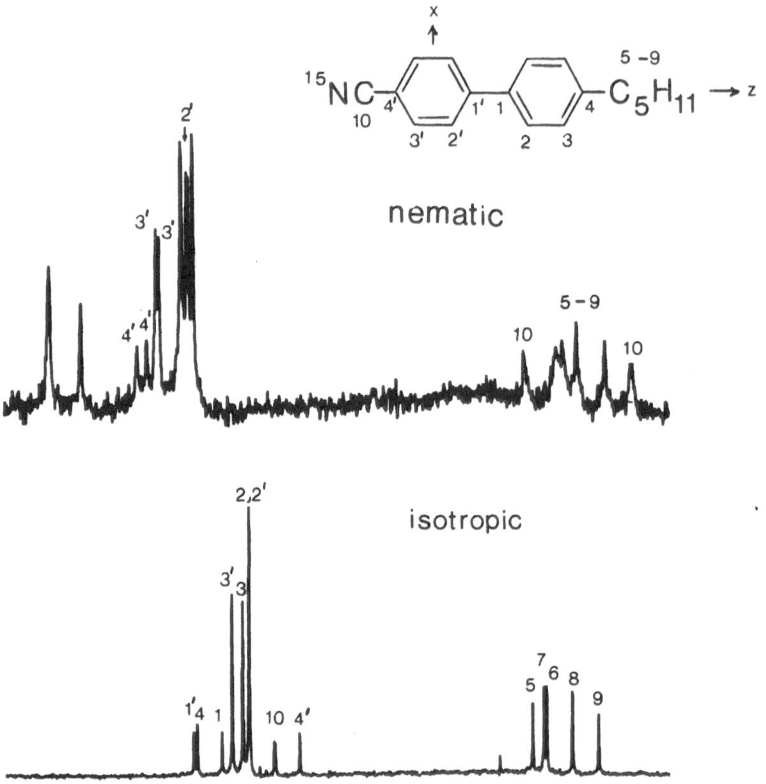

Figure 19. *Carbon-13 proton decoupled spectra of 5CB-N-15 in the nematic and isotropic phases.*

Carbon-13 spectra are normally recorded with complete proton spin decoupling, and it is extremely useful to incorporate a third kind of magnetic nucleus into the molecule so that the spectrum shows C-13 resonances split by dipolar coupling. The dipolar couplings obtained from such spectra can be used to monitor structure or orientational ordering (32). For example, Fig. (19) shows the C-13 spectrum of 5CB labelled with ^{15}N. In this case the dipolar coupling to the nitrile carbon measures the order parameter S_{zz} (see Fig. (19) for location of the axes). Now, the value of δ_{CN} is,

$$\delta_{CN} = \tfrac{2}{3} S_{zz} \{\sigma_{zz}^{CN} - \tfrac{1}{2} (\sigma_{xx}^{CN} + \sigma_{yy}^{CN})\} + \tfrac{1}{3}(S_{xx} - S_{yy})(\sigma_{xx}^{CN} - \sigma_{yy}^{CN}) \quad (36)$$

where σ^{CN} is the shielding tensor for the nitrile carbon, and S is the local ordering matrix for the phenyl ring substituted by the CN group; xyz are principal axes in this case for S.

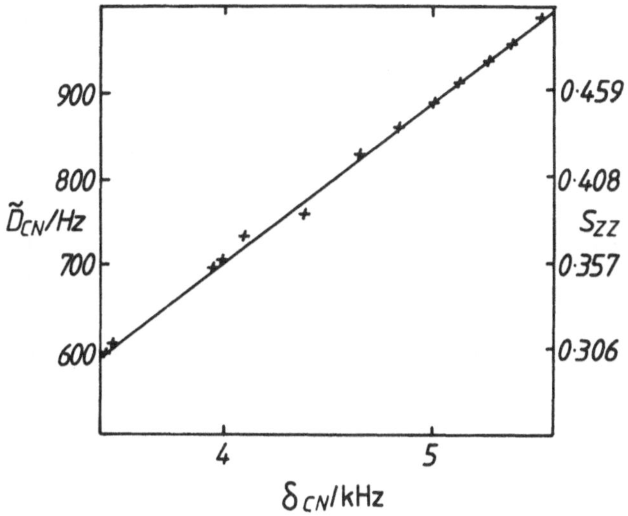

Figure 20. *Variation of* S_{zz} *with* δ_{CN} *for 5CB.*

It is found experimentally that δ_{CN}/S_{zz} is constant throughout the nematic phase, as shown in Fig. (20), hence the effect of the term involving $(S_{xx} - S_{yy})$ in Eq. (36) is negligibly small. This means that δ_{CN} can be taken as a direct measure of S_{zz} in 5CB. Moreover, it is reasonable to assume that changing the chain length in a series of alkylcyanobiphenyls will have little effect on σ^{CN} and hence the proportionality between δ_{CN} and S_{zz} can be taken to apply for the whole series. It is therefore possible to use δ_{CN} to monitor the value of S_{zz} in the alkyl, and probably also alkyloxycyanobiphenyl series. It is particularly interesting to compare S_{zz} at T_{NI} for each member of the series as shown in Fig. (21). The calculated values of S_{zz} at T_{NI} shown in Fig. (21) are obtained by the mean field theory developed by Counsell et al (34).

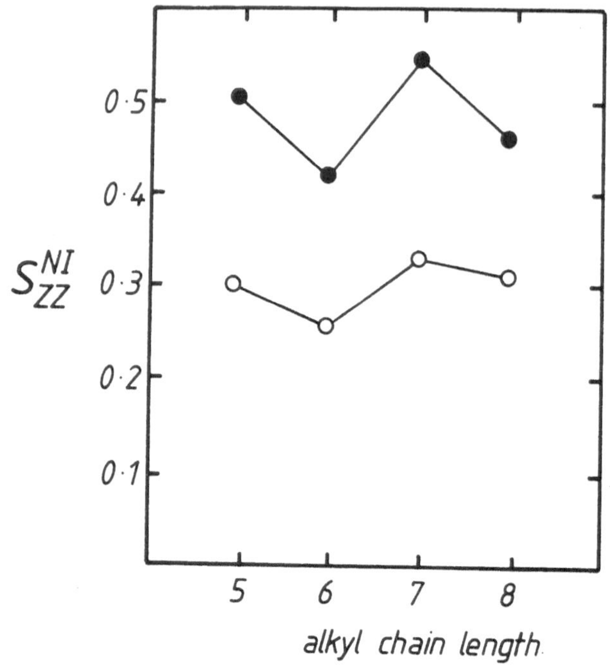

Figure 21. *The order parameter at the nematic-isotropic transition for n-alkylcyanobiphenyls. The observed (○) values are from C-13 measurements whilst those calculated ● are from the mean field theory of Counsell et al (34).*

CONCLUSION

It now is possible to reproduce the order parameter profiles for uniaxial liquid crystals which have only single alkyl chains, and extension of the calculations to include two chains poses no fundamental problems. The immediate challenges before us are to develop theories to explain the magnitude and temperature dependence of the interaction parameters X_a, X_{cc} and X_{cd}, and to obtain more data in order to critically test the existing theory. NMR spectroscopy has the capability to provide large data sets for individual mesogens although involving considerable effort and patience. No other method of studying the molecular aspects of liquid crystal phases has this same capability and our understanding of liquid crystal behaviour has much to gain by a continuing application of NMR spectroscopy to characterizing orientational order.

REFERENCES

1. J.W. Emsley and G.R. Luckhurst, *Mol. Phys.*, **40**, 19 (1980).

2. E.E. Burnell and C.A. de Lange, *J. Chem. Phys.*, **76**, 3474 (1982).

3. J.G. Snijders and C.A. de Lange, *J. Chem. Phys.*, **77**, 5386 (1982).

4. D. Catalano, C. Forte and C.A. Veracini, to be published.

5. J.W. Emsley, G.R. Luckhurst and G.N. Shilstone, unpublished results.

6. D.M. Ellis and J.L. Bjorkstam, *J. Chem. Phys.*, **46**, 4460 (1967).

7. G.R. Luckhurst, C. Zannoni, P.L. Nordio and A. Segre, *Mol. Phys.*, **30**, 1345 (1975).

8. J.W. Emsley, R. Hashim, G.R. Luckhurst, G.N. Rumbles and F. Viloria, *Mol. Phys.*, **49**, 1321 (1983).

9. D. Catalano, C. Forte, C.A. Veracini and C. Zannoni, *Israel J. Chem.*, to be published.

10. E.E. Burnell and C.A. de Lange, *J. Magn. Reson.*, **39**, 461 (1980).

11. E.E. Burnell and C.A. de Lange, *Chem. Phys. Letters*, **76**, 268 (1980).

12. J.W. Emsley, G.R. Luckhurst, G.W. Gray and A. Mosley, *Mol. Phys.*, **35**, 1499, (1978).

13. S. Sinton and A. Pines, *Chem. Phys. Lett.*, **76**, 263.

14. J.W. Emsley, G.R. Luckhurst and C.P. Stockley, *Mol. Phys.*, **44**, 565, (1981).

15. A.G. Avent, J.W. Emsley and G.R. Luckhurst, *Mol. Phys.*, **49**, 737 (1983).

16. P. Diehl and A.S. Tracey, *Mol. Phys.*, **30**, 1917 (1975).

17. A.G. Avent, J.W. Emsley and D.L. Turner, *J. Magn. Reson.*, **52**, 57 (1983).

18. J.W. Emsley, S.K. Khoo and G.R. Luckhurst, *Mol. Phys.*, **37**, 959 (1979).

19. N. Boden, R.J. Bushby and L.D. Clark, *Mol. Phys.*, **38**, 1683.

20. N. Boden, L.D. Clark, R.J. Bushby, J.W. Emsley, G.R. Luckhurst and C.P. Stockley, *Mol. Phys.*, **42**, 565.

21. S. Hsi, H. Zimmermann and Z. Luz, *J. Chem. Phys.*, **69**, 4126.

22. J.W. Emsley and D.L. Turner, *J. Chem. Soc. Faraday 2*, **77**, 1493 (1981).

23. V.W. Miner, P.M. Tyrell and J.H. Prestegard, *J. Magn. Reson.*, **55**, 438 (1983).

24. J.W. Emsley and D.L. Turner, *Chem. Phys. Lett.*, **82**, 447 (1981).

25. N. Boden, L.D. Clark and R.J. Bushby, *Chem. Phys. Lett.*, **64**, 519 (1979).

26. P.A. Beckmann, J.W. Emsley, A.M. Giroud, G.R. Luckhurst and D.L. Turner, to be published.

27. S. Marcelja, *J. Chem. Phys.*, **60**, 3599.

28. J.W. Emsley, G.R. Luckhurst and C.P. Stockley, *Proc. Roy. Soc. London*, **A381**, 117 (1982).

29. C.J.R. Counsell, J.W. Emsley and G.R. Luckhurst, *Mol. Phys.*, submitted for publication.

30. C.J.R. Counsell, J.W. Emsley, G.R. Luckhurst, B.A. Timimi and F.R. Viloria, to be published.

31. C.J.R. Counsell, *Ph.D Thesis*, Southampton (1984).

32. A. Pines and J.J. Chang, *Phys. Rev.*, **A10**, 996 (1974).

33. A. Hohener, L. Muller and R.R. Ernst, *Mol. Phys.*, **38**, 909 (1979).

34. C.J.R. Counsell, J.W. Emsley, G.R. Luckhurst and C.P. Stockley, unpublished results.

DETERMINATION OF BIAXIAL STRUCTURES IN LYOTROPIC MATERIALS BY DEUTERIUM NMR

J. William Doane

Department of Physics and Liquid Crystal Institute,
Kent State University,
Kent, Ohio 44242.

INTRODUCTION

It is an unusual circumstance when NMR can be used to determine the physical shape and dimension of structures which occur in condensed matter. Such is the case, however, in lyotropic materials where aggregates and modulated structures of unique shapes can be obtained (1). Structural information from ^{2}H-NMR comes about because of self-diffusion of an oriented molecule over the structure during the time scale of the measurement (~length of the free induction decay) (2). A diffusing molecule will time average the quadrupole interactions of its deuterium spins which will then reflect the structural shape and dimension. Of course, the size of the structures to be measured are limited to the range, R, a diffusing molecule will, on the average, travel. This range is $R \sim \sqrt{D\tau}$ where D is the diffusion constant and τ the length of the free induction decay. Normally in liquid crystals $D \sim 10^{-6}$ cm^2/sec and $\tau \sim 10^{-4}$ sec in which case $R \sim 1000$ Å. This number may be smaller or larger by roughly an order of magnitude depending upon the system.

Lyotropic phases with aggregates or modulated structures of this dimension are those phases which might be called mediating phases. These are short temperature (or concentration) range phases that are sandwiched in between two more commonly occurring phases, such as, for example, the lamellar and hexagonal phase. One such mediating phase which has been reported to occur between the lamellar and hexagonal phase is the so-called ribbon phase (2) which is illustrated in Fig. 1. The recorded ^{2}H-NMR spectra patterns are shown along with the illustrated structures, and it will be explained how one can

J. W. Emsley (ed.), Nuclear Magnetic Resonance of Liquid Crystals, 413–419.

obtain the actual dimension of the ribbon (the ratio d/r in Fig. 1) from these spectral patterns (2), to an accuracy which is in some ways better than that determined from x-rays (3) or freeze fracture microscopy (4).

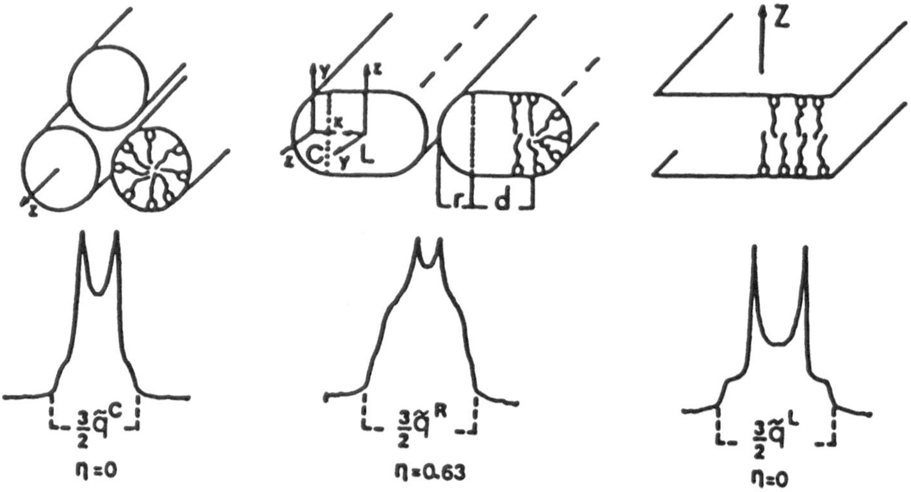

Figure 1. *Schematic view of the cylindrical, ribbon and lamellar aggregates along with their corresponding deuterium NMR spectral patterns recorded for potassium palmitate-d_3 in a mixture with 7 wt% potassium laurate and 30 wt% water.*

Another illustration of such a biaxial structure is the well known ripple phase (P_β phase) reported to occur in some phospholipid bilayer membranes (5). This phase appears in a narrow temperature interval between the lamellar L_α phase (lamellar phase of melted hydrocarbon chains) and the $L_{\beta'}$ phase (lamellar phase of tilted and stiffened hydrocarbon chains). A simple triangular representation of this model along with the spectral pattern for deuteriated water is shown in Figure 2. Like in the previous example, dimensions associated with the ripple can be obtained by NMR (6). From the spectral pattern one can obtain the ratio A/λ (see Fig. 2). From freeze-fracture studies one can determine a value for the periodic length, λ, which when combined with the ^{2}H-NMR measurement one obtains a value for the ripple amplitude, A. Again these measured values appear as reliable as x-rays and certainly more easily obtained (6).

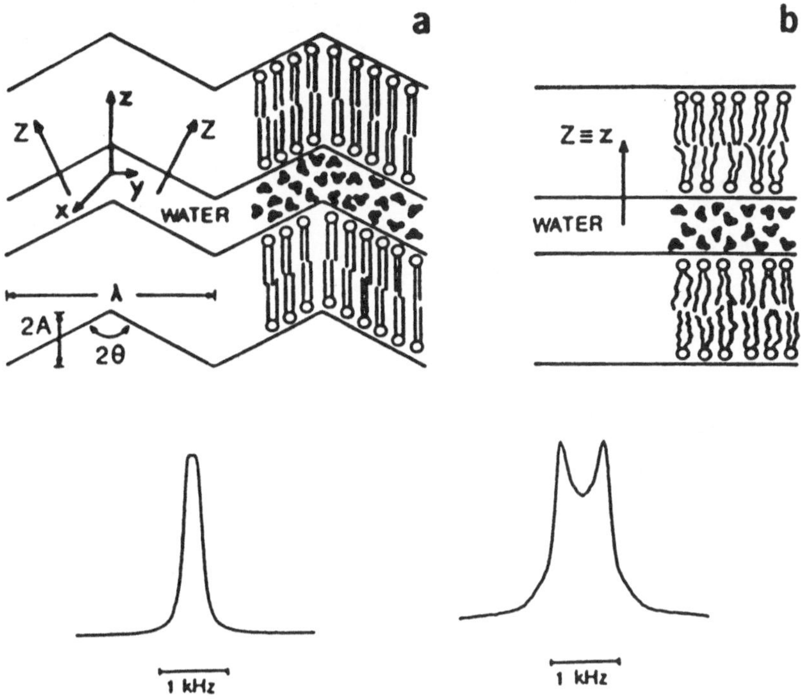

Figure 2. *Schematic view of the ripple and L phase of phosphatidylcholene bilayers with the corresponding deuterium spectral patterns for D_2O.*

ANALYSIS OF THE SPECTRAL PATTERNS

In our analysis we shall assume the fast motion regime. In this regime it is assumed that the molecule and hence deuteriated site has had ample time to sample all relevant orientations of the aggregate or structure during time t. In this case, one has a spectral pattern with a characteristic shape (2) which depends on the value of the time averaged coupling constant $\tilde{q} = eQ\tilde{V}_{zz}/h$ and the motionally induced asymmetry parameter, $\tilde{\eta} = (\tilde{V}_{xx}-\tilde{V}_{yy})/\tilde{V}_{zz}$. Note that ~ above a symbol denotes a tensor component in a space fixed reference frame. One often needs to add a line broadening parameter which can result from any one of a variety of mechanisms, usually in these cases believed to be a non-uniformity in the dimension of the modulated structures.

Following the usual NMR convention of $|\tilde{V}_{xx}| \leq |\tilde{V}_{yy}| < |\tilde{V}_{zz}|$ for the time averaged electric field gradient tensor maintains the condition $0 \leq \tilde{\eta} \leq 1$. These components of the time averaged field gradient tensors can be calculated from the value $\tilde{V}_{ZZ}$ in an unmodulated lamellar phase from the usual transformation equations as (7):

$$\begin{aligned}\tilde{V}_{xx} &= \tilde{V}_{ZZ}(\tfrac{3}{2}\sin^2\theta\cos^2\phi - \tfrac{1}{2})\\ \tilde{V}_{yy} &= \tilde{V}_{ZZ}(\tfrac{3}{2}\sin^2\theta\sin^2\phi - \tfrac{1}{2}) \qquad (1)\\ \tilde{V}_{zz} &= \tilde{V}_{ZZ}(1 - \tfrac{3}{2}\sin^2\theta),\end{aligned}$$

where x,y,z are the principal axes of the time averaged electric field gradient, usually determined by symmetry, and X,Y,Z is the principal axis frame in the lamellar, which is axially symmetric so that $\tilde{V}_{XX} = \tilde{V}_{YY} = -\tilde{V}_{ZZ}/2$. The angles θ and ϕ give the orientation of Z in the x,y,z frame. $\tilde{V}_{ZZ}$ also results from a time average but $\tilde{V}_{xx}$, $\tilde{V}_{yy}$ and $\tilde{V}_{zz}$ are further time averaged by self-diffusion of the deuteriated molecule over the modulated structures or aggregates. I now consider some special cases.

Hexagonal Phase (Elongated Cylindrical Aggregate)

This is the simplest of aggregates from which to calculate $\tilde{V}^C_{zz}$. Defining by symmetry the axis of the cylinder in Figure 1 to be the z axis then $\theta = 90$ in Eq. (1) and self-diffusion of the molecules around the cylinder time averages $\langle\sin^2\phi\rangle$ to a value of 1/2 whereby $\tilde{V}^C_{zz} = -\tilde{V}^L_{ZZ}/2$ or $\tilde{q}^C = -\tilde{q}^L/2$ and $\tilde{\eta} = 0$. Here $\tilde{q}^L$ and $\tilde{q}^C$ are the coupling constants of the lamellar and cylindrical phases respectively. This result occurs if there is no change in the degree of order of the molecule between the lamellar and cylindrical aggregate. Spectral patterns of Fig. 1 show that the splitting between the edge singularities of the uniaxial pattern in the cylindrical are not exactly half that of the lamellar phase indicating that the degree of order is slightly larger for the cylindrical aggregate.

Ribbon Phase

As the molecule diffuses around the cross section of the ribbon there is an exchange in the orientation of the principal axis frame. In this phase we have two possibilities for the resulting time averaged principal axis frame: that of the original lamellar or that associated with the ends of the ribbon which we model as one half of a cylinder. If we let W represent the probability that the molecule is in the lamellar, L, region then we can express the averages as (2):

$$\tilde{V}^{R}_{ii} = W\tilde{V}^{L}_{ii} + (1-W)\tilde{V}^{C}_{ii} \quad (2)$$

for the tensoral components of the ribbon, R, lamellar, L, and cylindrical, C, phases all expressed in the same frame. The result of this time average gives (2):

$$\tilde{q}^{R} = \begin{cases} \frac{3W+1}{4}\tilde{q}^{L} & \text{for } \frac{1}{3} \leq W \leq 1 \\ -\tilde{q}^{L}/2 & \text{for } 0 \leq W \leq \frac{1}{3} \end{cases}$$

$$\tilde{\eta}^{R} = \begin{cases} \frac{3-3W}{1+3W} & \text{for } \frac{1}{3} \leq W \leq 1 \\ 3W & \text{for } 0 \leq W \leq \frac{1}{3} \end{cases} \quad (3)$$

The accuracy of this model is easily checked experimentally as there is a unique relationship between $\tilde{\eta}^{R}$ and $\tilde{q}^{R}$. This relationship has been found to hold remarkably well for deuteriated water and deuteriated lipids (2).

The dimensions of the ribbon have been obtained by relating W to the relative cross-sectional areas of the cylindrical ends and lamellar central region whereby $W = (1+\pi r/2d)^{-1}$. Measurements (2) show d/r to be temperature dependent and vary between a value of one and two.

Ripple Phase

It is interesting that the ripple phase of Figure 2 modulates and hence averages only the water signal (6). The signal from the deuteriated phospholipids are not modulated probably for two reasons: (i) The diffusion rate of a phospholipid is $\sim 10^{-11}$ cm²/sec and too slow to diffuse over the ripple period on the time scale of the NMR measurement; (ii) the lipids are oriented parallel in this structure and their orientation is not modulated by the structure.

As illustrated in Figure 2, diffusion of the water over the bilayer surface causes an exchange in the orientation of Z and hence averages the tensoral components in accordance with Eqs. (1). When expressed in terms of the ripple angle β the coupling constant $\tilde{q}$ and asymmetry parameter $\tilde{\eta}$ for the ripple phase are:

(6)

$$\tilde{q}/\tilde{q}^L = \frac{3}{2}\sin^2\beta - \frac{1}{2}$$

$$\tilde{\eta} = \frac{3/2\cos^2\beta}{3/2\sin^2\beta - 1/2}$$

for $54 \leq \beta \leq 90$ and where $\tilde{q}^L$ is the lamellar value of the coupling constant. The ratios of the ripple amplitude, A, to the periodicity, λ, is given by $\tan\beta = A/2\lambda$.

Like the former case, there is a unique relationship between $\tilde{q}/\tilde{q}^L$ and $\tilde{\eta}$ which can be used to check the model experimentally. Experiment shows that in some cases (membranes containing cholesterol or other impurity) the above model does not work. The reason it does not work in these cases is that the above model assumes that the direction of the ripple is uniform throughout the region a molecule diffuses. A distribution in ripple direction over the diffusing range of the molecule is easily accounted for, and when Eq. (4) is modified to account for this feature, experiment fits well with this simple theory (6).

DISCUSSION

The above examples illustrate how simple ^{2}H-NMR techniques can be applied to provide valuable insights into modulated aggregate structures. In fact, this technique is not only considerably simpler and easier to apply than traditional techniques such as x-rays and freeze-fracture microscopy. The precision of measurement may even be considerably better than these other methods.

ACKNOWLEDGEMENTS

The author of these lecture notes acknowledges his students, Mr. L. Strenk and Dr. Z. Yaniv and his colleagues, Dr. G. Chidichimo, Dr. P. Westerman and Dr. N. Vaz, who contributed to the research on these materials. Research was supported in part by the National Science Foundation Solid State Chemistry grant DMR82-04342.

REFERENCES

1. V. Luzzati and A. Tardieu, *Ann. Rev. Phys. Chem.* **25**, 79 (1974).

2. G. Chidichimo, N.A.P. Vaz, Z. Yaniv and J.W. Doane, *Phys. Rev. Lett.* **49**, 1950 (1982).

3. J. Stamatoff, B. Feuer, H.J. Guggenheim, G. Tellez and T. Yamane, *Biophys. J.* **38**, 217 (1982).

4. R.R. Balmbara, O.A. Bucknall and J.S. Clunie, *Mol. Cryst. Liq. Cryst.* **11**, 173 (1970).

5. E. Sackmann, D. Ruppel and C. Gebhardt, in *Liquid Crystals of One- and Two-Dimensional Order*, edited by W. Helfrich and G. Heppke (Springer-Verlag, Berlin, 1980), pp. 309-326.

6. L.M. Strenk, P.W. Westerman, N. Vaz and J.W. Doane (to be published).

7. A. Abragam, *The Principles of Nuclear Magnetism* (Oxford University Press, London, 1961) p. 475.

PHASE BIAXIALITY IN THE CHOLESTERIC AND BLUE PHASES

J. William Doane

Department of Physics and Liquid Crystal Institute,
Kent State University,
Kent, Ohio 44242.

INTRODUCTION

In the previous chapter we saw some simple examples where self-diffusion of the molecules over biaxially ordered regions induced an asymmetry parameter, $\tilde{\eta}$, in the time averaged quadrupole interaction. Biaxiality was evident by a spatial grouping of the molecules which require two axes to describe the orientational order of the groups or aggregates.

In the cholesteric phase the source of biaxiality is not so evident. In fact, it will not be until chapter 19 where I introduce and show the measurement of specific order parameters that we shall see the precise mechanism of biaxiality in the cholesteric phase. For now it is only necessary to see that the cholesteric phase is biaxial because of the symmetry (1). The nematic phase is (except in unusual circumstances) uniaxial in that its order is completely defined by an axis, N, referred to as a director. Imposing a helical twist upon the nematic forms the cholesteric and breaks this uniaxial symmetry. An additional direction, the direction of the axis of the helix (pitch axis), P, is necessary to fully describe the orientational ordering. This biaxiality shows up on the ^{2}H-NMR signal as an asymmetry parameter but in this case this asymmetry is imposed on the signal by the effect that the pitch axis has on the orientational ordering of the molecule. In this case, unlike in the previous chapter, self-diffusion is not required to observe the asymmetry parameter. In fact, we shall see that self-diffusion impedes its measurement but does not prevent us from studying it.

J. W. Emsley (ed.), Nuclear Magnetic Resonance of Liquid Crystals, 421–430.

^{2}H-NMR SPECTRAL PATTERNS

There are basically three different types of ^{2}H-NMR spectral patterns that can be obtained from cholesteric materials depending upon how the pitch axis aligns in the magnetic field, B_0: (i) P and B_0 parallel, a case studied by Collings et al. (2) and Luz et al. (3) which gives a two line spectrum with a splitting of about one-half that obtained in an untwisted nematic at the same reduced temperature; (ii) P and B_0 perpendicular, the case to be discussed here, which yields a spectral pattern which consists of two sets of edge singularities (see Fig. 1) which correspond to those positions on the helix where N and B_0 are parallel (outer set) and perpendicular (inner set) (4); (iii) P randomly ordered in B_0 to give a pattern with one set of singularities like (in the absence of diffusion) the usual powder pattern in a solid.

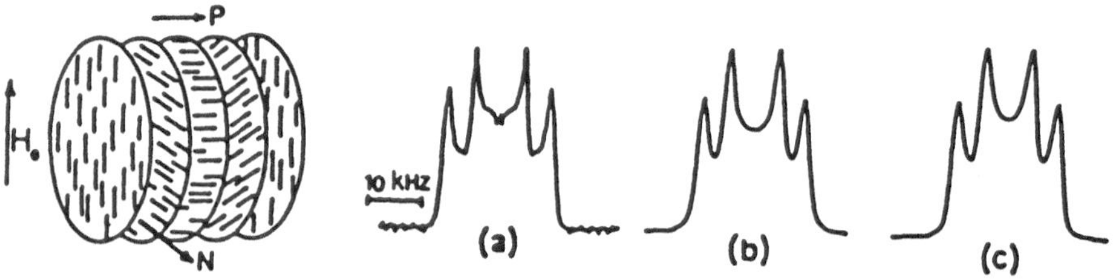

Figure 1. *Cholesteric structure with P and B_0 perpendicular and spectral patterns for (a) experimental spectral pattern for 40.7-d_4 twisted by chiral CB-15 (see reference 5), (b) calculated spectral pattern from Eq. (5), and (c) calculated spectral under the approximation of an average rather than time dependent spectral frequency as described in the text. Reproduced from reference 5.*

If it were not for self-diffusion the calculation of the spectral pattern shown in Fig. 1 would be trivial by starting with the equation for the frequencies of the two spectral lines $\nu^{\pm}$ of an untwisted nematic (5):

$$\nu^{\pm}=\pm\frac{3}{16}\tilde{q}[(1+\tilde{\eta}\cos 2\phi_1)+(3-\tilde{\eta}\cos 2\phi_1)\cos 2\theta_1] \qquad (1)$$

where $\tilde{q}$ is the time averaged coupling constant, $\tilde{\eta}$ the motionally induced asymmetry parameter, and θ_1, ϕ_1 the polar angles giving the direction of B_0 in the x,y,z time averaged principal axis frame with z parallel to N in Fig. 1. For the case where P and B_0 are perpendicular, there is a uniform distribution of θ_1 which yields a pattern like those of Fig. 1. For this case, self-diffusion of the deuteriated molecule along the pitch axis has the effect of further time averaging the quadrupolar interaction to alter the shape of the pattern (5). Since in this case we are concerned about diffusion in only one dimension it is convenient to account for this effect in terms of the probability function $W(\theta_1, \theta_0, \tau)$ for the probability that a molecule at orientation θ_1 finds itself at a new angle θ in a time τ (5):

$$W = \frac{1}{\sqrt{4\pi D\tau}} \exp \frac{P^2(\theta_1-\theta_0)^2}{16\pi^2 D\tau} , \qquad (2)$$

where D is the diffusion constant along the pitch axis and τ is the time.

Averaging Eq. (1) over θ_1 gives us an average frequency which is time dependent:

$$\bar{\nu}^{\pm}(\theta_o,\tau) = \pm \frac{3}{16} \tilde{q}[1-\tilde{\eta})+(3+\tilde{\eta})\cos 2\theta_o \exp [\frac{-16\pi^2 Dt}{P_o^2}]], \qquad (3)$$

where the constraint $\phi_1 = \pi/2$ is imposed by the choice of x as the pitch axis. It is noted that Eq. (3) has the proper limits for D = 0 and ∞.

One-half of the quadrupolar echo, can now be expressed in terms of the ensemble average (6),

$$G(t) = G_o \int_0^{\pi/2} [\cos \int_0^t \bar{\nu}(\theta_o t')dt'] \times R(\theta_o, t+2\tau)U(\theta_o)d\theta_o, \qquad (4)$$

where G_o is a constant and θ_o is distributed according to the function $U(\theta_o)$. In the case of the undistorted helix $U(\theta_o) = 1$. $R(\theta_o t + 2\tau)$ is a relaxation function which accounts for line broadening due to magnetic dipole-dipole interactions and magnetic field inhomogeneities. The time τ is that associated with the quadrupole echo. The damping function $R(\theta_o, t)$ was taken to be:

$$R(\theta_o, t) = \exp[-t^2\sigma^2(\theta_o)/2],$$

where

$$\sigma(\theta_o) = W' + W''[|P_2(\cos\theta_o)|-1],$$

where W' and W'' are fitting parameters. Performing the integration on dt' in Eq. (4) we obtain for the free-induction decay (5):

$$G(t) = G_o \int_0^{\pi/2} \{\cos[\Omega(\theta_o,t)t]\} \times R(\theta_o,t+2\tau)d\theta_o, \quad (5a)$$

where

$$\Omega(\theta_o,t) = \frac{3}{16}\tilde{q}[(1-\tilde{\eta})+(3+\tilde{\eta})K\cos 2\theta_o], \quad (5b)$$

and where

$$K = \frac{p^2}{16\pi^2 Dt}[1-\exp[\frac{-16\pi^2 Dt}{p^2}]]. \quad (5c)$$

The spectral pattern can be obtained by the Fourier transformation of Eqs. (5). Such a pattern so calculated is shown in Fig. 1b.

Eq. (5) must be evaluated numerically and the computation even on a fast computer is time consuming. There are, however, some useful features and simplifications that are worth noting. If we regard t_m as the length of the free-induction decay, then $\Omega(\theta_o,t_m)$ of Eq. (5) can be considered to be the average frequency over the time interval from $t = 0$ to $t = t_m$. This approximation considerably reduces the computation time to an almost identical pattern is shown in Fig. 1. This approximation further allows us to obtain a useful expression for the position of the characteristic singularities which arise in the spectral pattern. Ignoring the line-broadening function and changing the variable of integration in Eq. (5) from θ_o to Ω, the Fourier transform of Eq. (5) is trivially calculated to obtain the splittings $\delta\nu_1$ between the outer-edge singularities and $\delta\nu_2$ between the inner set which can also be obtained experimentally with high precision. From the Fourier transform of Eq. (5) with R = 1 they are calculated to be (4,5):

$$\delta\nu_1 = \tfrac{3}{8}\tilde{q}[(1-\tilde{\eta})+(3+\tilde{\eta})K],$$

$$\delta\nu_1 = \pm\tfrac{3}{8}\tilde{q}[1-\tilde{\eta})-(3+\tilde{\eta})K]. \quad (6)$$

In the expression for $\delta\nu_2$ the plus sign is to be considered whenever the self-diffusion is sufficiently large to force the singularities that define $\delta\nu_2$ to cross over zero, i.e., $(1-\tilde{\eta})<(3+\tilde{\eta})K$. Otherwise, $\delta\nu_2$ is negative. Including the line-broadening term has only the effect of broadening the singularities and Eq. (6) remains valid. It is seen from Eq. (6) that the effect of diffusion can be eliminated by considering the quantity:

$$\Delta = 2(|\delta\nu_1|)-(|\delta\nu_2|) = \frac{3}{2}\,\tilde{q}\,(1+\tilde{\eta}). \tag{7}$$

Eq. (7) has been used to measure values of $\tilde{\eta}$ where the value of $\tilde{q}$ can be obtained from samples containing low enough

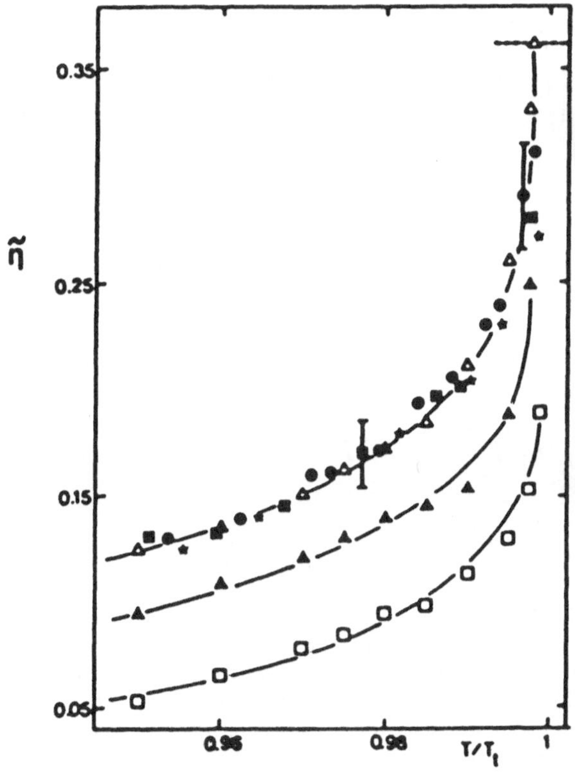

Figure 2. *Experimental values of $\tilde{\eta}$ versus reduced temperature T/T_T where T_T is the transition temperature (K) cholesteric to either the isotropic or blue phase, whichever occurs. The symbols correspond to different pitch lengths, p_o. Reproduced from reference 7.*

concentrations of chiral material to be untwisted by the magnetic field where Eq. (1) with $\theta_o = 0$ can be applied. Figure 2 shows that the values of $\tilde{\eta}$ obtained in the binary mixture of MBBA-d_4 and with different quantities of chiral MBMBA to create materials with different pitch lengths, p_o.

Eq. (5) can also be used to show the effects of self-diffusion upon the spectral shape. These effects are illustrated in the calculated spectral patterns shown in Fig. 3. In Fig. 3c it is seen, as one would expect, that in the limits of rapid diffusion rates, the pitch axis would time average to the axial principal axis to yield a two line spectrum.

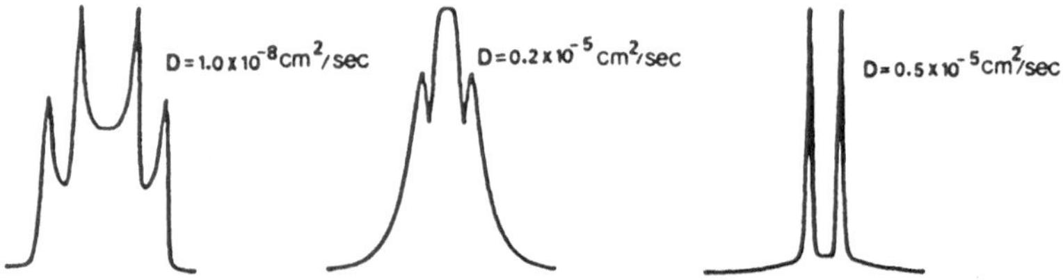

Figure 3. *Calculated spectral patterns from Eq. (5) illustrating the effect of molecular self-diffusion along the pitch axis in a sample where the helix is undistorted by the magnetic field. The spectral patterns were calculated for the values of the diffusion constant D indicated in the figure and for $\tilde{\eta} = 0$. Reproduced from reference (5).*

DISTORTION AND SPATIAL MODULATED DIFFUSION

For suitably large pitch lengths the cholesteric helical structure which aligns such the P and B_o are perpendicular will be distorted by the magnetic field and, if p_o is large enough, the field will completely unwind the helix to form a nematic. These features have been described theoretically by de Gennes. We shall show here that they can be readily observed by deuteron NMR.

When the helix becomes distorted the fraction of molecules which align parallel to B_o increases with respect to those aligned perpendicular to B_o. As a consequence, the relative intensities of the 0° and 90° singularities are dramatically affected as illustrated in Fig. 4.

Applying the theory of de Gennes (9) it is possible to modify Eq. (2) to calculate the spectra of Fig. 4 where (8):

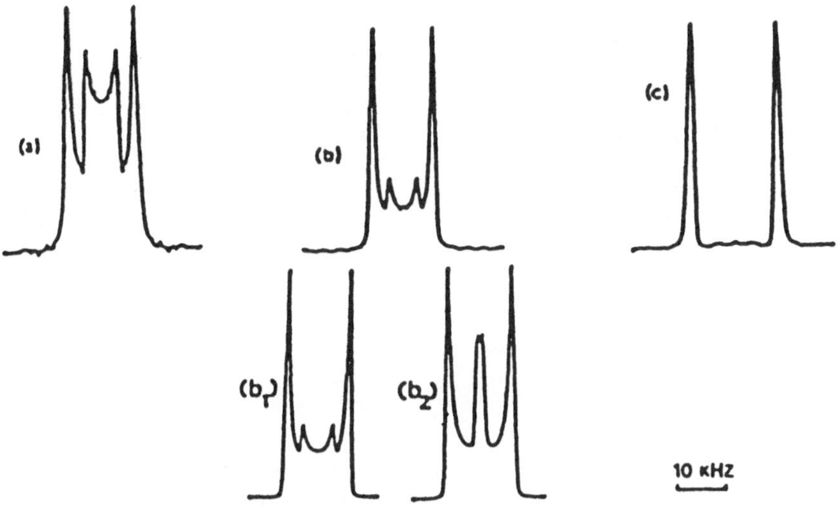

Figure 4. *Spectral patterns obtained at temperatures: (a) at T = 277.23K which show small effects of distortion; (b) at T = 296.23K which shows large distortion effects; and (c) at T = 301.30K where the distortion is total (nematic structure); (b_1) fitted spectral pattern assuming a space modulated self-diffusion constant along the pitch axis; (b_2) fitted spectral pattern assuming a constant value for the self-diffusion constant illustrative of the poor agreement that is obtained under this assumption. Reproduced from reference (8).*

$$W(\theta_o,\theta,t) = \frac{1}{\sigma\sqrt{2}} \frac{R(k)}{(1-k^2\cos^2\theta)^{\frac{1}{2}}} \exp\left[\frac{R(k)}{2}\int_{\theta_o}^{\theta}\frac{d_\varepsilon}{(1-k^2\cos^2\varepsilon)^{\frac{1}{2}}}\right],$$

$$R(k) = pE(k)/\pi^2,$$

$$\sigma = \sqrt{2Dt},$$

$$U(\theta_o) = pE(k)[\pi^2(1-k^2\cos^2\theta_o)^{\frac{1}{2}}]^{-1}. \qquad (8)$$

G_0 is a normalization constant and E(k) is the complete elliptic integral of the second kind. The distortion parameter k is a measure of the degree of distortion in the sense that if $B_0 = 0$ then k = 0 and the structure is undistorted; for successively stronger magnetic fields the value of k increases and, when the magnetic field reaches the critical value B_c then k = 1 and the cholesteric structure is totally unwound ($p = \infty$).

If Eqs. (5) and (8) are used assuming a constant diffusion constant along the pitch axis it is found to be impossible to fit the data [Fig. (b_2)]. The best fit found is shown in Fig. (b_1). The reason for this poor fit is that the diffusion constant is impeded by the tightly twisted regions. We have found, however, that a good fit to the data can be obtained by assuming that the diffusion constant is related in a linear fashion to the rate of twist along the pitch axis by the simple relationship (8):

$$D = D_N(1 - C\partial\alpha/\partial x) , \tag{9}$$

where D_N is the diffusion in the untwisted sample (nematic), C is a proportionality constant and the angle of rotation of the director along the pitch axis. Fig. (b_1) shows the data can be well fitted when Eq. (9) is incorporated into Eqs. (5) and (8).

BLUE PHASES

It was first discovered by Collings and McColl (10) and subsequently by Samulski et al. (3) that the blue phases, although optically isotropic, are not isotropic in the usual NMR sense in that the strong spin interactions such as the quadrupolar interaction are not averaged to zero. The idea here is that even though the phase may be cubic, as strongly implied by light scattering, the lattice parameter (~ wavelength of light) is too large for diffusion to average out the spin interaction. The blue phases give a rather featureless spectral pattern as illustrated in Fig. 5, which also shows a calculated pattern using Eq. (5) modified with $\tilde{\eta} = 0$ and an isotropically ordered pitch axis. Fig. 5, therefore, applies to an unaligned blue phase sample. Probe heads with extreme temperature homogeneity are required to obtain accurate patterns as the phases are less than 1°C wide.

DISCUSSION

Deuteron NMR is a technique used to study the cholesteric phase because it is sensitive to features of the phase which are not

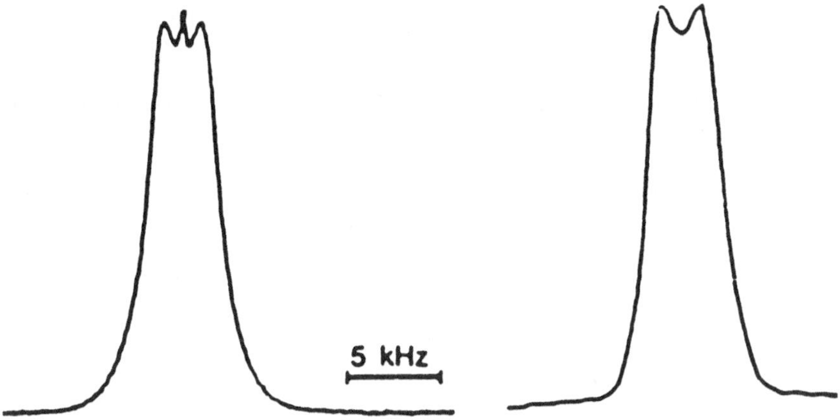

Figure 5. *Experimental and calculated blue phase spectrum of MBBA-d_4 in 70 wt% chiral MBMBA.*

readily observable by optics or other methods. The feature of biaxiality is perhaps the most significant as it appears both from these results and theoretically to be an essential feature of the phase. Furthermore, the blue phases may be biaxially induced. The feature that diffusion along a helix is affected by the pitch of the helix is a totally new concept (at least to liquid crystals). Also, there are probably few other techniques that could observe space modulated diffusion.

More experimental work is required to obtain better blue phase patterns and more theoretical work is necessary to calculate those patterns. At this time, however, it does not appear that the patterns contain enough features to be able to precisely define the blue phase structure.

ACKNOWLEDGEMENTS

The author acknowledges his student Dr. Z. Yaniv who worked on the cholesteric phase as part of his dissertation. Dr. N. Vaz and Dr. G. Chidichimo also contributed to the author's understanding of the cholesteric phase. Finally, the author acknowledges the support of NSF Solid State Chemistry grant DMR82-04342.

REFERENCES

1. R.M. Hornreich and S. Shtrikman, in *Liquid Crystals in One-*

and Two-Dimensional Order, edited by W. Helfrich and G. Heppke (Springer-Verlag, Berlin, 1980), p. 185.

2. P.J. Collings, S. Goss and J.R. McColl. *Phys. Rev. A* **11**, 684 (1975).

3. Z. Luz, R. Poupko and E.T. Samulski, *J. Chem. Phys.* **74**, 5825 (1981).

4. Z. Yaniv, N.A.P. Vaz, G. Chidichimo and J.W. Doane, *Phys. Rev. Lett.* **47**, 46 (1981).

5. G. Chidichimo, Z. Yaniv, N. Vaz and J.W. Doane, *Phys. Rev. A* **25**, 1077 (1982).

6. A. Abragam, *The Principles of Nuclear Magnetism* (Oxford University Press, London, 1961).

7. Z. Yaniv, M.E. Neubert and J.W. Doane (submitted for publication).

8. N. Vaz, G. Chidichimo, Z. Yaniv, *Phys. Rev. A* **26**, 637 (1982).

9. P.G. de Gennes, *Solid State Commun.* **6**, 163 (1968).

10. P.J. Collings and J.R. McColl, *J. Chem. Phys.* **69**, 3371 (1978).

PHASE BIAXIALITY IN SOME OF THE SMECTIC PHASES

J. William Doane

Department of Physics and Liquid Crystal Institute,
Kent State University,
Kent, Ohio 44242.

INTRODUCTION

Probably the first liquid crystalline phase to be recognized as being biaxially ordered is the smectic C, S_C, phase (1). Biaxial ordering in this phase is clearly evident from conoscopic observation under a polarizing microscope. This phase is probably also the first liquid crystalline phase in which biaxiality was studied by NMR. Unlike the optical observation, however, the NMR (2-5) observation of biaxial ordering is a difficult experiment to perform in this phase. Not only is the biaxial ordering weak but also suitable alignment of the sample in the magnetic field in order to observe this feature requires either an imposed electric field (2) or slowly spinning sample (3,4).

It is now generally recognized by workers in the field of liquid crystals that most smectic phases are biaxially ordered. The smectic A, S_A, and non-tilted smectic B, S_B, are probably the only examples of uniaxial smectic phases. While other smectic phases such as S_H, S_F, S_G, S_I etc. are biaxial, the mechanisms responsible for the biaxiality are not well known and the NMR of these phases has not been thoroughly studied. In this chapter I will primarily survey NMR experiments that only observe biaxial ordering. Many useful NMR measurements wait to be performed.

GENERAL THEORETICAL CONSIDERATIONS

Unlike the liquid crystalline phases considered in the previous

J. W. Emsley (ed.), Nuclear Magnetic Resonance of Liquid Crystals, 431–439.

two chapters we can normally ignore self-diffusion as a mechanism for imposing averaging in the smectic phases. In this case it is helpful to write a general expression for the splitting of a spin 1 deuterium quadrupole interaction in a magnetic field relative to some reference frame fixed to the sample. Often one principal axis is known. If this is the y-axis then for a rotation about that axis a two line spectrum of frequencies $\nu^{\pm}$ will follow the equation (6):

$$\nu^{\pm} = \pm\tfrac{3}{4}\,\tilde{q}[A(\tfrac{3}{2}\cos^2\theta_o - \tfrac{1}{2} + B\sin^2\theta_o\cos 2(\phi_o+\phi_1) + C\sin 2\theta_o - \cos(\phi_o+\phi_1)]$$

$$A = [(\tfrac{3}{2}\cos^2\theta_1 - \tfrac{1}{2}) + \tfrac{\tilde{\eta}}{2}\sin^2\theta_1]$$

$$B = [\tfrac{\tilde{\eta}}{2} + (\tfrac{3-\tilde{\eta}}{4})\sin^2\theta_1]$$

$$C = (\tfrac{3-\tilde{\eta}}{4})\sin 2\theta_1 \qquad (1)$$

where $\tilde{q} = eqV_{zz}/h$, $\tilde{\eta} = (\tilde{V}_{xx}-\tilde{V}_{yy})/\tilde{V}_{zz}$, and the Euler angles ϕ_1,θ_1,ψ_1 give the orientation of the newly oriented principal axis (x,y,z) frame in the sample (a,b,c) frame. The sample frame is defined such that c is parallel to B when the sample was cooled into the smectic phase. With the smectic sample so prepared it is then reoriented in the magnetic field, at the polar angle θ_o,ϕ_o in the a,b,c frame. Note that as in previous chapters ~ denotes a partially averaged tensor component relative to a space fixed reference frame.

When $\theta_1=\phi_1=0$ the above equation reduces to the commonly used expression

$$\nu^{\pm} = \pm\tfrac{3}{4}\,\tilde{q}\{(\tfrac{3}{2}\cos^2\theta_o - \tfrac{1}{2}) + \tfrac{\tilde{\eta}}{2}\sin^2\theta_o\cos 2\phi_o\}. \qquad (2)$$

The above equations can only be applied directly if one has a uniformly aligned liquid crystal which is often not the case. In non-uniformly aligned samples one must usually consider distributions in ϕ_1, θ_1 or ψ_1. One simple example is the S_G phase as follows.

S_G Phase

Probably the simplest form of biaxial ordering is that observed in a magneto-aligned sample of the material 4-n-pentoxy-benzylidene-4-n'-heptylaniline (50.7). ^{2}H-NMR spectra of

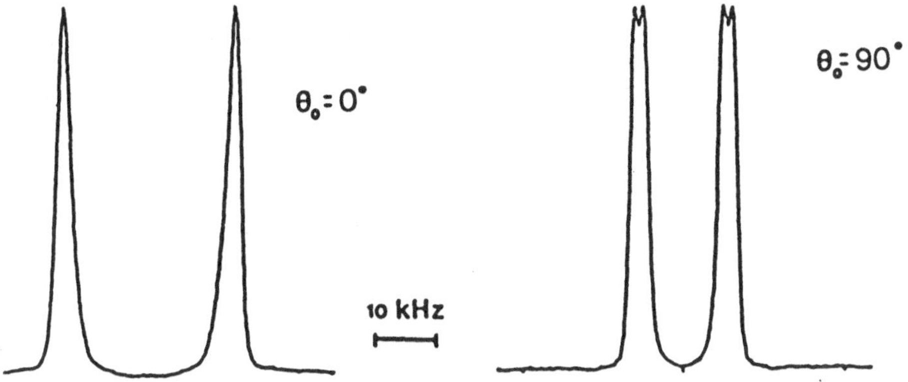

Figure 1. *Experimental angular dependence of the ^{2}H-NMR spectrum in the S_G phase of 50.7-d_4 at 31°C. Each single line breaks up into two edge singularities at θ_o = 90° (from reference (8)).*

50.7-d_4 selectively deuteriated on one aromatic ring is shown in Fig. 1. Biaxial ordering is evidenced by the splitting of the lines at the 90° orientation of the sample relative to the direction of the magnetic field. The splitting at this orientation results from $\tilde{V}_{xx} \neq \tilde{V}_{yy}$ and consequently a finite value of $\tilde{\eta}$. When the sample is prepared by cooling from the uniaxial nematic in the presence of the magnetic field the z principal axis remains aligned, whereas the x and y axes are randomly aligned in a plane normal to the direction of the applied field direction. In this case the angle ϕ_o in Eq. (2) is randomly distributed.

In order to calculate the shape of this spectral pattern it is convenient to express Eq. (2) in terms of dimensionless quantity $\omega = 2\nu/3\tilde{q}$ whereby

$$\omega = \pm(\alpha + \beta \cos 2\phi_o) \tag{3}$$

where $\alpha = (1 + 3\cos 2\theta_o)/8$ and $\beta = \tilde{\eta}(1 - \cos 2\theta_o)/8$. Since the x and y axes are randomly aligned, there is a uniform distribution in ϕ_o. The spectral pattern $G(\omega)$ in such a case can be calculated from the equation

$$G(\omega) = G_o d\phi_o/d\omega = \frac{G_o}{2\beta}\left[1 - \left(\frac{\omega-\alpha}{\beta}\right)^2\right]^{-\frac{1}{2}} \quad (4)$$

where G_o is a constant. As seen from Eq. (4) the pattern will consist of edge singularities at values of $\omega = \pm|\alpha + \beta|$ and $\pm|\alpha - \beta|$. The spectral pattern is illustrated in Fig. 2 which shows the positions of the edge singularities for $\theta_o = 90°$

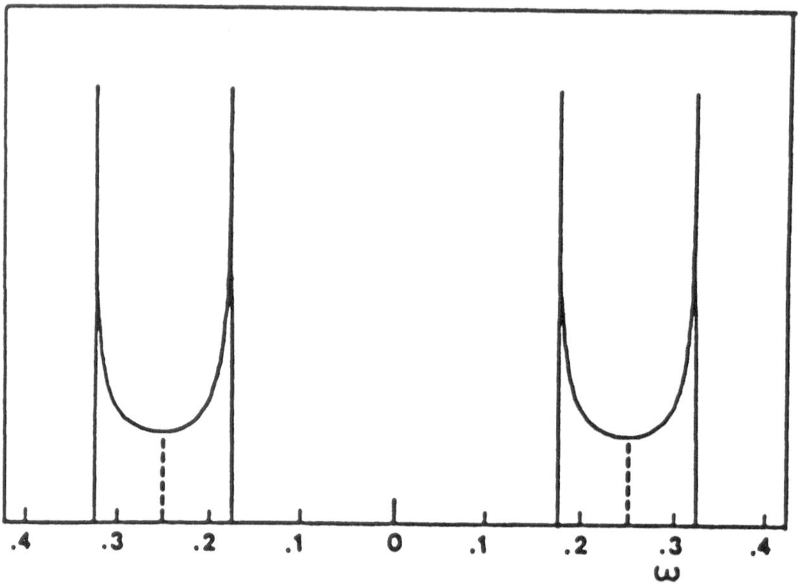

Figure 2. *Biaxial spectral pattern illustrated for the model outlined in the text where $\tilde{\eta} = 0.3$ and $\theta_o = 90°$.*

and $\tilde{\eta} = 0.3$. The separation between the two edge singularities of each line $\delta\omega$ is given by:

$$\delta\omega = \tilde{\eta}(1 - \cos 2\theta_o)/2 \quad (5)$$

which is seen to be maximum at $\theta_o = 90°$.

The appearance of the singularities in Fig. 1 at $\theta_o = 90°$ is clear and corresponds to a value of $\tilde{\eta} \simeq 0.2$.

Such a smectic G pattern was first observed with the solute $CDCl_3$ dissolved in the material 60.6 by Barbara et al (7).

It should be mentioned here that the biaxiality in all reported S_G phases does not appear to be the same (8). It is suspected that there is more than one S_G phase.

S_C Phase

Smectic samples which are rigid in the magnetic field like those described above are relatively simple to study compared to the S_C phase. The S_C phase is in some respects like the nematic phase in that the alignment of the z principal axis is easily affected by the magnetic field of the NMR experiment making it difficult to alter θ_0 in Eq. (1) in a controlled manner in order to study biaxiality. We have found two solutions to this problem: 1) to apply electric fields (2) in order to compete with the magnetic field for sample alignment, and 2) to spin the sample in the magnetic field (3,4). By spinning a nematic or smectic C sample about an axis perpendicular to the direction of B at a suitable rate it is possible to obtain a random distribution of the z axes in a plane which contains the direction of the B. Such a distribution produces a spectral pattern from which to measure the biaxiality.

1. Application of an Electric Field. The material n-heptyloxyazoxybenzene (HOAB) has a negative dielectric anisotropy. Application of an a.c. electric field in a direction perpendicular to the direction of the magnetic field has the effect of aligning the director in the same direction as the magnetic field.

By cooling HOAB from the nematic to the smectic C phase in the presence of both a magnetic and a.c. electric field, we found it to be possible to make a monodomain smectic C sample. The a.c. field has the effect of aligning the material so that the x and y axes are also ordered.

This monodomain feature of the sample can be observed with NMR. If the sample is polydomain as has been observed by cooling in the presence of a magnetic field alone, then the spectral lines become distributed when the sample is reoriented in the magnetic field. In a monodomain sample, on the other hand, the spectral lines do not distribute but either shift or remain unchanged depending upon which axis the sample is reoriented.

The left-hand spectrum of Fig. 3 shows the unsplit spectral lines of one of the deuteriated methyl groups of HOAB-d_{30} in the smectic C phase of $\theta_0 = 0$. The temperature of the sample has been adjusted to the temperature where the spectral lines are not split at all (see Fig. 3 of reference (2) where $\delta\nu_1 = 0$). These lines were recorded in a monodomain sample created in an 8 kV/cm, 1kHz electric field (defining a y-axis) and in an 8 kG magnetic field (defining a z-axis).

After the monodomain was formed, the electric field was turned off. When the sample was reoriented about the x-axis in the magnetic field, the spectral lines did not appear to change and remained as shown in the left-hand spectrum of Fig. 3. No variation in the splittings were observed for angles as large as 40°. This we interpreted as due to the fact that the director could follow the field when rotated about this axis. If, on the other hand, the sample was reoriented about the y-axis, then the lines split as shown in the middle and right-hand spectrum of Fig. 3.

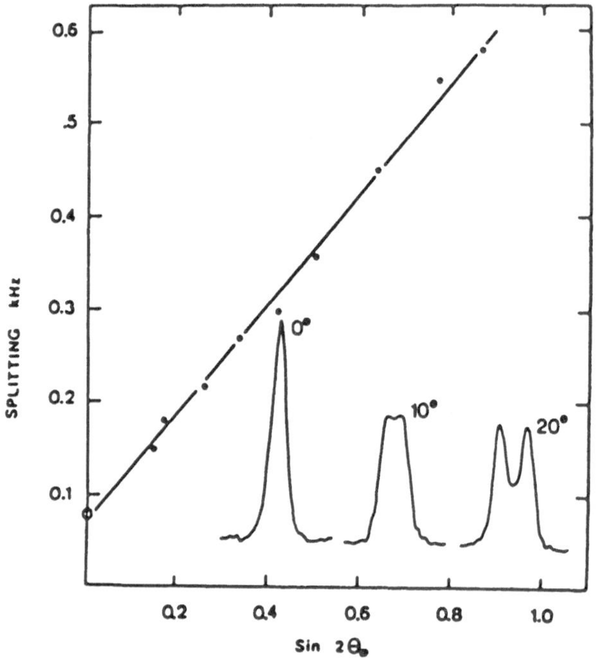

Figure 3. *Recorded deuteron spectra of one of the methyl lines of HOAB-d_{30} in the smectic C phase at the temperature where there is no splitting at $\theta_0 = 0$ (left spectrum). The other spectra are recorded for reorientation of the sample about the y-axis. The solid dots on the graph show splitting values for y-axis reorientation (from reference (2)).*

The splitting of the central line can only come from

biaxial order. This can be seen from Eq. (1). The first term (uniaxial term) in Eq. (2) is zero for the central line because the angle θ_1 is near the magic angle.

Upon the adjustment of the angle θ_o, the second and third biaxial terms, if present, will split the line. The relative contribution of each term can be determined through the angular dependence of the splitting as is shown in Fig. 3 where the $\sin 2\theta_o$ term is shown to be dominant at small angles.

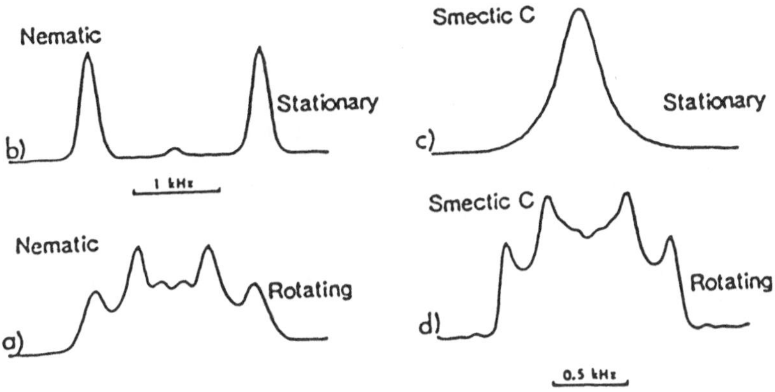

Figure 4. *Deuterium spectra for the methyl group of 4-n-heptyl-d_{15}-oxybenzoic acid-d_1 (7OBA-d_{16}) for: a) spinning nematic sample; b) non-spinning nematic sample; c) non-spinning smectic C sample after the sample had been cooled into the smectic C phase from the aligned nematic; d) spinning smectic C sample (from reference (3)).*

2. Spinning Sample Technique. Figures 4a and 4b show the spectra for a spinning and aligned nematic phase respectively of the deuteriated methyl group of 7OBA. Here the spectra are quite normal in that the spectral lines of the aligned nematic correspond as expected, to the outer singularities of the spectral pattern of the spinning sample. In the smectic C phase of this material, the spectral lines for the deuteriated methyl sites in the aligned sample are partially overlapped such that the splitting is near zero (~100 Hz) as indicated in Fig. 4c. When the sample is spun, a typical spectral pattern with two sets of singularities is observed but one in which the outer

singularities are far removed from the near zero splitting of the aligned sample (Fig. 4d). This abnormal behaviour occurs only for the methyl group in the S_C phase. The spectral lines from the other deuteriated sites on 7OBA follow a uniaxial-like behaviour where the resulting outer and inner singularities in the smectic C phase as in the nematic phase can easily be identified with the corresponding aligned spectra for $\theta_o = 0$ and $\theta_o = 90°$ (one-half that of the $\theta_o = 0$ splittings), respectively.

This result implies that all of the deuteriated sites (with the exception of the methyl site) have their principal z axes nearly parallel to each other and aligned parallel to B in the stationary sample. The fact that they are not all parallel in the smectic C phase is an indication that the molecule is not, on the average, rotating about its long axis and that all of the deuteriated sites do not average such that the rotation axis becomes a common principal axis as in the nematic phase.

It is possible to calculate the angle between the z axis of the methyl site and the direction of B in the aligned sample. If the z axis of the methyl site in the aligned sample is oriented at an angle of θ_1 with respect to B, then the 100 Hz splitting of Fig. 4c must follow the equation:

$$\delta\nu = \nu^+ - \nu^- = \tfrac{3}{2}\tilde{q}^{Me}[P_2(\cos\theta_1)] = 100 \text{ Hz}. \tag{6}$$

From the separation between the outer singularities $\delta\nu'$ of Fig. 4d however, we obtain $\delta\nu' = \frac{3}{2}\tilde{q}^{Me} = 1200$ Hz yielding a value of $\theta_1 = 51° \pm 3$.

We now must answer the question as to why the principal z axis of the methyl site is oriented at such a large angle relative to B in the aligned sample while the z axes of the other sites higher up on the hydrocarbon chain are nearly parallel to B. In order to explain this strange effect, we must resort to the use of orientational order parameters which will be introduced in the next chapter, where we will return to this problem and show that if the rotation about the long axis is only slightly biased to one orientation (an effect sometimes labelled as "partial rotational freeze-out"), then those deuteriated sites which have the smallest values of $\tilde{q}^i$, such as the methyl group, are the ones most strongly affected. On the average, the sites with the smallest values of $\tilde{q}^i$ will experience the largest shift in the orientation of their principal z axis.

DISCUSSION

Both of the samples presented demonstrate biaxial ordering. The first example of the S_G phase, however, does not show the mechanism responsible for the phase biaxiality, which is also the case of the cholesteric phase (see chapter 18). The S_C experiment, on the other hand, qualitatively shows the biaxial mechanism of partial rotational freeze-out for that phase but the magnitudes of the differences in the orientation of the principal z axes were not suitably explained.

In order to determine the mechanisms as well as quantitize these features, it is useful to introduce the concepts of order parameters to be discussed in the next chapter.

ACKNOWLEDGEMENTS

The author acknowledges Dr. H. Schmiedel of Leipzig, DDR and Dr. R. Dong of Brandon, Manitoba, Canada during their visits to Kent State University. NSF Solid State Chemistry grant DMR82-04342 is acknowledged.

REFERENCES

1. T. Taylor, J. Fergason and S. Arora, *Phys. Rev. Lett.* **24**, 359 (1970).

2. P.J. Bos, J. Pirs, P. Ukleja, M.E. Neubert and J.W. Doane, *Mol. Cryst. Liq. Cryst.* **40**, 59 (1977).

3. D.J. Photinos, P.J. Bos, J.W. Doane and M.E. Neubert, *Phys. Rev. A* **20**, 2203 (1979); *Phys. Rev. A* **23**, 3346 (1981).

4. P.J. Collings, D.J. Photinos, P.J. Bos, P. Ukleja and J.W. Doane, *Phys. Rev. Lett.* **42**, 996 (1978).

5. D.J. Photinos and J.W. Doane, *Mol. Cryst. Liq. Cryst.* **76**, 159 (1981).

6. Z. Yaniv, Ph.D. Dissertation (Kent State University, Kent, Ohio 44242) 1982.

7. T. Barbara and B.P. Dailey, *Mol. Cryst. Liq. Cryst.* **88**, 311 (1982).

8. Z. Yaniv, H. Schmiedel, R.Y. Dong, N.A.P. Vaz and J.W. Doane, *Mol. Cryst. Liq. Cryst.* (to appear, Proceedings of the Ninth International Liquid Crystal Conference, Bangalore, (1982).

ORIENTATIONAL ORDER PARAMETERS AND MECHANISMS OF PHASE BIAXIALITY

J. William Doane

Department of Physics and Liquid Crystal Institute,
Kent State University,
Kent, Ohio 44242

INTRODUCTION

In the lyotropic phases discussed in chapter 16, it was possible to model the phases in order to describe the ^{2}H-NMR spectral patterns. In these phases the biaxial structures were clear from the models. In the cholesteric and smectic phases discussed in chapters 17 and 18, however, the mechanisms responsible for the observed phase biaxiality were not clear. In this case it is useful to introduce orientational order parameters which are time averages of specific functions in order to examine biaxiality.

ORDERING IN UNIAXIAL PHASES

As discussed earlier, one only measures $\tilde{V}^j_{zz}$ or $\tilde{q}^j$ (jth) molecular site in a uniaxially ordered phase, i.e., $\tilde{\eta} = 0$. In the liquid crystal, the time averaged $\tilde{q}^j$ is related to the non-averaged solid state value $\tilde{q}^s$ by the expression (1):

$$\tilde{q}^j = \tilde{q}^s \langle \tfrac{3}{2} \cos^2\sigma_j - \tfrac{1}{2} \rangle \qquad (1)$$

where σ_j is the time dependent angle between the jth C-D bond (principal axis in the solid) and principal z axis of the liquid crystal. Often, however, one is more interested in how the molecule or one of its segments is ordered rather than a particular C-D bond. In such a case it is convenient to introduce a reference frame fixed to the molecule and refer Eq. (1) to that frame. Such a transformation is discussed by Zannoni

J. W. Emsley (ed.), Nuclear Magnetic Resonance of Liquid Crystals, 441–448.

in chapter 1, and with the appropriate choice of frame, Eq. (1) can be expressed as,

$$\tilde{q}^{j} = \tilde{q}^{s}\,(s^{j}_{0,0}S_{0,0} + \tfrac{3}{4}\,s^{j}_{0,-2}S_{0,-2}) \tag{2}$$

where

$$s^{j}_{0,0} = (\tfrac{3}{2}\cos^{2}\beta_{j} - \tfrac{1}{2}),\quad S_{0,0} = S_{ZZ} = \overline{D}_{0,0} = \langle\tfrac{3}{2}\cos^{2}\theta - \tfrac{1}{2}\rangle \tag{3a}$$

$$s^{j}_{0,2} = (\sin^{2}\beta_{j}\cos 2\alpha_{j}),\quad S_{0,2} = \tfrac{2}{3}\,(S_{XX} - S_{YY}) = \tfrac{2}{3}(\overline{D}_{0,2} + \overline{D}_{0,-2}) = \langle\sin^{2}\theta\cos 2\psi\rangle\ , \tag{3b}$$

where for convenience we have also expressed the order parameters in terms of the elements of the Saupe order matrix[2,3] $S_{\lambda\mu} = \tfrac{1}{2}\langle 3\cos\theta_{\lambda}\cos\theta_{\mu} - \delta_{\lambda\mu}\rangle$, where $\mu,\lambda = X,Y,Z$ and the $\cos\theta_{\lambda}$ are the direction cosines of the orientation of z in the molecular X,Y,Z frame, and $\delta_{\lambda\mu}$ the Kronecker symbol. The polar angles α_j, β_j give the orientation of the C-D bond in the molecular X,Y,Z frame and the polar angles θ,ψ give the orientation of z in the X,Y,Z frame, and $\overline{D}_{0,m}$ are partially averaged Wigner rotation matrices. Our director in this case is z, the time averaged principal axis of the field gradient. The axes x and y are undefined in the uniaxial phases.

If the C-D bonds of the jth sites in question are rigid relative to each other and the molecular frame, then the conformational parameters $s^{j}_{0,\ell}$ ($\ell = 0,1,2$) are constants.

Often Eq. (2) is further simplified in that (3) $S_{0,2} \ll S_{0,0}$. The parameter $S_{0,2}$ expresses anisotropic fluctuations of the molecule relative to the uniaxial director and is ~0.05 whereas $S_{0,0} \sim 0.6$.

To the author's knowledge, the complete set of Meier-Saupe order matrices for a molecule which forms a liquid crystal phase has been determined in only one case. This is because the molecules are flexible and contain only simple rigid segments (see, for examples, chapters 2 and 15).

An important distinction in nomenclature should be noted here. Often a finite value of $S_{0,2} = \tfrac{2}{3}\,(S_{XX} - S_{YY}) = \langle\sin^{2}\theta\cos 2\psi\rangle$ is referred to as molecular biaxiality. We distinguish this from phase biaxiality which we discuss next.

ORDERING IN BIAXIAL PHASES

One parameter which expresses phase biaxiality is the motionally induced asymmetry parameter $\tilde{\eta}$ which is a time average that can be expressed by the equation (1):

$$\tilde{\eta}^j = \frac{3}{2}\frac{\tilde{q}^S}{\tilde{q}^j} \langle \sin^2\sigma_j \cos 2\xi_j \rangle \tag{4}$$

where σ_j and ξ_j are the polar angles giving the orientation of the jth C-D bond direction in the principal axis frame of the time averaged field gradient tensor.

Like in the previous case of uniaxial phases, it is convenient to refer this quantity to a molecular frame. This problem is normally regarded as complicated but we shall show here that it need not be, particularly if one has at hand some useful information regarding the orientation of the principal axes of the phase and of the molecule. We shall start by assuming that the orientation of principal axes of the field gradient tensors which we label x,y,z are known relative to the sample. We shall also assume here that the principal axes of the Meier-Saupe order matrix X,Y,Z are also known. In this case the asymmetry parameter can be expressed in terms of two molecular order parameters (1,4):

$$\tilde{\eta}^j = \frac{3}{2}\frac{\tilde{q}^S}{\tilde{q}^j} (s^j_{0,0}S_{2,0} + s^j_{0,2}S_{2,2}) \tag{5a}$$

where

$$S_{2,0} = \tfrac{3}{8}(\overline{D}_{2,0} + \overline{D}_{-2,0}) = \langle \sin^2\theta \cos 2\phi \rangle \tag{5b}$$

$$S_{2,2} = \tfrac{1}{2}(\overline{D}_{2,2} + \overline{D}_{2,-2} + \overline{D}_{-2,-2} + \overline{D}_{-2,2})$$
$$= \tfrac{1}{2}\langle (1 + \cos 2\theta) \cos 2\phi \cos 2\psi - 2\cos\theta \sin 2\phi \sin 2\psi \rangle \tag{5c}$$

and $\tilde{q}^j$ follows Eq. (3). The order parameter $S_{2,0}$ expresses anisotropic fluctuations of this molecular long axis (Z axis) relative to the x,y,z principal axis of the field gradient. Here the Euler angles ϕ, θ, ψ give the orientation of the X,Y,Z frame in the x,y,z frame. The parameter $S_{2,2}$ expresses "birotational freeze-out." It is noted here that if the long molecular axis is well ordered ($\theta \cong 0$) then $S_{2,2} \cong \langle \cos 2(\phi+\psi) \rangle$; hence the name "birotational freeze-out" as this average is non-zero for rotations about the long axis partially restricted under a two-fold rotation axis. We note here the distinction between $S_{2,0} = \langle\sin^2\theta\cos2\phi\rangle$ and $S_{0,2} = \langle\sin^2\theta\cos2\psi\rangle$. The former parameter expresses phase biaxiality in that an x axis associated with the phase is required to define the angle ϕ. The orientation of x and the angle ϕ are not defined in a uniaxial phase. On the other hand ψ and consequently $S_{0,2}$ can be defined for a uniaxial phase and as such $S_{0,2}$ expresses a feature of uniaxial ordering. This parameter, however, requires a biaxial-shaped molecule and is sometimes said to express molecular biaxiality even though it is a uniaxial order parameter with regard to the phase.

It is further noted that for a cylindrical molecule then $S_{0,2} = S_{2,2} = 0$ and from Eqs. (3) and (5) the phase biaxiality is (5):

$$\tilde{\eta} = \tfrac{3}{2}\, S_{2,0}/S_{0,0} \tag{6}$$

The interesting feature in this case is that the value $\tilde{\eta}$ is independent of the deuteriated site, a very useful feature. It is perhaps instructive at this point to consider an application of Eqs. (5) and (6).

BIAXIAL MECHANISM OF THE CHOLESTERIC PHASE

In chapter 17 we saw how the value of $\tilde{\eta}^j$ could be measured for a cholesteric phase. In this chapter we are now prepared to make use of the measured values of $\tilde{\eta}^j$ to determine the mechanism for biaxial ordering in the cholesteric phase.

It is noted from Eq. (5) that in order to distinguish between $S_{2,0}$ and $S_{2,2}$, we need to measure a value for $\tilde{\eta}$ from two inequivalent deuteriated sites. To do this we have made use of the deuteriated materials 4-methoxybenzylidene-4'-butylaniline (MBBA) selectively deuteriated in the position (MBBA-α-d_2) and on one aromatic ring (MBBA-2',3',5',6'-d_4) (5). To each of these materials was added 4-methoxybenzylidene-4'-[(+)-2-methylbutyl]-aniline (MBMBA) in various quantities to obtain samples of different pitch lengths. Binary mixtures of MBBA-α-d_2 with 11.3 wt% MBBA and of MBBA-2',3',5',6'-d_4 with 20.0 wt% MBMBA (Fig. 1) were found optically to both have the same pitch length of 1.4 m. Values of the asymmetry parameter for these two mixtures were then measured. These values are shown in Fig. 2. We see from Fig. 2 that the measured value of $\tilde{\eta}$ is independent of the deuteriated site. As described above, this feature implies that there is no measurable ordering of the short molecular axis (i.e., $S_{2,2} \cong 0$) and that the ^{2}H-NMR biaxiality is a result of anisotropic fluctuations (i.e., $S_{2,0}$) of the long molecular axis

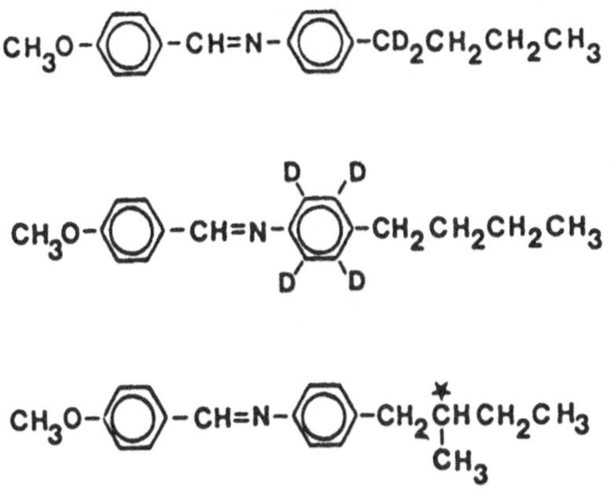

Figure 1. *Compounds: MBBA-α-d_2 (a); MBBA-2',3',5',6'-d_4 (b); chiral MBMBA (c).*

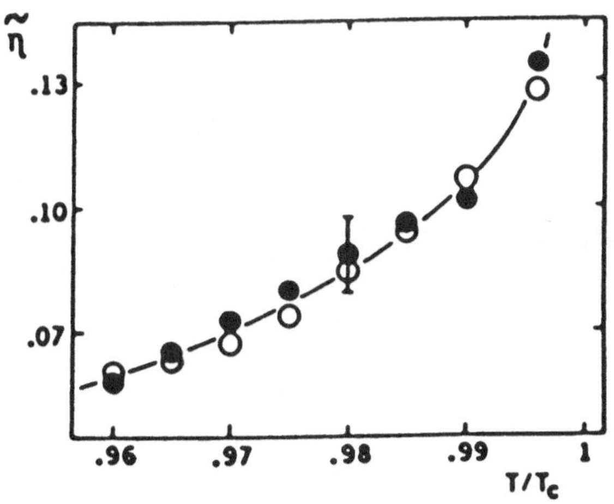

Figure 2. *Reduced temperature plot of the asymmetry parameter for binary mixtures of MBBA with MBMBA to yield a pitch length of 1.4 μm where ● = MBBA-α-d_2 and ○ = MBBA-2',3',5'6'-d_4.*

relative to the pitch axis of the helix. In fact it has been determined that the long molecular axis is least ordered relative to the pitch axis direction which then establishes this axis as the x axis under the convention $|V_{xx}| < |V_{yy}| < |V_{zz}|$ (6).

BIAXIALITY FROM ROTATIONAL FREEZE-OUT PARAMETERS

In chapter 18 it was shown that in the S_C phase the principal z axis of the methyl group in two different materials was shifted away from the z axis of the other deuteriated sites which were parallel to the direction of B in a sample prepared from cooling from the nematic phase. Furthermore, it was shown that in that case the third term of Eq. (1) of chapter 18 was the dominant term. It was left unclear in chapter 18, however, why the z axis of the methyl site was more shifted than the other sites and why it was the methyl splitting which followed the $\sin 2\theta_o$ dependence.

To answer that question we now turn to the coefficient C of the $\sin 2\theta_o$ term of Eq. (1) of chapter 18. The coefficient C in that term can, like $\tilde{\eta}$ and $\tilde{q}$, be related to orientational order parameters whereby (1,4):

$$C^j = \frac{3}{2}\frac{\tilde{q}^S}{\tilde{q}^j} \langle \sin 2\sigma_j \cos\xi_j \rangle \tag{7}$$

where the polar angles σ_j and ξ_j give the orientation of the jth C-D bond in the a,b,c frame. Like before, Eq. (7) can be expressed in terms of a molecular frame which in the case of a well ordered molecule (1,4) [i.e., $\theta(t) \cong 0$ or $S_{0,0} \cong 1$]

$$\langle \sin 2\sigma_j \cos\xi_j \rangle \cong s^j_{0,1}\, S_{1,1}$$

where

$$S_{1,1} = \tfrac{1}{2}[\overline{D}^2_{1,1} + \overline{D}^2_{1,-1} + \overline{D}^2_{-1,1} + \overline{D}^2_{-1,-1}]$$

$$= \langle \cos\theta \cos\phi \cos\psi \; - \cos 2\theta \sin\phi \sin\psi \rangle \cong \langle \cos(\phi+\psi) \rangle \,. \tag{8}$$

Here, $S_{1,1}$ represents "rotational freeze-out" and $s^j_{0,1}$ is as defined earlier. Note here that $S_{1,1}$ describes biasing in one orientation whereas $S_{2,2} \cong \langle \cos 2(\phi+\psi) \rangle$ shows biasing in two orientations about a two-fold rotation axis.

From the third term of Eq. (1) of chapter 18 we note that, for $\tilde{\eta} = 0$, $C^j = \frac{3}{4} \sin 2\theta^j_1$ and we have from Eqs. (7) and (8):

$$\sin 2\theta^j_1 \quad 2\, \frac{\tilde{q}^S}{\tilde{q}^j}\; s^j_{0,1} \langle \cos(\phi+\psi) \rangle \tag{9}$$

for the case where the long axis of the molecule is well ordered but there is a biasing toward one orientation as the molecule rotates about its long axis. From Eq. (9) we can now see why small values of $\tilde{q}^J$ yield larger values of θ_1 for a fixed biasing parameter $\langle\cos(\phi+\psi)\rangle$ for the molecule. In the smectic C experiment (7) in chapter 18 it is clear why the methyl site of 70BA-d_{16} which has a very small $\tilde{q}^J \cong 100$ Hz exhibited a large value of $\theta_1 \approx 51°$, whereas the other sites with $\tilde{q}^J \gtrsim 50$ kHz showed no measurable value of θ_1. It is also clear why the effect was also strong on the one methyl site of HOAB-d_{30}.

It is unfortunate, however, that in these samples a value of $\langle\cos(\phi+\psi)\rangle$ could not be measured since we do not have a value for the conformation parameter $s^j_{0,1}$. Unknown wiggling or conformational motion of the hydrocarbon chain prevents us from obtaining a value for this parameter. We know, however, that $\langle\cos(\phi+\psi)\rangle$ must be very small in that we only observe its effect on the site where $\tilde{q}^J \lesssim 100$ Hz. The molecule in the smectic C phase must therefore be only weakly hindered as it rotates about its long axis. A suitable experiment still waits to be done for the S_C phase in order to measure a value for $\langle\cos(\phi+\psi)\rangle$.

DISCUSSION

In this chapter I have attempted to illustrate the utility of order parameters in determining mechanisms for biaxial ordering. It is seen that many experiments wait to be done to measure these parameters and better understand molecular ordering in the liquid crystalline phases.

ACKNOWLEDGEMENTS

The author acknowledges his colleagues Dr. N. Vaz and Dr. Photinos who contributed toward the order parameter concepts. The work was supported in part by the National Science Foundation Solid State Chemistry grants DMR82-04342 and DMR82-44468.

REFERENCES

1. D.J. Photinos and J.W. Doane, *Mol. Cryst. Liq. Cryst.* **76**, 159 (1981).

2. See, for example, J.W. Emsley and J.C. Lindon, *NMR Spectroscopy Using Liquid Crystal Solvents* (Pergamon Press, Oxford, 1975) or N.A.P. Vaz and J.W. Doane, *Phys. Rev. A* **22**, 2238 (1980).

3. J.W. Doane, in *Magnetic Resonance of Phase Transitions*, edited by F.J. Owens, C.P. Poole, Jr., and H.A. Farach (Academic Press, New York, 1979), pp. 171-246.

4. D.J. Photinos, P.J. Bos, J.W. Doane and M.E. Neubert, *Phys. Rev. A* **20**, 2203 (1979).

5. Z. Yaniv, M.E. Neubert and J.W. Doane, in *Liquid Crystals and Ordered Fluids*, Vol. 4, edited by A.C. Griffin and J.F. Johnson (Pergamon Press, to appear).

6. Z. Yaniv, N.A.P. Vaz, G. Chidichimo and J.W. Doane, *Phys. Rev. Lett.* **47**, 46 (1981).

7. D.J. Photinos, P.J. Bos, M.E. Neubert and J.E. Doane, *Phys. Rev. A* **23**, 3346 (1981).

AMPHIPHILIC MOLECULES IN LYOTROPIC LIQUID CRYSTALS AND MICELLAR PHASES

J. Charvolin and Y. Hendrikx

Laboratoire de Physique des Solides*,
Université Paris-Sud.

INTRODUCTION

The behaviour of the paraffinic chains of amphiphilic molecules such as soaps, detergents and lipids, in aggregates of the liquid crystalline and micellar phases has been studied in detail with NMR. These studies demonstrated the dynamical disorder of the chains, particularly that introduced by the conformation changes of their carbon-carbon skeleton. In the most studied case, that of the lamellar phases with flat amphiphile/water interfaces, they also gave direct access to the disorder profile of a chain, making possible a description of its anisotropic behaviour around the local normal to the interface. A very limited number of studies only has been devoted to hexagonal and micellar phases, with cylindrical and spherical interfaces, where one might expect, on purely intuitive grounds, important changes in the disorder profile imposed by the curvatures of the interfaces.

We shall examine and compare in this respect a series of NMR data obtained when studying potassium soaps in the different aggregates of their various phases: infinite lamellae and cylinders of the ordered anisotropic lamellar and hexagonal phases, small spheroïds of the disordered micellar phases whose study is rather delicate because of their isotropy, small oblate spheroïds of the new nematic phases whose anisotropy makes the study of small aggregates easier. The surprising result is that the disorder profile of the chains is weakly dependent on the shape of the aggregate, i.e. on the interfacial curvature. We

* Laboratoire associé au CNRS, L.A.n° 2.

J. W. Emsley (ed.), Nuclear Magnetic Resonance of Liquid Crystals, 449–471.

shall briefly discuss this information and comment on it in relation with some thermodynamical approaches of these liquid crystals.

It is generally considered that the rich polymorphism of the so-called lyotropic liquid crystals is associated with the conformational polymorphism of the paraffinic chains of amphiphilic molecules. For instance at low temperature the chains are predominantly all trans, therefore can be regarded as

a) $CH_3(CH_2)_{16}CO_2K$

b) $CH_3(CH_2)_9OSO_3Na$

Figure 1. *Two typical amphiphilic molecules, potassium stearate (a), sodium decyl sulphate (b).*

rigid cylindrical rods, and they can aggregate in lamellar sheets only. At high temperatures, as they are deformed and disordered by isomeric rotations around their carbon-carbon bonds, the chains can aggregate in various shapes, such as the lamellae, cylinders and spheres shown in Fig. (2). Here it is expected that different states of disorder, or weighting of the conformations, would correspond to different aggregates as the chains adapt themselves to the packing constraints imposed by the interfacial curvature. Clearly this is an important point for understanding the structures of the aggregates, hence those of the lyotropic liquid crystals, and we shall consider it here in detail.

The first information about the disorder of the chains in the high temperature liquid crystals was obtained from X-ray scattering studies: a diffuse band is observed at large angles which looks like the one observed with melted paraffins of similar chain lengths (1). The chains therefore appear to be in a chaotic state which might be described as liquid. However this comparison can not be drawn too far because the two states are not quite analogous: in the liquid state the chain ends, which are sources of disorder, are uniformly distributed whereas in the liquid crystalline state half of them, the polar heads, are localized at the interface. Indeed NMR experiments, which are more powerful than X-ray ones to describe some aspects of local disorder, show that the two states are drastically different: the orientational order of the chain segments is never absent in a liquid crystalline lamellar phase as it is in the true liquid,

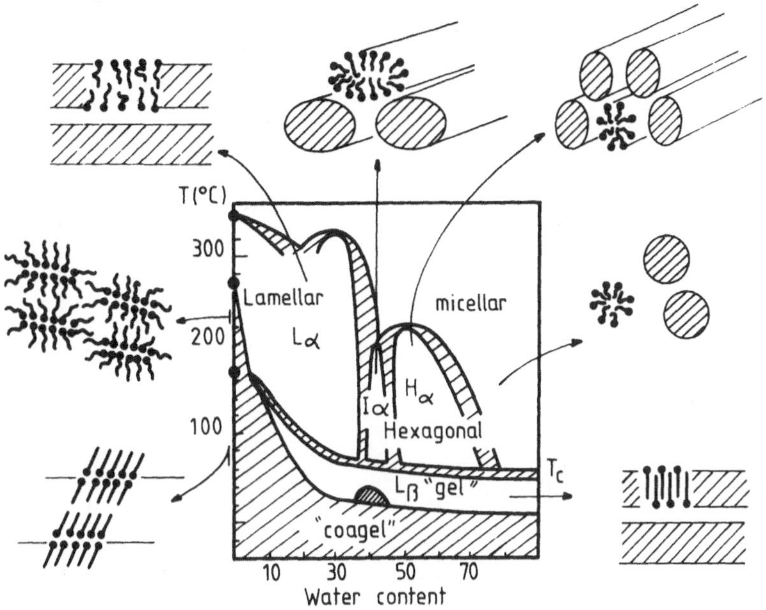

Figure 2. *Schematic representation of the phase diagram of the system potassium stearate (C_{18})/water. The concentration is expressed in weight per cent. The hatched regions are biphasic. The structure of one intermediate phase (I_α) only is drawn.*

as shown in Fig. (3) (2). Several models have been proposed to explain the slow decrease of the order observed for chains in lamellae with flat interfaces (3-8). Their fundamental common point is that the lateral steric constraints between chains, arising from the rather high density of polar heads at the interface, restrict the disordering of the chains and force them to span the monolayer along the normal to the interface in a somewhat stretched configuration (3,9). In a lamellar phase the methyl ends of the chains are therefore localized in the centre of the aggregate and the probability for them to come to the interface is negligible. Obviously we can not transfer this picture to phases with curved interfaces, hexagonal and micellar, without important modifications. In particular the methyl groups have to be expelled from the centres of the aggregates to prevent a divergence of the density caused by the splay of the normals. Indeed it has been demonstrated recently that in spherical micelles the methyl ends of the chains have a

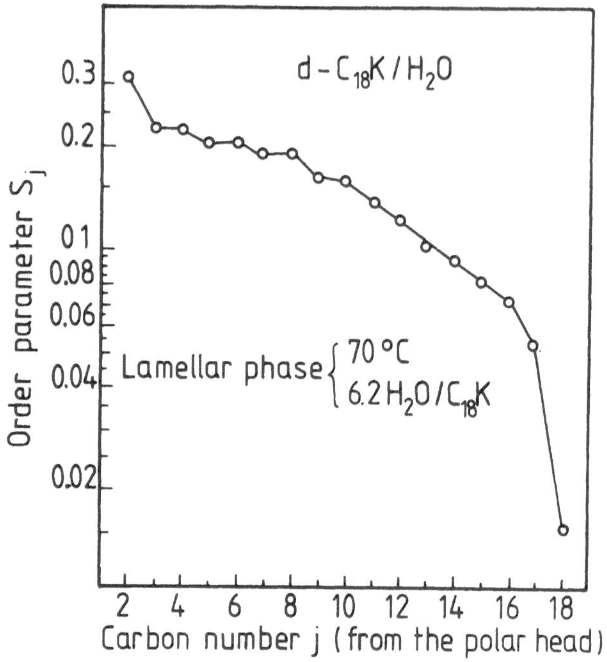

Figure 3. *Order parameters of the CD bonds of perdeuteriated potassium stearate chains (C_{18}) in a lamellar phase. The order parameter is measured relative to the lamella normal by DMR. The mean area A per molecule at the interface is 37Å^2.*

non negligible probability to be at the interface (10). The question now is how the chains exhibit such configurations: decreasing their stretching along the normal, i.e. increasing their conformational disorder, or/and stretching along directions different from the normal.

To answer this question we shall examine and compare the NMR data obtained with phases of increasing interfacial curvatures: lamellar, hexagonal, nematic, micellar. We shall develop particularly the study of the behaviour of the chains in the oblate spheroïds of the nematic phases (11,12). These are spheroïds of weak geometrical anisotropies which align spontaneously in concentrated solutions. They are therefore quite close to micellar aggregates but, whereas the study of a local orientational order is difficult in a micellar solution because it is isotropic, the anisotropy of the nematic solutions makes this approach easier. In order to develop such an analysis it is first necessary to recall a few particularities of the study of lyotropic liquid crystals with NMR (13,14).

RELEVANT ASPECTS OF NMR

We shall consider only dipolar and quadrupolar interactions whose effect on the spectrum of an oriented sample has been discussed by Veracini in chapter 6. For the systems of interest here the motions which may affect the orientations of chain segments, CH_2 and CH_3 have been discussed by Charvolin and Rigny (13). In Fig. (4) are shown the axes which are used to describe the motions present in a lyotropic sample according to what is known about their dynamics at the moment.

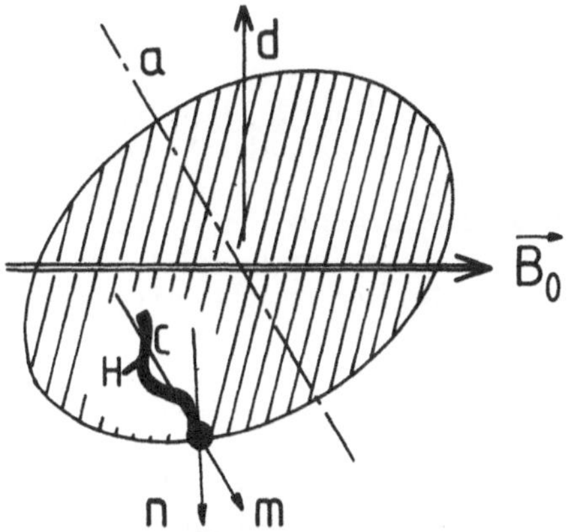

Figure 4. *Schematic representation of a section of an aggregate with axis* a *in a lyotropic phase of director* d *in a magnetic field* B_0. *The local directions are the normal* n *to the interface, the stretching axis* m *of the molecule, and the CH bond direction.*

(i) Rapid conformational changes ($\tau_c \sim 10^{-9}$s) which deform the molecules around a mean stretching axis *m*, and overall rotation about this axis.

(ii) Fluctuations of *m* about the local normal to the interface *n* (10^{-9}s $< \tau_f <$ 10^{-6}s); in a lamellar phase these motions might be collective in character, particularly the slowest (13).

(iii) Modulation of the orientation of the local normal *n* relative to the aggregate axis *a* experienced by the molecule as it diffuses along a curved interface

($\tau_d \sim \langle \ell^2 \rangle / D$ where $D \sim 10^{-6}$ to 10^{-7} $cm^2 s^{-1}$).

(iv) Fluctuation of the aggregate axes a relative to the director d of the phase (this occurs in the nematic phase but the dynamics are unknown).

We shall consider the molecules to be effectively of cylindrical symmetry about m, and for there to be the same symmetry about a, n, and d. The motions will be considered to be uncoupled so that it is possible to relate $\tilde{A}_B$, the partially averaged component of a dipolar or quadrupolar tensor along the magnetic field to A_I, the component along the symmetry axis of the interaction tensor in the molecular frame, by using the spherical harmonic addition theorem (appendix).

$$\tilde{A}_B = A_I \langle P_2(CH|m)\rangle \langle P_2(m|n)\rangle \langle P_2(n|a)\rangle \langle P_2(a|d)\rangle P_2(d|B_0). \quad (1)$$

Here (CH|m) represents the cosine of the angle between the CH direction and m, and $\langle P_2(CH|m)\rangle$ represents the effect of averaging over the conformational changes in an alkyl chain. Similarly $\langle P_2(m|n)\rangle$ represents averaging over fluctuations of chain axes m relative to interface normals n; $\langle P_2(n|a)\rangle$ represents the effect of translational diffusion and $\langle P_2(a|d)\rangle$ that of fluctuations of a relative to the director and is the order parameter of the phase. Finally, $P_2(d|B_0)$ allows for the angle between the director and the magnetic field, which from now on we will assume to be zero so that $P_2(d|B_0)$ is unity.

The nuclear relaxation times are also affected by the motions discussed above (see chapters 9 and 11); an exact description of relaxation for such complicated motion is difficult and we shall use an approximate one where it is considered that the motions fall into two categories, (i) and (ii) being fast (f) and the others slow (s). In this case we write the spin-lattice relaxation rate R_1 as,

$$R_1 = K^2(1-(\langle P_2(CH|m)\rangle \langle P_2(m|n)\rangle))^2 f^f(\omega)$$

$$+ K^2(\langle P_2(CH|m)\rangle \langle P_2(m|n)\rangle (1-\langle P_2(n|a)\rangle \langle P_2(a|d)\rangle))^2 f^s(\omega) \quad (2)$$

The value of K depends upon whether dipolar or quadrupolar interactions are responsible for relaxation, and $f^f(\omega)$ and $f^s(\omega)$ are functions of the appropriate spectral densities. This separation is a bit arbitrary in the general case. We shall use it only when interpreting relaxation times for nuclei in micellar samples where the symmetry of the phase makes the study of relaxation the only approach to obtaining orientational order parameters.

We can illustrate this formal presentation by considering the two extreme cases of the lamellar and micellar phases. The lamellar phase is a periodic organisation along one dimension of flat interfaces so that *d* and *a* are coincident and their common orientation relative to the magnetic field, B_0 is constant then

$$\tilde{A}_B = A_I <P_2(CH|m)> <P_2(m|n)> \quad (3)$$

and gives access to the local behaviour, $S = <P_2(CH|m)><P_2(m|n)>$, around the normal *n* to the interface directly. The order parameter curve of Fig. (3) is obtained from a direct measurement of $\tilde{A}_B$. On the other hand the micellar phase is a disordered solution of quasi-spherical aggregates: there is no director *d*, the axes of the aggregates *a* are in isotropic rotation, then

$$\tilde{A}_B = 0$$

and no direct information about the orientational order can be obtained in the micellar phase. Fortunately the study of the nuclear relaxation rate R_1 can provide some indirect information. Indeed, using the formula for R_1 with $<P_2(a|d)> = 0$ and extrapolating $f^f(\omega)$ and $f^s(\omega)$ from the studies of the ordered phases one may deduce from the measured R_1 values some estimate of S the local orientational disorder.

We shall now examine the data obtained in the different phases.

ORDER PARAMETER CURVES

Infinite Lamellae in Lamellar Phases

The simplest method of measuring the orientational order of the methylene groups of the chains in a lamellar phase consists of performing a nuclear magnetic resonance study of chains whose methylene groups have been deuteriated. The resonance of the deuterons of the jth methylene group is split by the residual quadrupolar interaction left by the anisotropy of the phase. The splitting $\Delta\tilde{\nu}$ is a direct measurement of S_j,

$$\Delta\tilde{\nu} = \frac{3}{2} q_{CD} S_j \quad (4)$$

The quadrupolar coupling constant q_{CD} is known and hence we have a direct access to the absolute value of the order parameter of the jth CD bond relative to the normal (14)

$$S_j = \langle P_2((CD_j)|m)\rangle \langle P_2(m|n)\rangle \ . \tag{5}$$

The variation of S_j along a potassium stearate chain (C_{18}) in a lamellar phase was shown in the introduction. We present here, in Fig. (5), a set of curves obtained in lamellar phases of potassium laurate (C_{12}) for different values of the polar head density at the interface (A^{-1}). The following analysis of this

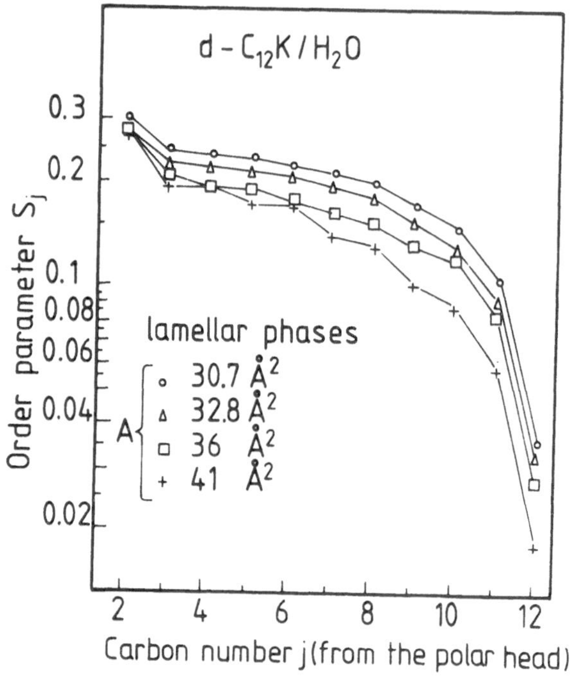

Figure 5. *Order parameters S_j, relative to the normal to the interface, of the $(CD)_j$ bonds of perdeuteriated potassium laurate chains in lamellar phases. The water content and the temperature determine the mean area per molecule at the interface A.*

set of curves was proposed: the slow decrease of S_j for $j<9$ is due mainly to the steric constraints introduced by the rather dense packing of the chains at the interface, the decrease becomes more rapid for $j>9$ because these constraints are partially relieved by the presence of chain ends in the central region of the bilayer. The hypothesis of the role of the steric interactions is confirmed by the fact that the disorder of the chain increases when their density at the interface, A^{-1}, decreases. But the disorder does not increase uniformly all along the chain: the later methylenes get more rapidly disordered than the first ones. This might be seen as a stiffening of the chain in the vicinity of its polar head or of the interface but there is also the possibility of an inertial effect introduced by the higher mass of the carboxylate group. The following step in this description would consist in separating the contributions of the conformational changes along the chain $\langle P_2((CD)_j(m)\rangle$ from that of the overall fluctuations of the chain $\langle P_2(m|n)\rangle$. In principle the conformational changes contribute to the distribution of the disorder along the chain and to its average level simultaneously whereas the overall fluctuations contribute to the average level of disorder only. Unfortunately we believe that the existing data are not sufficient to go further along this direction at the moment.

Infinite Cylinders in Hexagonal Phases

Here the molecules are aggregated in infinite cylinders packed along a two dimensional hexagonal lattice, the axes a of the cylinders are parallel to each other and define the director d, the normal to the interface is everywhere perpendicular to d. The deuteron NMR study gives the absolute value of the product of order parameters

$$\langle P_2((CD)_j|m)\rangle \langle P_2(m|n)\rangle \langle P_2(n|d)\rangle = S_j \langle P_2(n|d)\rangle$$

The last term can be easily calculated. It corresponds to the rotation of the local normal as the molecule diffuses along the interface, $\langle P_2(n|d)\rangle = -0.5$ (appendix). Therefore, here also we have direct access to S_j. The variation of S_j along a potassium laurate chain in an hexagonal phase with a mean area per molecule at the interface A = 52 Å^2 is shown in Fig. (6) (15). It is to be compared with those of the lamellar phases. If we choose a lamellar phase with A = 36 Å^2 it is very striking that the gradients of disorder are very similar in spite of so important changes of A and interfacial curvature. The S_j curve for the lamellar phase is only a factor 1.25 higher than that of the hexagonal phase. One may be tempted to say that the conformational states of the chain are similar in the two phases and that the amplitude of the overall fluctuations are a bit larger in the hexagonal phase. We shall come back to this in the discussion. Potassium laurate is not an unique case in this respect, the same behaviour was recorded in the hexagonal phase of a soap with shorter chain length (16).

Spheroïds in Micellar Phases

As was emphasized in the preceding paragraph about NMR, the

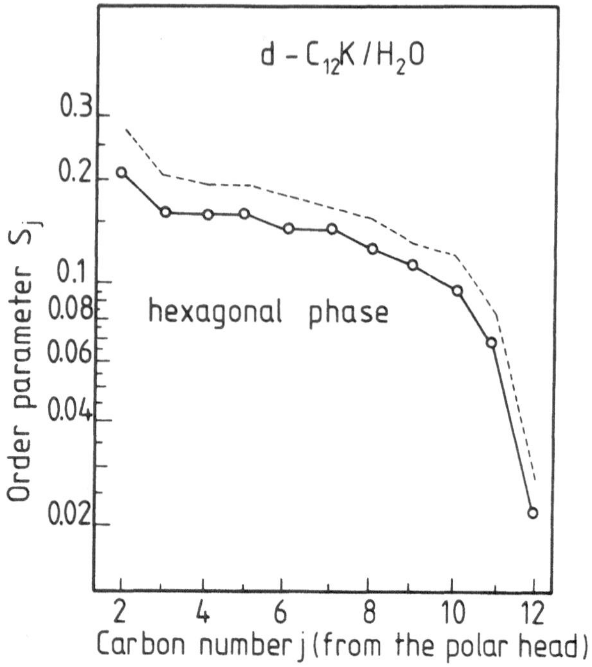

Figure 6. *Order parameters S_j, relative to the normal to the interface, of the $(CD)_j$ bonds of perdeuteriated potassium laurate chains in an hexagonal phase. The mean area per molecule at the interface is 52 Å² (60% weight H_2O, 50°C). The dotted curve is that of a lamellar phase with A = 36 Å².*

spectra do not provide direct access to the local anisotropy in an isotropic phase, but fortunately, some information can be gained from relaxation time measurements at different frequencies. This was done by studying the C^{13} relaxation of the chain skeleton induced by the modulation of the H^1-C^{13} dipolar couplings (17). The order parameters, S_j, obtained that way in a micellar phase of sodium octanoate with an estimated A of 80 $Å^2$ fall in the range 0.3-0.1, the first carbon having the highest value and the last one the lowest. This is quite similar to what we have observed in infinite lamellae and cylinders. Once again there is no drastic change of the gradient of orientational disorder in spite of the larger A and interfacial curvatures. It might be argued that this information is deduced with a model, the distinction between slow and fast motions as suggested by the studies of the ordered phases with infinite aggregates, whose appropriateness to disordered phases of finite aggregates might be questioned. It would therefore be interesting to find micellar aggregates in anisotropic situations to have access to S_j through direct measurements of $\Delta\tilde{\nu}$. This is possible with the so-called lyotropic nematic phases.

Oblate Spheroïds in Nematic Phases

Lyotropic nematic phases are concentrated aqueous solutions of aggregates of amphiphilic molecules with long range orientational order. They are phases of intermediate degree of order between the phases with long range translational order, such as the lamellar and hexagonal phases, and the totally disordered phases, such as the micellar phases. Their structural study is just at its beginning and we can picture the aggregates at the moment as oblate spheroïds, of disk-like shape, or prolate spheroïds, of rod-like shape, according to temperature and composition. In the solution the axes of the aggregate, a, are nearly parallel, they fluctuate about an average direction which is the director, d, of the phase. We shall consider here a nematic phase of oblate aggregates, or discotic phase N_D, in the vicinity of the transition to a phase of prolate spheroïds. Here the aspect ratio of the aggregates is rather low, about 2.8 and the area A = 60 $\mathring{A}^2$ (18,19). A drawing of the phase is given in Fig. (7). One can see that such a structure is well suited for our purpose: its aggregates are very close, if not identical, to the spheroïds of the micellar phases considered above, it is anisotropic so that DMR measurements may give access to the orientational order of the chains.

The DMR study gives the absolute value of the following products

$$\langle P_2((CD)_j|m)\rangle\ \langle P_2(m|n)\rangle\ \langle P_2(n|a)\rangle\ \langle P_2(a|d)\rangle = S_j \langle P_2(n|a)\rangle \langle P_2(a|d)\rangle$$

The two factors $\langle P_2(n|a)\rangle$ and $\langle P_2(a|d)\rangle$ depend on the structure of the aggregate and the degree or order of the phase respectively. The first one $\langle P_2(n|a)\rangle$ expresses the effect of the modulation of the orientation brought in by the translational diffusion of the molecule along the interface of the aggregate. It is obvious that its value depends on the percentages of flat and curved interfaces, which can be estimated from structural determinations by X-ray (18) and neutron (19) scattering. The second term $\langle P_2(a|d)\rangle$ is the order parameter of the phase, a lower limit of which can be extracted from the same scattering studies. Such a set of data, DMR (20) and scattering (19) ones, has been collected for a nematic phase of oblate spheroïds. The DMR data provide the product

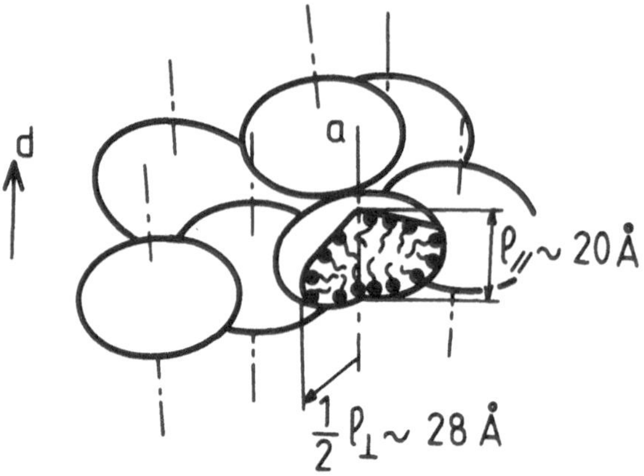

Figure 7. *Schematic representation of a nematic phase of oblate aggregates. The composition by weight is 30.35% potassium laurate, 7.04% decanol, 62.61% D_2O, the temperature is 17°C. (Erratum: $\ell_{//}$~23 Å and $\frac{1}{2}$ $\ell_{\perp}$~32 Å).*

S_j $\langle P_2(n|a)\rangle$ $\langle P_2(a|d)\rangle$ and from the scattering data (19) we can estimate $\langle P_2(n|a)\rangle$ following the calculation given in the appendix to be ~0.427 and the limits for the order parameter of the phase to be $1 > \langle P_2(a|\mathbf{d})\rangle > 0.7$. Therefore we can draw the upper and lower limit S_j curves in a nematic phase of oblate spheroïds. They are shown in Fig. (8). It is clear now that in quasi-spherical aggregates the gradient of orientational disorder along a C_{12} chain is close to those observed in infinite lamellae (Fig. (3,5)) and cylinders (Fig. (6)).

DISCUSSION

We have studied the orientational order relative to the normal to the interface of the CH_2 and CH_3 groups of soap molecules to detect some relations between the conformational states of their chains and the very rich polymorphism of their aggregates in lyotropic liquid crystals. After having described the state of the chains in lamellar phases, with flat interfaces and rather low mean area A per chain at the interface, from 30 to 40 Å^2, we moved to hexagonal phases, with cylindrical interfaces and larger area, A~52 Å^2, and then to micellar and nematic phases, with quasi-spherical interfaces and high A~60-70 Å^2. Doing so we expected to observe dramatic changes in the degree of

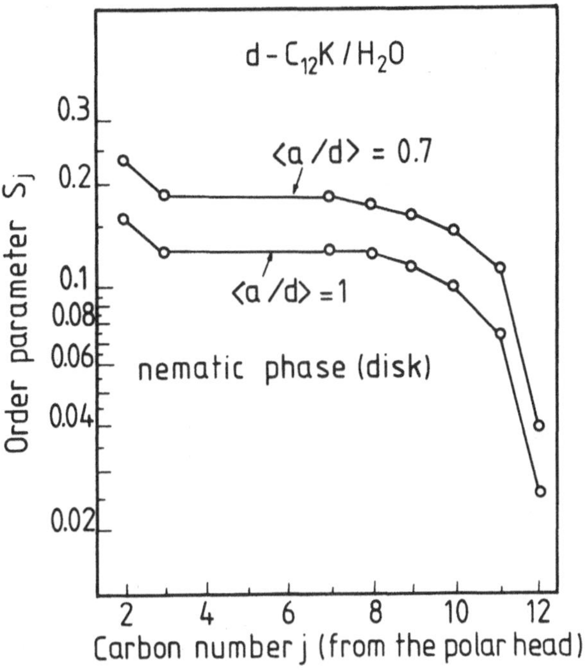

Figure 8. *Order parameters S_j, relative to the normal to the interface, for the $(CD)_j$ bonds of a perdeuteriated potassium laurate chain in the nematic phase of oblate spheroids of Fig. (7). The two curves correspond to the extreme values of the phase order parameter $\langle a|d\rangle$ ($\equiv \langle P_2(a|d)\rangle$).*

orientational disorder and its distribution along the chain, for instance a change of sign for the gradient of order as the mean area per chain at the interface A increases and the interface becomes concave on the chain side. Indeed we observe no important qualitative change, the general shapes of the curves and their magnitude are quite comparable, only small quantitative changes are observed which are summarized in Fig. (9).

We might invoke three possibilities to account for these observations (i) the method, DMR, is not sensitive to presumed differences in the chain disorder (ii) different average conformations of the chain around the stretching axis correspond to states of orientational disorder which are quite similar (iii) the "universal" behaviour does not result from a

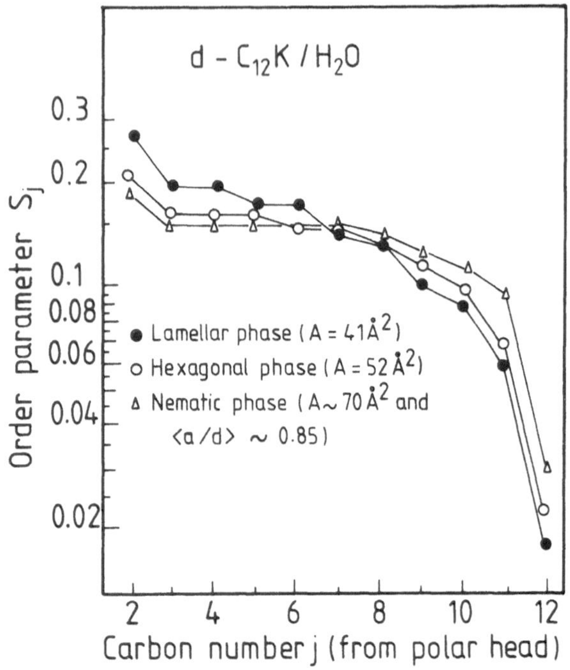

Figure 9. *Order parameters S_j, relative to the normal to the interface, of the $(CD)_j$ bonds of perdeuteriated potassium laurate chains in a lamellar phase (●), an hexagonal phase (○), a nematic phase of oblate spheroïds (Δ) which has been given an intermediate value of 0.85 for the phase order parameter $\langle a|d\rangle$ ($\equiv\langle P_2(d|d)\rangle$).*

limitation of the method. The first two points may be eliminated: point (i) because the sensitivity of the method to changes of mean area A is apparent in Fig. (5) and the disordering associated with a variation of A from 35 $Å^2$ to 60 $Å^2$ should be detected, point (ii) is improbable considering the number of aggregate shapes investigated and the corresponding average conformations which we might think compatible with these shapes. We focus therefore on the third possibility and look for the meaning of such an observation.

In Fig. (9) we have chosen to compare the results in the hexagonal and nematic phases with that of a lamellar phase having A = 41 $Å^2$. The latter is the largest value of A possible with a flat interface in this system, i.e. it is the reference to detect some effect associated with the curvature. While it is clear that the conformational disorder does not change dramatically one may notice however that the gradient of orientational disorder along the chain slowly decreases when the curvature increases from that of a lamella to that of a spheroïd. The first links become disordered, the last ones more ordered. Coming back to Fig. (5), from which we may extract the dependance of S_j with A, it is clear that none of the first segments of the chains becomes as disordered as it should be when A increases to 52 Å^2 and 60 Å^2, in the hexagonal and nematic phases respectively, and that none of the segments of the tails experience the expected decrease of their lateral area associated with the convergence of the normals to the centre. The curvature has therefore a weak effect. Even though there are indications that some more freedom is given to the links in the vicinity of the interface, in agreement with the increase of A, and that the tails experience a small decrease of lateral area in the centre of the aggregate, in agreement with the splay of the normals associated with the curvature, it shows that the chains suffer only small changes in their conformational states. This is to be related to the conclusion of previous structural studies by X rays and NMR of lamellar phases composed of mixtures of molecules with different chain lengths which suggests that the ability of the chains to increase their disorder far from what it is in a lamellar phase is limited. Indeed when short chains are put among long ones they introduce defects of density which are to be filled by an increased disorder of the tails of the long chains for the structure to remain lamellar. This is true for a low concentration of short chains but at high concentration the lamellar structure is modified in a way which shows that the long chains are reluctant to be more disordered (2,21). To end with the analysis of the curves of Fig. (5) and (9) it is interesting to notice that the relative disordering of the first three links hardly changes as previously observed in the lamellar phase.

Although not quantitative yet this discussion strongly suggests that the chains are somewhat reluctant to change their conformational state from what it is in the lamellar phase. Of course this imposes constraints on the packing of the chains in their aggregates. Thus if in hexagonal and micellar phases the chains want to remain almost as stretched around their average axis as they are in the lamellar phase, all of them can not converge simultaneously to the axis or the centre of the aggregates and some disorientation from the normal is needed in order to decrease the pressure at the centre and to fill in the unoccupied volume at the interface. This picture, which requires a few chains to have their methylene and methyl group at the contact with water, is supported by a recent NMR study of micelles labelled with paramagnetic ions (10). Also the probability for a chain to be along the interface should not be too high as the magnitudes of the curves keep rather close. This probability might be estimated through modelling, such as that of a recent mean field calculation of the chain behaviour in micelles which shows S_j curves quite similar to the experimental ones (22).

COMMENTS ABOUT THE THERMODYNAMICS

This discussion introduces two questions:

i) why the chains of amphiphilic molecules have such a behaviour when at an interface?

ii) does this behaviour play a role in the structure?

To answer the first question we may come back to the models which have been developed to understand the NMR data in the lamellar phases. Only a few of them consider variation of A sufficiently large for our purpose. The model (3) we have mainly referred to in our introduction is indeed valid only for small values of A. It is a model of flexible chains in steric repulsive interactions only but, when A increases, two new energy terms appear which can not be ignored. One is the intra-chain term related to the increase of the number of gauche conformations as the chains become more disordered: it decreases the intrinsic flexibility of the chains. The second is the attractive inter-chain term which becomes effective as the distance between the chains increases. We know two approaches which take these terms in account and provide the free energy of chains for sufficiently large variation of A. One is a computer simulation of chains adsorbed on a flat interface (23). We reproduce the free energy curves calculated with this model in Fig. (10). These curves present three minima for rather well defined values of A: 25,35 and 50 $Å^2$. The minimum at $Å^2$ = 25 A is an absolute minimum at 20°C, the chains are not deformed by gauche conformations: this state corresponds to that observed in the low temperature "gel" phase described in Fig. (2). The minimum at A = 35$Å^2$ becomes an absolute minimum when the temperature increases, the chains are deformed by gauche conformations and a few $g^{\pm}tg^{\pm}$ sequences, or "kinks"; this is the state of the lamellar phase and indeed the calculated and experimental S_j curves are in good agreement. Then, with increasing A, the free energy increases up to a barrier beyond which a third minimum appears for A ≅ 50 $Å^2$. There is some suspicion about the physical significance of this third minimum which might be an artefact introduced by the basic assumption of the computation, i.e. the chains are all constrained to have the same conformation. This model suggests that if we pull the

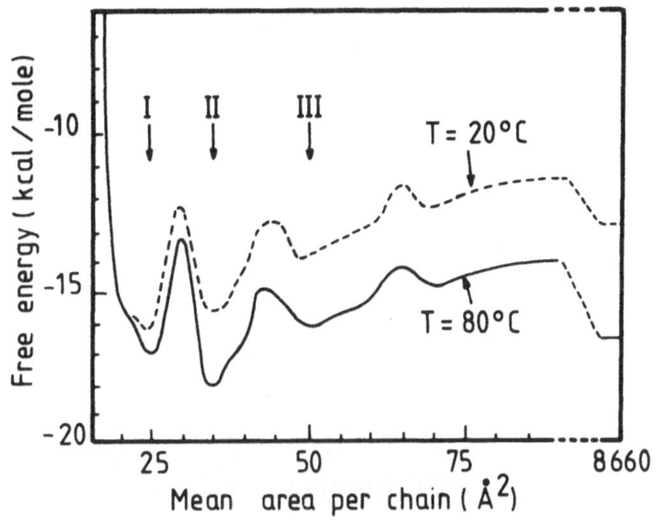

Figure 10. *Free energy of tetradecane chains adsorbed on a flat interface as their mean area A at the interface increases (23).*

chains apart from the value A ~ 35 $Å^2$ their free energy increases. A subsequent development of this model (24) shows that this increase may be depressed by the introduction of a slight interfacial curvature because in the compressed part of the chain the energetic gain is higher than the entropic loss. However this model does not consider the very important curvatures of the liquid crystals. The second model is a more recent mean field calculation of the free energy cost to pack dodecane (C_{12}) into bilayers, cylinders and spheres (25). The result is shown in Fig. (11). The energy scale is different from that of the preceding model but here also we observe a minimum for the flat interface of the lamellar phase in the same range of A. When A increases the free energy in the lamellar phase increases rapidly and the cylinders of the hexagonal phase appear more favourable. Also the order parameter curves deduced from this second model agree qualitatively with the experimental ones shown in Fig. (9).

Therefore these two models tell us that if we pull the polar heads apart along a flat interface, from A~35-40 $Å^2$, by increasing the degree of hydration, the chains will resist. This antagonistic behaviour of the polar and paraffinic layers may

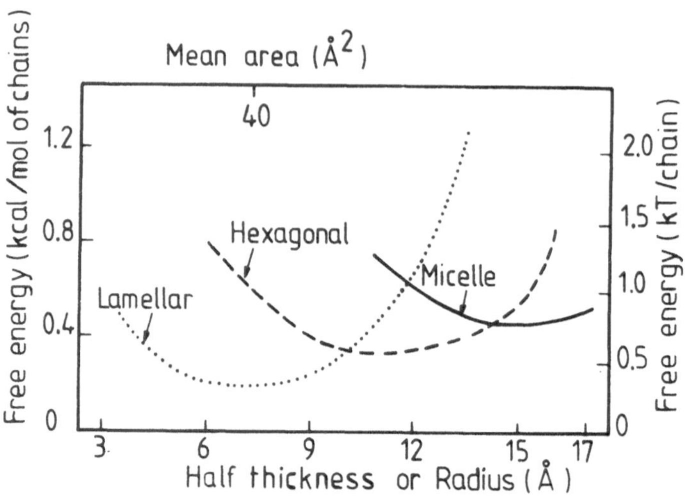

Figure 11. *Free energy cost of transferring the C_{12} chains from a bulk n-alkane environment to the aggregate (25).*

lead to a curvature of the interface which, in the case discussed above, should become concave on the paraffinic side. But, as we have deduced from our data that the conformational state of the chains hardly changes, this mechanism therefore introduces an energy term of curvature elasticity associated with the splay of the chain mean axes. This is a term which is not often included in the thermodynamics approach to the structures of lyotropic liquid crystals where the chains are treated as passive, fitting the volume imposed by the polar layer without strains (26,27). The possible roles of this term have been discussed recently in the context of bilayer stability (28) bilayer defects (29) and rod/disk coexistence in micellar solutions (30).

CONCLUSION

In this chapter we emphasized the fact that NMR data of lyotropic liquid crystals can be analyzed and interpreted within the frame of the precise structural descriptions provided by scattering experiments only. We illustrated this discussing the behaviour of soap molecules in the different phases of the potassium laurate/water system for which a relatively large quantity of structural information is available. By doing so we could approach the important point of the relation between the molecular behaviour and the structure of the phase which is not yet clear in the existing models. The striking fact is that the averaged molecular conformation, as seen by DMR, is hardly dependent on the structure, more precisely it hardly changes with the mean area per polar head at the interface and the curvature of the interface. This result suggests the necessity of including an energy term of elastic curvature associated with a splay of the mean molecular axes when considering the thermodynamics of the mesophases of this system. In view of the very close resemblances between the phase diagrams of different amphiphilic molecules it may be expected that this conclusion, deduced from the study of one particular system, is general.

As far as the particular aspect of the behaviour of alkyl chains is concerned we think it appropriate to quote here some very recent results which may provide new insights into this problem. They show that what we called a "universal" behaviour in lyotropic liquid crystals is indeed not limited to this field. Alkanes and the alkyl chains of soaps and lipids exhibit the same order parameter curves as the ones discussed up to now when simply solubilized in nematic solvents (31). Such solvents build no interface and are characterized by a uniaxial field only. Therefore these results emphasize the role of the latter, as it is associated with the existence of an interface, and minimize the role of any specific mode of anchoring of the molecule at the amphiphile/water interface. It is also clear now that much care must be exercised when using DMR alone as a tool for structural characterizations in lipidic systems.

REFERENCES

1. V. Luzzati, in *Biological Membranes*, p.71, edited by D. Chapman, Academic Press (1968).

2. B. Mély and J. Charvolin, in *Physico-Chimie des Composés Amphiphiles*, p.41, edited by R. Perron, Colloque National du CNRS n° 938, CNRS (1979).

3. P.G. de Gennes, *Phys. Lett.* **A47**, 123 (1974).

4. S. Marcelja, *Biochim. Biophys. Acta* **367**, 165 (1974).

5. P. Bothorel, J. Belle and B. Lemaire, *Chem. Phys. Lipids* **12**, 96 (1974).

6. F. Jähnig, *J. Chem. Phys.* **70**, 3279 (1979).

7. K.A.Dill and P.J.Flory, *Proc.Natl.Acad.Sci.* **77**, 3115 (1980).

8. D.W.R. Gruen, *Biochim. Biophys. Acta* **595**, 161 (1980).

9. This may be so whatever the chain length, S. Alexander, *J. de Physique* **38**, 983 (1977).

10. B. Cabane, *J. de Physique* **42**, 847 (1981).

11. B.J. Forrest and L.W. Reeves, *Chemical Reviews* **81**, 1 (1981).

12. J. Charvolin and Y. Hendrikx, in *Liquid Crystals of One and Two-Dimensional Order*, edited by W. Helfrich and G. Hepke, Springer Serie in *Chemical Physics* **11**, 265 (1980).

13. J. Charvolin and P. Rigny, *J. Chem. Phys.* **58**, 3999 (1973).

14. J. Charvolin and A. Tardieu, *Solid State Physics* **suppl. 14** edited by F. Seitz and D. Turnbull, p. 209, Academic Press (1978).

15. B. Mély, J. Charvolin and P. Keller, *Chem. Phys. Lipids* **15**, 161 (1975).

16. I. Henriksson, L. Odberg and J.C. Ericsson, *Mol. Cryst. Liq. Cryst.* **30**, 73 (1975).

17. T. Ahlnäs, H. Walderhaug and O. Söderman, *International Symposium on Surfactants in Solution*, Lund (1982), K.L. Mittal and B. Lindman editors, Plenum Press (1984).

18. J. Charvolin, A.M. Levelut and E.T. Samulski, *J. de Phys. Lett.* **40**, L-587, (1979).

19. Y. Hendrikx, J. Charvolin, M. Rawiso, L. Liébert and M. Holmes, *J. Phys. Chem.*, **87**, 3991 (1983).

20. F.Y. Fujiwara and L.W. Reeves, *J. Phys. Chem.* **84**, 653 (1980). B.J. Forrest and L.W. Reeves, *J. Am. Chem. Soc.* **103**, 1641 (1981). Y. Hendrikx, unpublished results.

21. J. Charvolin, *J. de Chimie Physique* **80**, 15 (1983), (special issue on Polymorphism in Liquid Crystals).

22. D.W.R. Gruen, *J. Colloïd Interface Sci.* **84**, 281 (1981).

23. J. Belle and P. Bothorel, *Nouveau Journal de Chimie* **1**, 265 (1977).

24. B. Lemaire and P. Bothorel, *Macromolécules* **13**, 311 (1980).

25. D.W.R. Gruen and E.H.B. de Lacey, *International Symposium on Surfactants in Solution*, Lund (1982), K.L. Mittal and B. Lindman editors, Plenum Press (1984).

26. J.N. Israelachvili, D.J. Mitchell and B.W. Ninham, *J. Chem. Soc. Faraday Trans II* **72**, 1525 (1976).

27. B. Jönsson and H. Wennerström, *J. Coll. Interface Sc.* **80**, 482 (1981) and *International Symposium on Surfactants in Solution*, Lund (1982), K.L. Mittal and B. Lindman editors, Plenum Press (1984).

28. W. Helfrich in *Physique des défauts, Les Houches, Session XXXV*, 1980, R. Balian Editor, North Holland (1981).

29. A.G. Petrov, M.D. Mitov and A.I. Derzhanski in *Advances in Liquid Crystal Research and Application*, Ed. L. Bata, Pergamon Press, Vol. 2, p. 695 (1980).

30. W.E. McMullen, A. Ben-Shaul and W.M. Gelbart, *J. Coll. Interface Sci.*, **98**, 523 (1984).

31. E.T. Samulski, private communication, and *J. Israel Chem. Soc.*

APPENDIX

Motional Averaging for a CD Bond in Rotation

The quadrupolar splitting for a CD bond at an angle θ from the magnetic field B_0 can be written

$$\Delta\nu = \frac{3}{2}\,\frac{e^2q\,Q}{h}\,\langle P_2\cos\theta)\rangle = \frac{3}{2}\,\frac{e^2q\,Q}{h}\,\langle\frac{3\cos^2\theta-1}{2}\rangle\ .$$

When the CD bond has a rapid motion of uniaxial rotation around

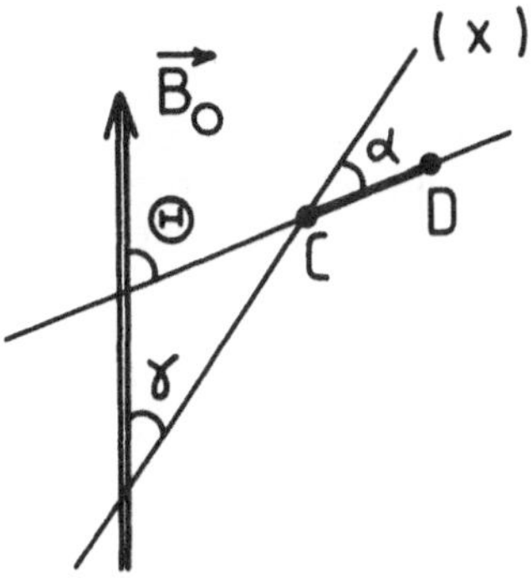

an axis x it is possible to express $\Delta\nu$ in a simple way using the theorem of addition of spherical harmonics

$$\Delta\nu = \frac{3}{2}\,\frac{e^2 q\,Q}{h}\,\left\langle\frac{3\cos^2\alpha-1}{2}\right\rangle\left\langle\frac{3\cos^2\gamma-1}{2}\right\rangle$$

in the text we choose to write the spherical harmonics indicating which axis is referenced with respect to which other one, therefore $\Delta\nu$ becomes

$$\Delta\nu = \frac{3}{2}\,\frac{e^2 q\,Q}{h}\,\langle P_2(CD|x)\rangle\,\langle P_2(x|B_0)\rangle$$

and the value of the static quadrupolar coupling constant, e^2qQ/h is determined by the nature of the C-D bond, as discussed in Chapter 6.

Averaging Effect of the Translational Diffusion

The translational diffusion of a molecule along the curved surface of an aggregate will modulate the orientation of the average local axis, the normal to the interface in our model, relative to the magnetic field. This motion is the cause of an averaging of the quadrupolar interaction which contributes to reducing the splitting of the lines by a factor

$$\langle P_2(n|a)\rangle = \left\langle\frac{3\cos^2(n,a)-1}{2}\right\rangle$$

where, following our conventions, n is the normal to the interface, a the axis of symmetry of the aggregate and $(n|a)$ their angle.

In lamellar phases n and a are both normal to the lamella so that $(n,a) = 0$ and $\langle P_2(n|a)\rangle = 1$, in hexagonal phases n the normal and a the axis of the cylinder are perpendicular so that $(n,a) = \pi|2$ and $\langle P_2(n|a)\rangle = -\frac{1}{2}$. For the finite aggregates of the micellar and nematic phases the angle (n,a) is no longer constant as above but varies along the interface and the value of $\langle P_2(n|a)\rangle$ has to be obtained through an integration over the interface. The result is obvious for perfect spheres, $\langle P_2(n|a)\rangle = 0$, but the calculation must be done for the oblate spheroïds shown in Fig. (7).

As their exact shape is not yet precisely known we can approximate them also as disks with rounded rims for a simpler calculation.

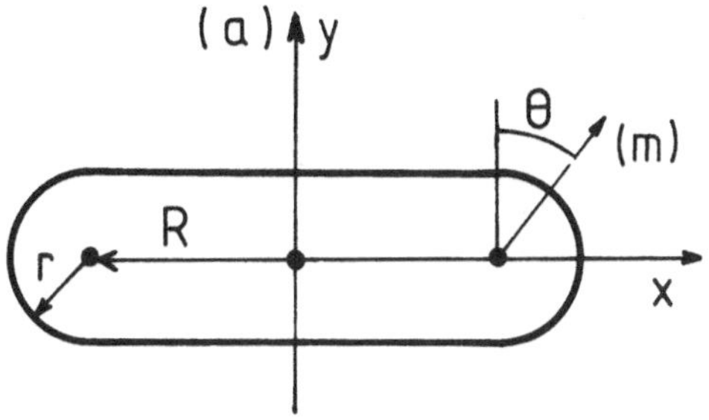

$$\langle P_2(n|a)\rangle = \int_{\text{(interface)}} \frac{3\cos\ \ -1}{2}\ ds \ / \int_{\text{(interface)}} ds$$

$$\int_{\text{(interface)}} ds = 2\pi R^2 + 2\pi\int_0^{\pi} (R+r\sin\theta)r\,d\theta = 2\pi(R^2+\pi Rr + 2r^2)$$

$$\int_{\text{(interface)}} \frac{3\cos^2\theta-1}{2} = 2\pi R^2 + 2\pi \int_0^{\pi} \left(\frac{3\cos^2\theta-1}{2}\right)(R+r\sin\theta)r\,d\theta = 2\pi\left(R^2+\frac{\pi}{4}Rr\right)$$

and if we write R r = ρ−1

$$\langle P_2(n|a)\rangle = \tfrac{1}{4}\ (4(\rho-1)+\pi)\left((\rho-1)+\frac{2}{(\rho-1)}+\ \right)^{-1}$$

as the anisotropy ratio $\rho = 2.8$ for the aggregates studied here then $\langle P_2(n|a)\rangle = 0.427$.

STRUCTURE OF THE HYDROPHOBIC-HYDROPHILIC INTERFACE IN LYOTROPIC LIQUID CRYSTALS: COUNTERION BINDING STUDIES

N. Boden and S.A. Jones

Department of Physical Chemistry,
The University,
Leeds LS2 9JT, England.

INTRODUCTION

The macroscopic structure of the mesophase which obtains for a particular surfactant-water mixture is governed by the size and shape of the amphiphilic aggregate. The latter is, in turn, determined by the molecular interactions, especially those at the aggregate-water interface and which control the surface curvature (1). These can be probed by studying the interaction and binding of both counterions and water at this interface using NMR methods. Arguably, the most informative quantities to measure are quadrupolar splittings of the alkali metal ions and ^{2}H and ^{17}O in labelled water. The object of this chapter is to critically assess the information obtainable from such measurements.

The lamellar phase, which consists of bimolecular lamellae separated by layers of water, has invariably been used as a model system for counterion binding studies. In particular, the quadrupolar splitting of ^{23}Na in the lamellar phase of the sodium octanoate-decanol-water system (2) has been extensively studied (3-5). The results of these studies will be briefly reviewed. Then new and more extensive measurements on the more practicable potassium oleate-water system will be described. The effects of temperature, hydration and solubilizates on the quadrupolar splitting of ^{39}K will be examined. The influence of solubilizates on counterion binding is of especial interest as it will reflect the manner in which the solubilizate perturbs the structure of the bilayer-water interface. Such measurements alone are, however, insufficient to establish a molecular model for the solubilization process. In the following chapter it will

J. W. Emsley (ed.), Nuclear Magnetic Resonance of Liquid Crystals, 473–496.

be shown how these measurements when combined with complementary ^{2}H NMR measurements on both solubilizate and amphiphile and X-ray diffraction measurements of the bilayer dimensions can lead to quite detailed information about the solubilizate-bilayer interaction.

THEORY OF COUNTERION QUADRUPOLAR SPLITTINGS

To date, NMR studies of alkali metal ions have focussed mainly on ^{23}Na (3-15) and to a lesser extent on ^{7}Li (3,15,16), ^{133}Cs (3,15,17), and $^{85/87}Rb$ (3). Despite the diversity of the mesophases formed by potassium soaps, ^{39}K has not hitherto been studied because it is among the least NMR sensitive of all atomic nuclei (18) and the use of a very high field spectrometer is essential. The quadrupolar splittings of the halide ions seem to be rather large and they have been little studied (19).

The most striking feature of the spectrum of a counterion in a lamellar phase or indeed in any other lyotropic mesophase is that it is dominated by the nuclear quadrupole-electric field gradient interaction. Moreover, the exchange of counterions among their accessible sites is fast compared to the inverse of the quadrupolar interaction so that the observed spectrum represents an ensemble average. For a nucleus with spin $I \geq 1$ in a macroscopically aligned uniaxial mesophase this spectrum will, to first order, consist of 2I equally spaced lines corresponding to the various $m \leftrightarrow m + 1$ transitions each with intensity roughly proportional to $[I(I+1) - m(m-1)]$. The separation of the lines, referred to as the quadrupole splitting, is given by [20,21]

$$\widetilde{\Delta\nu}(\phi) = \frac{3}{2I(2I-1)} |\tilde{q}_{zz}| P_2(\cos\phi) \tag{1}$$

with

$$\tilde{q}_{zz} = \sum_n p_n \chi_n \{S_{cc}^{\ n} + \eta_n (S_{aa}^{\ n} - S_{bb}^{\ n})\} . \tag{2}$$

In equation (1), ϕ is the angle between the mesophase director and the direction of the applied magnetic field. $\tilde{q}_{zz}$ is the partially averaged value of the nuclear quadrupole-electric field gradient coupling measured parallel to the director: it represents an ensemble average over all of the n sites that the counterion accesses in times less than the inverse of the spread of the quadrupole splittings over these sites. The ^{39}K spectrum observed in a macroscopically aligned sample of the N_c^{+} nematic phase of potassium dodecanoate-KCl-D_2O (22) and shown in Fig. (1a) confirms that such an averaging does indeed occur. For ^{39}K,

$I = \frac{3}{2}$, and we see from equation (1) that the first order spectrum consists of three equally spaced lines of relative intensities 3:4:3 and with separation $\Delta\tilde{\nu}(\phi=0°) = |\tilde{q}_{zz}|/2$ as is observed.

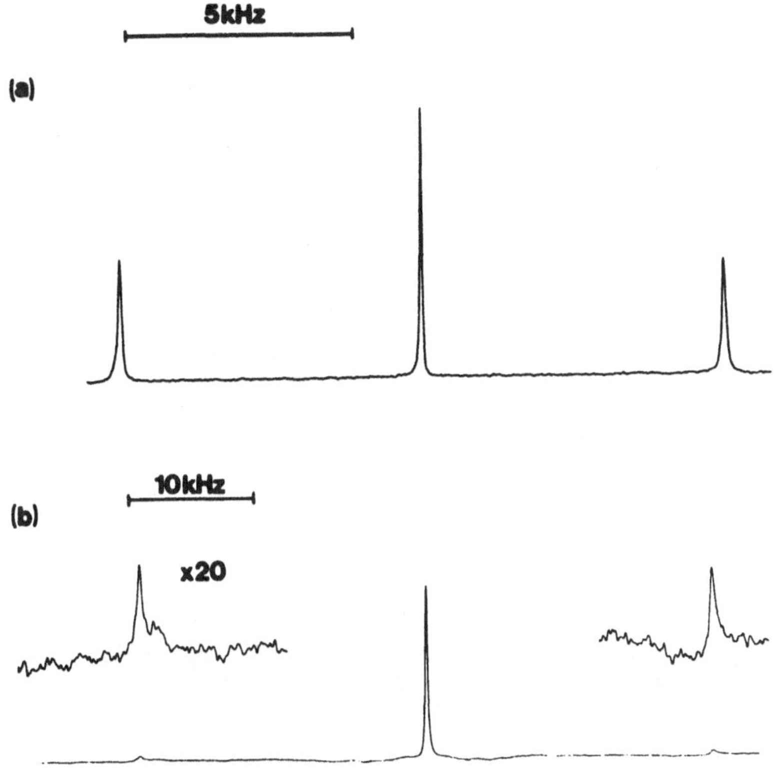

Figure 1. *^{39}K spectra for (a) an aligned N_C^+ nematic mesophase of potassium dodecanoate-KCl-D_2O (34.0:2.25:63.75 wt %) and (b) an unaligned lamellar phase of potassium oleate-water (70:30 wt % ≡ $R_{W/A}$ = 7.6). In (a) 3094 scans were accumulated with a recycle time of 0.04 s, total experiment time 2.06 min, whilst in (b) 155591 scans were accumulated with a recycle time of 0.02 s, experimental time 51.8 min.*

The S_{ij}^n in equation (1) are the elements of the Saupe ordering matrix (23) for the principal axes (a,b,c) of the nuclear quadrupolar interaction tensor at the nth site with statistical weight p_n. $\chi_n = (\frac{e^2qQ}{h})_n$ is the quadrupolar coupling constant and η_n the asymmetry parameter. The values of these latter two quantities will not be known; they will depend on the detailed structure of the binding site, especially on the distortion of the hydration shell of the counterion from spherical symmetry. The coupling of the nucleus with the electric field gradients which are due to charged groups in the surface does not contribute significantly to the observed quadrupolar splitting.

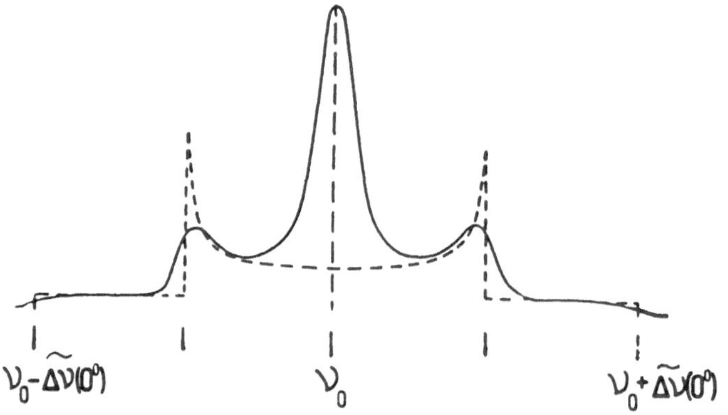

Figure 2. *First order powder spectrum for $I = \frac{3}{2}$ in a uniaxial mesophase (21). The dashed curve assumes that only quadrupolar interactions contribute, whilst the solid curve includes broadening due to spin relaxation or unaveraged dipolar coupling.*

The first order powder spectrum predicted for an unaligned uniaxial phase is illustrated in Fig. (2). This is to be compared with the ^{39}K spectrum (Fig. (1b)) observed for an unaligned sample of the lamellar phase of the potassium oleate-water system. The central component is seen to be located symmetrically with respect to the outer components and does not show any broadening caused by either second order effects or chemical shielding anisotropy (21). The outer components correspond to the singularities of the powder spectrum which occur at $\nu_0 \pm \frac{1}{2}\Delta\nu(0°)$; thus, their separation is equal to the quadrupole splitting. For the spectrum given in Fig. (1b), $\widetilde{\Delta\nu}(0°)$ is 46.212 kHz and it was measured using a Bruker WH-400 spectrometer operating at 18.67 MHz. The pulse width was 60 μs which corresponds to a uniform excitation spectral width ($\Delta\nu_s \approx 1/\pi\, t_p$) of 5.3 kHz. It is, therefore, quite reasonable to suppose that the amplitudes of the singularities are smaller than their expected values because of the finite bandwidth of the spectrum of the pulse. Since lamellar phases are difficult to align, spectra are usually measured on unaligned samples.

SODIUM-23 STUDIES

Nonionic Surfactants

Fig. (3) shows the ^{23}Na splitting for various concentrations of sodium chloride in the lamellar phase of the 1-monooctanoin-water system (5). The splitting increases with

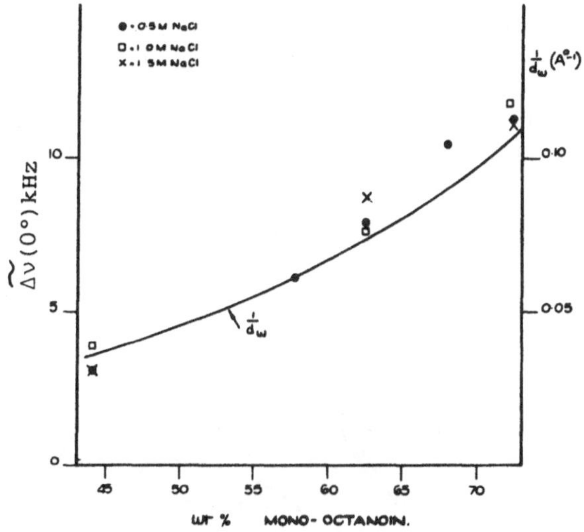

Figure 3. *Dependence of the ^{23}Na quadrupolar splitting $\tilde{\Delta\nu}(0°)$ on salt and water content in the lamellar phase of 1-monooctanoin-water system. The continuous line represents the inverse of the water layer thickness d_w (5).*

decreasing water content and is practically independent of salt concentration and temperature. This behaviour is explained by assuming the observed splitting arises from distortion of the hydration sheath of Na^+ ions adsorbed on the bilayer surface. If d is the thickness of this adsorbed layer and there is a uniform distribution of ions across the water layer of thickness d_w, then

$$\widetilde{\Delta\nu} = 2(d/d_w)\ \widetilde{\Delta\nu}_{bound} \tag{3}$$

which is consistent with the observed splitting. Note that the presence of charged surfactants is not necessary to produce splittings.

Charged Surfactants

The composition dependence of the ^{23}Na quadrupolar splittings as observed (5) in the sodium octyl sulphate-decanol-water system is summarized in Fig. (4) and for the sodium octanoate-decanol-water system in Fig. (5). In the former and also in the latter system at high water contents (>60% by wt. water) the splitting is essentially independent of temperature and hydration. This behaviour can be explained in terms of the ion-condensation model developed by Wennerström and co-workers (24).

At the high water contents involved the hydration of the interface is complete and addition of water causes a one-dimensional swelling (2). The only structural parameter which changes is the thickness of the water lamellae. For the system in Fig. (4) this thickness changes from 20 to 85 Å. The ion condensation model (24) predicts that when the thickness (2a) of the water lamellae exceeds a critical value which

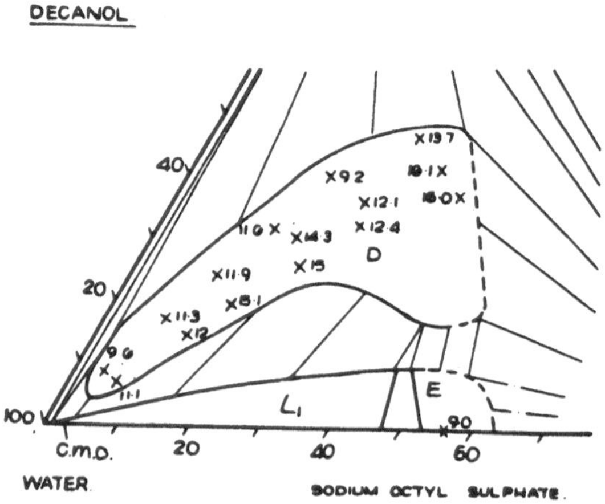

Figure 4. *Dependence of the ^{23}Na quadrupolar splitting $\widetilde{\Delta\nu}(0°)$ on composition for sodium octyl sulphate-decanol-water system (5). Splittings are in kHz.*

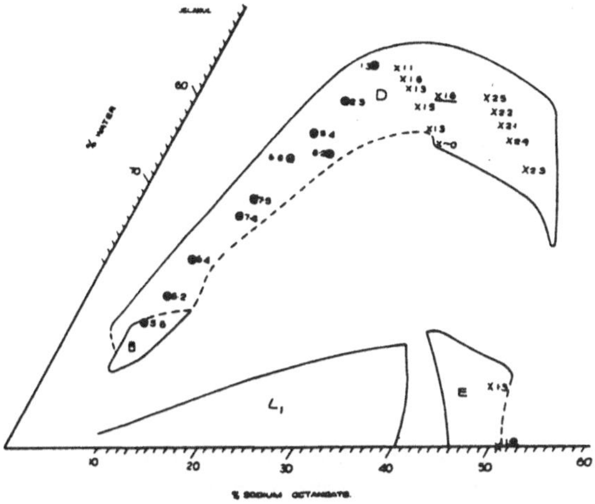

Figure 5. *Dependence of the ^{23}Na quadrupolar splitting on composition in the lamellar phase of the sodium octanoate-decanol-water system;* ⊗ *: splitting is either invariant with temperature or increases with temperature. X: splitting decreases, passes through zero, and then increases with increasing temperature (5).*

depends upon the surface charge density σ according to

$$(|\sigma|a)_{crit} = 1.31\ \pi\varepsilon\varepsilon_0\ kT\ |z|/e \tag{4}$$

a fraction (≈0.5) of the counterions, which depends upon σ, condense on to the surface. For values of $(|\sigma|a)$ larger than the critical value the number of bound ions is predicted to be independent of water content, temperature and addition of a simple salt. For the octyl sulphate system σ is one unit charge per 82 $Å^2$ which gives a critical surface separation of 7.5 Å. This is much less than the actual separation and explains why the observed splitting is independent of water content and temperature. The effect on the splitting of the addition of a salt which contains the counterion of the surfactant is also consistent with this model (25) and the assumption that the splitting is proportional to the fraction population p_b of a single 'bound' site, viz.

$$\Delta\tilde{\nu} = p_b \, |\Delta\nu_b| \tag{5}$$

The ^2H quadrupolar splitting of D_2O, as observed at these high water contents, can also be accounted for in terms of the above expression (26). Assuming the splitting vanishes for 'unbound' water molecules, then

$$\Delta\tilde{\nu}_{D_2O} = \frac{nx_A}{x_{D_2O}} \, |\Delta\tilde{\nu}_b| \tag{6}$$

where $\Delta\tilde{\nu}_b = \frac{3}{2}x_D\langle S_{DD}\rangle$

(7)

and $\langle S_{DD}\rangle = \sum_n p_n\{S^n_{cc} - \eta_n(S^n_{aa} - S^n_{bb})\}$.

The splittings observed for the lamellar phases of alkali metal octanoate-decanol-water systems are plotted versus x_A/x_{D_2O} in Fig. (6) and are seen to be consistent with equation (5).

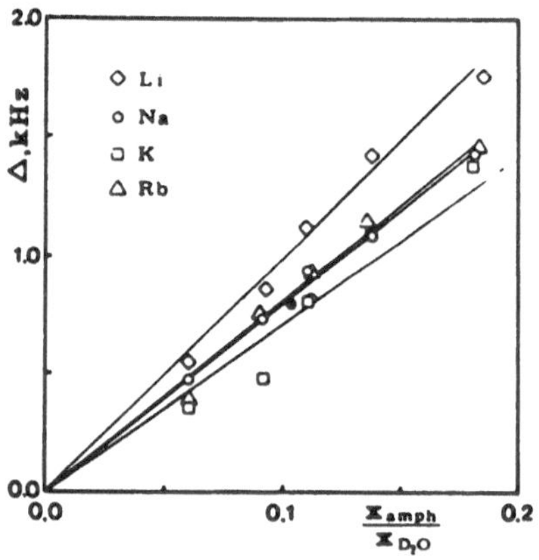

Figure 6. ^2H *quadrupolar splitting of* D_2O *in alkali metal octanoate-decanol-water systems as a function of the mole ratio of total amphiphile-to-water at constant mole ratio of soap to total amphiphile of 0.27 (26).*

The distinction between 'bound' and 'unbound' water molecules is seemingly pertinent; moreover, the order parameter of the latter is zero implying no long-range ordering of the water molecules by the amphiphilic lamellae. The upper limit of the linearity of these plots gives a value of about 5 for the hydration number n for all the alkali metals and this, when it is used with x_D = 220 kHz, gives $|S_{cc}|$ = 0.01.

At low water contents, where all of the water is bound to the counterions or polar head groups, the observed splittings are a complex function of composition and temperature (see Fig. (5)). The sodium octanoate-water-decanol system is not the most amenable for studies of the composition dependence because a long chain alcohol is needed for the formation of a lamellar phase and this severely limits accessible compositions.

POTASSIUM-39 STUDIES

The potassium oleate-water system is a very convenient system to study. Phase diagrams are available for both the binary system (27) and the ternary systems formed on solubilizing decanol (21) (Fig. (7)) and p-xylene (2) (Fig. (8)). The binary system exhibits a lamellar phase extending from 18 to 32% by weight of water at 21°C; this phase is stabilized up to approximately 90% by weight of water on addition of decanol, whilst p-xylene has

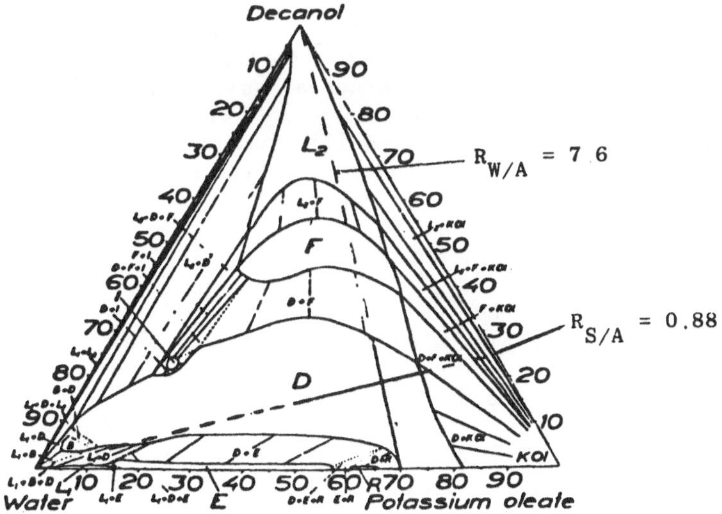

Figure 7. *Potassium oleate-decanol-water system at 20°C (21).*

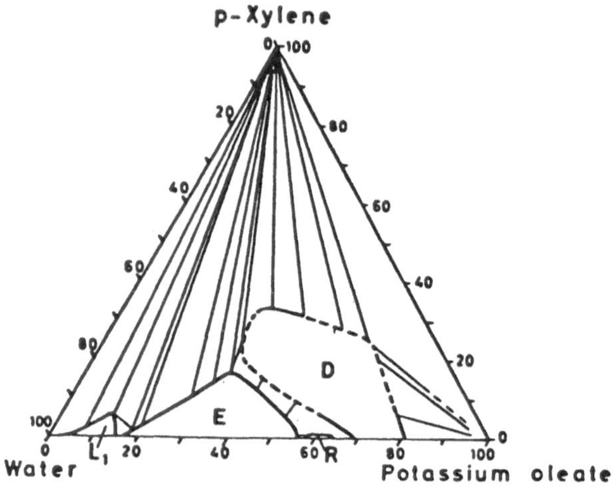

Figure 8. *Potassium oleate-p-xylene-water system at 20°C (2).*

comparatively little effect. These two phase diagrams will be used as models for the solubility of other alcohols and hydrocarbons.

The effects of water content, temperature and solubilization on the ^{39}K quadrupolar splitting in the low water content, non-swelling lamellar phase will be systematically examined. A simple chemical equilibrium model will be shown to rationalize the results in contrast to the behaviour observed in the swelling region of the mesophase.

Effects of Water Content and Temperature

The variation of the ^{39}K quadrupolar splitting with water content is given in Fig. (9a) and with temperature in Fig. (9b). Note how as the water content is increased the splitting at first falls rapidly and then tends to level out at a plateau value. Samples with mole ratio of water-to-amphiphile $R_{W/A}$ between 5 and 8 are in the non-swelling limit and the effect of addition of water is to increase the area per amphiphile at the bilayer surface (28). From Fig. (9b), we can see that for a sample with $R_{W/A}$ = 7.6, the splitting increases roughly linearly with temperature up to about 60°C after which the rate of increase rapidly falls off.

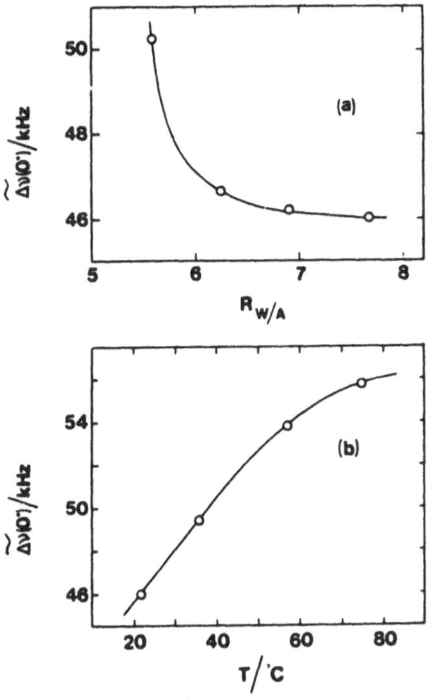

Figure 9. *Variation of the* ^{39}K *quadrupolar splitting* $\widetilde{\Delta\nu}(0°)$, *as measured in the lamellar phase of potassium oleate-water, with (a) the mole ratio of water-to-amphiphile* $R_{W/A}$ *at a fixed temperature of 21°C and (b) temperature at fixed* $R_{W/A}$ = 7.6.

The observed temperature dependence can be understood in terms of the classical double layer model of a micelle illustrated in Fig. (10). Two distinct binding sites are envisaged. In site I, the K^+ ions are specifically 'bound' between neighbouring carboxyl groups, whilst in site II, though essentially still bound, they are distributed in a diffuse layer, i.e. the binding is 'non-specific'. For ions in site II, $\eta = 0$, by symmetry, and the major axis *c* of the principal axes system of the electric field gradient tensor is parallel to the director; thus $S_{cc} > 0$ and $\Delta\nu > 0$ provided $\chi > 0$. Conversely, for an ion in site II, the *c* axis will be included at an angle θ which is expected to be greater than 54.7°,

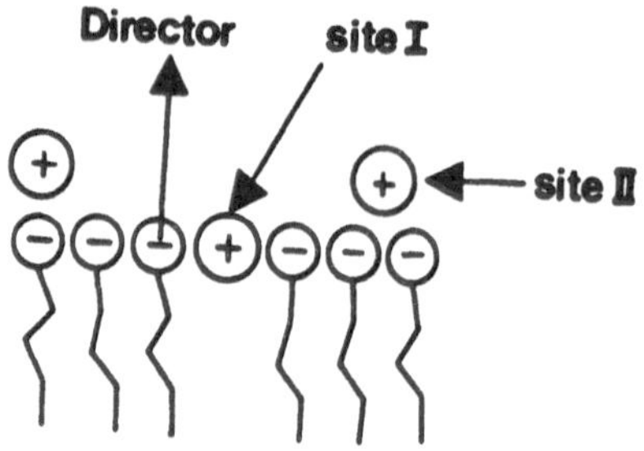

Figure 10. *Two site model for counterion binding at a bilayer surface.*

thereby making $S_{CC} < 0$ and $\Delta\nu < 0$. It is anticipated that site I will be energetically more favourable than site II, so that an increase in temperature will be accompanied by an increase in the population of ions in site II at the expense of site I. This will cause an increase in the magnitude of $\widetilde{\Delta\nu}$ if initially it is positive and a decrease if it is negative (4,5,20). Thus, the increase in splitting with temperature as seen in Fig. (9b) is consistent with a positive quadrupolar splitting.

The fractional population of site II is also expected to increase with hydration (4), yet the splitting is seen (Fig. (9d)) to decrease. The most plausible explanation of this paradoxical observation is that the quadrupolar coupling constant decreases with increasing hydration and brings about a greater reduction in the splitting than the increase due to any change in site populations. The electric field gradient at the K^+ ion has its origin in the distortion of the symmetry of the hydration sphere; a 'hydration number' of 4 to 6 is predicted (29) as for Na^+, though for the former ion the ion-water interaction is weaker owing to its larger radius. At the low water content end of the accessible composition range, there is insufficient water to fully hydrate both the potassium ions and the carboxyl groups of the amphiphiles, the partially hydrated potassium ions are, therefore, strongly bound to the carboxyl head groups, either directly or through a hydrogen bond involving a water molecule. Added water interacts strongly with the potassium ions, increasing the symmetry and thereby reducing the electric field gradient at the nucleus. The observed behaviour is consistent with this model: following the marked initial decrease in splitting, addition of water beyond $R_{W/A} \geqq 6$ has only a relatively small effect. Thus, the decrease in splitting brought about by the reduction in quadrupolar coupling constant upon increasing hydration dominates any increase resulting from changes in the site populations.

Effects of Solubilization of Alcohols (Amphiphiles)

The solubilization of an alcohol into the lamellar phase of the potassium oleate-water system has a marked effect on the ^{39}K quadrupolar splitting. Fig. (11) shows a plot of splitting as a function of $R_{S/A}$, the mole ratio of octanol (solubilizate)-to-potassium oleate (amphiphile), at a constant mole ratio of water-to-amphiphile $R_{W/A}$ of 7.6 (see Fig. (7)) and at a fixed

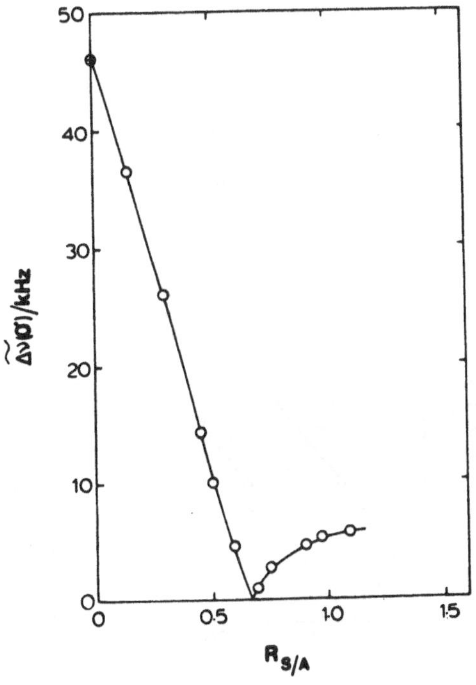

Figure 11. *Variation of the ^{39}K quadrupolar splitting $\widetilde{\Delta\nu}(0°)$ with $R_{S/A}$ the mole ratio of octanol (solute)-to-potassium oleate (amphiphile) measured in the lamellar phases of potassium oleate-octanol-water mixtures with $R_{W/A}$ = 7.6 and at 21°C.*

temperature of 21°C. The splitting is seen to fall to zero and then to increase. This behaviour is in marked contrast to that observed in the sodium octanoate-decanol-water system where the effect of variation of alcohol content is relatively small, presumably because the range of accessible compositions is restricted to the neighbourhood of the zero point.

The measurements presented in Fig. (11) imply that the quadrupolar splitting changes from a positive to a negative value as octanol is titrated into the lamellar phase. This is an interesting result and it is confirmed by the temperature dependences of the splittings which are quite different for samples on either side of the zero point (see Fig. (12)). For samples with mole ratios $R_{S/A}$ less than that of the zero point the splitting increases with increasing temperature, as for the two component system, whilst for samples with greater mole ratios the splitting first falls to zero and then increases.

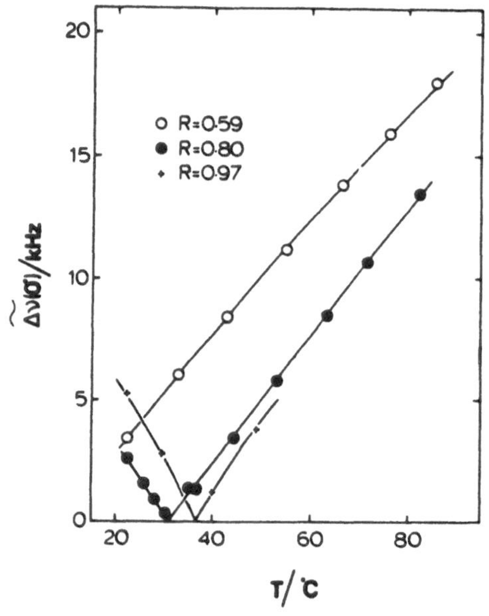

Figure 12. *Variation of the* ^{39}K *quadrupolar splitting* $\tilde{\Delta\nu}(0°)$ *with temperature in the lamellar phase of various potassium oleate-octanol-water mixtures all with* $R_{W/A} = 7.6$.

Figure 13. *Modification of two site model to incorporate the effects of solubilization of octanol followed by addition of water.*

The effect of octanol on counterion binding can be rationalized in terms of an extension of the simple two site model as illustrated schematically in Fig. (13). The presence of alcohol gives rise to an increase in the population of ions in specifically bound sites, albeit those involving an alcohol hydroxyl group should be distinguished from the sites in the pure potassium oleate-water system.

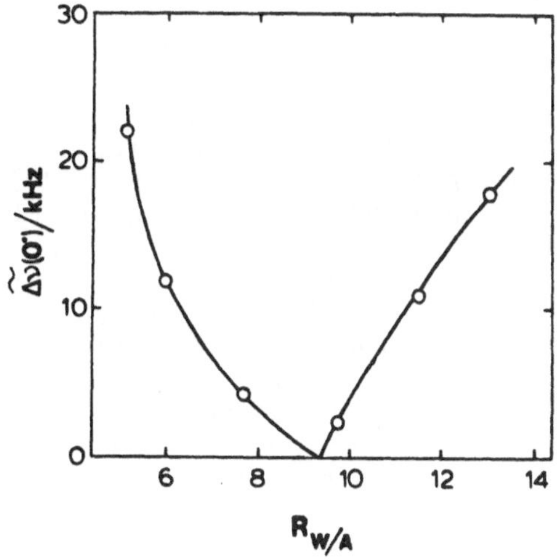

Figure 14. *Variation of the ^{39}K quadrupolar splitting $\Delta\tilde{\nu}(0°)$ with mole ratio of water to amphiphile in lamellar potassium oleate-octanol water mixtures with $R_{S/A}$ = 0.88 and at a constant temperature of 21°C.*

Fig. (14) shows the effect on the quadrupole splitting of the addition of water to a lamellar phase with a high octanol content. The mole ratio of octanol-to-potassium oleate is held constant at 0.88 and that of water-to-potassium oleate is varied from 5 to 13 (see Fig. (7)). On addition of water the splitting first decreases to zero and then increases implying the sign of the quadrupolar splitting changes from negative at low water content to positive at high water content. This observation suggests that the effect of water is to displace ions from the specific to the diffuse binding site (see Fig. (13)). This is not unreasonable as there is now sufficient water to satisfy the hydration requirements of both the counterion and head groups of the amphiphiles. The binary system responded quite differently to increasing the water content (Fig. (9a)): in this case the displacement of ions into the diffuse binding sites triggers the transition to the normal hexagonal phase. It seems that in the presence of an alcohol the electrostatic repulsion between neighbouring carboxylate groups is suppressed and the transition quenched.

Higher values of the mole ratio $R_{W/A}$ than are shown in Fig. (14) are accessible (values up to 70 are possible for $R_{S/A}$ = 0.88), but we found that the singularities of the powder spectra could not be detected for these samples. This phenomenon has been noticed for ^{23}Na in the sodium octanoate-decanol-water system and ascribed to exchange of sodium ions between domains with different director orientations (5): when the exchange rate becomes comparable with the difference between the quadrupolar splittings in adjacent domains the singularities become too broad to detect. An alternative explanation is that the lamellae undergo large scale undulations at high water contents where the lamellar phase undergoes one-dimensional swelling (30). The fluctuations in the quadrupolar splitting associated with the fluctuations in the orientation of the local lamellar director will similarly give rise to an exchange broadening of the singularities when the rate of these fluctuations is comparable with the magnitude of the quadrupolar splitting.

Benzyl alcohol behaves quite differently than the n-alcohols. On titrating this alcohol into the binary system, the ^{39}K splitting falls to a minimum value and then rises again without passing through zero (Fig. (15)). Thus, benzyl alcohol does not bring about a change in the sign of the splitting, a conclusion supported by the observation, for samples to both sides of this minimum, that the splitting increases with temperature.

Effects of Hydrocarbons (Lipophiles)

Hydrocarbon solubilizates have a much less pronounced effect on

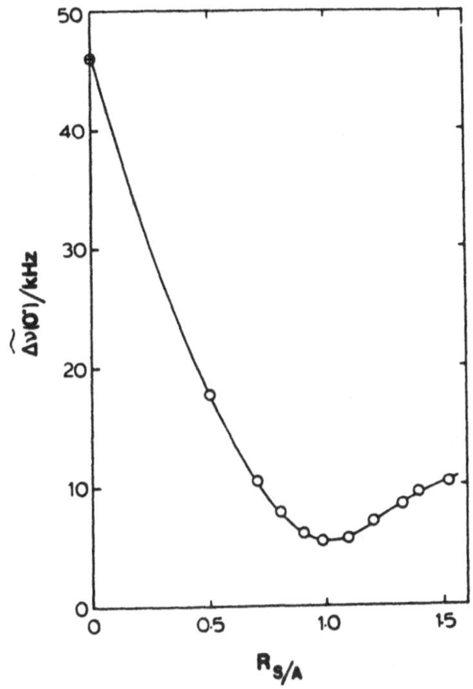

Figure 15. *Variation of the potassium quadrupolar splitting $\Delta\nu(0°)$ with the mole ratio of benzyl alcohol (solute) to potassium oleate (amphiphile) in the lamellar phase of potassium oleate-benzyl alcohol-water with $R_{W/A}$ = 7.6 and at a constant temperature of 21°C.*

the ^{39}K quadrupolar splitting than the alcohols. The variation of splitting with mole ratio of solubilizate-to-potassium oleate for benzene and cyclohexane is shown in Fig. (16). Both of these solutes have a similar effect; they bring about a reduction of approximately 35% in the splitting when $R_{S/A} \cong 1.5$, the solubility limit. There seems to be a slight difference between the aromatic and aliphatic solutes, benzene causing a slightly greater reduction in the splitting than cyclohexane. On the basis of the two site model, the small reduction in the splitting brought about by these solutes corresponds to a small increase in the population of the specific binding site.

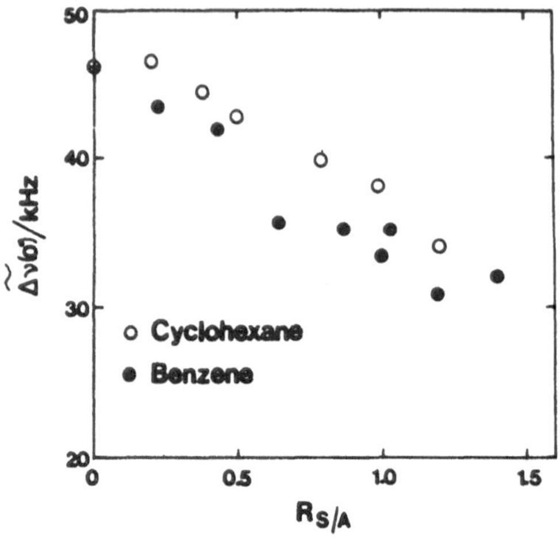

Figure 16. *Variation of the K quadrupolar splitting with mole ratio of solute (cyclohexane or benzene)-to-amphiphile* $R_{S/A}$ *in lamellar potassium oleate solute water mixtures with* $R_{S/A}$ = *7.6 and at a constant temperature of 21°C.*

The variation of the ^2H quadrupolar splitting for D_2O with the mole ratio of hydrocarbon-to-potassium oleate is shown in Fig. (17). The splitting increases by about 30% over the composition range investigated and there is no apparent difference between the effects of benzene and cyclohexane.

CHEMICAL EQUILIBRIUM MODEL OF COUNTERION BINDING

Let us examine the implications of the two site equilibrium model as invoked in the interpretation of the ^{39}K quadrupolar splittings. Obviously, it is a drastic oversimplification. Nevertheless, it is instructive to develop it further and to attempt to extract from the measurements values for the apparent equilibrium constant and its associated thermodynamic parameters. These quantities, though model dependent, ought to give a qualitative indication of the energetics of counterion binding.

Assuming that the observed quadrupolar splitting is a weighted average over the two sites, we can write

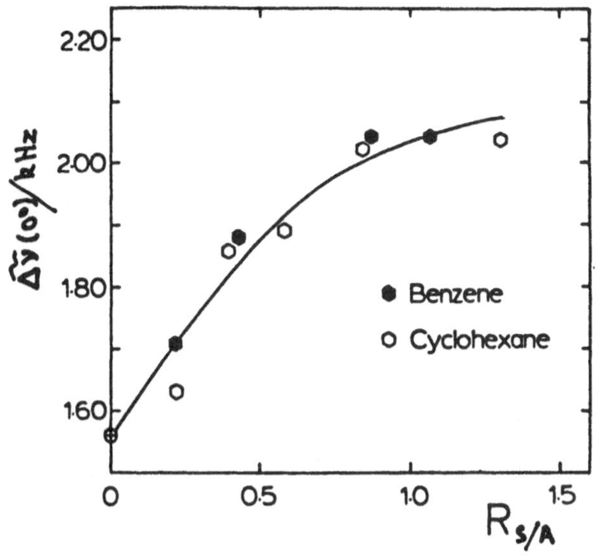

Figure 17. *Variation of the* ^2H *quadrupolar splitting of* D_2O *as a function of the mole ratio of solute-to-potassium oleate at constant* $R_{W/A}$ *= 7.6 and at 21°C.*

$$\widetilde{\Delta\nu} = p_s \widetilde{\Delta\nu} + p_d \widetilde{\Delta\nu}_d \quad (8)$$

where p_s and p_d are the fractional populations and $\widetilde{\Delta\nu}_s$ and $\widetilde{\Delta\nu}_d$ the quadrupolar splittings of ions in the specific (negative splitting) and diffuse (positive splitting) binding sites, respectively. An apparent equilibrium constant K is defined by

$$\ln K = p_s/p_d = -\Delta G/RT = -(\Delta H - T\Delta S)/RT \quad (9)$$

We will consider the sample with $R_{W/A}$ = 7.6. To obtain values for K it is necessary to have values for the splittings in the two sites. Values of $\widetilde{\Delta\nu}_d$ = 60 kHz and $\widetilde{\Delta\nu}_s$ = -7 kHz will be used; the former corresponds to the splitting obtained by a high temperature extrapolation of the temperature dependence shown in Fig. (9b), whilst the latter corresponds to the limiting splitting at high alcohol content (Fig. (11)). The values of K presented in Fig. (18) were obtained using these values in equation (8) to give the p_n. The slopes of all of the plots are

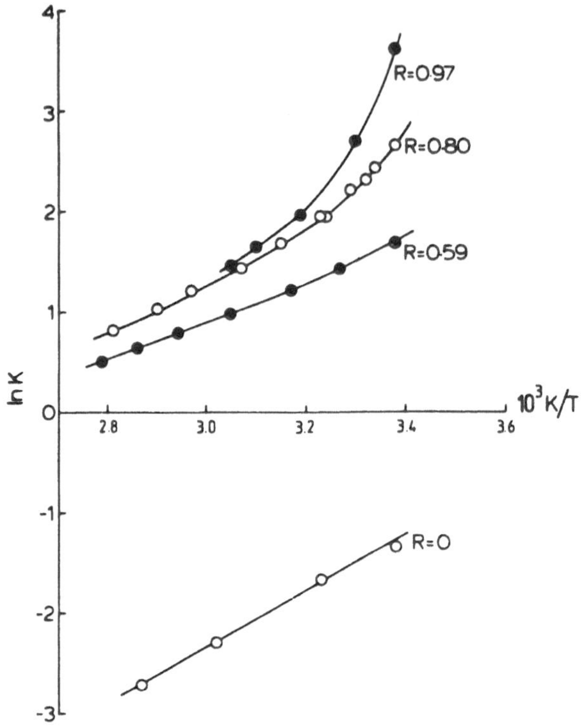

Figure 18. *Plots of the logarithm of K, the apparent equilibrium constant for the exchange of potassium ions between the 'specific' and 'diffuse' binding sites of the two site model, versus 1/T for lamellar potassium oleate-octanol-water mixtures all with* $R_{W/A} = 7.6$.

positive which indicates that ΔH is negative, i.e. the specific binding site is of lower energy than the diffuse one. This is exactly what one would expect and what has been assumed thus far. The value of K at 21°C varies from 0.3 for $R_{S/A} = 0$ to 66.7 for $R_{S/A} = 0.97$ illustrating the anticipated effect of alcohol on ion binding. The values observed for ΔH are summarized in Fig. (19). We see that ΔH is constant, independent of temperature, only in the case of the binary system. In the presence of octanol, ΔH is a function of both alcohol content and temperature, implying a distribution of binding sites. In particular, ΔH increases both with alcohol content and on

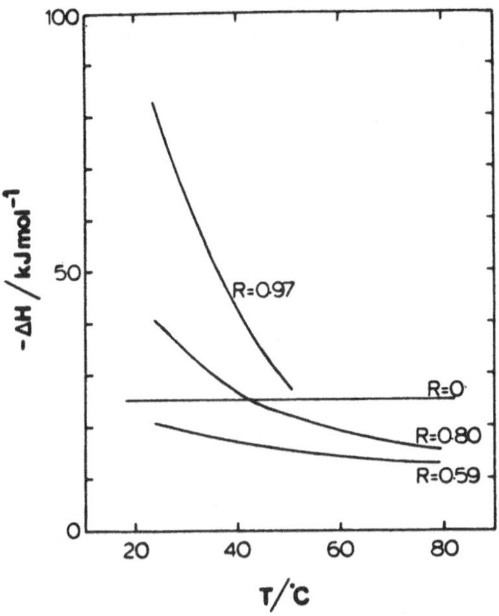

Figure 19. *Temperature dependence of the apparent enthalpy difference ΔH between the 'specific' and 'diffuse' binding sites of the potassium ion in lamellar potassium oleate-octanol-water mixtures for various mole ratios of octanol-to-potassium oleate, all with* $R_{W/A} = 7.6$.

lowering the temperature, i.e. with increase in the population of specific ion binding site. Thus, it seems that as the population of the specifically bound site increases its energy decreases. Indeed, it is not unreasonable to expect that an arrangement in which the counterions are interposed between carboxylate head groups will minimize the coulombic energy. The value of ΔS, estimated at 303 K, varies from -80 J K^{-1} mol^{-1} for the sample with $R_{S/A} = 0$ to -190 J K^{-1} mol^{-1} for the sample with $R_{S/A} = 0.97$. Despite the undoubtedly large errors involved, the negative value indicates that the specifically bound site is of lower entropy than the diffuse one.

The two site model, despite its serious limitations, gives results which are, qualitatively, of physical significance. The results should not, however, be taken to have any quantitative significance.

The variation of the ^{39}K quadrupolar splitting on addition of benzyl alcohol is different from that observed with the long chain alcohols. In terms of the two site model, the results suggest that the diffuse binding site is first depopulated and then repopulated as benzyl alcohol is added. In the following chapter it will be shown that the minimum splitting which occurs at $R_{S/A} = 1.0$ coincides with a maximum in the composition dependence of the orientational ordering of both the benzyl alcohol molecule and the oleate chain.

COMMENT

It is pertinent to comment on the potential use of ^{39}K NMR for studying counterion binding in lyotropic mesophases. A high magnetic field spectrometer is essential and a wide band pulse spectrometer is preferable. The quadrupolar splittings observed can be as large as 50 kHz, which means the powder spectrum has a width of 100 kHz. This is far too broad to be detected without serious distortion of the lineshape using a narrow band, high resolution spectrometer such as the one employed in this study. The maximum quadrupolar splittings measured for the potassium oleate-octanol-water system tend to be larger than those reported for ^{23}Na (50 kHz c.f. 10 kHz) in the sodium octanoate-decanol-water system (4,5). The electric quadrupole moment of ^{39}K ($0.11 \times 10^{-24} cm^2$) is smaller than that for ^{23}Na ($0.15 \times 10^{-24} cm^2$) and it is suggested that the difference in splittings is due to a larger electric field gradient at the ^{39}K than at the ^{23}Na nucleus as a consequence of the more easily deformable hydration shell of the K^+ ion.

REFERENCES

1. J.N. Israelachvili, D.J. Mitchell and B.W. Ninham, *J.C.S. Faraday 2*, **72**, 1525 (1976).

2. P. Ekwall, in *'Advances in Liquid Crystals'*, ed. G.H. Brown (Academic Press, New York, 1975), Vol. 1, p.1.

3. G. Lindblom and B. Lindman, *Mol. Cryst. Liq. Cryst.*, **22**, 45 (1973).

4. G. Lindblom, B. Lindman and G.J.T. Tiddy, *Acta Chem. Scand., Ser. A*, **29**, 876 (1975).

5. G. Lindblom, B. Lindman and G.J.T. Tiddy, *J. Amer. Chem. Soc.*, **100**, 2299 (1978).

6. A. Johansson and B. Lindman, *'Liquid Crystals and Plastic Crystals'*, ed. G.W. Gray and P.A. Winsor (Ellis Horwood, Chichester, U.K., 1974), Vol. 2, p.192.

7. M. Shporer and M.M. Civan, *Biophys. J.*, **12**, 114 (1972).

8. D.M. Chen and L.W. Reeves, *J. Amer. Chem. Soc.*, **94**, 4384 (1972).

9. H.J.C. Berendsen and H.T. Edzes, *Ann. New York Acad. Sci.*, **204**, 459 (1973).

10. D.M. Chen, K. Radley and L.W. Reeves, *J. Amer. Chem. Soc.*, **96**, 5251 (1974).

11. K. Radley, L.W. Reeves and A.S. Tracey, *J. Phys. Chem.*, **80**, 174 (1976).

12. J. B. Rosenholm and B. Lindman, *J. Colloid and Interface Sci.*, **57**, 362 (1976).

13. K. Abdolall, E.E. Burnell and M.I. Valic, *Chem. Phys. Lipids*, **20**, 115 (1977).

14. G.J.T. Tiddy, G. Lindblom and B. Lindman, *J.C.S. Faraday 1*, **74**, 1290 (1978).

15. O. Soderman, S. Engström and H. Wennerström, *J. Colloid and Interface Sci.*, **78**, 110 (1980).

16. E. Everiss, G.J.T. Tiddy and B.A. Wheeler, *J.C.S. Faraday 1*, **72**, 1747 (1976).

17. N.O. Persson and G. Lindblom, *J. Phys. Chem.*, **83**, 3015 (1979).

18. B. Lindman and S. Forsen, in *'NMR and the Periodic Table'*, eds R.K. Harris and B. Mann, (Academic Press, New York, 1978) p.129.

19. G. Lindblom, H. Wennerström and B. Lindman, *Chem. Phys. Letters*, **8**, 489 (1971).

20. H. Wennerström, G. Lindblom and B. Lindman, *Chem. Scr.*, **6**, 97 (1974).

21. M.N. Cohen and F. Reif, in *'Solid State Physics'*, eds. F. Seitz and D. Turnbull (Academic Press, New York, 1957), Vol. 5, p.321.

22. R.C. Long, *J. Magn. Reson.* **12**, 216 (1973).

23. A. Saupe, *Z. Naturforsch*, **19a**, 161 (1964).

24. S. Enström and H. Wennerström, *J. Phys. Chem.*, **82**, 2711 (1978).

25. H. Wennerström, B. Lindman and G. Lindblom and G.J.T. Tiddy, *J.C.S. Faraday Transactions 1*, **75**, 663 (1979).

26. N. Person and B. Lindman, *J. Phys. Chem.*, **79**, 1410 (1975).

27. F. Husson, H. Mustacchi, A.E. Skoulios and V. Luzzati, *Acta Cryst.*, **13**, 660 (1960); F. Husson, H. Mustacchi and V. Luzzati, *ibid.*, **13**, 668 (1960).

28. P. Ekwall, L. Mandell and K. Fontell, *J. Colloid and Interface Sci.*, **31**, 508 (1969).

29. J.E. Enderby and G.W. Neilson, *Rep. Prog. Phys.*, **44**, 38 (1981).

30. J.M. di Meglio, M. Dvolaitzky, R. Ober and C. Taupin, *C.R. Acad. Sc. Paris*, **296**, 405 (1983).

SOLUBILIZATE-BILAYER INTERACTIONS IN LAMELLAR MESOPHASES

N. Boden and S.A. Jones

Department of Physical Chemistry,
The University,
Leeds, LS2 9JT, England.

INTRODUCTION

In the previous chapter we learned how information about the binding of K^+ ions at the bilayer-water interface in the lamellar phase of the potassium oleate-water system could be deduced from measurements of the ^{39}K quadrupolar splitting. The magnitude of this splitting is dependent upon the molecular structure and the amount of a third component (solubilizate). In particular, it appears to reflect the manner in which the solubilizate interacts with and perturbs the surface of the bilayer. In the present chapter we wish to pursue this problem of solubilizate-bilayer interaction further. Two additional pieces of experimental data will be considered. First, X-ray diffraction measurements of the bilayer structure and, secondly, ^2H NMR measurements on both the solubilizate and soap (amphiphile). The object of the ^2H studies will be to obtain empirical information about the structure of the bilayer and, in particular, the effects of perturbants, not to investigate models for the conformational and reorientational motions of the amphiphiles about the bilayer director. We will see that examination of the correlations between the X-ray and NMR data gives an insight into the molecular mechanism of the solubilization process and of the way the solubilizate affects the mesophase stability.

J. W. Emsley (ed.), Nuclear Magnetic Resonance of Liquid Crystals, 497–532.

SOLUBILIZATION OF ALCOHOLS

X-ray Diffraction Measurements

From X-ray diffraction measurements the thickness and surface area of the bilayer can be calculated (see Appendix I for details). The surface area of the bilayer is a cooperative property of all of the components in the mesophase. A precise prescription is, therefore, required for apportioning the surface area amongst the various components. This can be done, as described in some detail in Appendix I, by defining a quantity called the partial molecular surface area which for the ith component is

$$A_i = (\partial A/\partial N_i)_{T,p,N_{c\neq i}} \tag{1}$$

Knowledge of the variation of this quantity with composition for each component in a three-component system gives direct insight into the way the individual components determine the surface area of the bilayer (equation A5). To illustrate this point we shall consider the composition dependences of the A_i of all three components in the potassium oleate-decanol-water system as shown in Fig. (1).

The ratio of $R_{W/A}$ is held constant at 7.6 and corresponds to the composition line shown in Fig. (7) of the previous chapter. The partial molecular surface area of the water (the increase in surface area brought about by the addition of one water molecule at a given composition) falls from 1.6 Å^2 in the two component system to almost zero at high alcohol concentration. This corresponds to a transition from a situation in which added water interacts with and causes an increase in the area of the bilayer to one in which all added water is intercalated between the bilayers causing an increase in the bilayer separation but no lateral expansion, i.e. a transition from a 'non-swelling' to a one dimensional 'swelling' system (1).

The partial molecular surface area of decanol at infinite dilution, i.e. when $R_{S/A}=0$, typifies the solubilizate-bilayer interaction. The value in this case at $x_S = 0$ is approximately 12.5 Å^2 which is considerably less than the cross-sectional area of a decanol molecule. Yet it is generally accepted (1) that alcohols are solubilized with their hydroxyl groups located at the bilayer-water interface and with the alkyl chain moiety extending into the bilayer. A plausible explanation of this observation is that the expansion of the surface caused by addition of a decanol molecule is largely offset by an increase in the ordering of the soap chains and the commensurate reduction in their cross-sectional area. The value of A_S is seen to increase roughly in proportion to the concentration of decanol, suggesting that the ordering effect of the latter decreases with increasing concentration.

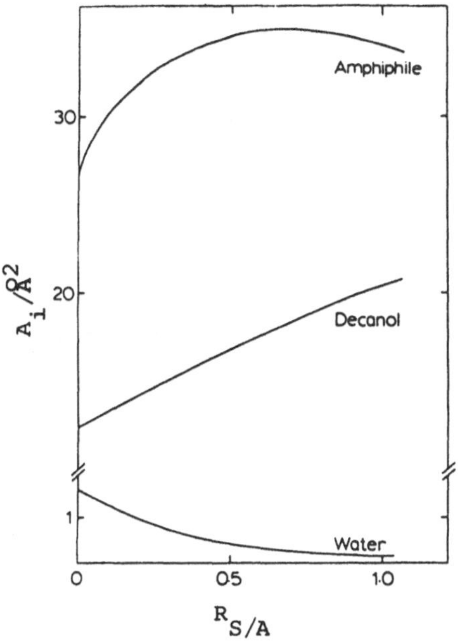

Figure 1. *Partial molecular surface areas in the potassium oleate-decanol-water system with $R_{W/A}$ = 7.6 and T = 21°C.*

The behaviour of the partial molecular surface area of the soap A_A is more complex than that of either of the other two components, there being an initial increase of about 7 $\AA^2$ followed by a small decrease. In the absence of decanol $A_A \approx 26.5\ \AA^2$ which is very close to the limiting value of 25-26 $\AA^2$ for the mean interfacial area per polar group in the swelling region (2) and is similar to the cross-sectional area of an unsaturated hydrocarbon chain in a monolayer (3). The decrease in A_A at high concentrations is probably attributable to ordering of the chains, but the initial increase is harder to explain. It must be associated with the effect of decanol on the binding of the potassium ion. That this is so can be seen by comparison with the behaviour of A_A in the egg lecithin-water-cholesterol system as illustrated in Fig. (2).

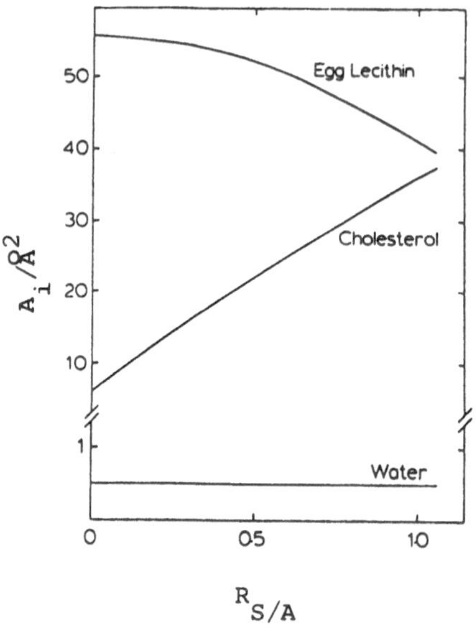

Figure 2. *Partial molecular surface areas in the egg lecithin-cholesterol-water system with* $R_{W/A}$ = 25 *and* T = 20°C, *as calculated from data taken from reference 4.*

This interpretation of the partial molecular surface measurements can be tested by studying the ordering of the soap and alcohol using ^2H NMR.

^2H NMR Measurements

To study the orientational ordering of the soap chains one would ideally use perdeuteriated potassium oleate. This substance cannot, however, be synthesized due to the instability of the double bond. Perdeuteriated potassium stearate is a very suitable probe molecule. A typical powder spectrum is shown in Fig. (3). It consists of seventeen superimposed 'Pake' lineshapes. The separation of a particular doublet (singularities) corresponds to

$$\Delta\tilde{\nu} = \tfrac{3}{4}\chi_D\{S_{cc} + (S_{aa} - S_{bb})\}$$

$$\approx \tfrac{3}{4}\chi_D S_{CD} \qquad (2)$$

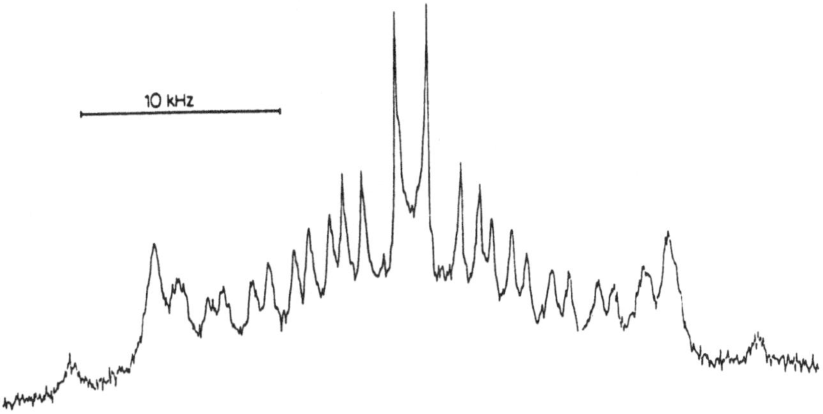

Figure 3. *^2H spectrum of potassium stearate-d_{35} solubilized in the lamellar phase of the potassium oleate-water system.*

for a $>CD_2$ or a $-CD_3$ group since $\eta \approx 0$. The splittings cannot be assigned to the individual methylene segments. We therefore assume that they decrease monotonically from the carboxylate group, albeit an alternation in splittings for the first four carbons is expected (5). The ensuing analysis does not, however, depend on a rigorous assignment of splittings. Consider first the effect of hydration on chain order. Increasing water content has little effect on the general shape of the order parameter profile (plot of S_{CD} versus chain segment number) (Fig. (4)) it simply causes a reduction in the segmental order parameters. Fig. (5) shows a plot of the average order parameter $\langle S_{CD} \rangle$ versus the surface area per amphiphile as determined by X-ray diffraction. This correlation is also observed in other potassium soap-water systems (6) and implies that the ordering of the chains is coupled to the surface area per soap molecule.

The effect on the chain order of addition of decanol at constant mole ratio of water-to-amphiphile $R_{W/A} = 7.6$ is illustrated in Fig. (6). The ordering of the top few segments of the chain is seen to increase roughly in proportion to the amount of decanol added, but for the remainder the effect is very much dependent upon the location of the segment. This is closely illustrated in Fig. (7) in which the S_{CD} for each segment is normalized with respect to its value in the absence

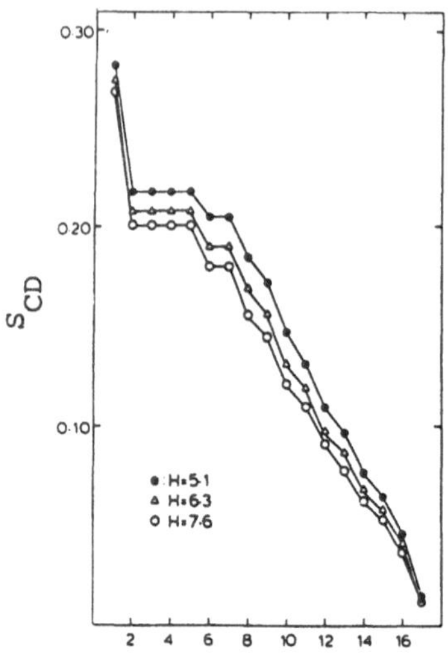

Figure 4. *C-D bond order parameters of potassium stearate-d_{35} in the lamellar phase of potassium oleate-water ($H = R_{W/A}$).*

of decanol. Note that the ordering is greatest at segment ten and also how the terminal four segments of the chain after initially increasing in order start to disorder at the higher decanol contents. A further point of interest is that the order parameter of the first segment increases only slightly with concentration and becomes smaller than that of the second segment at the highest concentration. Very similar behaviour is observed with octanol as shown in Fig. (8), except its ordering effect is not quite as great as that of decanol. The maximum is now at segment nine and the disordering of the end of the chain at high concentrations is more pronounced.

The corresponding order parameter profile for octanol-d_{17} is shown in Fig. (9). The general pattern of behaviour is very similar to that of the soap. Increases in order parameter ratios are very similar for each segment with the exception of segment one whose behaviour closely resembles that of the corresponding segment in the soap. Thus, the observed ordering effect seems to correspond with a reduction in the whole-body reorientational motion. The effect of alcohol is

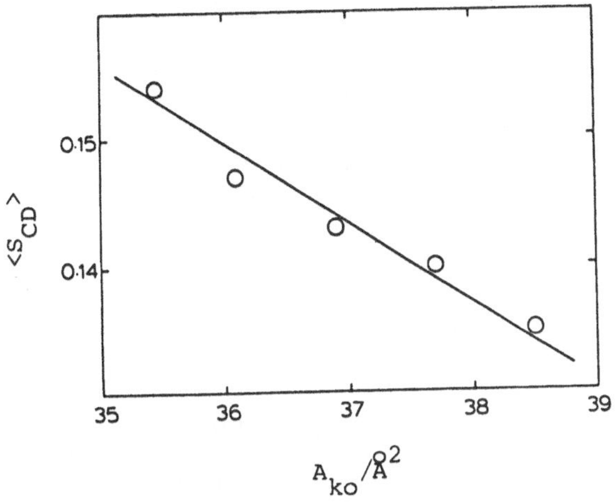

Figure 5. *Average C-D bond order parameter of potassium stearate in the lamellar phase of the potassium oleate-water system plotted as a function of the surface area per soap molecule as obtained by X-ray diffraction.*

exactly the opposite to that of water: it causes a contraction of the surface which damps the collective orientational fluctuations of the amphiphiles (7). The enhancement in the order at segments eight-ten of the soap corresponds to the location of the disordered terminus of the alcohol chains. The disordering of segments from ten downwards must be caused by the presence of an excess of gauche ± defects introduced to increase the effective thickness of the chain in order to take up the free volume created on solubilization of the alcohol.

A value for the average length of the stearate chains can be estimated from the segmental order parameters by the method outlined in Appendix 2. The thickness of the hydrocarbon sandwich d_{hc} taken as twice the average length of the stearate chain is compared with the X-ray thickness d_b in Fig. (10). The X-ray value is about 3 Å smaller. This discrepancy is most likely due to differences in the extent to which the stearate and oleate chains extend into the bilayer rather than in the method employed in the calculation of d_{hc}. Interestingly, the two sets of data converge as the concentration of solubilizate increases, suggesting that as the bilayer becomes more ordered the difference between the stearate and oleate chains becomes smaller.

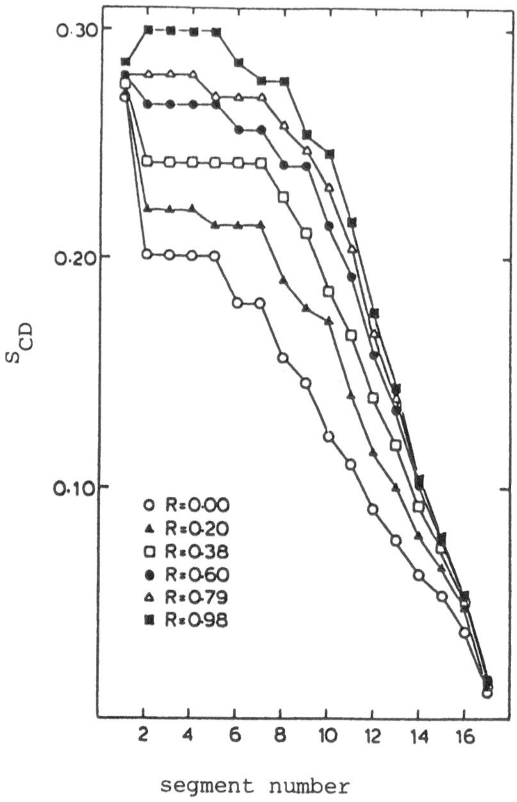

Figure 6. *C-D bond order parameters of potassium stearate-d_{35} in the lamellar phase of the potassium oleate-decanol-water system ($R = R_{S/A}$) with $R_{W/A} = 7.6$.*

The average length and cross-sectional area of the hydrocarbon chain as calculated from the S_{CD} of octanol-d_{17} are plotted as a function of $R_{S/A}$ in Fig. (11). The cross-sectional areas are seen to be very similar to those calculated for the stearate chain and given in Fig. (12).

The above results explain the paradoxically small partial molecular surface area of decanol (Fig. (1)). The increase in the surface area brought about by the addition of one alcohol molecule, i.e. its partial molecular surface area, is substantially smaller than the cross-sectional area projected on to the bilayer surface (Fig. (11)). This is due to the

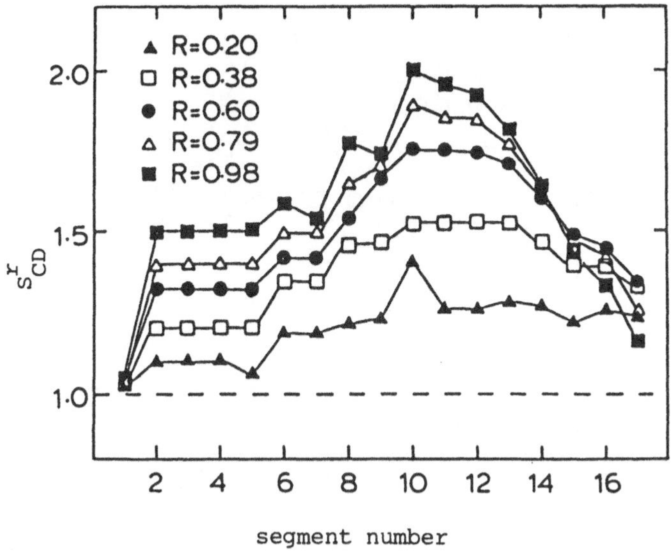

Figure 7. *Relative C-D bond order parameters of potassium stearate-d_{35} in the lamellar phase of the potassium oleate-decanol-water system for various $R_{S/A}$ (≡R) at $R_{W/A}$ = 7.6.*

pronounced ordering effect of the alcohol on the soap chains. Note also how the partial molecular surface area of decanol increases with concentration (Fig. (1)) whilst its actual cross sectional area decreases (Fig. (11)).

A further point of interest is the dependence of the partial molecular surface area of the alcohol on the size of the hydrocarbon moiety (Fig. (13)); longer alcohols such as oleyl alcohol cause considerably greater increases in surface area than smaller alcohols such as benzyl alcohol. This suggests that steric repulsions between the amphiphile and solute chains play a role in determining the surface area. Steric repulsions between the chains probably contribute to the attenuation in the ordering effect of the alcohol as its concentration is increased.

The change in the bilayer thickness brought about by the addition of alcohol also depends on the chain length (Fig.

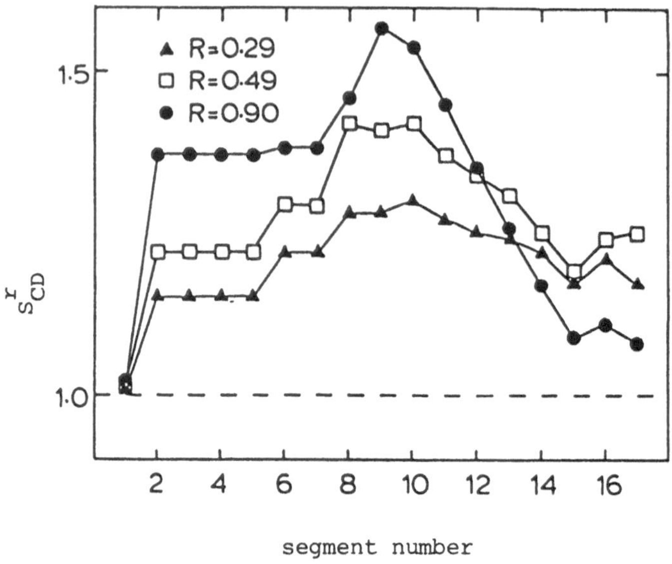

Figure 8. *Relative C-D bond order parameters of potassium stearate-d_{35} in the lamellar phase of the potassium oleate-octanol-water system at various $R_{S/A}$ (≡R) at $R_{W/A}$ = 7.6.*

(14)). A long chain alcohol such as decanol produces quite a large increase in bilayer thickness, whereas heptanol causes only a very small increase and pentanol actually produces a slight thinning of the bilayer.

The complex behaviour of the partial molecular surface area of potassium oleate can also now be explained. The decrease at high alcohol concentration is due to ordering of the soap chains which causes a reduction in their cross-sectional area. A similar decrease with increasing solute concentration is observed in the partial molecular surface area of the amphiphile molecule in the egg lecithin-water-cholesterol system (Fig. (2)) and this is also caused by the cholesterol induced ordering of the amphiphile chain (8). In this case, however, there is no initial rise in the partial molecular surface area. This suggests that this feature of the behaviour of the partial molecular surface area must be due in some way to the behaviour of the counter-ion. The variation of the ^{39}K quadrupolar splitting on addition of decanol (Fig. (11), previous chapter)

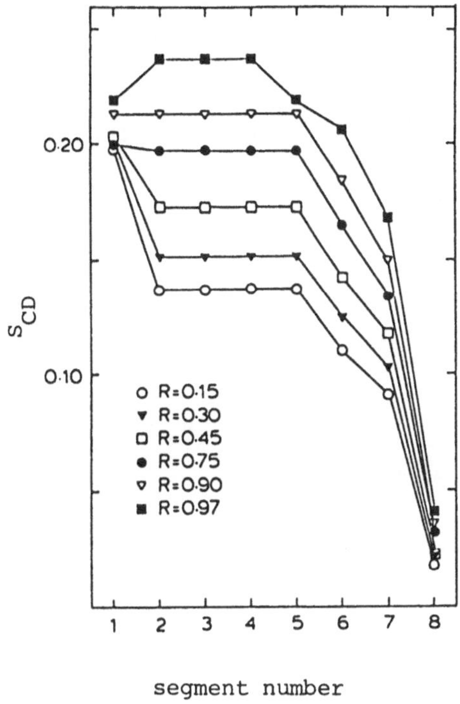

Figure 9. *C-D bond order parameters of octanol-d_{17} in the lamellar phase of the potassium oleate-octanol-water system at various $R_{S/A}$ (≡R) and with $R_{W/A}$ = 7.6. Segments 1 and 2 were assigned by measurements with 1,1-d_2 and 2,2-d_2-decanol.*

showed that there was an increase in the population of the counter-ion binding site with negative order parameter, i.e. an increase in the number of counter-ions located between the carboxylate groups. Since the partial molecular surface area of the amphiphile will contain contributions from both the carboxylate anion and the counter-ion an increase in the number of counter-ions, which will be hydrated and therefore quite bulky, interposed between the head-groups will cause an increase in the partial molecular surface area. There are, therefore, two competing processes influencing the behaviour of the partial molecular surface area of the amphiphile as decanol is added: a chain ordering effect which tends to reduce the partial molecular surface area and an increase in the number of

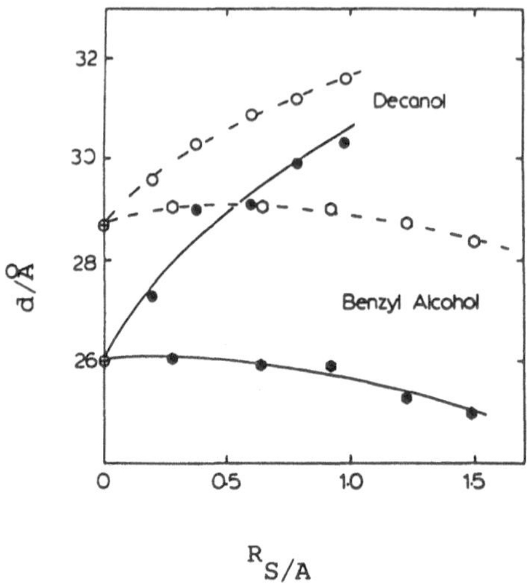

Figure 10. *Thickness of bilayer as calculated from C-D bond order parameters ○ (decanol) and ⬡ (benzyl alcohol) and X-ray diffraction measurements ● (decanol) and ⬢ (benzyl alcohol) as a function of $R_{S/A}$ in the lamellar phases of potassium oleate-solute-water systems with $R_{W/A}$ = 7.6.*

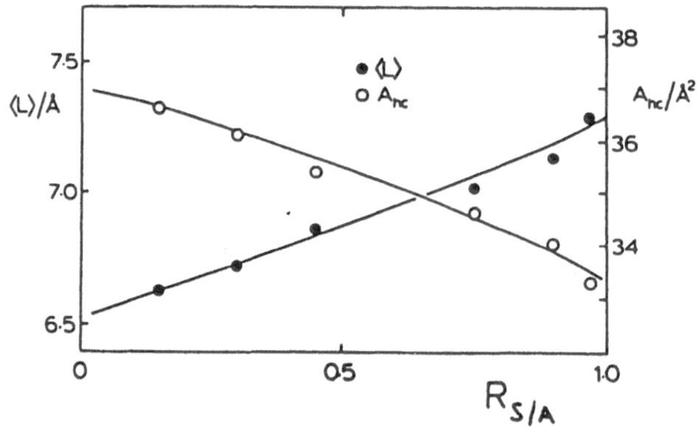

Figure 11. *The average length ⟨L⟩ and cross-sectional area A_{hc} of octanol-d_{17} as a function of $R_{S/A}$ in the lamellar phase of potassium oleate-octanol-water with $R_{W/A}$ = 7.6. Values for A_{hc} obtained from $A_{hc} = V_{hc}/\langle L \rangle$ using $V_{hc}/Å^3 = 27.4 + 26.9n_c$.*

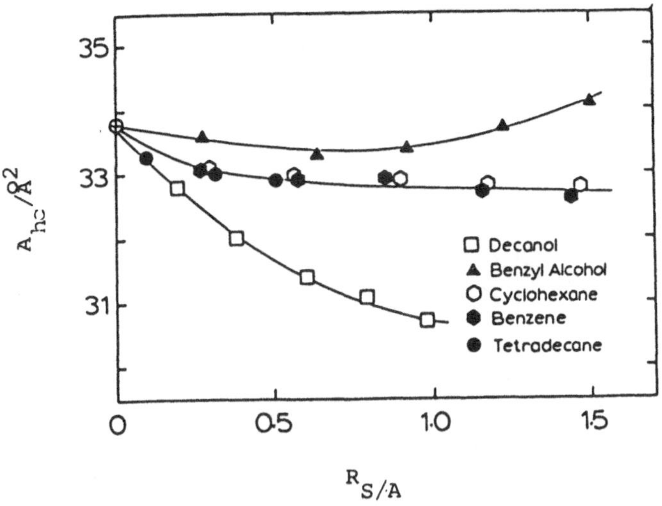

Figure 12. *Cross-sectional area of stearate chains as a function of $R_{S/A}$ in the lamellar phase of potassium oleate-solute-water with $R_{W/A}$ = 7.6.*

potassium ions in the surface which tends to increase it.

The effect of hydration on the ordering of the octanol molecule is shown in Fig. (15), and we see that for $R_{W/A} > 10$ further addition of water has little effect.

Benzyl alcohol behaves differently from the long chain alcohols octanol and decanol. The variations in the partial molecular surface areas as benzyl alcohol is added are shown in Fig. (16). The partial molecular surface area of the solubilizate follows the same trend as those of the other alcohols. The partial molecular surface area of the amphiphile again increases at first and then decreases though in this case the relative magnitudes of the two effects are very different to those observed as decanol was added, the initial increase being smaller and the reduction at higher solute concentration more pronounced.

Up to a mole ratio of benzyl alcohol-to-potassium oleate of about one the order parameters measured by ^2H NMR for both the benzyl alcohol molecule (Fig. (17)) and the top part of the amphiphile chain (Fig. (18)) increase with increasing alcohol

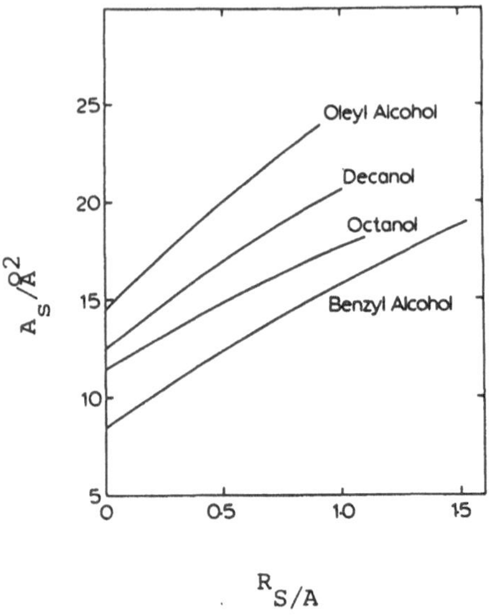

Figure 13. *Partial molecular surface areas of solutes as a function of $R_{S/A}$ in the lamellar phases of potassium oleate-solubilizate-water systems with $R_{W/A}$ = 7.6.*

concentration. The lower part of the chain, however, disorders. This behaviour can be easily explained. Benzyl alcohol is located with its hydroxyl group at the interface and it therefore orders the top segments of the amphiphile chains in the same way as octanol. However, the benzyl alcohol molecule is comparatively short and the ordering effect does not extend very far into the bilayer. Rather, free volume is generated underneath the solubilizate which must be taken up by the amphiphile chains. The simplest way in which this can be done is by the generation of gauche defects leading to a disordering of the chains and a reduction in their effective lengths.

At mole ratios of benzyl alcohol-to-soap of more than about one the order parameters for the benzyl alcohol molecule and the top of the chain start to decrease. The ^{39}K studies (Fig. (15), previous chapter) also suggested that at these solute concentrations the specific ion binding site begins to be depopulated. This effect, rather than the continued ordering of the chains, must be the cause of the rapid fall in the partial molecular surface area of the amphiphile at high concentrations. The partial molecular surface area of the water goes through a

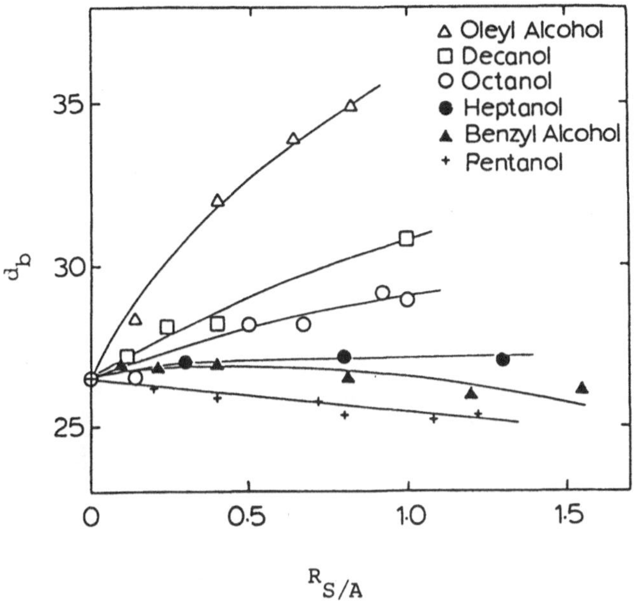

Figure 14. *Bilayer thickness d_b as a function of $R_{S/A}$ for various solubilizates in lamellar phases of potassium oleate-solubilizate-water with $R_{W/A}$ = 7.6.*

minimum at approximately the same mole ratio of solute-to-soap as the maximum in the order parameters of the benzyl alcohol molecules and the top of the amphiphile chains and the minimum in the ^{39}K splitting. For all the solutes studied there seems to be a correlation between A_W and the ^{39}K quadrupolar splitting: as the splitting falls and the potassium ions occupy the specifically bound sites the partial molecular surface area of the water also falls. It is as if there are a limited number of specific surface binding sites which can be occupied by the counter-ions or by water. Water only perturbs the surface when there are sufficiently few potassium ions interposed between the head-groups.

SOLUBILIZATION OF HYDROCARBONS

X-ray Diffraction Measurements

Three characteristically different types of behaviour are

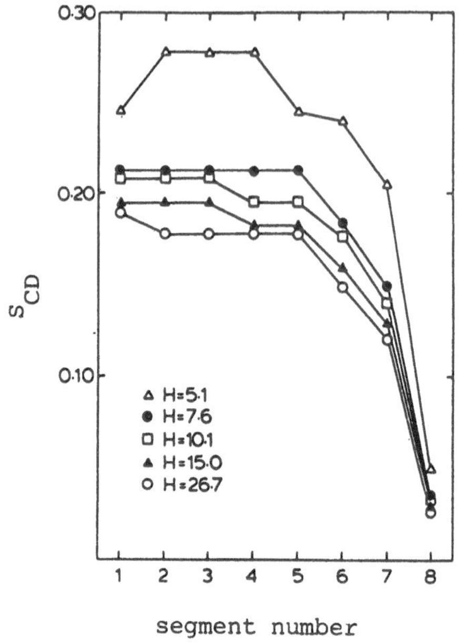

Figure 15. *Hydration dependence of the C-D bond order parameters of octanol-d_{17} solubilized in the lamellar phase of the potassium oleate-water system with $R_{S/A} = 0.88$ ($H \equiv R_{W/A}$).*

observed depending on the molecular structure of the solubilizate. Examples of each are given in Fig. (19). The partial molecular surface area of the water behaves similarly in all three cases, but the behaviour of the solubilizate and amphiphile are quite different. The results for a variety of solubilizates are compared in Fig. (20a) and the following three types of behaviour are evident.

1. For benzene the partial molecular surface area is constant at a value of approximately 4.8 Å^2. This suggests that the distribution of benzene through the bilayer is independent of composition.

2. The smaller alkanes and toluene all behaved similarly, the partial molecular surface area being approximately 4.5 Å^2 at infinite solution and falling as the mole ratio of solubilizate to soap increases. This suggests that at low concentrations the solubilizate is distributed throughout

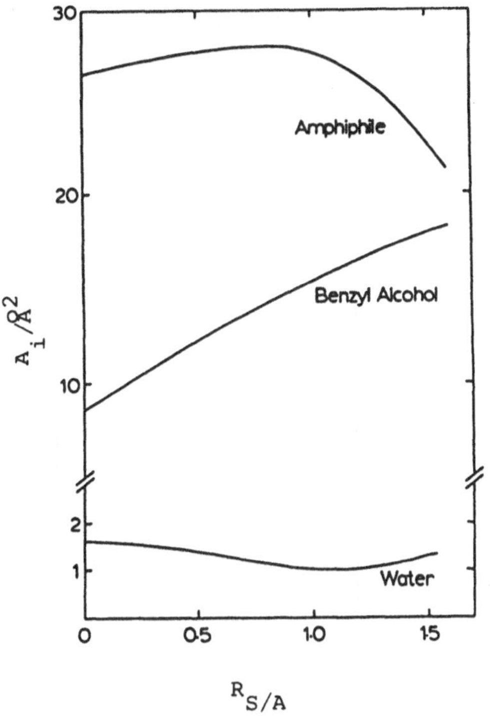

Figure 16. *Partial molecular surface areas as a function of $R_{S/A}$ in the lamellar phase of the potassium oleate-benzyl alcohol-water system with $R_{W/A}$ = 7.6.*

the bilayer, whilst at higher concentrations it partitions into the centre of the bilayer so that further addition causes little change in surface area.

3. For tetradecane the partial molecular surface area is initially close to zero, but then increases rapidly, indicating that the solubilizate is initially located mainly in the centre of the bilayer but as the concentration builds up it is forced to go between the chains, causing a pronounced increase in surface area.

The above interpretations of the partial molecular surface areas are substantiated by the behaviour of the bilayer thickness. Fig. (21) shows the bilayer thickness as a function of the mole ratio of solubilizate-to-soap for three different

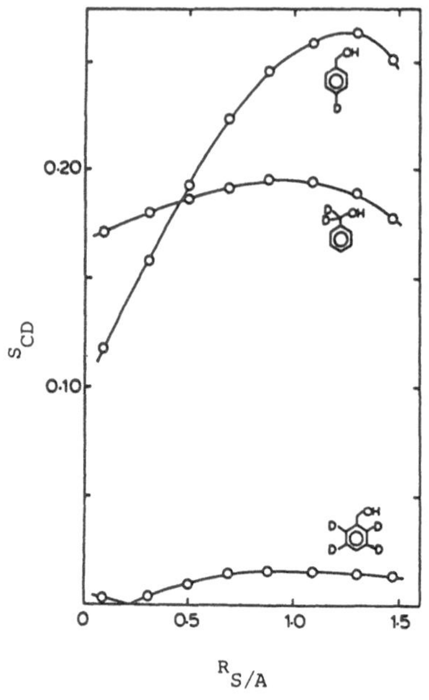

Figure 17. *C–D bond order parameters of benzyl alcohol as a function of $R_{S/A}$ in the lamellar phase of the potassium oleate-benzyl alcohol-water system with $R_{W/A}$ = 7.6.*

solutes: benzene, cyclohexane and tetradecane. Each plot also shows, as a dashed line, the bilayer thickness which would be measured if the solute molecules were located entirely in the centre of the bilayer and caused no change in the surface area, but simply forced apart the two opposing monolayers of amphiphile. The deviation of the experimental points from this line is a measure of the extent to which the solutes intercalate between the amphiphile chains.

For benzene the bilayer thickness increases linearly with the mole ratio of solubilizate-to-amphiphile but suggesting a fairly uniform distribution of benzene across the bilayer

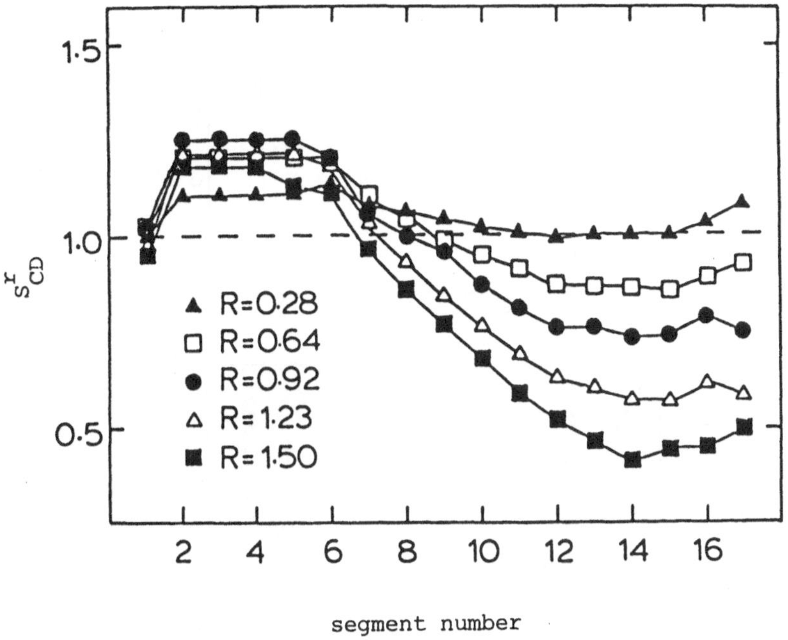

Figure 18. *Relative C-D bond order parameters of potassium stearate-d_{35} in the lamellar phase of the potassium oleate-benzyl alcohol-water system with $R_{W/A}$ = 7.6.*

independent of composition. For cyclohexane the bilayer thickness initially increases less rapidly than the theoretical line showing that the solubilizate penetrates between the chains to a certain extent. At higher mole ratios of solute the experimental curve becomes almost parallel to the theoretical line consistent with the solubilizate partitioning into the centre of the bilayer. For tetradecane the behaviour is the opposite of that observed for cyclohexane: at very low concentrations the data follow the theoretical line very closely, suggesting that the solubilizate is located entirely in the centre of the bilayer, but at higher concentrations the bilayer thickness is markedly less than that predicted by the model and there must, therefore, be some penetration of the tetradecane molecules into the amphiphile layers.

The behaviour of the partial molecular surface areas of the amphiphiles are compared in Fig. (20b) and are seen to be quite

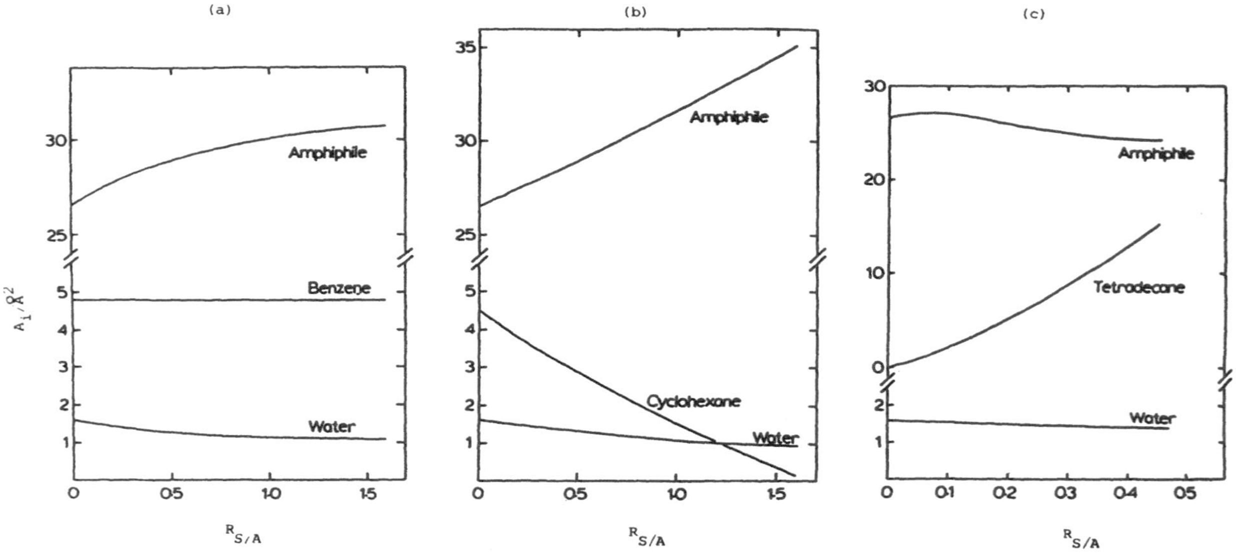

Figure 19. *Partial molecular surface areas as a function of $R_{S/A}$ in the lamellar phases of potassium oleate-water-solubilizate systems where the solubilizate is (a) benzene, (b) cyclohexane and (c) tetradecane. $R_{W/A}$ = 7.6.*

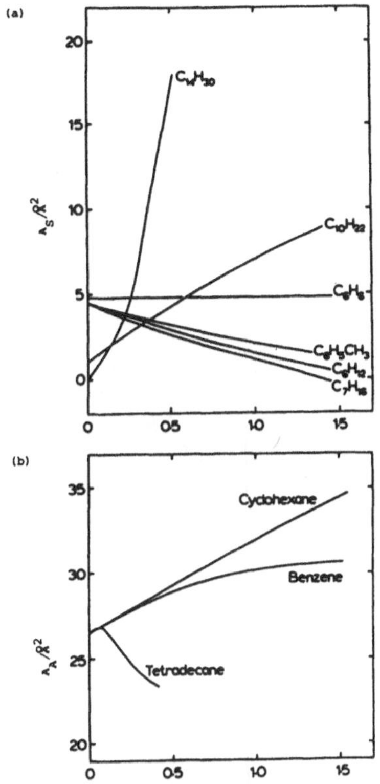

Figure 20. *Partial molecular surface areas of (a) solubilizate and (b) amphiphile in the lamellar phases of potassium oleate-water-solubilizate systems with* $R_{W/A}$ *= 7.6 and at 21°C.*

different for the three solubilizates. These results can be explained similarly to those for the alcohols in terms of a competition between ordering of the amphiphile chains, which tends to reduce the partial molecular surface area of the amphiphile, and transfer of potassium ions to sites between the carboxylate groups, which tends to increase it. These two effects can be monitored by ^2H and ^{39}K NMR, respectively.

NMR Measurements

We see in Fig. (16) of the previous chapter that the ^{39}K

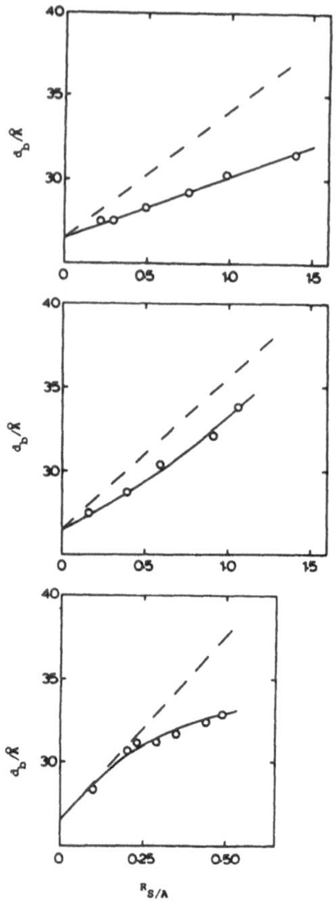

Figure 21. *Bilayer thickness as a function of $R_{S/A}$ in lamellar phases of potassium oleate-water solubilizate systems where the solubilizate is in (a) benzene, in (b) cyclohexane and in (c) tetradecane. $R_{W/A}$ = 7.6. The dashed line represents the behaviour if the solubilizate were located solely in the centre of the bilayer.*

quadrupolar splitting exhibited very similar behaviour for both benzene and cyclohexane, though the former caused a slightly larger reduction in the splitting than the latter suggesting that more K^+ ions enter the surface of the bilayer in the case of benzene than cyclohexane. At higher mole ratios the ordering of the top of the amphiphile chains brought about by the addition of benzene (Fig. (22a)) counteracts the increase in the

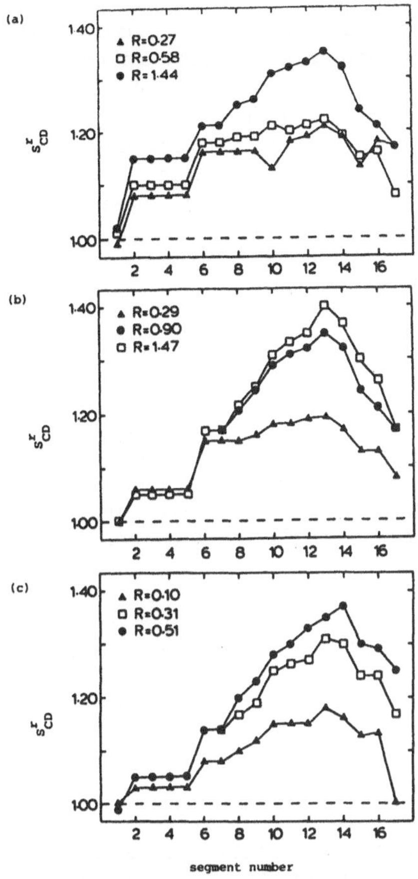

Figure 22. *Relative C-D band order parameters of potassium stearate-d_{35} in the lamellar phases of potassium oleate-water-solubilizate where the solubilizate is in (a) benzene, in (b) cyclohexane and in (c) tetradecane. $R_{W/A}$ = 7.6.*

partial molecular surface area. For cyclohexane the top segments of the chain become more ordered at first, but further addition of solute causes no further change in the order parameters (Fig. (22b) so that the partial molecular surface area of the soap continues to increase with concentration of solubilizate. No ^{39}K measurements were made on the potassium oleate-water-tetradecane system, but the reduction in the partial molecular surface area of the amphiphile can only be partially accounted for in terms of ordering of the chains (Fig. (21c)). The cross-sectional area calculated from the segmental order parameters decreases by only about 1 Å^2 (Fig. (12)). Other factors such as displacement of K^+ ions from 'specific' to 'diffuse' binding sites might be involved; if so, they should be evident in the ^{39}K splittings.

In marked contrast to the behaviour of the amphiphiles, the order parameters of the solutes all decrease with increase in concentration of solubilizate as shown in Fig. (23). The interpretation of these measurements is, however, complicated by the fact that the observed splitting is an average over all of the solubilization sites, which can have order parameters of opposite sign.

Comparison of Bilayer Thickness as Calculated from X-ray Diffraction and ^{2}H NMR Measurements

The thicknesses obtained by the two methods are compared in Fig. (24). There is a marked discrepancy between the two sets of measurements. The NMR measurements give the penetration depth of the hydrocarbon chain of the stearate probe molecule which varies very little, but similarly with composition for all three solutes. The X-ray diffraction data gives the actual bilayer thickness which is seen to increase rapidly with concentration of solubilizate: this is because it includes the contribution from hydrocarbon located between the opposing monolayers. The difference between the two values is greatest for cyclohexane which is consistent with the proposed models.

Distribution of Hydrocarbon Solubilizates in the Bilayer

Incorporation of hydrocarbon solubilizates into the bilayers of the potassium oleate-water lamellar phase causes the bilayer thickness to increase. This is, at low concentrations, roughly in proportion to the molar volume of the solubilizate (Fig. (25)). Despite this universal response of the bilayer structure to hydrocarbon solubilizates the actual distribution of solubilizate across the bilayer is dependent upon its molecular structure. Benzene tends to be fairly evenly distributed across the bilayer but with a slightly greater tendency to pack into the centre. Short chain alkanes such as pentane and cyclohexane, too, tend to be distributed across the bilayer at low concentrations and to partition mainly in the centre at high concentrations. Long chain alkanes such as tetradecane though they partition preferentially into the centre of the bilayer at very low concentrations, at higher concentrations they are located between the amphiphile chains. For long chain alkanes solubilized into phosphatidyl choline bilayers it has been shown by X-ray diffraction measurements that the solubilizate molecules are located within the opposing monolayers, aligned parallel to and alongside the amphiphile chains (10).

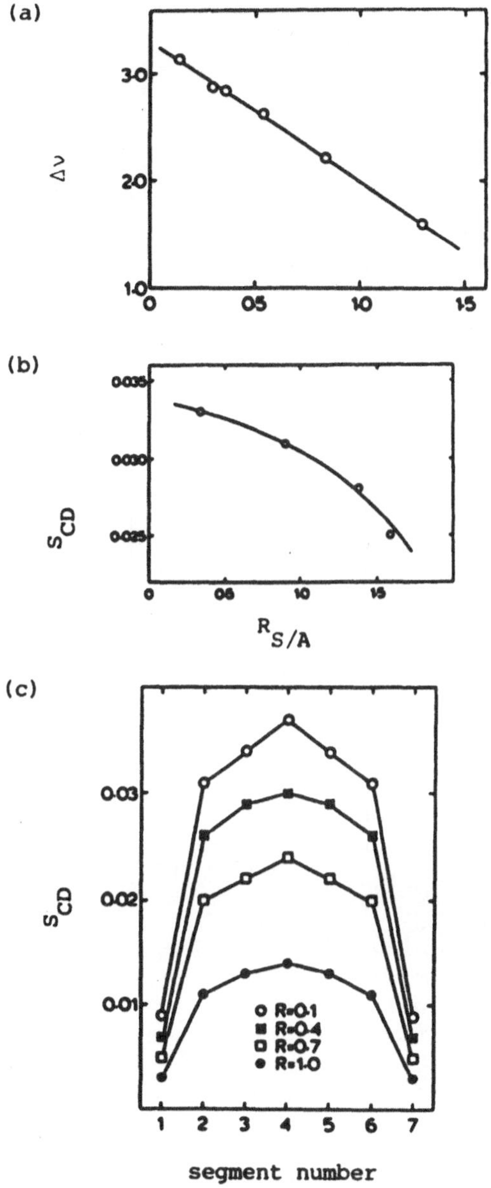

Figure 23. *(a) 'Apparent' ^2H quadrupolar splitting of cyclohexane-d_{12} in the lamellar phase of potassium oleate-water-cyclohexane - the spectrum is broadened by ring inversion (9). (b) and (c) show C-D bond order parameters for (b) benzene and (c) pentane in lamellar phases of potassium oleate-water-solubilizate systems $R_{W/A}$ = 7.6.*

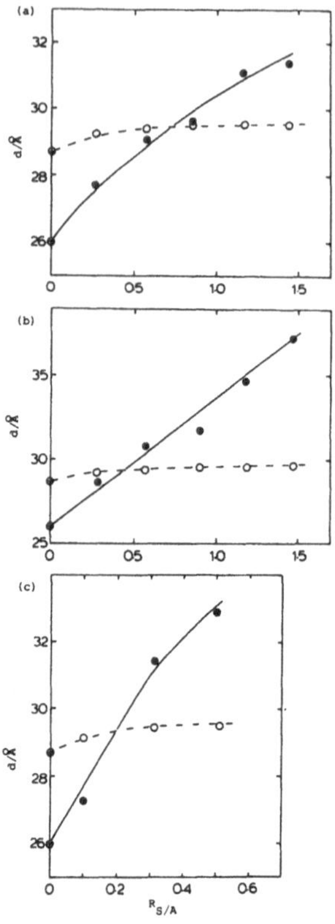

Figure 24. *Bilayer thickness as calculated from the C-D bond order parameters of potassium stearate-d_{35} (○) and X-ray diffraction measurements (●) in the lamellar phases of potassium oleate-water-solubilizate systems where the solubilizate is in (a) benzene, in (b) cyclohexane and in (c) tetradecane.* $R_{W/A} = 7.6$.

The main factors influencing the distribution of hydrocarbon solutes in the bilayer are as follows (11).

1. An increase in hydrocarbon-water contact which is very unfavourable.

2. The internal entropy of the hydrocarbon molecule is reduced by the lateral pressure of the amphiphile chains. This constraint is strongest in the ordered part of the bilayer

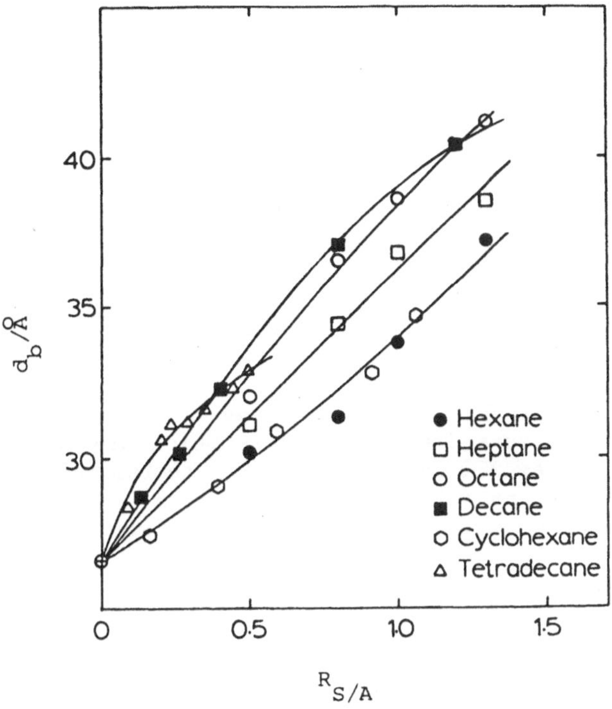

Figure 25. *Bilayer thickness as obtained from X-ray diffraction measurement in the lamellar phases of potassium oleate-water-solubilizate systems with* $R_{W/A} = 7.6$.

near the interface and weakest in the centre.

3. The adsorption of solubilizate into the outer regions of the bilayer necessitates a straightening of the amphiphile chains. This is unfavourable in terms of the free energy of the chains, but has the entropic advantage of giving a more even distribution of solubilizate across the bilayer.

Factors 1 and 3 suggest that energetic considerations would favour the solubilization of hydrocarbons in the geometric centre of the bilayer. Distribution of the solubilizate through the bilayer must, therefore, be entropy driven. An obvious difference between benzene and a small aliphatic hydrocarbon, such as heptane or cyclohexane, is that the former is rigid while the latter are flexible and undergo considerable internal motion. Solubilization of a flexible molecule in the ordered region of the bilayer, while gaining a better distribution of solute through the bilayer, results in a loss of internal entropy which is unfavourable. Benzene has no internal entropy to lose and the only entropic factor governing its solubilization is the entropy of mixing. This argument provides a plausible explanation for the fact that cyclohexane and heptane tend to partition into the centre of the bilayer while benzene is more evenly distributed across it.

Tetradecane is solubilized initially entirely in the centre of the bilayer, presumably because there are a huge number of possible conformers available to this molecule and the loss of conformational entropy necessary for it to enter the ordered part of the bilayer, where it would be restricted to those conformers which were essentially parallel to the lipid chains, is prohibitive. As the concentration is increased, however, the lateral pressure due to the ordering of the amphiphile chains strongly disfavours those conformers which are parallel to the bilayer surface. The solubilizate is, therefore, forced to go between the chains although this causes an increase in the area of hydrocarbon-water contact which severely limits its solubility.

RESPONSE OF THE BILAYER STRUCTURE TO SOLUBILIZATE

The structural changes brought about by the addition of solubilizate to the potassium oleate-water lamellar phase are as follows.

1. The addition of hydrocarbons gives rise eventually to a dispersion of lamellar phase in an isotropic solution. This occurs at $R_{S/A} \simeq 1.5$ to 1.8 with the exception of tetradecane whose solubility limit is $\simeq 0.55$.

2. The addition of alcohols induces a transition to reversed hexagonal phase; exceptions to this are pentanol and benzyl alcohol which produce dispersions. The maximum mole ratios of the various alcohols to soap are listed in Table 1.

These observations can be rationalized in terms of current theories (3,12) for the structure into which amphiphiles aggregate in water. The size and shape of the aggregate that obtains for a particular amphiphile system is determined by the 'packing parameter' $v/a_0\ell_c$ where a_0 is the optimal surface area per amphiphile, ℓ_c the maximum length the chain can assume and v its volume. Spherical micelles are formed if $v/a_0\ell_c < \frac{1}{3}$, cylindrical micelles if $\frac{1}{3} < v/a_0\ell_c < \frac{1}{2}$, bilayers if $\frac{1}{2} < v/a_0\ell_c < 1$ and inverted structures if $1 < v/a_0\ell_c$. For the lamellar phases of all of the potassium oleate-water-solubilizate systems, the value of the packing parameter is approximately unity when a_0 is taken to be the surface area per soap molecule, ℓ_c to be one-half of the bilayer thickness (d_b) and v to be the total hydrocarbon volume per soap molecule.

Table 1. *Solubilities of alcohols in the lamellar phase of potassium oleate-water ($R_{W/A}$ = 7.6) at 21°C.*

Solute	Mole Ratio
Pentanol	1.6 ± 0.1
Benzyl Alcohol	1.5 ± 0.1
Octanol	1.1 ± 0.1
Decanol	1.0 ± 0.1
Oleyl Alcohol	0.9 ± 0.1

The solubilization of alcohols is favoured by the entropy of mixing and by the reduction in the Coulombic repulsions between the carboxylate groups due to the increase in the number of potassium ions located between them which accompanies the addition of alcohol. The latter process is likely to be dominant and to easily outweigh the unfavourable effect of the ordering of both the amphiphile and alcohol chains. The addition of a long chain alcohol clearly increases the hydrocarbon chain volume of the bilayer and also the surface area. However, the surface area does not increase as rapidly as the hydrocarbon chain volume so that the bilayer thickness increases, preserving the value of the packing parameter (Fig. (14)). As the alcohol concentration is increased the energy required to order the chains further becomes greater and at some point the value of the packing parameter becomes greater than one and a transition to reversed hexagonal phase occurs. This transition takes place at progressively lower mole ratios of alcohol to soap as the chain length of the alcohol, and hence its ordering effect on the amphiphile chains, becomes greater.

In general, as hydrocarbons are added to the lamellar phase the solubilizate partitions largely into the centre of the bilayer and the ordering of the amphiphile chains is not as pronounced as that produced by alcohol solubilizates. One might expect that it would be possible to add hydrocarbon indefinitely, forcing the two opposing monolayers of amphiphile apart and forming a layer of essentially liquid-like hydrocarbon in the centre of the bilayer. This is found, however, not to be the case. The reason for this is that once the centre of the bilayer is truly liquid-like further addition of hydrocarbon would represent the transfer of solubilizate from one bulk liquid to another at a higher pressure due to the van der Waals attraction across the bilayer of the two water layers (11).

In the case of the long-chain alkane, where the solute is located to a large extent between the amphiphile chains, it may be that the increased hydrocarbon/water contact at the interface places an upper limit on the solubility.

APPENDIX 1

X-ray Diffraction Measurements

The classic model of a lamellar phase is one which consists of bimolecular lamellae (bilayers) separated by layers of water (Fig. (1A)). The bilayer thickness d_b is related to the interlayer separation d_0 by

$$d_b = \phi_b d_0 \tag{A1}$$

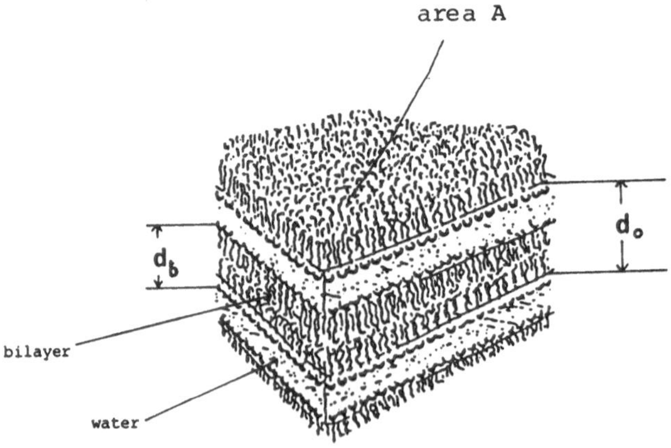

Figure 1A. *Model of lamellar bilayer.*

where ϕ_b is the volume fraction of the bilayer. Values for d_0 can be obtained by measuring the diffraction angle 2θ and using the Bragg equation

$$n\lambda = 2d_0 \sin\theta. \tag{A2}$$

If all of the solubilizate is within the bilayer

$$\phi_b = \frac{x_A V_A + x_S V_S}{x_A V_A + x_W V_W + x_S V_S}, \tag{A3}$$

where x_i and V_i are, respectively, the mole fraction and partial molar volume of component i, and the subscripts A, W and S refer to amphiphile, water and solubilizate.

Consider the volume element A x d_0 of bilayer in Fig. (A1) which contains N_A, N_W and N_S molecules of, respectively, amphiphile, water and solubilizate. The mean area per molecule at the bilayer-water interface is

$$A_m = 2A/(N_A + N_W + N_S).$$

But, since $N_i = N_{Av} A d_0 \phi_i / V_i$,

$$A_m = \frac{2}{N_{Av} d_0} \sum_i x_i V_i \quad (i = A, W, S), \tag{A4}$$

It has been customary (1) to express the interfacial area as the area per amphiphile, i.e. $2A/N_A$. However, as the surface area is a communal or cooperative property of all the components comprising the mesophase, it would be preferable to be able to express it in a way that separates the contribution from each component in a mathematically precise manner. This can be achieved by defining a quantity called the 'partial molecular surface area' which apportions the interfacial area among the various components in an analogous manner to the relationship between the total volume and the partial molar volumes of the components in liquid mixtures.

The Partial Molecular Surface Area

The surface area of the bilayer is an extensive property of a lamellar phase and can be expressed as a function of temperature T, pressure p and the amount n_i of each component. Applying Euler's theorem gives

$$A = \sum_i n_i \left(\frac{\partial A}{\partial n_i}\right)_{T,p,n_c \neq i} \qquad (i = A, W, S)$$

$$= \sum_i N_i A_i \tag{A5}$$

where

$$N_i = N_{Av} n_i$$

and

$$A_i = \left(\frac{\partial A}{\partial N_i}\right)_{T,p,N_{c \neq i}}.$$

The quantity A_i defines the partial molecular surface area of component i.

In a three component system the mean molecular surface area is

$$A_m = \frac{A}{N_A + N_W + N_S} = x_A A_A + x_W A_W + x_S A_S. \tag{A6}$$

Since $x_A + x_W + x_S = 1$ there are two independent composition variables, but fixing the ratio of the amounts of any two of the three components, say $n_W/n_A = x_W/x_A = R_{W/A}$, then only the mole fraction x_S is independent. Differentiating A_m with respect to x_S at constant T and p and recognizing that

$$\sum_i x_i (\partial A_i/\partial x_S)_{T,p,x_A/x_W} = 0,$$

we obtain

$$\frac{\partial A_m}{\partial x_S} = A_A \frac{\partial x_A}{\partial x_S} + A_W \frac{\partial x_W}{\partial x_S} + A_S \frac{\partial x_S}{\partial x_S}$$

$$= - \frac{A_A}{1 + R_{W/A}} - \frac{R_{W/A}}{1 + R_{W/A}} A_W + A_S . \tag{A7}$$

Since

$$x_A = \frac{1 - x_S}{1 + R_{W/A}} \quad \text{and} \quad x_W = \frac{R_{W/A}}{1 + R_{W/A}} (1 - x_S) .$$

Multiplying equation (A7) by x_S and combining the result with equation (A6) we obtain

$$A_m = x_S \frac{\partial A_m}{\partial x_S} + \frac{1}{1 + R_{W/A}} (A_A + R_{W/A}A_W). \qquad (A8)$$

Similarly, multiplying equation (A7) by $(1 - x_S)$ and combining the result with equation (A6) gives

$$A_m = A_S - (1 - x_S) \frac{\partial A_m}{\partial x_S} . \qquad (A9)$$

Equations (A8) and (A9) tell us that the tangent to the curve of A_m versus x_s at $x_s = x_s'$ has intercepts $A_S(x_S')$ at $x_S = 1$ and

$$\frac{1}{1 + R_{W/A}} (A_A + R_{W/A}A_W)(x_S') \text{ at } x_S = 0 .$$

Alternatively, choosing x_W as the independent variable and fixing the ratio $n_S/n_A = x_S/x_A = R_{S/A}$ we obtain

$$A_m = x_W \frac{\partial A_m}{\partial x_W} + \frac{1}{1 + R_{S/A}} (A_A + R_{S/A}A_S) \qquad (A10)$$

$$A_m = A_W - (1 - x_W) \frac{\partial A_m}{\partial x_W} . \qquad (A11)$$

Proceeding as above, values for both A_W and the second term in equation (A10) corresponding to any value x_W' of x_W can be obtained. Combining the results of the two experiments yields values for A_A', A_W' and A_S' corresponding to the composition x_A', x_W', x_S'.

APPENDIX 2

The Calculation of the Effective Length of a Hydrocarbon Chain from its Order Profile

The deuterium quadrupolar splittings obtained from fully deuteriated amphiphile chains in lamellar liquid crystals can be used to estimate the effective chain length and hence the bilayer thicknesses (13). The reasoning behind this is as follows. The average orientation of a C-D bond is essentially perpendicular to the bilayer normal $\mathbf{n}$ and so it is reasonable to suppose that the order parameter S_{CD} is negative. It is also possible to define a segmental order parameter

$$S_{mol} = \tfrac{1}{2} \langle 3 \cos^2\theta - 1 \rangle ,$$

where θ is the instantaneous angle between the normal to the plane of the segment and **n**. This is approximately related to S_{CD} by

$$S_{mol} = -2\, S_{CD}$$

for a methylene segment, and by

$$S_{mol} = -3\, S_{CD}$$

for a methyl group. Assume that the angular fluctuations of the hydrocarbon chains arise from rotational isomerization around carbon-carbon bonds and that the only segment orientations allowed are as follows:

(a) parallel to the bilayer normal (which we call A segments), and

(b) inclined at 60° to the bilayer normal (called B segments).

The jump frequency between the various conformational states is very much faster than the time scale of the NMR experiment and consequently only the time average of the splitting in A and B orientations is observed. If the probabilities of a segment being in the states A and B are p_A and p_B, respectively, then the time average of the various orientations is given by

$$S_{mol} = p_A \tfrac{1}{2} (3 \cos 0° - 1) + p_B \tfrac{1}{2} (3 \cos^2 60° - 1)$$

and since

$$p_A + p_B = 1$$

it follows that

$$p_B = (1 - S_{mol})/1.125$$

For a segment in the A state the projected length onto the bilayer normal is ℓ = 1.25 Å which corresponds to the effective length of a carbon-carbon bond in an all-trans chain. In the B state the projected length is ℓ cos 60°. The average length of the ith segment is therefore

$$\langle \ell_i \rangle = p_{iA}\ell + p_{iB}\ell \cos 60° = \ell(1 - 0.5p_{iB}) \quad \text{(A12)}$$

and the total length of an n-carbon chain is

$$\langle L \rangle = \sum_{i=1}^{n} \langle \ell_i \rangle = \ell(n - 0.5 \sum_{i=1}^{n} p_{iB}) . \quad \text{(A13)}$$

Equation (A13) has been found to be generally applicable, i.e. it is insensitive to the model chosen for the motion of the chains.

REFERENCES

1. P. Ekwall, in *'Advances in Liquid Crystals'*, ed. G.H. Brown (Academic Press, New York, 1975), Vol. 1, p.1.

2. P. Ekwall, L. Mandell and K. Fontell, *J. Colloid and Interface Sci.*, **31**, 508 (1969); **31**, 530 (1969).

3. C. Tanford, *'The Hydrophobic Effect'*, 2nd Edn., (John Wiley & Sons, New York, 1980).

4. H. Lecuyer and D.G. Dervichian, *J. Mol. Biol.*, **45**, 39 (1969).

5. J. Seelig and W. Neiderberger, *Biochemistry*, **13**, 1585 (1974).

6. B. Mely, J. Charvolin and P. Keller, *Chem. Phys. Lipids*, **15**, 161 (1975); B. Mely and J. Charvolin, *Chem. Phys. Lipids*, **19**, 43 (1977).

7. P. van der Ploeg and H.J.C. Berendsen, *J. Chem. Phys.*, **76**, 3271 (1982).

8. G.L. Turner and E. Oldfield, *Nature*, **277**, 669 (1979).

9. R. Poupko and Z. Luz, *J. Chem. Phys.*, **75**, 1675 (1981).

10. T.J. McIntosh, S.A. Simon and R.G. McDonald, *Biochemica et Biophysica Acta*, **597**, 445 (1980); T.J. McIntosh and M.J. Castello, *Biochimica et Biophysica Acta*, 1981, **645**, 318 (1981).

11. D.W.R. Gruen, *Biophys. J.*, **33**, 149 (1981); D.W.R. Gruen and D.A. Haydon, *Biophys. J.*, **33**, 167 (1981).

12. J.N. Israelachvili, D.J. Mitchell and B.W. Ninham, *J. Chem. Soc., Faraday 2*, **72**, 1525 (1976).

13. A. Seelig and J. Seelig, *Biochemistry*, **13**, 4839 (1974); A. Schindler and J. Seelig, *Biochemistry*, **14**, 2283 (1975).

NMR OF LIQUID CRYSTALLINE LIPIDS IN BIOLOGICAL MEMBRANES

Ian C.P. Smith

Division of Biological Sciences,
National Research Council,
Ottawa, Canada K1A 0R6.

SOME CHARACTERISTICS OF BIOLOGICAL MEMBRANES

Every living cell is surrounded by a membrane containing lipid, protein, and often carbohydrate. The membrane serves, at the very least, to distinguish the outside from the inside world of the cell, thus conferring identity on the cell. The membrane can regulate the vectorial transport of nutrients, waste, and exogenous materials such as drugs, as well as provide a site for cell-specific antigens and hormone receptors. In the cells of higher organisms, we find more specialized membranes such as those of nuclei, mitochondria, or endoplasmic reticulum.

A distinctive component of membranes is lipid, which can have a wide variety of chemical structures - phospholipid, glycolipid, sphingolipid, sterol, sterol ester, etc. All such lipids are amphiphilic, having water-loving and water-hating ends. It has been known for many years that simple aqueous dispersions of these lipids form lamellar arrays of (presumably) highly-ordered lipids. Given also that membrane proteins contain both hydrophobic and hydrophilic regions, a model has evolved for the intact biological membrane in which lipids are arranged in bilayers with various possible intercalations of protein, Fig. (1). The similarity of the fatty acyl or alkyl chains of the membrane lipids to those in the various lyotropic mesophases suggests that membranes may have liquid crystalline properties. This has turned out to be true in almost all cases studied to date. The structure of the lipid components resembles those of the various smectic mesophases, and in some systems hexagonal and cubic phases have been inferred.

J. W. Emsley (ed.), Nuclear Magnetic Resonance of Liquid Crystals, 533–566.

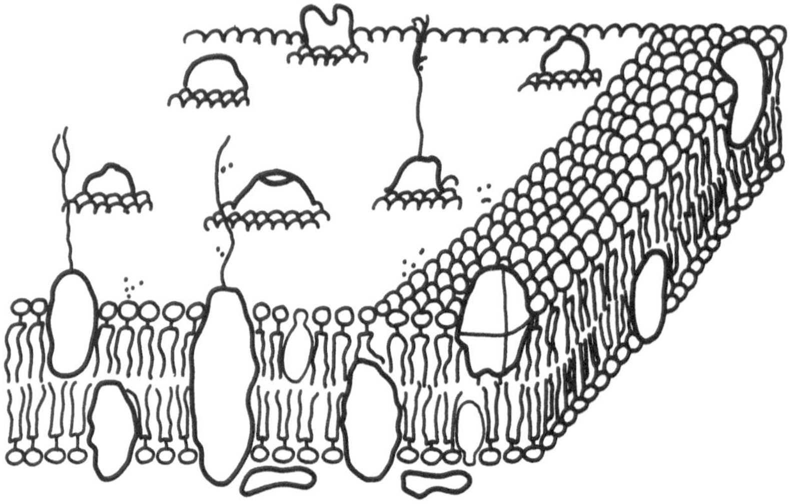

Figure 1. *Schematic slice through a biological membrane composed of a lipid bilayer into which proteins of various types are intercalated. The circular parts of the lipids represent the hydrophilic headgroups, and the wiggly lines the hydrophobic hydrocarbon chains. Figure courtesy of Dr. E.J. Dufourc.*

In view of the above, it is clear that many of the concepts and physical methods used in the study of liquid crystals should also be applicable to biological membranes. The situation is complicated by the multiplicity of components, and by the need to study these systems while they are alive. Nonetheless, significant progress has been made, and NMR has in large part been responsible for this progress. Although some use has been made of ^{1}H, ^{13}C, and ^{14}N, the major nuclei at present are ^{2}H and ^{31}P.

DEUTERIUM NMR SPECTRA OF MEMBRANES

Fig. (2) shows the fatty acyl chains of a phospholipid as they are thought to occur in liquid crystalline lipid bilayers. The carbon-carbon single bonds allow the possibility at each position of a saturated chain for one trans and two gauche conformers. The relative populations of these conformers may vary greatly with position along the chain. The all-trans state of the chains is the longest, thinnest, and most-easily packed; however, a single gauche conformer causes a large change in the length, effective width, and packing ability of the chains. This change can be partly overcome by cooperative conformational combinations, such as the gauche$^+$-trans-gauche$^-$ kink shown in Fig. (2). The ensemble average distribution of gauche and trans

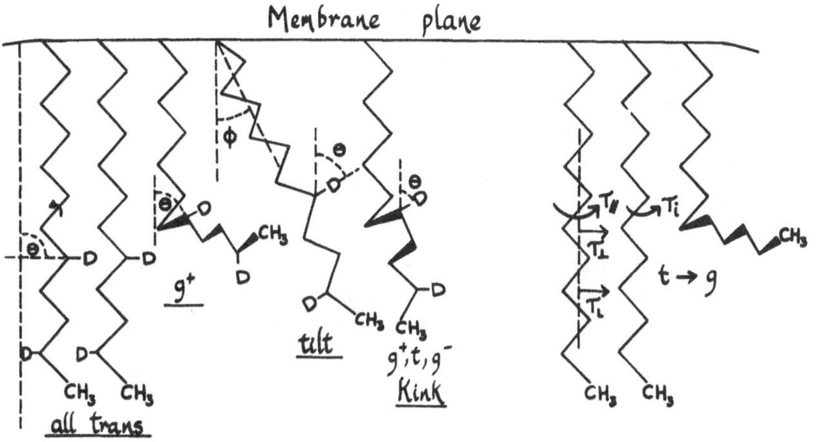

Figure 2. *Representation of some of the states of the fatty acyl chains of membrane lipids. The symbols g and t refer to the gauche and trans conformers with respect to rotation about a C-C bond. Other symbols are defined in the text.*

conformers thus provides a measure of the state of molecular organization of the fatty acyl chains - this can be estimated from the bond order parameter, S_{CD}, which is the ensemble and time average of the function $(3\cos^2\theta-1)/2$, where θ is the angle between a carbon-deuterium bond and the axis of ordering. When the appropriate transformation is made to account for the angle between the C-D bond and the instantaneous director (long axis) of the molecule, the resulting parameter, S_{mol}, has the value 1 for the trans conformer, and 0 for no conformational preference. This simple picture must be modified somewhat if the long molecular axis makes a net tilt (angle ϕ) with respect to the membrane normal (Fig. (2)). In addition, if rapid axial motion about the long molecular axis does not take place, or if the symmetry of the labelled group is low, two or more order parameters may be needed to describe the average orientations of the chains.

A complementary aspect of the membrane state involves the various motions of the chains. The order parameter gives an average picture of the degree of organization, but does not provide insight into the rates of interconversion between gauche and trans conformers at the various positions (τ_i), the rate of modulation of tilt angle ($\tau_\perp$), the rate of overall axial rotation of the chain ($\tau_\parallel$), or the rate of lateral diffusion of the entire phospholipid molecule (τ_L). These rates describe the mobilities of the lipid molecules. The rather vague concept of "fluidity" usually heard in descriptions of membrane properties involves both the degree of organization and the rates of movement of the lipids. These must be determined separately if the properties of the lipids are to be described unambiguously.

Due to its relatively small quadrupole moment, deuterium is an ideal probe of membrane lipids. The underlying physics of these systems has been presented in various reviews (1-9). Since there are two allowed transitions for the ^2H nucleus of spin = 1, an oriented C-^2H fragment gives rise to two resonances separated by a quadrupolar splitting. The magnitude of this splitting, $\Delta\tilde{\nu}$, depends upon the angle between the C-^2H bond and the applied magnetic field, Fig. (3). In a randomly oriented solid all possible angles are present, and the resulting powder spectrum is the sum of all the corresponding subspectra. In contrast to the chemical liquid crystals discussed so far, where the director of ordering is fixed by the external magnetic field, the situation in a membrane is more similar to that of a solid powder. The membranes do not orient significantly in the magnetic field, and their rates of overall rotation are very slow on the ^2H NMR time scale (any rate less than 3×10^5 sec^{-1} is slow). Thus, if there were no motion *within* the membrane, the resulting spectrum would have a D_q of 128 kHz, corresponding to the spectrum of a rigid randomly-oriented powder. If rapid motion occurred about the long axis of the fatty acid molecule (dashed axis in Fig. (3)), but with no disordering of the chain segments, a quadrupolar splitting (corresponding to $\theta = 90°$ in Fig. (3)) of 64 kHz would result. Disordering of the chains, with rapid interconversion between disordered states, leads to a further reduction in the quadrupole splitting D_q. The magnitude of D_q is directly related to the order parameter of the C-^2H fragment in question; $|S_{CD}| = \frac{4}{3}\frac{D_q}{C}$, where C is the quadrupolar coupling constant for ^2H in the particular chemical site (usually 170 kHz for attachment to an sp^3-hybridized carbon). This relationship assumes axial symmetry of the deuterium quadrupolar coupling tensor (usually true for attachment to saturated carbons), and axially-symmetric motion of the C-^2H fragment (usually true in the liquid-crystalline state of lipids, but often untrue for gel state lipids). The sign of the order parameter cannot be determined by a single measurement (if $|S_{CD}| \leq \frac{1}{2}$). Motions of

ORIGIN OF ^{2}H POWDER SPECTRUM

$$\Delta\nu = \frac{3}{4} \frac{e^2 qQ}{h} (3\cos^2\theta - 1)$$

$\theta = 90°$

$\theta = 54°$

$\theta = 0°$

Σ all
possible
θ

D_q

$2D_q$

H

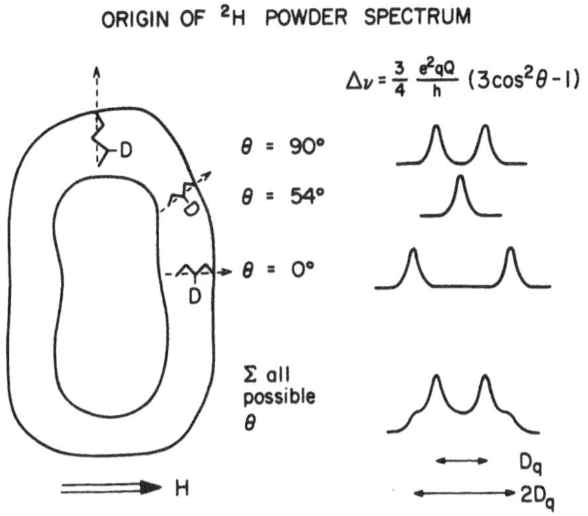

Figure 3. *Origin of the powder spectrum seen in ^{2}H NMR of membranes. The angle θ is that between the applied magnetic field, H, and the axis of motional averaging (dashed). The symbol D_q is used for the splitting derived from the powder pattern, which corresponds to the $\Delta\nu$ for $\theta = 90°$.*

intermediate rate and/or low symmetry lead to complex lineshapes, the most common of which has no resolvable peaks, and has a humplike form. In the case of axially asymmetric motion, a humplike spectrum similar to that observed when the quadrupolar splitting tensor of deuterium is axially asymmetric may result. This is known as an $\eta = 1$ lineshape (9).

Motional properties of the lipid chains may be monitored by the various relaxation times. The spin-lattice relaxation time, T_1, is sensitive to motions near the resonance frequency (15-60 MHz), whereas the spin-spin relaxation time, T_2, is more sensitive to slow motions. Both may show a dependence on the degree of molecular ordering, as well as on the angle between the director and the applied magnetic field. Attempts are currently underway to establish a sound theoretical basis for their interpretation (10, 11).

Special instrumental techniques are required to observe ^{2}H

NMR spectra of membranes. The wide spectra necessarily dictate the use of strong radiofrequency pulses, since the excitation of all resonances across the entire spectrum must be uniform to avoid spectral distortion. This is often not achieved for the broader patterns, and corrections must be applied if a detailed lineshape analysis is desired (12). Improved detection sensitivity can be achieved by operation at high magnetic fields (currently up to 11.5 T), and the design of probes especially for the ^{2}H NMR experiment. The latter involves optimization of filling factor, good tuning versatility to accommodate the lossy samples, and relatively fast electronic response. Wide spectral windows are achieved by high speed analog-to-digital conversion (500 kHz to 2 MHz). This is also necessary to provide adequate time resolution of the rapidly decaying ^{2}H NMR signals. The principal drawback in early ^{2}H NMR studies of membranes was this very fast signal decay. Fig. (4) gives a view of the problem.

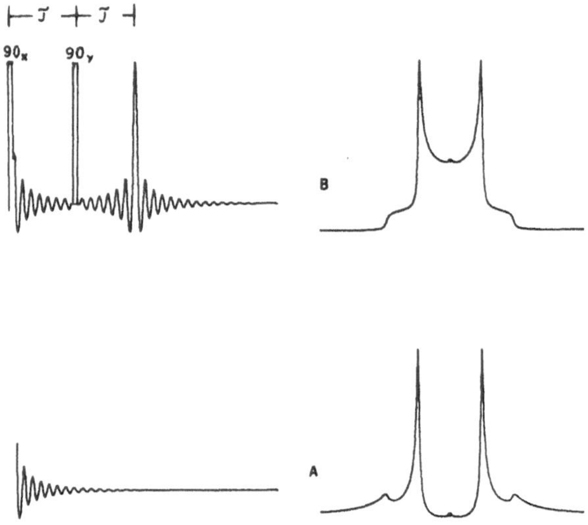

Figure 4. *Comparison of behaviour of the spin system, and the resulting ^{2}H NMR powder spectrum, from a quadrupolar echo (top) or a single 90° pulse (bottom). Figure courtesy of Dr. H.C. Jarrell.*

Excitation of the spin system also excites the detection coil and the probe circuitry. This hardware excitation decays with a characteristic time determined by probe design. The higher the quality factor Q, the longer the decay time. The hardware decay, if observed, introduces spurious signals into the detector. To avoid this aberration, we have to insert a delay between the pulse and the commencement of data acquisition. If this delay is comparable to the decay time of the NMR signal, very poor baselines and a weak distorted NMR signal will be obtained. These problems have now been largely circumvented by the use of the quadrupole echo technique (13), Fig. (4), and a phase-alternated acquisition sequence. The quadrupole echo sequence involves two successive 90° pulses, differing in phase by 90°; the second 90° pulse leads to formation of an echo at a time τ. If τ can be adjusted to be longer than the hardware ringdown time, no interference from the spurious signal occurs, and the full intensity of the signal may in principle be obtained from the echo. Attention must be paid to the relative magnitudes of the interval τ and the effective T_2 values of the spectrum, to avoid other types of spectral distortion (14). Finally, if the echo is transformed from its true maximum, no phasing of the frequency domain spectrum is required.

Cycling of the phases of the excitation, combined with the echo technique, yields a spectrum of high signal-to-noise ratio with a very flat baseline. An excellent comprehensive review of the experimental methods is available (4).

ACHOLEPLASMA LAIDLAWII B: A SIMPLE BIOLOGICAL MEMBRANE

The biological membrane which has been most thoroughly studied to date is the plasma membrane of the microorganism *Acholeplasma laidlawii*, strain B. *Acholeplasma* are minute prokaryotic organisms approaching the larger viruses in size. However, unlike viruses, they are capable of autonomous growth and reproduction in cell-free media. They are not only the smallest self-replicating organisms known so far; they are also the simplest in ultrastructure. They have no cell walls nor intracellular membranous structures, and are bounded by a single plasma membrane (15). By carefully depleting the medium of fatty acid, and then supplementing it with that desired, incorporation of particular fatty acids to levels of 100% of total fatty acid can be accomplished. Cholesterol, although not essential for growth, can be incorporated up to levels of 40 mole % of total lipid.

Isolated membranes of *A. laidlawii* can be prepared by the simple and mild process of osmotic lysis. They may then be freeze-dried and stored. Intact cells yield spectra very similar to those of freshly-prepared or reconstituted freeze-dried membranes (16). This is a particular advantage when one is subject to the vagaries of NMR instruments.

Our initial effort along these lines dealt with the popular, saturated fatty acid, palmitic acid (C16:0). The spectrum of 13-$^{2}H_2$-C16:0 in *A. laidlawii* membranes, taken using the quadrupole echo technique, is shown in Fig. (5). From

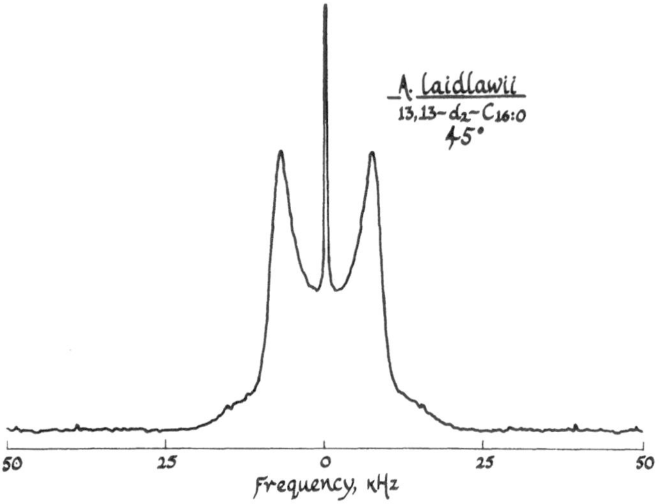

Figure 5. *The ^{2}H NMR spectrum of membranes of A. laidlawii enriched in 13-$^{2}H_2$-palmitic acid, 45°C.*

spectra such as this the profile of order versus position shown in Fig. (6) was constructed (17). Note that three quadrupolar splittings were measured for the 2-position. This is due to the different environments experienced by the sn-1 and sn-2 chains, and to the inequivalence of the two deuterons at carbon-2 of the sn-2 chain. Proceeding down the chains we find that the degree of ordering is relatively high and constant down to carbon-10, but decreases dramatically with position thereafter. Thus, the regions of the acyl chains near the edges of the bilayers are well organized, but those at the centre are extremely disorganized. This has important implications for the permeation through membranes, and the intramembrane location of, membrane-active drugs. Translating to the molecular order parameter (intrinsic molecular ordering), the plateau has a value of ca. 0.4. Nature has retained the two options of greater or lesser degrees of order in this region in response to interactions or perturbations.

Turning now to the same measurement for the membranes

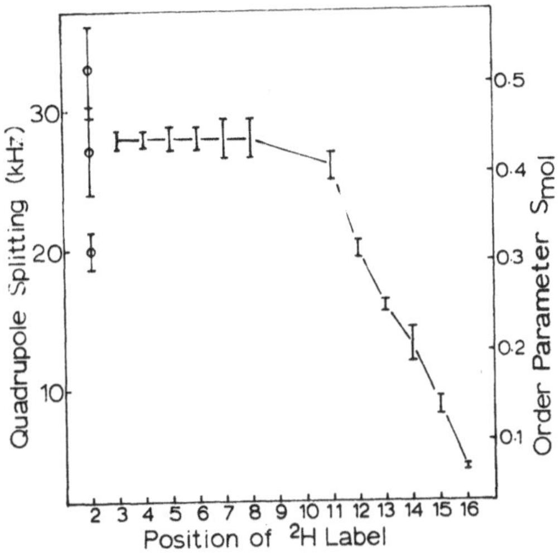

Figure 6. *Quadrupole splitting (or molecular order parameter) versus position of labelling for A. laidlawii membranes enriched in specifically deuteriated palmitic acid. Data from reference 17.*

enriched in oleic acid (18), Fig. (7), we see both similarities and differences on comparison with the data of Fig. (6). The presence of the cis double bond at position nine results in a severe drop in quadrupole splitting, as expected. This is somewhat deceptive, since the large decrease is due mainly to the particular geometry of the cis double bond. The low symmetry at this position leads to a requirement for at least three order parameters (and therefore three measurable parameters) for complete characterization. This has been attempted for the oleoyl chain of 1-palmitoyl-2-oleoyl-phosphatidylcholine, for which infrared linear dichroism data were available (19), and a cyclopropane-containing lipid in which more quadrupole splittings were available (20). In these cases it was shown that the double bond and the cyclopropane ring have degrees of order similar to or greater than that of the preceding methylene groups. The striking similarity of the data in Fig. (7) to those of Seelig and Waespe-Šarčevič (19) suggests that a similar conclusion holds for C18:1 in *A. laidlawii* membranes.

At positions removed from the cis double bond, the variation of D_q with position looks very similar to that already

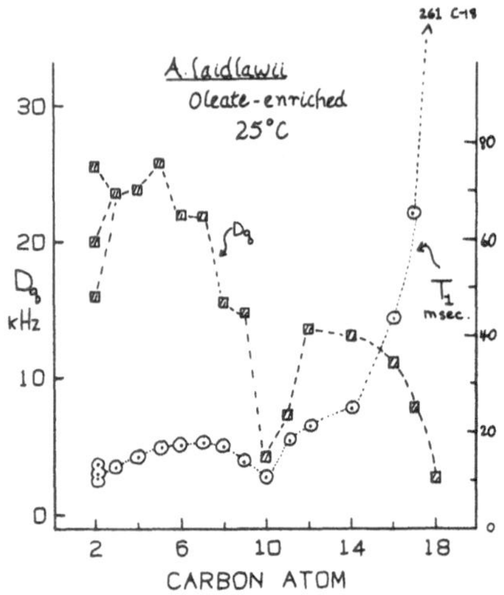

Figure 7. *Quadrupolar splitting and spin-lattice relaxation time (T_1) versus position of labelling for A. laidlawii membranes enriched in specifically deuteriated oleic acid. The T_1 data are of M. Rance, K.R. Jeffrey and I.C.P. Smith, unpublished.*

discussed for C16:0, including the three splittings for position-2 of the chains. Thus, despite the geometric disruption in packing due to the cis double bond, the ordering properties of saturated and unsaturated acyl chains are remarkably similar.

Fig. (7) superimposes measurements of spin-lattice relaxation time on the positional dependence of order parameter for the oleate-enriched membranes (18). The spin-lattice relaxation time is mainly sensitive to the rates of motions near the resonance frequency, although it has a weak dependence on order parameter (10). In its simplest interpretation, the T_1 value can be considered as an indicator of the rate of motion of the deuteriated carbon. Fig. (7) shows a modest dependence of T_1 on position, with a significant dip at the position of the double bond, and a large increase in the rates of rapid motion for the last three positions. Note that the terminal methyl group has a T_1 value of 261 ms, well off the upper limit of the scale for Fig. (7). Comparable data (21) for multilamellar vesicles of dipalmitoyl phosphatidylcholine at 45°C are: C3-C12, 27-38 ms; C-13, 40ms; C-14, 46 ms; C-15, 56 ms; C-16, 275 ms.

The "fluid" membranes of *A. laidlawii* have T_1 values that are everywhere shorter than those of the model membrane with saturated chains, despite the fact that the "fluid" membranes were measured at a temperature further removed from that of the liquid crystal to gel transition. Thus, the "fluidizing" effect of the cis double bond is negligible from the mobility point of view; in fact, for the 9- and 10-positions a significant decrease from the average rate of motion of the chains is seen (as might be expected since rotation about the C9-C10 bond is not possible). The "fluidizing" effect may be on the neighbouring chains, which sense the bulkiness of the cis double bond, or simply a lowering of the liquid crystal to gel transition temperature (*vide infra*). The only direct experimental proof of the former assertion comes from comparison of the data for the model systems dipalmitoyl phosphatidylcholine (22) and 1-palmitoyl-2-oleoyl phosphatidylcholine (19). The order parameters for the palmitoyl chains were all significantly lower than those of the former compound.

When more than one quadrupolar splitting is present in a powder spectrum, accurate measurement becomes difficult due to overlaps. Bloom and coworkers have demonstrated a method, known as de-Pake-ing, which reduces the spectrum to that expected for a particular orientation of the director of ordering with respect to the magnetic field (23). The lower part of Fig. (8) shows the ^2H NMR spectrum of *A. laidlawii* membranes enriched to 47% in the cyclopropane-containing fatty acid, dihydrosterculic acid, enriched in deuterium at the 2-position. As remarked upon earlier for other systems, the C-2 position of the fatty acid at sn-2 apparently yields two quadrupolar splittings due to spatial inequivalence of the two deuterons. However, a surprising result is obtained if the data are de-Pake-ed. This is seen in the upper part of Fig. (8) where one-half of the symmetrical de-Pake-ed pattern is shown on an expanded frequency scale. Not only the two expected components for the C-2 deuterons are seen - there are minor components on either side of the principal components. The relative intensities of the two patterns correlate well with the relative proportions of two principal lipid headgroup classes, monoglucosyl and diglucosyl-diglyceride, with the former group having the greater population. The de-Pake-ing procedure was very helpful in elucidating the order parameter profile for this peculiar fatty acid in *A. laidlawii* membranes (24).

Many model and biological membranes are known from calorimetric studies to undergo a thermally-induced transition from a highly-ordered and relatively immobile state of the fatty acyl chains (gel state lipid) to a less-ordered and highly mobile state of the chains (liquid crystalline state). In model membranes the temperature range over which this transition takes

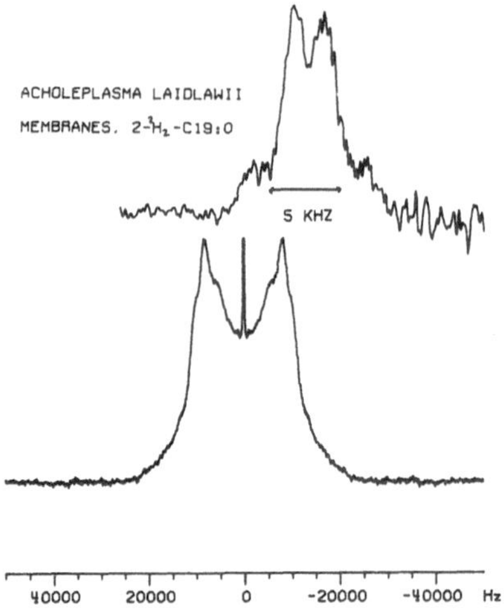

Figure 8. *^{2}H NMR spectrum of A. laidlawii membranes enriched in 2- H -dihydrosterculic acid (bottom), and one-half of the de-Pake-ed spectrum (top) showing the enhanced resolution of components.*

place is very narrow (25), whereas it is considerably broader for biological membranes such as those of *A. laidlawii.*

Passage into the gel state leads to large changes in the ^{2}H NMR spectra of membrane lipids. In the early 1970's this usually resulted in the complete disappearance of the spectrum, because the spectrometers were not capable of observing the rapidly decaying free induction signals (26). The instrumental developments discussed earlier led ultimately to observation of the spectra of the gel state of the lipids in *A. laidlawii* (17,27), and of dipalmitoyl phosphatidylcholine (21). Fig. (9) shows some of these improved spectra. Note the broad subspectrum present in the 37° spectrum - it is now clearly 60 kHz in width and contributes about 50% of the total intensity. The 60 kHz width is that expected for an acyl chain that is all-trans as far as the labelled position, with retention of rapid motion about the long axis of the chain. By 25°C, below which *A. laidlawii* does not grow efficiently on C16:0, all lipid is essentially in this state. Cooling the membranes further results in an even broader ^{2}H NMR spectrum; at 3°C intensity spreads over a range of ca. 120 kHz. This latter spectrum does not have

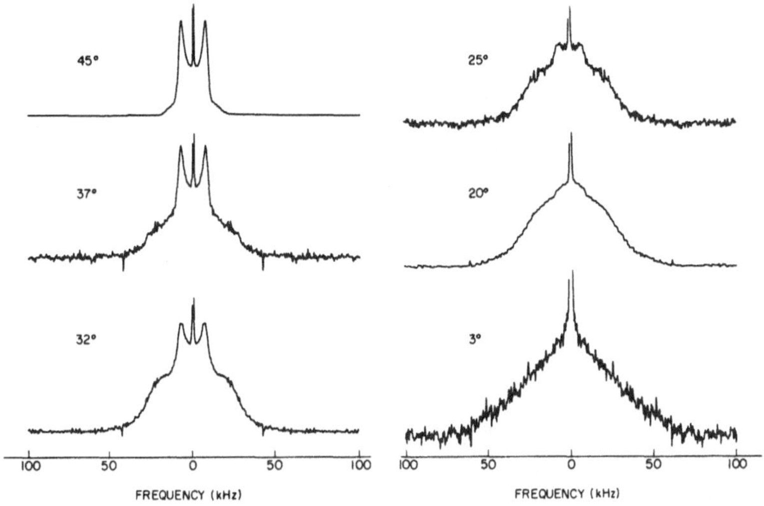

Figure 9. *^2H NMR spectra of A. laidlawii membranes enriched in 13-$^2H_2$-palmitic acid. The growth temperature of the organism was 37°C (27).*

a reliable lineshape since the 90° pulses of 9 μs duration are not short enough to cover the entire width of the spectrum with equal radiofrequency power. However, a spectrum obtained with pulses of 3 μs showed discernible shoulders separated by 110 kHz (27). For no rapid motion whatsoever of the chains, a splitting of 128 kHz is expected. Thus, by 3°C the acyl chains are almost totally immobile on the ^2H NMR time scale.

Up to now in our discussion of ^2H NMR spectra of membranes we have limited ourselves to quadrupolar splittings, overall pattern widths, and relaxation times. From the hundreds of spectra that have now been obtained, it is clear that information is also available in the lineshapes. The problem is how to make use of it. On the theoretical side one can attempt to simulate lineshapes in terms of motional models with the inclusion of several modes and rates, as has been done for the tuque-like spectra often obtained at low temperatures (28,29). On the experimental side one can measure the moments of the spectra, which have been used by solid state NMR practitioners for many years (30).

The n'th moment of a spectrum which is symmetrical about its midpoint is given by:

$$M_n = \frac{\int_0^\infty x^n F(x)\ dx}{\int_0^\infty F(x)\ dx},$$

where F(x) is the spectral lineshape, $x = \omega - \omega_0$, and ω_0 is the central frequency. The integrations are performed from the centre of the spectra in order to conserve the odd moments, which otherwise should be identically zero due to the symmetry of the spectra. As the moments give extra weight to the wings of

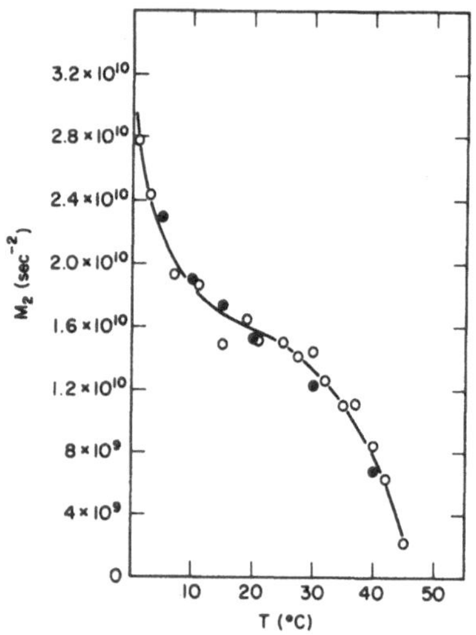

Figure 10. *Temperature dependence of the second moment of the spectra shown in Fig. (9). (27).*

the spectra, they are particularly sensitive to the rates of motions which partially average the components of the quadrupolar splitting tensor. The second moment is often used as a monitor of molecular motion. Fig. (10) shows the temperature dependence of the second moment for the spectra of Fig. (9). The onset of the phase transition is seen in the large increase in

M_2 from 45 to 25°. If all motion had ceased at 25°, there would be only slight changes in M_2 with decreasing temperature; on the contrary, it increases even more steeply over the lower range, indicating the decrease in rate of motion about the long axes of the acyl chains referred to earlier. However, even at the lowest temperature in Fig. (10), molecular motion is still occurring at a significant rate on the ^{2}H NMR time scale, as M_2 is far below the rigid-lattice limit of 1.28 x 10^{11} sec^{-2}. Part of this deviation from the rigid-lattice limit may also be due to the difficulty of obtaining the full lineshape when its extremities approach separations of 200 kHz.

Further insight into the information contained in the ^{2}H NMR powder spectra may be gained from a moment analysis proposed by Bloom and coworkers (31). They showed that the moments of the order parameter distribution, S_n, can be simply related to those of the ^{2}H NMR spectrum, M_n. Thus, the mean order parameter is simply related to the first moment: $S_{CD} = 2\sqrt{3}\, M_1/(e^2qQ/h)$, where e^2qQ/h is the quadrupolar coupling constant. This is useful in cases where large component linewidths obscure accurate measurement of the order parameter from the apparent quadrupole splitting. Another very useful parameter is defined as:

$$\Delta_2 = \frac{\langle S_2\rangle - \langle S_1\rangle^2}{\langle S_1\rangle^2} = \frac{M_2}{1.35\, M_1^2} - 1$$

where

$$S_n = \int_0^\infty S^n P(S)\,dS,$$

and P(S) is the probability of a given order parameter. Thus, the Δ_2 parameter is a direct measure of the mean square deviation of the order parameter, or the degree of homogeneity of the acyl chain packing (when homogeneous line broadening effects can be ignored). Fig. (11) shows the temperature-dependence of Δ_2 for C14:0-ω-d_3 incorporated to 90% by the avidin technique into the membrane lipids of *A. laidlawii* (32). At temperatures above that of the phase transition (T_c), this parameter has a value near zero, indicating a narrow distribution for S and a high degree of homogeneity for the lipids. Over the range of the phase transition Δ_2 has a high value as a consequence of the phase heterogeneity. At lower temperatures Δ_2 returns to a lower value, suggesting a relatively high degree of homogeneity in the semi-rigid, strongly-ordered lipids (rapid axial motion has not been quenched at 30°C).

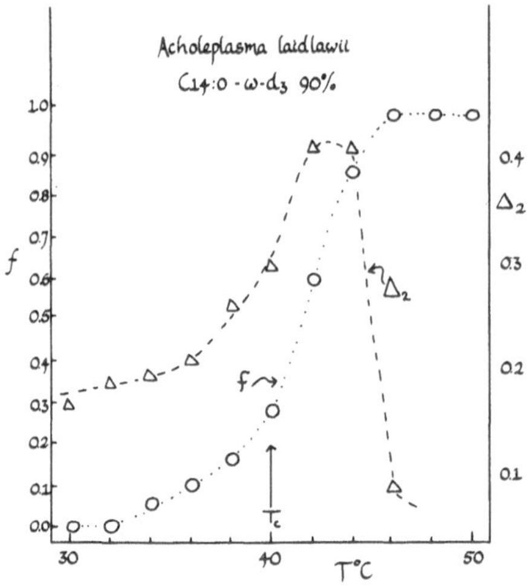

Figure 11. *Temperature dependence of the moment parameter,* Δ_2*, and of the fraction of fluid lipid, f, for A. laidlawii membranes enriched in 14-*$^2H_3$*-myristic acid (32).*

Spectral moments have also been very helpful in quantitating the fractions of gel and liquid crystalline lipid in spectra such as that for 37°C in Fig. (9). The strong overlap of the two types of spectrum makes deconvolution difficult. Jarrell et al. showed that the fraction of lipid in the liquid crystalline phase, f, can be determined from the moments of the spectra via:

$$M_n = f\, M_n^L + (1-f)M_n^G,$$

where M_n is the measured n'th moment of the composite spectrum, and M_n^L and M_n^G are the n'th moments of spectra taken at temperatures corresponding to the onset and completion of the phase transition, respectively (33). The method assumes that M_n^L and M_n^G are constant over the range of the phase transition. Fig. (11) shows the temperature dependence of f for membranes of *A. laidlawii* enriched in C14:0-ω-d_3. Note the asymmetric appearance of the dependence - it is steeper on the high temperature side.

A further use of spectral moments is to analyze the data for perdeuteriated fatty acyl chains in biological membranes. In model systems a considerable number of individual quadrupole splittings can be measured directly (21,34), whereas the greater linewidths found in biological membranes obscure much of this detail (17,32). By fitting the first four moments of the spectrum in terms of a polynomial, the dependence of the order parameter on position of labelling of the palmitate (35) or myristate (32) chains in *A. laidlawii* membranes could be estimated. This procedure, although approximate, gives a rough view of the profile with an enormous saving in time and effort, particularly if the effect of perturbants on the profile is to be determined.

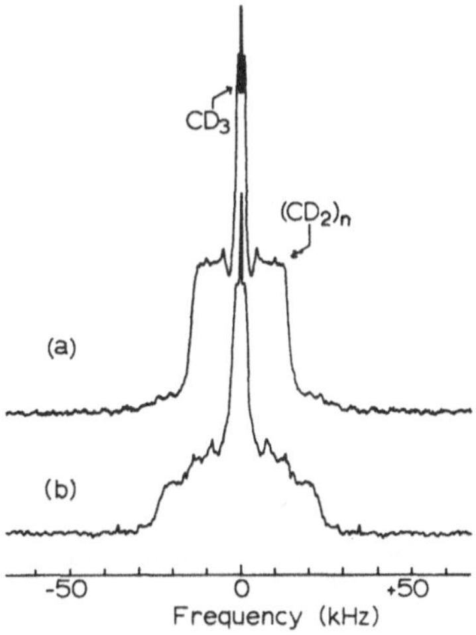

Figure 12. *^{2}H NMR spectra of A. laidlawii membranes enriched in perdeuteriated palmitic acid in the absence (top) and presence (bottom) of 40 mole % cholesterol (35).*

Fig. (12) shows the ^{2}H NMR spectra of *A. laidlawii* membranes containing perdeuteropalmitic acid, in the presence and absence of cholesterol. Due to unfavourable linewidths, only the quadrupole splittings of the terminal methyl and penultimate methylene groups can be distinguished, although the width of the pattern yields an estimate of the largest order parameter in the system, presumably due to the first 8-10 segments of the palmitoyl chains. As mentioned earlier, by analysis of the first four moments of the ^{2}H NMR spectra in terms of a polynomial

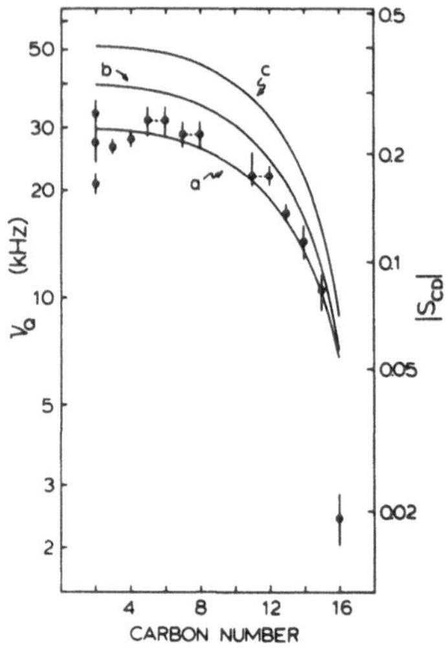

Figure 13. *Derived quadrupolar splittings (ν_Q) and bond order parameter (S_{CD}) for palmitic acid in A. laidlawii membranes. The solid curves are derived from moment analysis of the spectra from perdeuteriated palmitic acid; the dots are from specifically deuteriated acid: (a) 45°C, no cholesterol; (b) 45°C, 40% cholesterol (c) 22°C, 40% cholesterol (35).*

dependence of S on position, it was possible to delineate this dependence (35). Fig. (13) compares the dependence obtained from the moment analysis with that derived from direct measurement of the quadrupolar splittings of specifically-deuteriated palmitoyl chains in the cholesterol-free membranes. The high quality of the agreement in this case gave us confidence to apply the method to the membranes containing cholesterol. Curve b in Fig. (13) is for membranes containing 40 mole % of lipid as cholesterol, at 42°; curve c is for the same sample at 22°. Comparing curves a and b, we note that the largest absolute increases in molecular ordering of the acyl chains, due to cholesterol incorporation, occur for the first 10-12 carbon atoms of the chain, as has been observed earlier with egg phosphatidylcholine (34) and dimyristoyl phosphatidylcholine (36). This is expected in view of the known location of cholesterol in model membranes, the rigid nature of the A-D ring system, and the narrower cross sectional area and greater flexibility of the alkyl tail, as pointed out earlier (34). The degree of order for the cholesterol-enriched membrane at 22° (profile c) is even higher, but the main features of the profile are conserved.

The fused ring system of cholesterol is a case where only rigid body motion can take place. Recently it has been shown that analysis of the quadrupolar splittings from various positions on the ring system can lead to a determination of the location of the axis about which rapid axial rotation takes place, and of the order parameter for the ring system (37). The motional axis was found to make an angle of 79° with the C_3-H bond, as opposed to the 90° angle presumed in earlier studies. The order parameter for cholesterol in bilayers of egg lecithin containing 50 mole % cholesterol was found to be 0.87, close to the value of 1.0 corresponding to perfect order. This has recently been found to be true for cholesterol in the membranes of the human red blood cell (38).

Many drugs interact with membranes, if for no reason other than their amphipathic natures. In some cases this is thought to be the principal site of drug action. The great strides realized in spectrometer sensitivity have now made possible the study by ^{2}H NMR of deuteriated drugs in membranes and the effects of drugs on deuteriated membranes. Very little has been done to date on biological membranes, so I shall allude briefly to the studies reported so far on model systems.

It has been considered for the past 80 years that anaesthetics could act by dissolving in the hydrophobic regions of membranes and thus affecting their properties (39). Until the development of site-specific probes such as ^{2}H-labelled compounds, the molecular details of these interactions were unavailable. Boulanger et al. have made a thorough study of the interactions between the local anaesthetic tetracaine and multilamellar dispersions of phosphatidylcholine (40,41). When the anaesthetic was deuteriated, quadrupolar splittings of different magnitudes from various positions in the molecule were detected, indicating an environment for the anaesthetic with a detectable degree of order. Addition of unlabelled anaesthetic to labelled lipids resulted in changes in the lipid quadrupolar splittings. Fig. (14) shows the ^{2}H NMR spectra of phosphatidylcholine (PC) labelled at position 12 of the palmitoyl chain at the sn-2 position of the glyceryl moiety, and in the choline moiety, in the absence and presence of tetracaine (TTC). Increasing amounts of tetracaine led to decreasing quadrupolar splittings at the 12-position. The choline moiety yielded two splittings, barely distinguishable in Fig. (14d), of which one increased and the other decreased upon addition of tetracaine. The effects depended upon pH, since the dimethylamino group of tetracaine has an ionizable hydrogen

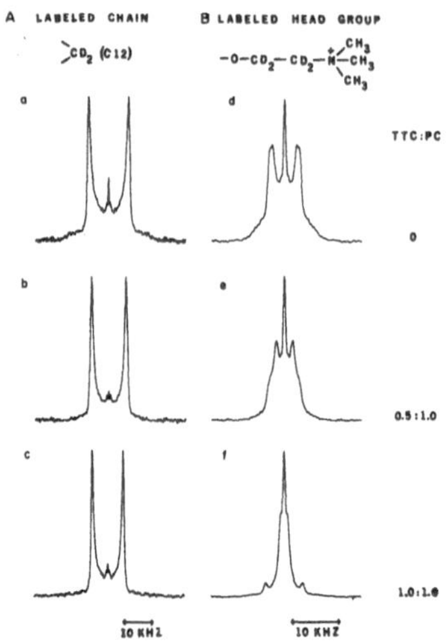

Figure 14. ^{2}H *NMR spectra of phosphatidylcholine membranes, labelled at carbon-12 of the sn-2 chain, or in the choline head group as shown. From top to bottom the molar ratio of bound tetracaine to lipid increases (41).*

(pK = 7.5). Taking all the quadrupolar splittings, their response to anaesthetic dose, and ^{31}P NMR of the headgroup phosphodiester, it was possible to construct a detailed molecular model for the anaesthetic-lipid interaction, including the location of the anaesthetic within the bilayer (41). Confirmation of the above in both phosphatidylcholine (42) and phosphatidylethanolamine (43,44) has appeared recently. In addition, data consistent with those for the model systems have been obtained for these anaesthetics interacting with the membranes of bovine spinal cord (45).

^{31}P NMR OF MEMBRANES

The ^{31}P chemical shift of a rigid phosphodiester, such as is found in membrane lipids (Fig. (15)), depends upon the orientation of the group with respect to the magnetic field of the spectrometer. This is represented in the left panel of Fig. (16), where the chemical shifts expected along the three principal directions are labelled σ_{11}, σ_{22}, and σ_{33}. For

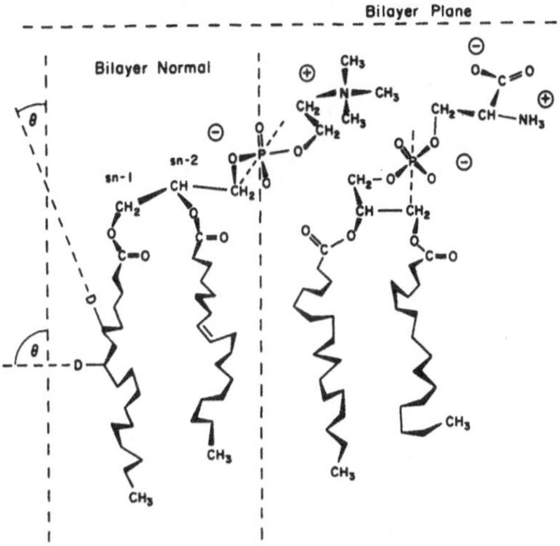

Figure 15. *Some membrane lipids as they might be arranged in a bilayer: left, phosphatidylcholine; right, phosphatidylserine.*

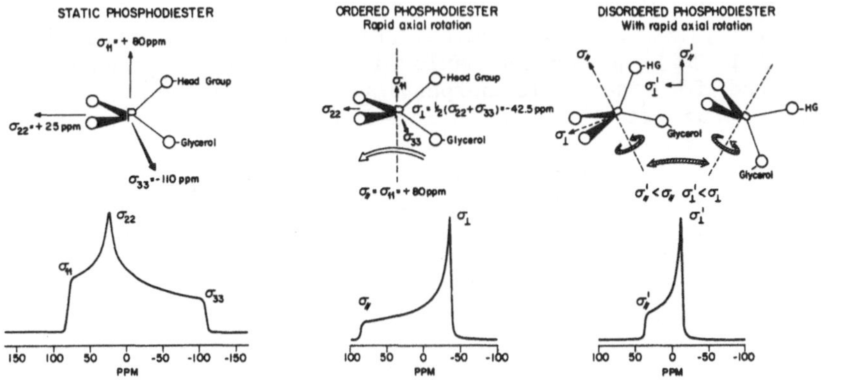

Figure 16. ^{31}P *NMR spectra (120 MHz) for some possible states of membrane phosphodiester groups.*

orientations other than along the three principal axes, the chemical shifts will lie somewhere between these three. The σ-values are the components of the chemical shift (or shielding) tensor for the interaction between the ^{31}P nucleus and the applied magnetic field. Knowledge of the orientation of the axis system, and of σ_{11}, σ_{22}, and σ_{33}, is required to calculate the chemical shift expected for any orientation. If one has a randomly oriented sample, all angles between the applied field and the principal axes are populated. The ^{31}P NMR spectrum is thus the sum of spectra for all possible orientations. Rather than a featureless lump, the so-called powder spectrum has several distinctive features: σ_{11} and σ_{33} define the outermost edges, and σ_{22} leads to a peak. The principal values of the chemical shift tensor can be estimated directly from the spectrum, as shown in Fig. (16); the directions of the principal axes in the molecular system can only be found by studying a single crystal of known structure. Note the values given in the left panel of Fig. (16) for σ_{11} and σ_{33} - the powder spectrum spans some 190 ppm. Principal tensor components for the phosphate moiety in a variety of compounds have been compiled (7).

From the case of the rigid solid, let us now consider an example closer to the situation found in membranes. We shall allow rapid anisotropic motion which can average some of the components of the chemical shift tensor. Suppose for simplicity that this motion is about the 1-axis, thus averaging σ_{22} and σ_{33}. The chemical shift will have the same value for the field anywhere in the 2-3 plane, but a different value when the field is perpendicular to the plane; the chemical shift tensor is axially symmetric. The chemical shift expected when the field is parallel to the unique axis is labelled $\sigma_{\parallel}$, whereas that expected for the field perpendicular to this axis is $\sigma_{\perp}$. The centre panel of Fig. (16) demonstrates this averaging, and the resultant powder spectrum. Note the characteristic shape with the buildup of intensity at $\sigma_{\perp}$. Both $\sigma_{\parallel}$ and $\sigma_{\perp}$ can be measured from the spectrum as shown, but this will involve some error whose magnitude depends upon the linewidths of the individual components of the powder lineshape (7).

Proceeding to the situation most relevant to membranes, let us now allow rapid motion of limited amplitude *of* the 1-axis of the phosphodiester moiety, retaining the rapid motion *about* the 1-axis, right panel of Fig. (16). The 1-axis now moves within a cone, and there is partial averaging of the former $\sigma_{\perp}$ and $\sigma_{\parallel}$, to yield new, smaller effective values, $\sigma_{\perp}'$ and $\sigma_{\parallel}'$. The effective tensor still has axial symmetry, but the total chemical shift anisotropy (CSA) represented by the spectrum, $\Delta\sigma = \sigma_{\parallel}' - \sigma_{\perp}'$, is reduced. The amount by which it is reduced is related to the allowed amplitude of the motion. Should the amplitude of this rapid motion become totally unrestricted, the motion is effectively isotropic, and the pattern collapses to a single line at the isotropic chemical shift. This value varies slightly from lipid to lipid, and with conditions such as pH, presence of ions, etc., but is in the region of ±5 ppm.

The spectrum shown on the right panel of Fig. (16) is typical of many seen for membrane systems; the observed $\Delta\sigma$ is determined in part by the amplitude of angular excursion during motional averaging, and by the orientation of the axis of motional averaging with respect to the principal components of the chemical shift tensor, σ_{11}, σ_{22}, and σ_{33}. Thus, one must be careful not to draw conclusions about the ordering of the phosphate moiety from only the width and shape of the powder pattern, as has been pointed out forcefully by Thayer and Kohler (46). Furthermore, it is often considered that the type of motional averaging, and the resultant powder spectrum we have just described above, is characteristic of a bilayer arrangement of lipids. This is not necessarily true, although most systems known to be bilayer have indeed yielded spectra similar to that shown on the right panel of Fig. (16). The rate of motion within a partially-ordered environment also influences the shape of the ^{31}P powder spectrum.

The previous section dealt with the spectra of systems with very slow or very fast rotational rates. In the intermediate region (for diffusion coefficients $10<R<10^7$ s^{-1}, and the spectral widths usually encountered with membranes) the ^{31}P NMR spectra are sensitive to both type and rate of the motion. Although the lipids of liquid crystalline membranes usually have motional rates in the fast limit region, this is often not so at lower temperatures. In such cases, the spectra can be used as a source of information about both the nature of the motion and the orientation of the headgroup.

Freed and coworkers have developed computational methods for simulating spectra for nuclei of spin ½ (47). They assumed that motion is very anisotropic, and that rotation about one axis is much faster than about the others. In such a case, an asymmetric rotation can be described by two diffusion coefficients $R_{\parallel}$ (fast motion) and $R_{\perp}$ (net slow motion).

Fig. (17) shows how the shapes of the spectra respond to changes in the rate of the fast motion ($R_{\parallel}$). This is similar to Fig. (5) in the original paper of Campbell et al. (47), but was generated using an improved version of the program (courtesy of Dr. C.F. Polnaszek), which yields correct simulations for very slow motions. Such simulations can provide information about the orientation of the rotational axes, and the rates of motion about these axes. However, one should be cautious in view of the

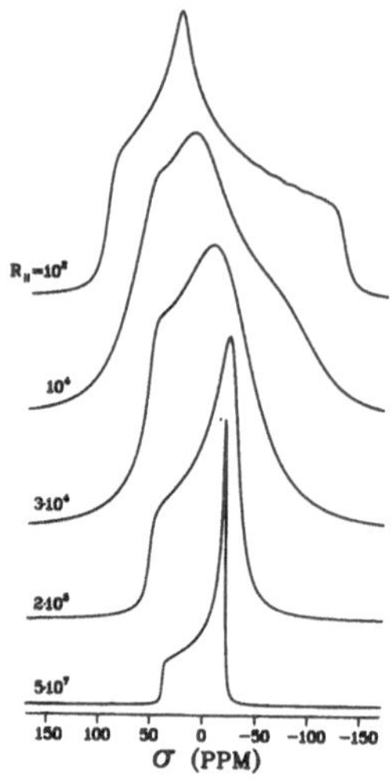

Figure 17. *Simulated ^{31}P NMR spectra for disordered phosphodiester undergoing various rates of motion within the allowed space: top to bottom, slow to fast motion (2).*

various assumptions made in such calculations. The simulated spectra will depend on the model of motion (Brownian motion was used in the calculations for Fig. (17), but others such as free diffusion or discrete jump diffusion are possible). Campbell et al. assumed in their calculations that the component line width, and hence the relaxation time T_2, was constant across the pattern (47). In our experience, this is only true in some cases and must be tested experimentally. The motion of the phosphate group in lipids is a superposition of internal motion and axial diffusion about the long molecular axis; more studies on the general problem of motion of lipid molecules in membranes are necessary before a complete interpretation of ^{31}P NMR spectra can be done with confidence. Furthermore, for proper simulations, good quality, undistorted experimental spectra are necessary. These are rather rare (*vide infra*).

Considerable insight can be gained by studies of proximal nuclei (^{13}C, ^{2}H, ^{1}H), and by measuring relaxation times of the various nuclei at different magnetic field values. The present simulations give a good idea of the behaviour expected for various models, but analysis in terms of these models is still ambiguous.

Thus far we have discussed only the effect of varying the rate of rotation about one axis of the phosphate motion. For relatively small membrane fragments (<10,000Å), the rate of overall isotropic diffusion of the vesicle leads to further averaging of spectral components (48). For relatively small, spherical vesicles two diffusion processes contribute to the averaging of the spectra: rapid Brownian tumbling of the entire vesicle (characterized by the diffusion coefficient D_t) and lateral diffusion of lipids around the vesicle (with diffusion coefficient D_{diff}). The correlation time, τ_c, is given by the equation:

$$\frac{1}{\tau_c} = \frac{6}{R^2}(D_t + D_{diff})$$

where R is the radius of a vesicle and $D_t = KT/8\pi r\eta$, where η is the viscosity (49). Using such a motional model, and Freed's theory for the motional dependence of lineshapes (50), Burnell et al. (48) were able to simulate experimental spectra of dioleoylphosphatidylcholine vesicles for different temperatures and viscosities of the medium, Fig. (18). As seen from the equation above, the effectiveness of the averaging process will increase with decreasing size of the vesicle. In the limiting case of very fast motion (very small vesicles), averaging of anisotropy is complete and a narrow line is observed at the isotropic chemical shift. Of course, even for large vesicles, partial averaging similar to that shown in Fig. (18) is possible by an increase of D_{diff}. For a proper interpretation of spectra such as that for E in Fig. (18) it is essential to estimate the size of the membrane fragments. Furthermore, measuring chemical shift anisotropy for such a spectrum can lead to quite a large error.

In the first part of this section we considered the type of spectrum expected for rapid anisotropic motion within a bilayer superstructure, Fig. (19), left panel. It has been shown that certain lipids will form a completely different type of structure in which the molecules project radially from the centre of a cylinder, Fig. (19), right panel (51). As mentioned earlier in this section, rapid lateral diffusion within a bilayer phase only affects ^{31}P NMR lineshapes if the lamellae enclose a particle of small radius. The cylinders in a hexagonal phase have a very small radius, and therefore lateral diffusion about the cylinder axis can cause further averaging of tensor components. The unique axis of the system now becomes the axis of the cylinder, and thus we label its chemical shift component $\sigma_{\parallel}^{H}$. However, noticing that along this axis the field would be roughly normal to the fatty acyl chains, we would expect the value of this chemical shift to be similar to $\sigma_{\perp}^{L}$. On

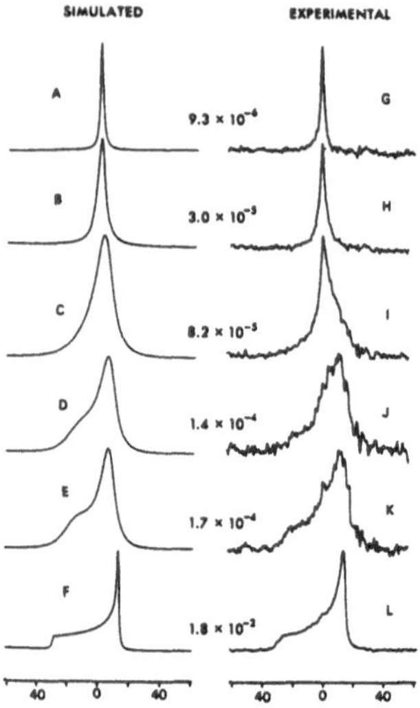

Figure 18. *Simulated and experimental (80 MHz) ^{31}P NMR spectra for vesicles of dioleoylphosphatidylcholine under conditions varied so as to change the rate of overall vesicle rotation; the correlation time for overall rotation, used in the simulations, is shown between the spectra (48).*

the other hand, $\sigma^H_\perp$ will be an average of $\sigma^L_\perp$ and $\sigma^L_\parallel$, due to rapid motion around the cylinder axis. Hence $\sigma^H_\perp = (\sigma^L_\perp + \sigma^L_\parallel)/2$, $\sigma^H_\parallel = \sigma^L_\perp$, and $\Delta\sigma^H = -\Delta\sigma^L/2$. The net result of this is that the ^{31}P powder pattern for the hexagonal phase has a $\Delta\sigma$ which is roughly half that of a corresponding lamellar phase, and a shape which has the build-up of intensity due to the axial component on the other side of the chemical shift zero. One might be tempted to use the breadth and handedness of the ^{31}P NMR patterns as diagnostic for the presence of either phase. This should never be done in the absence of other corroborating data, since a change in the location of the axis of motional averaging relative to the phosphodiester moiety, or a change in the headgroup conformation, with no change in the overall symmetry of the superstructure, can cause a transition from the type of spectrum on the left of Fig. (19) to that on the right (46).

Up to this point we have discussed the basis for, and the interpretation of, ^{31}P NMR spectra of membrane systems, but we

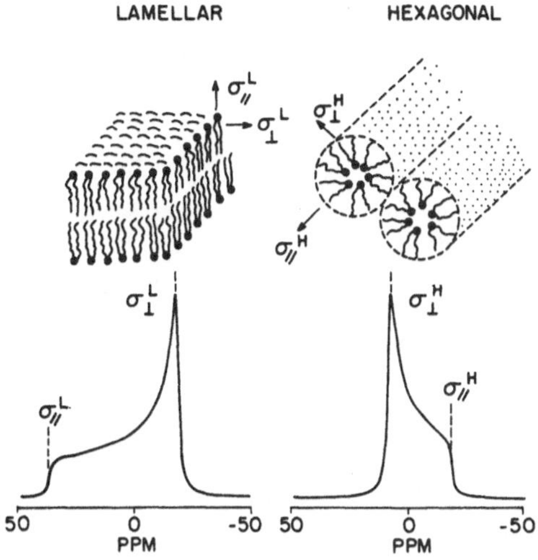

Figure 19. *Representation of the lamellar and hexagonal phases of lipids, and their usual ^{31}P NMR spectra.*

have not questioned the fidelity of the spectra themselves. For a rigid phosphodiester the ^{31}P spectra have widths of ca. 190 ppm (23.1 kHz for a spectrometer operating at 7.05 Telsa), the T_2 values can be quite short, and the T_1 values can be quite long. This combination implies that the powder spectra could be badly distorted by insufficient attention to the methods of spectral acquisition. Most of the membrane spectra reported until very recently have in fact suffered from greater or lesser degrees of distortion, in some cases leading to erroneous conclusions. The first problem arises from the spectral width. The effective strength of the radiofrequency pulse must be such as to excite equally the nuclei contributing to all frequencies within the powder spectrum; in practical terms this means that the 90° pulses should be of the order of 5-10 μsec. in duration. A second problem arises from the relatively short T_2 values expected in these spectra. Most NMR spectrometers are of the high resolution type, which implies probes of high Q factor and therefore long ringdown time after application of the radiofrequency pulse. To avoid spectral distortion due to probe and receiver ringdown, it is common in high resolution practice to insert a delay between excitation of the system and the beginning of the acquisition of the free induction decay. This procedure of necessity yields a loss of signal intensity and the need for a first order phase correction of the data. In the limit of short T_2 and long t_d one ends up with a wobbly baseline and no signal!

The problems referred to above were experienced earlier with ^2H NMR of membranes (*vide supra*), and to a large part resolved by means of a quadrupolar echo (13). Recently considerable improvement in the ^{31}P NMR spectra of membranes has been reported by use of the analogous Hahn echo based on residual chemical shift anisotropy (52). A 90° pulse excites magnetization which decays in amplitude according to T_2; a 180° pulse after a delay τ results in a refocussing of this magnetization to reach an echo maximum after a further time τ. If τ is longer than the ringdown time of the system, no spurious signal from this latter source will remain. If the exact maximum of the echo can be determined, and a Fourier transform done on half the echo starting from the time of maximum amplitude, no first order phase correction should be required on the transformed signal. Extensive phase cycling of the two pulses serves to minimize other aberrations due to mis-set pulses, channel imbalance, and residual baseline instability. To avoid saturation effects, some estimate of the T_1 values of the system must be made, and an appropriate cycle time employed.

Fig. (20) shows the experimental ^{31}P NMR spectra of a dry powder of dipalmitoyl phosphatidylcholine (DPPC) under a variety of conditions. The top spectrum was obtained under conditions similar to the best one might hope for from a high resolution spectrometer, using single 90° pulses and a preacquisition delay of 12 µsec. This is to be compared with spectrum 20c, obtained with the Hahn echo, and with that simulated for a rigid powder in Fig. (16), left panel. The single pulse method clearly causes extreme spectral distortion; attempts to measure the principal components of the chemical shift tensor would lead to overestimates by 20%. The proton-coupled spectrum in Fig. (20b) is shown to demonstrate that incomplete proton decoupling can also cause serious distortions; in this case it actually masks some of the distortions due to the use of single 90° pulses.

The purple membrane of *Halobacterium cutirubrum* is one of the simplest natural membranes, containing only a single protein - bacteriorhodopsin. The lipids of the purple membrane are very unusual; rather than the usual fatty acids, they all contain phytanyl chains with ether linkages (53). The major phospholipid, phosphatidylglycerophosphate (PGP), which constitutes 85% of the phospholipids, contains two phosphate groups, one mono- and one di-esterified. As a consequence, the ^{31}P NMR spectra of the purple membrane are a superposition of two patterns (Fig. (21)), which can be separated by computer simulation (dashed spectra). There are contradictory reports

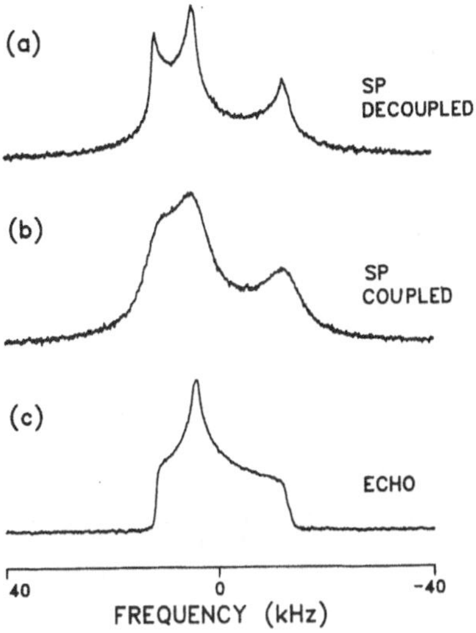

Figure 20. ^{31}P *NMR spectra (121 MHz) of a dry powder of dipalmitoyl phosphatidylcholine taken under the following conditions: (a) single 90° pulses, H-decoupled; (b) as in (a), but H-coupled; (c) CSA echo technique, H-decoupling (52).*

about the presence of a lipid phase transition between 5°C and 35°C in the purple membrane (54,55), and in lipids extracted from halobacteria (56,57). ^{31}P NMR studies of the temperature dependence of the chemical shift anisotropy showed no discontinuity for either the purple membrane or the major phospholipid, PGP (58). This is consistent with calorimetric results arguing against the presence of a phase transition (55,57).

The analysis of T_1 data for the highly ordered membrane is complex. However, an unequivocal value for the correlation time can be obtained if a minimum is observed in the temperature-dependence of T_1. Such data are shown in Fig. (22). For a minimum to occur, $\omega_0{}^2\tau_c^2 \cong 1$, and hence $\tau_c = 1 \times 10^{-9}$s for dioleoylphosphatidylcholine in reconstituted membranes of sarcoplasmic reticulum. Note that no minimum is observed in the 36.4 MHz data due to the dominance of dipolar relaxation at this field strength.

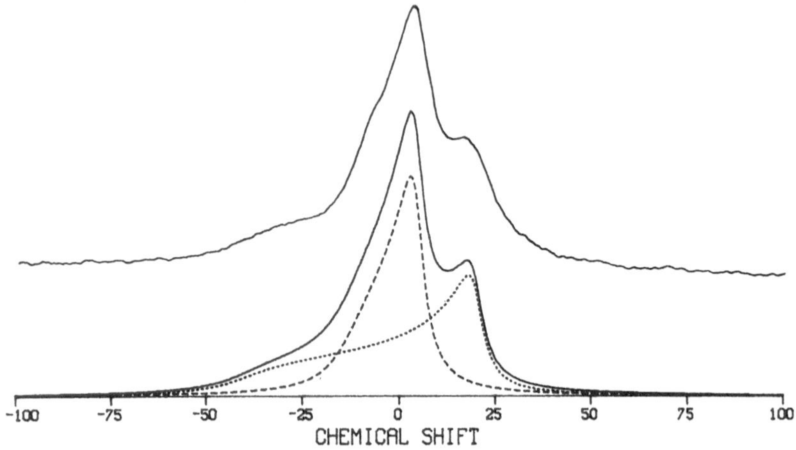

Figure 21. ^{31}P *NMR (121 MHz) spectra of the purple membrane from H. cutirubrum: upper, experimental spectrum; lower, simulation (solid line) in terms of two components (dashes) with lamellar-type spectra (58).*

A large number of ^{31}P NMR studies has been reported for model or biological membranes, particularly by the research groups of Cullis and de Kruijff. These were summarized recently in a review (60).

SYNOPSIS

We have seen that a variety of information can be derived from the ^{2}H and ^{31}P NMR spectra of membranes. Many of the techniques are derived from those used on the less complicated chemical liquid crystal systems. Analysis in terms of absolute and irrevocable parameters is sometimes difficult, due to broad lines, spectral overlap, poor signal-to-noise ratios, the superimposed effects of motion and order, etc. Nonetheless, considerable progress has been made, and life becomes easier with the ever-present technological improvements. There is no reason for anything but optimism with respect to future applications. It is excellent schools such as this one at San Miniato, that catalyze the transfer of expertise and interest from the physical to the biological sides, and vice-versa. The organizers are to be warmly thanked for a job well done.

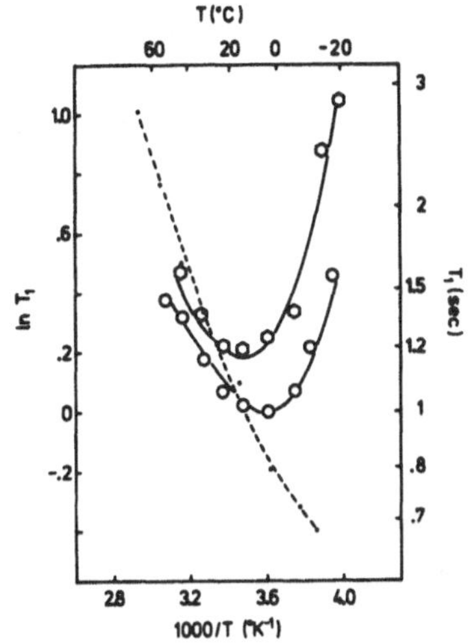

Figure 22. *Temperature dependence of the* ^{31}P *relaxation time* T_1 *(o) dioleoylphosphatidylcholine liposomes at 121 MHz; ---, same at 36 MHz (O) sarcoplasmic reticulum membrane enriched in dioleoylphosphatidylcholine, 121 MHz (59).*

REFERENCES

1. I.C.P. Smith and H.C. Jarrell, *Acc. Chem. Res.* **16**, 266 (1983).

2. I.C.P. Smith and I.H. Ekiel, in *"Phosphorus-31 NMR"*, ed. D. Gorenstein, pp. 448-475. Academic Press, New York, 1984.

3. I.C.P. Smith, in *"Biomembranes"*, Vol. 12, ed. L.A. Manson and M. Kates. Plenum Press, New York, 1984, in press.

4. J.H. Davis, *Biochim. Biophys. Acta* **737**, 117 (1983).

5. R.G. Griffin, in *"Methods in Enzymology"*, ed. J.M. Lowenstein, Vol. 72, part D, pp. 108-174. Academic Press, New York.

6. J. Seelig, *Quart. Rev. Biophys.* **10**, 353 (1977).

7. J. Seelig, *Biochim. Biophys. Acta* **515**, 105 (1978).

8. R.E. Jacobs and E. Oldfield, *Prog. NMR Spectr.* **14**, 113 (1981).

9. H.W. Spiess, in *"NMR Basic Principles and Progress"*, ed. P. Diehl, E. Fluck and R. Kosfeld, pp. 55-215. Springer, Berlin, 1978.

10. M.F. Brown, *J. Chem. Phys.* **77**, 1576 (1982).

11. D.Jaffe, R.L.Vold and R.R.Vold, *J.Chem.Phys.* **78**, 4852 (1983).

12. M. Bloom, J.H. Davis and M. Valic, *Can. J. Phys.* **58**, 1510 (1980).

13. J.H. Davis, K.R. Jeffrey, M. Bloom, M.I. Valic and T.P. Higgs, *Chem. Phys. Lett.* **42**, 390 (1976).

14. N. Boden and P.K. Kahol, *Mol. Phys.* **40**, 1117 (1980).

15. S. Razin, *Prog. Surf. Membr. Sci.* **9**, 257 (1975).

16. H.C. Jarrell, K.W. Butler, R. Deslauriers, I.H. Ekiel and I.C.P. Smith, *Biochim. Biophys. Acta* **688**, 622 (1982).

17. G.W. Stockton, K.G. Johnson, K.W. Butler, A.P. Tulloch, Y. Boulanger, I.C.P. Smith, J.H. Davis and M. Bloom, *Nature* **269**, 267 (1977).

18. M. Rance, K.R. Jeffrey, A.P. Tulloch, K.W. Butler and I.C.P. Smith, *Biochim. Biophys. Acta* **600**, 245 (1980).

19. J. Seelig and N. Waespe-Šarčevic, *Biochemistry* **17**, 3310 (1978).

20. E.J. Dufourc, I.C.P. Smith and H.C. Jarrell, *Chem. Phys. Lipids* **33**, 153 (1983).

21. J.H. Davis, *Biophys. J.* **27**, 339 (1979).

22. A. Seelig and J. Seelig, *Biochemistry* **13**, 4839 (1974).

23. M. Bloom, J.H. Davis and A.L. Mackay, *Chem. Phys. Lett.* **80**, 198 (1981).

24. H.C. Jarrell, A.P. Tulloch and I.C.P. Smith, *Biochemistry* **22**, 5611 (1983).

25. R.N. McElhaney, *Chem. Phys. Lipids* **30**, 229 (1982).

26. G.W. Stockton, K.G. Johnson, K.W. Butler, C.F. Polnaszek,

R. Cyr and I.C.P. Smith, *Biochim. Biophys. Acta* **401**, 535 (1975).

27. I.C.P. Smith, K.W. Butler, A.P. Tulloch, J.H. Davis and M. Bloom, *FEBS Lett.* **100**, 57 (1979).

28. T.H. Huang, R.P. Skarjune, R.J. Wittebort, R.G. Griffin and E. Oldfield, *J. Amer. Chem. Soc.* **102**, 7377 (1980).

29. M. Rance, K.R. Jeffrey, A.P. Tulloch, K.W. Butler and I.C.P. Smith, *Biochim. Biophys. Acta* **688**, 191 (1982).

30. A. Abragam, *"The Principles of Nuclear Magnetism"*, Clarendon Press, Oxford, 1961.

31. M. Bloom, J.H. Davis and F.W. Dahlquist, *Proc. XXth Collogue Ampère*, Tallinn, Estonia, 1978.

32. H.C. Jarrell, K.W. Butler, R. Deslauriers, I.H. Ekiel and I.C.P. Smith, *Biochim. Biophys. Acta* **688**, 622 (1982).

33. H.C. Jarrell, R.A. Byrd and I.C.P. Smith, *Biophys. J.* **34**, 451 (1981).

34. G.W. Stockton and I.C.P. Smith, *Chem. Phys. Lipids* **17**, 251 (1976).

35. J.H. Davis, M. Bloom, K.W. Butler and I.C.P. Smith, *Biochim. Biophys. Acta* **597**, 477 (1980).

36. R. Jacobs and E. Oldfield, *Biochemistry* **18**, 3280 (1979).

37. M.G. Taylor, T. Akiyama and I.C.P. Smith, *Chem. Phys. Lipids* **29**, 327 (1981).

38. E.C. Kelusky, E.J. Dufourc and I.C.P. Smith, *Biochim. Biophys. Acta* **735**, 302 (1983).

39. H. Myer, *Arch f. Exper. Path. u. Pharmakol.* **42**, 109 (1899).

40. Y. Boulanger, S. Schreier and I.C.P. Smith, *Can. J. Biochem.* **58**, 986 (1980).

41. Y. Boulanger, S. Schreier and I.C.P. Smith, *Biochemistry* **20**, 6824 (1981).

42. E.C. Kelusky and I.C.P. Smith, *Can. J. Biochem. Cell Biol.* (in press).

43. E.C. Kelusky and I.C.P. Smith, *Molec. Pharmacol.* (in press).

44. E.C. Kelusky and I.C.P. Smith, *Biochemistry* **22**, 6011 (1983).

45. M. Pass, K.W. Butler, E. Kelusky and I.C.P. Smith (unpublished).

46. A.M. Thayer and S.J. Kohler, *Biochemistry* **20**, 6831 (1981).

47. R.F. Campbell, E. Meirovitch and J.H. Freed, *J. Phys. Chem.* **83**, 525 (1979).

48. E.E. Burnell, P.R. Cullis and B. de Kruijff, *Biochim. Biophys. Acta* **603**, 63 (1980).

49. M. Bloom, E.E. Burnell, M.I. Valic and G. Weeks, *Chem. Phys. Lipids* **14**, 107 (1975).

50. J.H. Freed, G.V. Bruno and C.F. Polnaszek, *J. Phys. Chem.* **73**, 3385 (1971).

51. V. Luzzati and F. Husson, *J. Cell. Biol.* **12**, 207 (1962).

52. M. Rance and A. Byrd, *J. Mag. Res.* **52**, 221 (1983).

53. M. Kates and S.C. Kushwaha, in *"Energetics and Structure of Halpholic Microorganisms"*, ed. S.R. Caplan and M. Ginzburg, pp. 461-480. Elsevier/North Holland, Amsterdam, 1978.

54. H. Degani, D. Bach, A. Danon, H. Garty, M. Eisenbach and S.R. Caplan, *ibid.*, pp. 225-232.

55. M.B. Jackson and J.M. Sturtevant, *Biochemistry* **17**, 911 (1978).

56. W.Z. Plachy, J.K. Lanyi and M. Kates, *Biochemistry* **13**, 4096 (1974).

57. J.S. Chen, P.G. Barton, D. Brown and M. Kates, *Biochim. Biophys. Acta* **352**, 202 (1974).

58. I. Ekiel, D. Marsh, B.W. Smallbone, M. Kates and I.C.P. Smith, *Biochem. Biophys. Res. Commun.* **100**, 105 (1981).

59. J. Seelig, L. Tamm, L. Hymel and S. Fleischer, *Biochemistry* **20**, 3922 (1981).

60. P.R. Cullis, B. de Kruijff, M.J. Hope, A.J. Verkleij, R. Nayar, S.B. Farran, C. Tilcock, T.D. Madden and M.B. Bally, in *"Membrane Fluidity in Biology"*, Vol. 2, pp. 39-81. Academic Press, New York, 1982.

SUBJECT INDEX